Edited by
Gabriel Cristóbal,
Laurent Perrinet, and
Matthias S. Keil

Biologically Inspired Computer Vision

Related Titles

Hornberg, A. (ed.)

Handbook of Machine and Computer Vision.

The Guide for Developers and Users

2nd Edition
2016
Print ISBN: 978-3-527-41339-3; also available in electronic formats

Kwon, K., Ready, S.

Practical Guide to Machine Vision Software

An Introduction with LabVIEW

2015
Print ISBN: 978-3-527-33756-9; also available in electronic formats

Jeong, H.

Architectures for Computer Vision

From Algorithm to Chip with Verilog

2014
Print ISBN: 978-1-118-65918-2; also available in electronic formats

Kaschke, M., Donnerhacke, K., Rill, M.S.

Optical Devices in Ophthalmology and Optometry

Technology, Design Principles and Clinical Applications

2013
Print ISBN: 978-3-527-41068-2; also available in electronic formats

Gevers, T., Gijsenij, A., van de Weijer, J., Geusebroek, J.

Color in Computer Vision

Fundamentals and Applications

2012
Print ISBN: 978-0-470-89084-4; also available in electronic formats

Bodrogi, P., Khanh, T.

Illumination, Color and Imaging

Evaluation and Optimization of Visual Displays

2012
Print ISBN: 978-3-527-41040-8; also available in electronic formats

Osten, W., Reingand, N. (eds.)

Optical Imaging and Metrology

Advanced Technologies

2012
Print ISBN: 978-3-527-41064-4; also available in electronic formats

Cristobal, G., Schelkens, P., Thienpont, H. (eds.)

Optical and Digital Image Processing

Fundamentals and Applications

2011
Print ISBN: 978-3-527-40956-3; also available in electronic formats

Edited by Gabriel Cristóbal,
Laurent Perrinet, and Matthias S. Keil

Biologically Inspired Computer Vision

Fundamentals and Applications

The Editors

Professor Gabriel Cristóbal
Instituto de Optica (CSIC)
Serrano 121
28006 Madrid
Spain

Dr. Laurent Perrinet
Aix Marseille Université, CNRS
Institut de Neurosciences de la Timone
Faculté de Médecine, UMR 7289
Bâtiment Neurosciences
27, Bd Jean Moulin
13385 Marseille Cedex 05
France

Dr. Matthias S. Keil
Universidad de Barcelona
& Instituto de Investigación en Cerebro
Cognición y Conducta (IR3C)
Facultad de Psicología
Dpto. Psicología Básica
Passeig de la Vall d'Hebron, 171
Campus Mundet

Cover

Macro pictures of the book cover (Odontodactylus japonicus male (left) and Phidippus audax male I (right)) have been provided by Roy L. Caldwell and Thomas Shahan, respectively.

Library of Congress Card No.: applied for

British Library Cataloguing-in-Publication Data
A catalogue record for this book is available from the British Library.

Bibliographic information published by the Deutsche Nationalbibliothek
The Deutsche Nationalbibliothek lists this publication in the Deutsche Nationalbibliografie; detailed bibliographic data are available on the Internet at <http://dnb.d-nb.de>.

Print ISBN: 978-3-527-41264-8
ePDF ISBN: 978-3-527-68049-8
ePub ISBN: 978-3-527-68047-4
Mobi ISBN: 978-3-527-68048-1
oBook ISBN: 978-3-527-68086-3

Cover Design Grafik-Design Schulz
Typesetting SPi Global, Chennai, India
Printing and Binding Markono Print Media Pte Ltd, Singapore

Printed on acid-free paper

Contents

List of Contributors

Alexandre Alahi
Stanford University
Computer Science Dept.
353 Serra Mall
Stanford
CA 94305
USA

David Alleysson
Grenoble-Alpes University /
CNRS
Lab. Psych. NeuroCognition
BSHM
Domaine Universitaire
1251 Av. Centrale
BP 46
38402 Saint Martin d'Hères
cedex
France

Matthias Bethge
AG Bethge
Centre for Integrative
Neuroscience
Otfried-Müller-Str. 25
72076 Tübingen
Germany

N. Andrew Browning
Boston University
Center for Computational
Neuroscience and Neural
Technology (CompNet)
677 Beacon Street
Boston
MA 02215
USA

and

Scientific Systems Company Inc.
Active Perception and Cognitive
Learning Group (APCL)
500 West Cummings Park
Woburn
MA 01810
USA

Matthieu Cord
Université Pierre et Marie Curie
UPMC-Sorbonne Universities
LIP6/UPMC
4 Place Jussieu
75005 Paris
France

Gabriel Cristóbal
Instituto de Optica
Spanish National Research Council (CSIC)
Serrano 121
28006 Madrid
Spain

Emmanuel D'Angelo
Pix4D SA
EPFL Innovation Park
Lausanne
VD 1015
Switzerland

Holly E. Gerhard
AG Bethge
Centre for Integrative Neuroscience
Otfried-Müller-Str. 25
72076 Tübingen
Germany

Georges Goetz
Stanford University
Neuroscience Department
Stanford
CA 94305
USA

Nathalie Guyader
Grenoble-Alpes University/CNRS
Gipsa-lab
Domaine Universitaire
11 rue des mathématiques
BP 46
38402 Saint Martin d'Hères cedex
France

Matthias S. Keil
University of Barcelona
Institute for Brain, Cognition and Behaviour (IR3C)
Basic Psychology Department
Passeig de la Vall d'Hebron
171 Campus Mundet
08035 Barcelona
Spain

Tim Köhler
DFKI GmbH
Robotics Innovation Center
Robert-Hooke-Straße 1
28359 Bremen
Germany

Olivier Le Meur
University of Rennes 1/IRISA
SIROCCO
Campus universitaire de Beaulieu
35042 Rennes Cedex
France

Eduardo Magdaleno Castellò
Departamento de Ingeniería Industrial
Universidad de La Laguna
Cno San Francisco de Paula s/n
38271 LA LAGUNA, S.C. Tenerife
Spain

Henrik Malm
Lund University
Department of Biology
Sölvegatan 37
SE-223 62 Lund
Sweden

Matei Mancas
University of Mons (UMONS)
TCTS Lab, 31
Boulevard Dolez
B-7000 Mons
Belgium

Guillaume S. Masson
Institut de Neurosciences de la Timone, UMR 7289
Aix-Marseille Université & CNRS
27 Bd Jean Moulin
13385 Marseille Cedex 05
France

Anna Montagnini
Institut de Neurosciences de la Timone, UMR 7289
Aix-Marseille Université & CNRS
27 Bd Jean Moulin
13385 Marseille Cedex 05
France

Magnus Oskarsson
Lund University
Centre for Mathematical Sciences
Box 118
SE-221 00 Lund
Sweden

C. Alejandro Parraga
Centre de Visió per Computador/Comp. Sci. Dept.
Universitat Autònoma de Barcelona
Edifício O
Campus UAB (Bellaterra)
C.P.08193
Barcelona
Spain

Alejandro Pérez Nava
Universidad de La Laguna
Departamento de Estadística
Investigación Operativa y Computación
Cno San Francisco de Paula s/n
38271 LA LAGUNA, S.C.
Tenerife
Spain

Fernando Pérez Nava
Universidad de La Laguna
Departamento de Estadística
Investigación Operativa y Computación
Cno San Francisco de Paula s/n
38271 LA LAGUNA, S.C.
Tenerife
Spain

Laurent U. Perrinet
Institut de Neurosciences de la Timone, UMR 7289
Aix-Marseille Université & CNRS
27 Bd Jean Moulin
13385 Marseille Cedex 05
France

Christoph Posch
Sorbonne Universités
UPMC Univ Paris 06
UMR_S 968
Institut de la Vision
INSERM U968
CNRS UMR_7210
17 rue Moreau
75012 Paris
France

Florian Raudies
Boston University
Center for Computational
Neuroscience and Neural
Technology (CompNet)
677 Beacon Q8 Street
Boston
MA 02215
USA

Manuel Rodríguez Valido
Universidad de La Laguna
Departamento de Ingeniería
Industrial
Cno San Francisco de Paula s/n
38271 LA LAGUNA, S.C.
Tenerife
Spain

Mukul Sarkar
Department of Electrical
Engineering
Indian Institute of Technology
Delhi
Hauz Khas
New Delhi - 110016
India

Peggy Seriès
Institute for Adaptive and Neural
Computation Informatics
University of Edinburgh
10 Crichton Street
Edinburgh EH8 9AB
UK

Grigorios Sotiropoulos
Institute for Adaptive and Neural
Computation Informatics
University of Edinburgh
10 Crichton Street
Edinburgh EH8 9AB
UK

Lucas Theis
AG Bethge
Centre for Integrative
Neuroscience
Otfried-Müller-Str. 25
72076 Tübingen
Germany

Christian Thériault
Université Pierre et Marie Curie
UPMC-Sorbonne Universities
LIP6/UPMC
4 Place Jussieu
75005 Paris
France

Nicolas Thome
Université Pierre et Marie Curie
UPMC-Sorbonne Universities
LIP6/UPMC
4 Place Jussieu
75005 Paris
France

Reinhard Voelkel
SUSS MicroOptics SA
Rouges-Terres 61
2068 Hauterive
Neuchâtel
Switzerland

Eric Warrant
Lund University
Department of Biology
Sölvegatan 37
SE-223 62 Lund
Sweden

Foreword

Since the mid-twentieth century, our scientific world of information technology has become more and more multidisciplinary, combining the knowledge and experience of many disciplines such as biology, psychophysics, mathematics, and engineering.

The first attempts to widen their field of view came from engineers and mathematicians who were fascinated by the tremendous faculties of our brain to process and extract information from the noisy and changing world of our environment: would it be possible to mimic our brain to build a novel kind of machines for signal, images, and information processing? This opened the era of artificial neural networks.

Among the pioneers of this new approach are the seminal works of McCullogh and Pitts (1943) for their *Formal Neuron,* Von Neuman (1948) for his *Automata Networks,* Hebb (1949) for his famous *Synaptic Learning Rule,* Rosenblatt (1958) for his *Perceptron,* Widrow and Hoff (1960) for the *Adaline,* inspired from a combination of the perceptron, and Hebb's learning rule in the framework of linear predictive filters in signal processing.

Later, in the following decades, Minsky and Papert (1969) revisited the perceptron and its limitations, opening the way to the *multilayered perceptron* with a new learning rule: the *gradient back-propagation* suggested by Werbos (1974) and extended by Rumellart and McClelland (1986). We also cite the Hopfield networks (1982) derived from some statistical physics considerations. The common feature of all these networks is that they are taught to classify or recognize items according to an *ad hoc* learning rule.

In the mean time, a new concept of artificial neural networks, called "self-organization" emerged from the pioneering work by Kohonen (1984) with his *self-organizing maps.* In this framework, the neural network rather than to be taught what to do, was left able to evolve with respect to the input information structure, according to some general learning rule. With this concept, artificial neural networks acquired some scent of "intelligence" and many researchers worked around this idea of self-adaptive networks: among them, in nonlinear dimensionality reduction and data visualization Demartines and Herault (1999) with *curvilinear component analysis,* Tenenbaum (2000) with *isomap,* Lee and Verleysen (2004) with *curvilinear distance analysis.* We also cite the *sources separation network* also called *independent component analysis* (Herault, Ans, and Jutten, 1986) widely used as a signal processing technique to extract signals from unknown mixtures of them, provided that these signals are statistically independent and that the mixtures are different.

During this period, many international workshops and conferences concerning neural networks (ICANN, NIPS, IWANN, ESANN, etc.) were launched, gathering researchers mainly from mathematics and engineering. However, several biologists and psychophysicists progressively attended these conferences, often under the solicitations of the early participants.

Since the beginning of the 1980s, some researchers turned themselves toward more "realistic" models of neurons, among them Carver Mead (1988) who defined the concept of *neuromorphic circuits,* with his famous CMOS electronic model of the retina: he mimicked the retinal electric synapses by means of simple resistors and capacitors. He designed an analog electronic integrated circuit which exhibited the main properties of the vertebrate retina. This circuit was much more efficient than the usual digital approaches: fast, robust, continuous-time, and low energy consumption are the most important qualifying terms of this approach.

Maybe because this work was about the retina, it has motivated a renewal of interest in the biological model as an alternative – or rather as a complement – to computer vision. Hence, many researchers in engineering began to consider and study the biology of vision, asking for help from their colleagues in biology and psychophysics. This led to the creation of several multidisciplinary teams around the world, acting in the framework of cognitive science. A new era was thus born.

This book is the congregation of internationally renowned scientists in biology, psychology, mathematics, electronics, computer science, and information technology. They have made a number of major contributions in biological vision, computer vision, and cognitive science. The book addresses the most important aspects of the fields, from visual perception to technology at a high scientific level. Across its chapters, it offers a comprehensive view of biological motivations, theoretical considerations, as well as application suggestions.

On going through the chapters, the reader will discover in-depth studies of the state-of-the-art statements in various disciplines and cross-disciplinary topics related to vision and visual perception. Without any order of precedence, the following subjects are of interest: electronic neuromorphic circuits, retinal and color processing, insect vision, visual psychophysics, visual attention, and saliency, Bayesian inference in visual perception, detection of scale-invariant features, object recognition, spiking neurons, sparse coding, plenoptic cameras, motion detection, visual navigation.

Far from being a simple list of topics, this book offers the opportunity to consider a global approach to vision and invites the reader to make his (her) own point of view on a synthetic basis. It will be of great importance for researchers who are interested in the fundamental principles of biological vision and in computer vision, hardware, and autonomous robotics. Bringing together scientists of different disciplines, and bi- or three-disciplinary scientists, the book offers an example for future organization for academic and even industrial research teams.

Grenoble, Spring 2015 *Jeanny Herault*

Part I
Fundamentals

Biologically Inspired Computer Vision: Fundamentals and Applications, First Edition.
Edited by Gabriel Cristóbal, Laurent Perrinet, and Matthias Keil.

1
Introduction

Gabriel Cristóbal, Laurent U. Perrinet, and Matthias S. Keil

> *And first he will see the shadows best, next the reflections of men and other objects in the water, and then the objects themselves; then he will gaze upon the light of the moon and the stars and the spangled heaven; and he will see the sky and the stars by night better than the sun or the light of the sun by day?*
>
> *Plato, The Republic, Book VII*

As state-of-the-art imaging technologies become more and more advanced, yielding scientific data of unprecedented detail and volume, the need to process and interpret all this data has made image processing and computer vision also increasingly important. Sources of data that have to be routinely dealt with for today's applications include video transmission, wireless communication, automatic fingerprint processing, massive databanks, non-weary and accurate automatic airport screening, and robust night vision to name a few. These technological advances were closely followed by the increase in computer power of traditional, von-Neumann architectures. However, a major concern is that such architectures are efficient only for specific computations such as retrieving an item in a database or to apply a precise sequence of operations. These usually fail when the application comes closer to biological vision, such as the categorization of images independently of changes in lighting, movements, and the environment or the imaging device. Surprisingly, while computers can handle most complex tasks (such as computing hash numbers, logarithms, and linear algebra), most of the time, the "simpler" the task is for humans (such as "find the animal in the image"), the more complex it becomes to implement on a computing device. Multidisciplinary inputs from other disciplines such as computational neuroscience, cognitive science, mathematics, physics, and biology have had, and will have, a fundamental impact on the progress of imaging and vision sciences. One main advantage of the study of biological organisms is to devise very different types of computational paradigms beyond the usual von Neumann architecture, for example, by implementing a neural network with a high degree of local connectivity, which may be more easily adapted to solve the problems of modern imaging.

Biologically Inspired Computer Vision: Fundamentals and Applications, First Edition.
Edited by Gabriel Cristóbal, Laurent Perrinet, and Matthias Keil.

This book serves as a comprehensive but rigorous reference in the area of biologically inspired computer vision modeling. Biologically inspired vision, that is, the study of visual systems of living beings, can be considered as a two-way process. On the one hand, living organisms can provide a source of inspiration for new computationally efficient and robust vision models, and on the other hand, machine vision approaches can provide new insights into understanding biological visual systems. Over the different chapters, this book covers a wide range from the fundamental to the more specialized topics. This book often follows Marr's classical, three-level approach to vision (computational, algorithmic, and hardware implementation) [1], but also goes beyond Marr's approach in the design of novel and more advanced vision sensors. In particular, the last section of the book provides an overview of a few representative applications and current state of the art of the research in this area.

The scope of this book somewhat overlaps that of a few other books published in this area, most of them corresponding to conference proceedings, for example, Refs [2, 3]. More recently, several special sessions on this same topic were organized at different workshops such as the ones on "Biologically consistent vision" in conjunction with the 2011 Computer Vision and Pattern Recognition Conference and "Biological and Computer Vision Interfaces" in conjunction with the 2012 European Conference on Computer Vision. The first monograph on this topic was published more than 20 years ago [4] and therefore does not reflect the latest advances in the field. A very good reference in the field is the book by Frisby and Stone [5] that provides the foundation for computational and physiological research on vision. Another relevant reference is the book by Petrou and Bharath although it is more focused toward specialized hardware both at low power and high speed [6]. A more recent reference in the area is the book by Pomplun and Suzuki [7] which is more focused on specific aspects of visual function such as attention, binocularity, or cortical structures. At the time of writing, it is worth mentioning a special issue on bioinspired imaging which highlights recent progress in the domain of vision and biological optics [8]. As a consequence, this book is valuable for both undergraduate and graduate students and also for specialized researchers as it presents information that is usually spread out over different sources into a single and comprehensive monograph.

1.1 Why Should We Be Inspired by Biology?

A central question in this area is to understand why and how is it useful to build technical systems with a biological inspiration. Or, in other words, why should biologically-motivated studies be useful for constructing artificial systems? One answer is that living systems are engineered to perfection by evolution, and thus they provide a seemingly inexhaustible source of inspiration for engineering. In this book, to cover few examples of how such biological findings will be of direct benefit, for example, for engineering new devices and sensors. More generally, one objective of this book is to cover "two-way process" analogy previously mentioned. On the one hand, we have chapters where findings from biology, neurobiology, or psychophysics guide the development of computer vision algorithms. On the other hand, we have more technical chapters that are motivated by biology.

In general, we can argue that Nature has been continuously a source of inspiration not only for constructing artificial vision systems but also in other domains such as materials science. The term "biomimetics" has been coined for the field of science and engineering that seeks to understand and to use nature as a model for copying, adapting, and inspiring concepts. An amazingly early design of Nature can be observed in the *Erbenochile erbeni*, a Devonian trilobite with calcite-derived compound eyes, allowing the animal to see not only 360° on the horizontal plane but also to cover a significant vertical area. Even more intriguing is the fact that their eyes have eye shades for blocking glare, suggesting that trilobites were diurnal and not nocturnal (see Figure 1.1) [9]. Often, the visual system of non-human organisms surpasses the performance of our visual system. Examples include ultraviolet vision of honey bees, infrared vision of some snakes, or the high temporal resolution of the blow fly visual system. All species were optimized by evolution over large time scales. This has led to efficient

Figure 1.1 The *Erbenochile erbeni* eyes are distributed in a vertical half-cylinder surface with 18 or 19 lenses per vertical row. Note the striking similarity of this natural eye with the CurvACE sensor presented in Figure 22 of Chapter 7. Image credit: Wikipedia

solutions to many of the challenges they face. It is therefore not surprising that humans always make efforts to use Nature as a model for innovation and problem solving.

Another intriguing example are the eyes of the peacock mantis shrimp (*Odontodactylus japonicus*), which are capable of serial or parallel vision, trinocular vision, and depth perception; the color vision is extended to the ultraviolet and infrared ranges (see left cover image of this book). They also exhibit at least 16 different photoreceptors in addition to detection of polarized light and circular polarized light [10]. Inspired by the mantis shrimp's compound eye and polarized vision, an international team of researchers has recently built a miniature sensor that can detect subtle differences in early stage cancerous cells from *in vivo* endoscopy of the mouse colon [11]. The eyes of other animals use other strategies to solve predatory tasks. The jumping spider uses special staircase-shaped retinas to produce sharp and out-of-focus images simultaneously (see right cover image of this book). By comparing the focused and defocused images, the spider can estimate depth through unmoving eyes. Nagata *et al.* have developed a mathematical model that predicts the accuracy of the jumps for different wavelengths [12]. These recent promising results could open up new avenues both in mathematical modeling and signal processing for more challenging applications. In the context of animal vision studies, Land and Nilsson's book [13] provides a fascinating comparative account of the evolution and function of animal eyes with emphasis on the description of the optical system. Furthermore, a very recent book on the same topic is the one by Cronin *et al.* [14]. As a consequence, biologically inspired computer vision is not only about mimicking the structure of animals' visual systems but also to gain insights from the vast range of different solutions that have emerged through natural selection to provide efficient solutions to vision-based problems.

1.2 Organization of Chapters in the Book

This book contains 18 chapters that have been organized in four different parts:

- Fundamentals
- Sensing
- Modeling
- Applications

The cross-links between the different chapters have been sketched in Figure 1.2. This book aims at providing an overview of bioinspired computer vision, starting from fundamentals to the most recent advances and applications in the field. The three chapters selected in the section "Applications" are good representatives of how the transfer of ideas from biology to computer vision can be done in practice. Figure 1.3 shows a picture of a tag cloud that has been generated from the table of contents of the book.

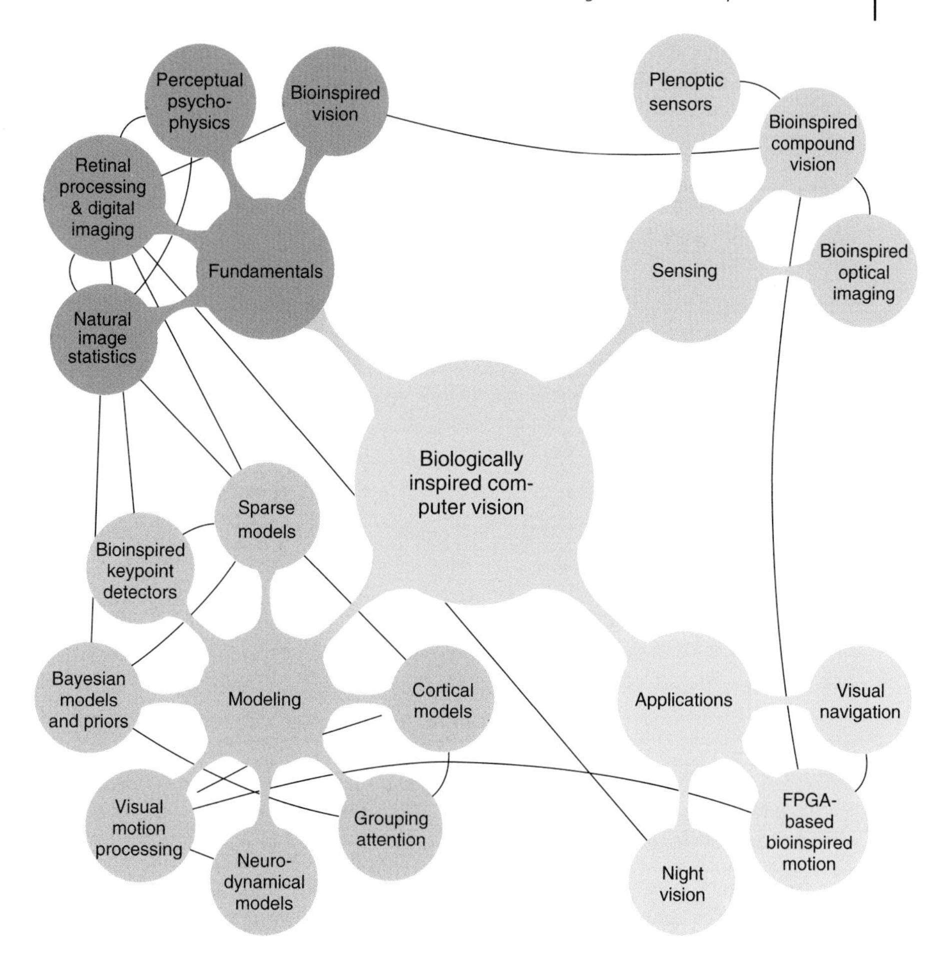

Figure 1.2 Mindmap of the book contents. Cross-links between chapters have been indicated as thin lines.

The structure of the book is as follows:

- Section 1: Fundamentals.
  - Chapter 2 describes basic bioinspired vision technology, which has the aim of outperforming conventional frame-based vision systems in many fields of application. It provides an overview of biosensors and neuromorphic retinas.
  - Chapter 3 describes how the retina is capable of much more complicated information processing than was initially thought.
  - Chapter 4 describes how natural image statistics can be exploited to effectively improve visual inference in computer vision systems.
  - Chapter 5 provides the basics of visual psychophysics, that is, how to measure the performance of observers in predetermined visual tasks.

Figure 1.3 Tag cloud of the abstracts and table of contents of the book. Credit: www.wordle.net

- Section 2: Sensing.
  - In Chapter 6, algorithms inspired by the compound eyes of insects, whose vision is based on sensing the polarization of light are described, illustrating how this can be modeled to enhance the visual perception of standard cameras.
  - Chapter 7 describes how natural concepts for miniaturization could be imitated for building computer vision systems with perfect adaptation to factors such as small size, special tasks, and specific applications.
  - Chapter 8 describes the basics of plenoptic sensing and how these new devices can extend the capabilities of current standard cameras.
- Section 3: Modeling.
  - Chapter 9 describes Bayesian models as a useful modeling approach for describing perception and behavior at the computational level.
  - Chapter 10 explains how neurodynamical models could be used not only as biologically inspired models for processing images but also for explaining perceptual phenomena.
  - Chapter 11 presents models of bottom-up visual attention and their applications.
  - Chapter 12 presents a review of several recent studies focusing on the integration of retinal and extra-retinal information for visual motion processing and human tracking behavior.
  - Chapter 13 describes cortical models for image recognition mainly based on the HMAX architecture of Riesenhuber and Poggio [15].
  - Chapter 14 describes how bioinspired approaches may be applied to computer vision problems using predictive coding schemes focusing on sparse models as simple and efficient instances of such techniques.

  - Chapter 15 describes methods for extracting and representing key points motivated from a biological standpoint.
- Section 4: Applications.
  - Chapter 16 describes how by mimicking neural processes of nocturnal animals, efficient computer vision algorithms can be devised.
  - Chapter 17 provides an overview of elementary motion detectors (EMDs) oriented to computer vision applications when resources available are limited (e.g., power consumption).
  - Finally, chapter 18 describes in detail a bioinspired model (ViSTARS) oriented to visually guided navigation in a cluttered world.

1.3
Conclusion

Biological vision shows excellence in terms of performance and robustness. Following one of the recommendations of the book referees, one of the aims of the book is to make it multidisciplinary, although perhaps in the future, the topic of biologically inspired computer vision could become a single discipline by itself. One of the reasons of the resurging interest in the topic of the book has been both the availability of massive computing power (e.g., cloud computing) and high-performant computing power (GPU, FPGA, etc). This has been illustrated in the Chapters 16 and 17 of this book.

For the reader's convenience, there is an accompanying website with supplementary material at bicv.weebly.com. It contains selected MATLAB and Python codes, testing images and errata.

Acknowledgments

We would like to express our appreciation for the quality of the chapters delivered by the authors and for their efforts to keep the chapter length within the limits given. This project could not have been achieved without the valuable contributions made by a significant number of experts in the field from both the academia and industry. We are grateful to their willingness to contribute to this groundbreaking resource. We would like to extend thanks to all the Wiley VCH members who have helped us manage the project; in particular, thanks are due to Val Moliere for her enthusiastic support. We want to express also our gratitude to Roy L. Caldwell and Thomas Shahan for providing us the macro pictures of the book cover (*Odontodactylus japonicus male* (left) and *Phidippus audax male* I(right)).

References

1. Marr, D. (2010) *Vision: A Computational Investigation into the Human Representation and Processing of Visual Information*, MIT Press.
2. Lee, S., Bultoff, H., and Poggio, T. (eds) (2000) *Biologically Motivated Computer Vision*, Lecture Notes in Computer Science, vol. 1811, Springer-Verlag.

3. Bultoff, H. *et al.* (eds) (2002) *Biologically Motivated Computer Vision*, Lecture Notes in Computer Science, vol. 2525, Springer-Verlag.
4. Overington, I. (1992) *Computer Vision: A Unified Biologically Inspired Approach*, Elsevier Science.
5. Frisby, J. and Stone, J. (2010) *Seeing*, 2nd edn, MIT Press.
6. Petrou, M. and Bharath, A. (eds) (2008) *Next Generation Artificial Vision Systems: Reverse Engineering the Human Visual System*, Artech House.
7. Pomplun, M. and Suzuki, J. (eds) (2012) *Developing and Applying Biologically-Inspired Vision Systems: Interdisciplinary Concepts*, IGI Global.
8. Larkin, W., Etienne-Cummings, R., and Van der Spiegel, J. (2014) Bioinspired imaging: discovery, emulation and future prospects. *Proc. IEEE*, **102** (10), 1404–1410.
9. Fortey, R. and Chatterton, B. (2003) A Devonian trilobite with an eyeshade. *Science*, **301**, 1689.
10. Horvath, G. (ed.) (2014) *Polarized Light and Polarization Vision in Animal Sciences*, 2nd edn, Springer.
11. York, T. *et al.* (2014) Bioinspired polarization imaging sensors: from circuits and optics to signal processing algorithms and biomedical applications. *Proc. IEEE*, **102**, 1450–1469.
12. Nagata, T. *et al.* (2012) Depth perception from image defocus in a jumping spider. *Science*, **335** (6067), 469–471.
13. Land, M. and Nilsson, D. (2012) *Animal Eyes*, 2nd edn, Oxford Animal Biology, Oxford University Press.
14. Cronin, T. *et al.* (2014) *Visual Ecoloy*, Princeton University Press.
15. Riesenhuber, M. and Poggio, T. (1999) Hierarchical models of object recognition in cortex. *Nat. Neurosci.*, **2**, 1019–1025.

2
Bioinspired Vision Sensing

Christoph Posch

2.1
Introduction

Nature has been a source of inspiration for engineers since the dawn of technological development. In diverse fields such as aerodynamics, the engineering of surfaces and nanostructures, materials sciences, or robotics, approaches developed by nature during a long evolutionary process provide stunning solutions to engineering problems. Many synonymous terms like *bionics*, *biomimetics*, or *bioinspired engineering* have been used for the flow of concepts from biology to engineering.

The area of sensory data acquisition, processing, and computation is yet another example where nature usually achieves superior performance with respect to human-engineered approaches. Despite all the impressive progress made during the last decades in the fields of information technology, microelectronics, and computer science, artificial sensory and information processing systems are still much less effective in dealing with real-world tasks than their biological counterparts. Even small insects outperform the most powerful man-made computers in routine functions involving, for example, real-time sensory data processing, perception tasks, and motor control and are, most strikingly, orders of magnitude more energy efficient in completing these tasks. The reasons for the superior performance of biological systems are only partially understood, but it is apparent that the hardware architecture and the style of computation are fundamentally different from what is state of the art in human-engineered information processing. In a nutshell, biological neural systems rely on a large number of relatively simple, slow, and imprecise processing elements and obtain performance and robustness from a massively parallel principle of operation and a high level of adaptability and redundancy where the failure of single elements does not induce any observable system performance degradation.

Biologically Inspired Computer Vision: Fundamentals and Applications, First Edition.
Edited by Gabriel Cristóbal, Laurent Perrinet, and Matthias Keil.

2.1.1
Neuromorphic Engineering

The idea of applying computational principles of biological neural systems to artificial information processing has existed for decades. An early work from the 1940s by Warren McCulloch and Walter Pitts introduced a neuron model and showed that it was able to perform computation [1]. Around the same time, Donald Hebb developed the first models for learning and adaptation [2]. In 1952, Alan Hodgkin and Andrew Huxley linked biological signal processing to electrical engineering in their famous paper entitled *A quantitative description of membrane current and its application to conduction and excitation in nerve* [3], in which they describe a circuit model of electrical current flow across a nerve membrane. This and related work earned them the Nobel Prize in physiology and medicine in 1963.

In the 1980s, Carver Mead and colleagues at Californian Institute of Technology (Caltech) developed the idea of engineering systems containing analog and asynchronous digital electronic circuits that mimic neural architectures present in biological nervous systems [4–6]. He introduced the term "neuromorphic" to name these artificial systems that adopt the form of, or morph, neural systems. In a groundbreaking paper on neuromorphic electronic systems, published 1990 in the *Proceedings of the IEEE* [5], Mead argues that the advantages (of biological information processing) can be attributed principally to the use of elementary physical phenomena as computational primitives, and to the representation of information by the relative values of analog signals, rather than by the absolute values of digital signals. He further argues that this approach requires adaptive techniques to correct for differences of nominally identical components and that this adaptive capability naturally leads to systems that learn about their environment. Experimental results suggest that adaptive analog systems are 100 times more efficient in their use of silicon area, consume 10 000 times less power than comparable digital systems, and are much more robust to component degradation and failure than conventional systems [5]. The "neuromorphic" concept revolutionized the frontiers of (micro-)electronics, computing, and neurobiology to such an extent that a new engineering discipline emerged, whose goal is to map the brain's computational principles onto a physical substrate and, in doing so, to design and build artificial neural systems such as computing arrays of spiking silicon neurons but also peripheral sensory transduction such as vision systems or auditory processors [7, 8]. The field is referred to as *neuromorphic engineering*.

Progressing further along these lines, Indiveri and Furber argue that the characteristics (of neuromorphic circuits) offer an attractive alternative to conventional computing strategies, especially if one considers the advantages and potential problems of future advanced Very Large Scale Integration (VLSI) fabrication processes [9, 10]. By using massively parallel arrays of computing elements, exploiting redundancy to achieve fault tolerance, and emulating the neural style of computation, neuromorphic VLSI architectures can exploit to the fullest potential the features of advanced scaled VLSI processes and future emerging technologies, naturally coping with the problems that characterize them, such as device inhomogeneities and imperfections.

2.1.2
Implementing Neuromorphic Systems

Neuromorphic electronic devices are usually implemented as VLSI integrated circuits or systems-on-chips (SoCs) on planar silicon, the mainstream technology used for fabricating the ubiquitous microchips that can be found in practically every modern electronically operated device. The primary silicon computational primitive is the transistor. Interestingly, when operated in the analog domain as required by the neuromorphic concept instead of being reduced to mere switches as in conventional digital processing, transistors share physical and functional characteristics with biological neurons. For example, in the weak-inversion region of operation, the current through an MOS transistor exponentially relates to the voltages applied to its terminals. A similar dependency is observed between the number of active ion channels and the membrane potential of a biological neuron. Exploiting such physical similarities allows, for example, constructing electronic circuits that implement models of voltage-controlled neurons and synapses and realize biological computational primitives such as excitation/inhibition, correlation, thresholding, multiplication, or winner-take-all selection [4, 5]. The light sensitivity of semiconductor structures allows the construction of phototransducers on silicon, enabling the implementation of vision devices that mimic the function of biological retinas. Silicon cochleas emulate the auditory portion of the human inner ear and represent another successful attempt of reproducing biological sensory signal acquisition and transduction using neuromorphic techniques.

2.2
Fundamentals and Motivation: Bioinspired Artificial Vision

Representing a new paradigm for the processing of sensor signals, one of the greatest success stories of neuromorphic systems to date has been the emulation of sensory signal acquisition and transduction, most notably in vision. In fact, one of the first working neuromorphic electronic devices was modeled after a part of the human neural system that has been subject to extensive studies since decades – the retina. The first so-called silicon retina made the cover of *Scientific American* in 1991 [11] and showed that it is possible to generate in an artificial microelectronics device, in real time, output signals that correspond directly to signals observed in the corresponding levels of biological retinas. Before going into more detail about biological vision and bioinspired artificial vision, and in order to appreciate how biological approaches and neuromorphic engineering techniques can be beneficial for advancing artificial vision, it is inspiring to look at the shortcomings of conventional image sensing techniques.

2.2.1
Limitations in Vision Engineering

State-of-the-art image sensors suffer from severe limitations imposed by their very principle of operation. The sensors acquire the visual information as a series of "snapshots" recorded at discrete points in time, hence time quantized at a predetermined rate called "frame rate." The biological retina has no notion of a frame, and the world, the source of the visual information we are interested in, works asynchronously and in continuous time. Depending on the timescale of changes in the observed scene, a problem that is very similar to undersampling, known from other engineering fields, arises. Things happen between frames and information gets lost. This may be tolerable for the recording of video data for a human observer (thanks to the adaptability of the human visual system), but artificial vision systems in demanding applications such as, for example, autonomous robot navigation, high-speed motor control, visual feedback loops, and so on, may fail as a consequence of this shortcoming.

Nature suggests a different approach: biological vision systems are driven and controlled by events happening within the scene in view, and not – like image sensors – by artificially created timing and control signals that have no relation whatsoever to the source of the visual information. Translating the frameless paradigm of biological vision to artificial imaging systems implies that control over the acquisition of visual information is no longer being imposed externally to an array of pixels but the decision making is transferred to the single pixel that handles its own information individually.

A second problem that is also a direct consequence of the frame-based acquisition of visual information is redundancy. Each recorded frame conveys the information from all pixels, regardless of whether or not this information – or part of it – has changed since the last frame had been acquired. This method obviously leads, depending on the dynamic contents of the scene, to a more or less high degree of redundancy in the acquired image data. Acquisition and handling of these dispensable data consume valuable resources and translate into high transmission power dissipation, increased channel bandwidth requirements, increased memory size, and postprocessing power demands.

Devising an engineering solution that follows the biological pixel-individual, frame-free approach to vision can potentially solve both problems.

2.2.2
The Human Retina from an Engineering Viewpoint

The retina is a neural network lining the back hemisphere of the eyeball and can be considered an extended and exposed part of the brain. Starting from a few light-sensitive neural cells, the photoreceptors, evolving some 600 million years ago, and further developed during a long evolutionary process, the retina is where the acquisition and first stage of processing of the visual information takes place. The retina's output to the rest of the brain is in the form of patterns of spikes produced by retinal ganglion cells, whose axons form the fibers of the optic nerve. These spike patterns encode the acquired and preprocessed visual information to be transferred to the visual cortex. The nearly 1 million ganglion cells in the retina compare signals received from groups of a few to several hundred photoreceptors, with each group interpreting what is happening in a part of the visual field. Between photoreceptors and ganglion cells, a complex network of various neuronal cell types processes the visual information and produces the neural code of the retina (Figure 2.1). As various features, such as light intensity, change in a given segment of the retina, a ganglion cell transmits pulses of electricity along the optic nerve to the brain in proportion to the relative change over time or space – and not to the absolute input level. Regarding encoding, there is a wide range of possibilities by which retinal ganglion cell spiking could carry visual information: by spike rate, precise timing, relation to spiking of other cells, or any combination of these [12]. Through local gain control, spatial and temporal filtering, and redundancy suppression, the retina compresses about 36 Gbit/s of raw high dynamic range (DR) image data into 20 Mbit/s spiking output to the brain. The retina's most sensitive photoreceptors, called "rods," are activated by a single photon, and the DR of processable light intensity exceeds the range of conventional artificial image sensors by several orders of magnitude.

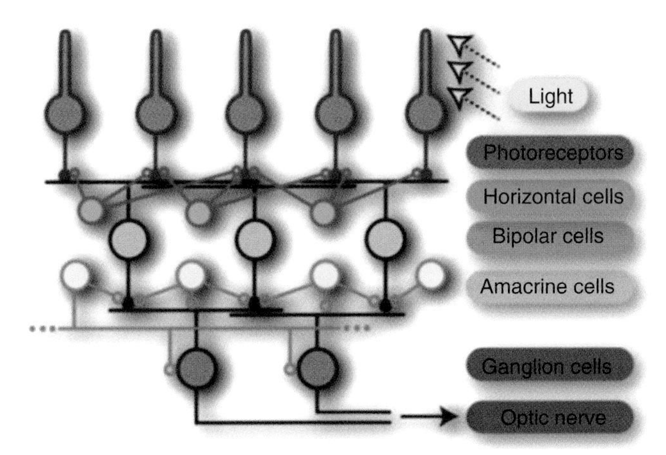

Figure 2.1 Schematic of the retina network cells and layers. Photoreceptors initially receive light stimuli and transduce them into electrical signals. A feedforward pathway is formed from the photoreceptors via the bipolar cell layer to the ganglion cells, which form the output layer of the retina. Horizontal and amacrine cell layers provide additional processing with lateral inhibition and feedback. Finally, the visual information is encoded into spike patterns at the ganglion cell level. Such encoded, the visual information is transmitted along their axons, forming the optic nerve, to the visual cortex in the brain. The schematic greatly simplifies the actual circuitry, which in reality includes various subtypes of each of the neuron types with different specific connection patterns. Also, numerous additional electrical couplings within the network are suppressed for clarity. (Adapted from Ref. [12].)

The retina has three primary layers: the photoreceptor layer, the outer plexiform layer (OPL), and the inner plexiform layer (IPL) [13, 14]. The photoreceptor layer consists of two types of cells, called *cones* and *rods*, which transform the incoming light into electrical signals, triggering neurotransmitter release in the photoreceptor output synapses and driving horizontal cells and bipolar cells in the OPL. The two major classes of bipolar cells, the ON bipolar cells and OFF bipolar cells, separately encode positive and negative spatiotemporal contrast in incoming light by comparing the photoreceptor signals to spatiotemporal averages computed by the laterally connected layer of horizontal cells. The horizontal cells are interconnected by conductive gap junctions and are connected to bipolar cells and photoreceptors in complex triad synapses. Together with the input current produced at the photoreceptor synapses, this network computes spatially and temporally low-pass filtered copies of the photoreceptor outputs. The horizontal cells feed back onto the photoreceptors to help set their operating points and also compute a spatiotemporally smoothed copy of the visual input. Effectively, the bipolar cells are driven by differences between the photoreceptor and horizontal cell outputs. In the yet more complex OPL, the ON and OFF bipolar cells synaptically connect to many types of amacrine cells and different types of ON and OFF ganglion cells in the IPL. Like the horizontal cells do with the photoreceptors and the bipolar cells, the amacrine cells mediate the signal transmission process between the bipolar cells and the ganglion cells, the spiking output cells of the retina network.

Bipolar and ganglion cells can be further subdivided into two different groups[1]: cells with primarily sustained and cells with primarily transient types of responses. These cells carry information along at least two parallel pathways in the retina, the magnocellular or transient pathway, where cells are more sensitive to temporal features (motion, changes, onsets) in the scene, and the parvocellular or sustained pathway, where cells are more sensitive to sustained features like patterns and shapes. The relevance of modeling these two pathways in the construction of bioinspired vision devices will be discussed further in Section 2.3.2.

In the following, a simplified set of characteristics and functions of biological vision that is feasible for silicon integrated circuit focal-plane implementation is summarized. As discussed previously, the retina converts spatiotemporal information contained in the incident light from the visual scene into spike trains and patterns, conveyed to the visual cortex by retinal ganglion cells, whose axons form the fibers of the optic nerve. The information carried by these spikes is maximized by the retinal processing, encompassing highly evolved adaptive filtering and sampling mechanisms to improve coding efficiency [15], such as:

- Local automatic gain control at the photoreceptor and network level to eliminate the retina's dependency on absolute light levels

1) In reality, this picture of a simple partition into sustained and transient pathways of course is too simple; there are many parallel pathways computing many views (probably at least 50 in the mammalian retina) of the visual input.

- Spatiotemporal bandpass filtering to limit spatial and temporal frequencies, such as reducing redundancy by filtering low frequencies and noise by filtering high frequencies
- Rectification in separate ON and OFF output cell types, perhaps to simplify encoding and locally reduce spike-firing rates

The varying distribution of different receptor types along with corresponding pathways across the retina, combined with saccades (precise rapid eye movements), elicits the illusion of high spatial and temporal resolution in the whole field of view. In reality, the retina acquires information in the retinal centre around the fovea at relatively low temporal resolution but with a high spatial resolution, whereas in the periphery, receptors are spaced at a wider pitch, but respond at much higher temporal resolution. In comparison to the human retina, a conventional image sensor sampling at the Nyquist rate requires transmitting more than 20 Gbit/s to match the human eyes' photopic range (exceeding 5 orders of magnitude), its spatial and temporal resolution, and its field of view. In contrast, by coding 2 bits of information per spike [16], the optic nerve transmits just about 20 Mbit/s to the visual cortex – a thousand times less.

To summarize, biological retinas have many desirable characteristics which are lacking in conventional image sensors – but inspire and drive the design of bioinspired retinomorphic vision devices. As will be further discussed in the remainder of this article, many of these advantageous characteristics have already been modeled and implemented on silicon. As a result, bioinspired vision devices and systems already outperform conventional, frame-based devices in many respects, notably wide DR operation, device-level video compression, and high-speed/low-data-rate vision.

2.2.3 Modeling the Retina in Silicon

The construction of an artificial "silicon retina" has been a primary target of the neuromorphic engineering community from the very beginning. The first silicon retina of Mahowald and Mead models the OPL of the vertebrate retina and contains artificial cones, horizontal cells, and bipolar cells. A resistive network computes a spatiotemporal average that is used as a reference point for the system. By feedback to the photoreceptors, the network signal balances the photocurrent over several orders of magnitude. The silicon retina's response to spatial and temporal changing images captures much of the complex behavior observed in the OPL. Like its biological counterpart, the silicon retina reduces the bandwidth needed to communicate reliable information by subtracting average intensity levels from the image and reporting only spatial and temporal changes [11, 17].

A next-generation silicon retina chip by Zaghloul and Boahen modeled all five layers of the vertebrate retina, directly emulating the visual messages that the ganglion cells, the retina's output neurons, send to the brain. The design incorporates both sustained and transient types of cells with adaptive spatial and temporal filtering and captures several key adaptive features of biological retinas. Light sensed by electronic photoreceptors on the chip controls voltages in the chip's circuits in a way that is analogous to how the retina's voltage-activated ion channels cause ganglion cells to generate spikes, in this way replicating responses of the retina's four major types of ganglion cells [7, 18].

2.3 From Biological Models to Practical Vision Devices

Over the past two decades, a variety of neuromorphic vision devices has been developed, including temporal contrast vision sensors that are sensitive to relative light intensity change, gradient-based sensors sensitive to static edges, edge-orientation sensitive devices, and optical-flow sensors [19]. Many of the early inventors and developers of bioinspired vision devices stem from the neurobiological community and saw their chips mainly as a means for proofing neurobiological models and theories and did not relate the devices to real-world applications. Very few of the sensors so far have been used in practical applications, yet in industry products. Many conceptually interesting pixel designs lack technical relevance because of, for example, circuit complexity, large silicon area, low fill factors, or high noise levels, preventing realistic application. Furthermore, many of the early designs suffer from technical shortcomings of VLSI implementation and fabrication such as transistor mismatch and did not yield practically usable devices. Recently, an increasing amount of effort is being put into the development of practicable and commercializable vision sensors based on biological principles. Today, bioinspired vision sensors are the most highly productized neuromorphic devices. Most of these sensors share the event-driven, frameless approach, capturing transients in visual stimuli. Their output is compressed at the sensor level, without the need of external processors, optimizing data transfer, storage, and processing, hence increasing power efficiency and compactness of the vision system. The reminder of this section reviews some of the recent developments in bioinspired artificial vision (Figure 2.2).

2.3.1 The Wiring Problem

As touched upon earlier, neuromorphic engineers observe striking parallels between the VLSI hardware used to implementing bioinspired electronics and the "hardware" of nature, the neural *wetware*. Nevertheless, some approaches taken by nature cannot be adopted in a straightforward way. One prominent challenge posed is often referred to as the "wiring problem." Mainstream VLSI technology does not allow for the dense three-dimensional wiring observed everywhere in biological neural systems. In vision, the optic nerve connecting

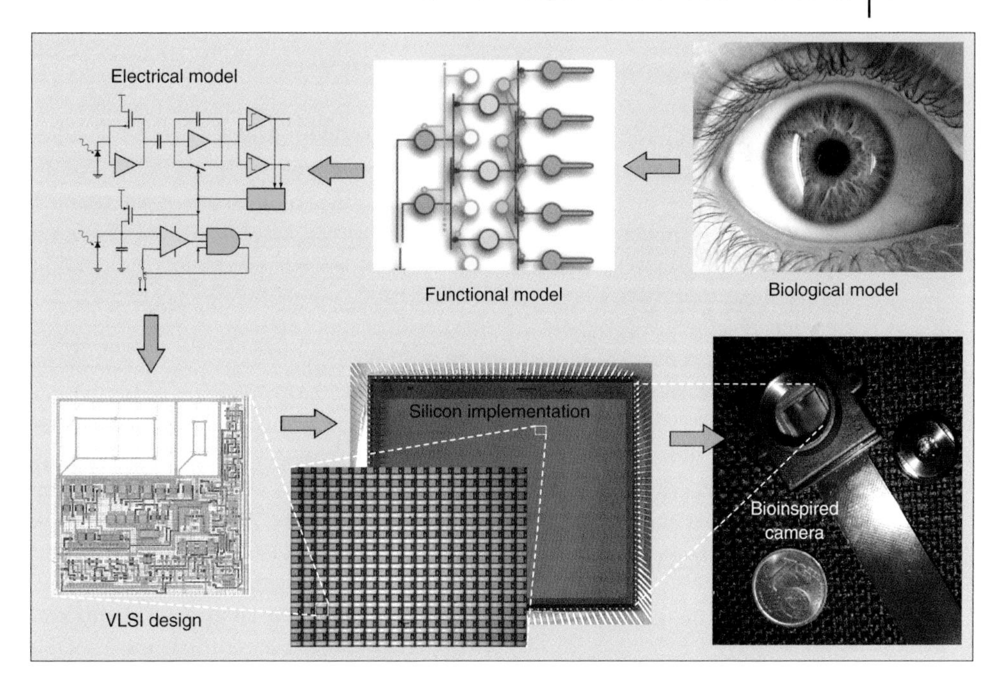

Figure 2.2 Modeling the retina in silicon – from biology to a bioinspired camera: ATIS "silicon retina" bioinspired vision sensor [20], showing the pixel cell CMOS layout (bottom left), microscope photograph of part of the pixel array and the whole sensor (bottom middle), and miniature bioinspired ATIS camera.

the retina to the visual cortex in the brain is formed by the axons of the about 1 million retinal ganglion cells, the spiking output cells of the retina. Translating this situation to an artificial vision system would imply that each pixel of an image sensor would have its own wire to convey its data out of the array. Given the restrictions posed by chip interconnect and packaging technologies, this is obviously not a feasible approach. However, VLSI technology does offer a workaround. Leveraging the 5 orders of magnitude or more of difference in bandwidth between a neuron (typically spiking at rates between 10 and 1000 Hz) and a digital bus enables engineers to replace thousands of dedicated point-to-point connections with a few metal wires and to time multiplex the traffic over these wires using a packet-based or "event-based" data protocol called address event representation (AER) [15, 16]. In the AER protocol, each neuron (e.g., pixel in a vision sensor) is assigned an address, such as its x, y-coordinate within an array. Neurons that generate spikes put their address in the form of digital bits on an arbitrated asynchronous bus. The bus arbiter implements a time-multiplexing strategy that allows all neurons to share the same physical bus to transmit their spikes. In this asynchronous protocol, temporal information is self-encoded in the timing of the spike events, whereas the location of the source of information is encoded in the form of digital bits as the "payload" of the event.

2.3.2 Where and What

The abstraction of two major types of retinal ganglion cells and corresponding retina pathways appear to be increasingly relevant with respect to the creation of useful bioinspired artificial vision devices. The magnocells are at the basis of the *transient* or *magnocellular pathway*. Magnocells are approximately evenly distributed over the retina; they have short latencies and use rapidly conducting axons. Magnocells have large receptive fields and respond in a transient way, that is, when changes – movements, onsets, and offsets – appear in their receptive field. The parvocells are at the basis of what is called the *sustained* or *parvocellular pathway*. Parvocells have longer latencies and the axons of parvocells conduct more slowly. They have smaller receptive fields and respond in a sustained way. Parvocells are more involved in the transportation of detailed pattern, texture, and color information [21].

It appears that these two parallel pathways in the visual system are specialized for certain tasks in visual perception: the magnocellular system is more oriented toward general detection or alerting and is also referred to as the "where" system. It responds with high temporal resolution to changes and motion. Its biological role is seen in detecting, for example, dangers that arise in the peripheral vision. Magnocells are relatively evenly spaced across the retina at a rather low spatial resolution and are the predominant cell type in the retinal periphery. Once an object is detected (often in combination with a saccadic eye movement), the detailed visual information like spatial details, color, and so on, seems to be carried primarily by the parvo system. It is hence called the "what" system. The "what" system is relatively slow, exhibiting low temporal, but high spatial resolution. Parvocells are concentrated in the fovea, the retinal center.

Practically, all conventional frame-based image sensors completely neglect the dynamic information provided by a natural scene and perceived in nature by the magnocellular pathway, the "where" system. Attempts to implementing the function of the magnocellular transient pathway in an artificial neuromorphic vision system has recently led to the development of the "dynamic vision sensor" (DVS). This type of visual sensor is sensitive to the dynamic information present in a natural scene and directly responds to changes, that is, temporal contrast, pixel individually, and near real time. The gain in terms of temporal resolution with respect to standard frame-based image sensors is dramatic. But also other performance parameters like the DR greatly profit from the biological approach. This type of sensor is very well suited for a plethora of machine vision applications involving high-speed motion detection and analysis, object tracking, shape recognition, 3D scene reconstruction, and so on [22–25]; however, it neglects the sustained information perceived in nature by the parvocellular "what" system.

Further exploitation of the concepts of biological vision suggests a combination of the "where" and "what" system functionalities in a bioinspired, asynchronous, event-driven style. The design of asynchronous, time-based image sensor (ATIS) [20, 26], an image and vision sensor that combines several functionalities of the biological "where" and "what" systems, was driven by this notion. Both DVS and ATIS will be further discussed in this chapter.

2.3.3
Temporal Contrast: The DVS

In an attempt to realize a practicable vision device based on the functioning of the magnocellular transient pathway, the "DVS" pixel circuit has been developed [27]. The DVS pixel models a simplified three-layer retina (Figure 2.3), implementing an abstraction of the photoreceptor – bipolar – ganglion cell information flow. Single pixels are spatially decoupled but take into account the temporal development of the local light intensity.

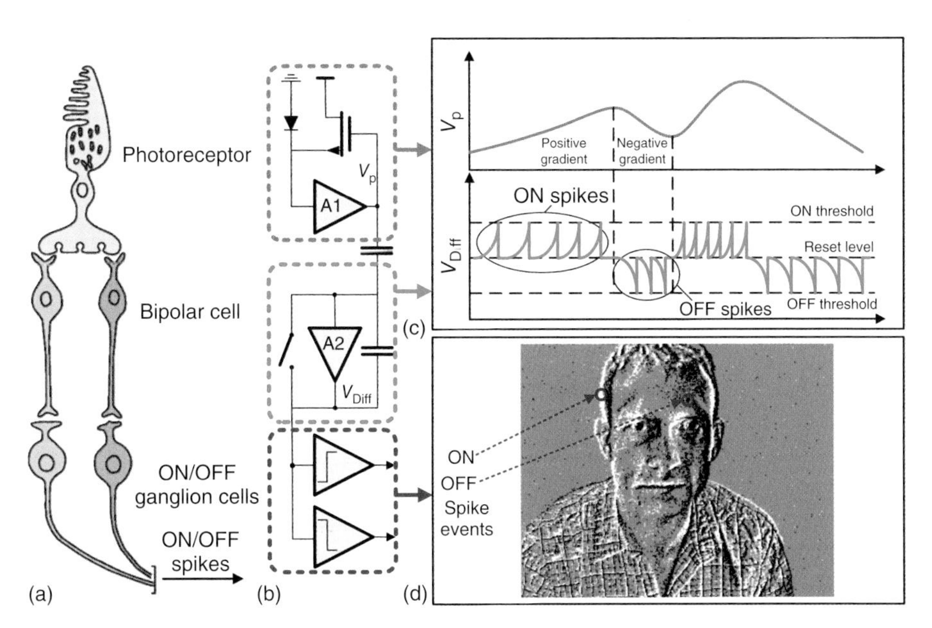

Figure 2.3 (a) Simplified three-layer retina model and (b) corresponding silicon retina pixel circuitry; in (c), typical signal waveforms of the pixel circuit are shown. The upper trace represents an arbitrary voltage waveform at the node V_p tracking the photocurrent through the photoreceptor. The bipolar cell circuit responds with spike events of different polarity to positive and negative gradients of the photocurrent while being monitored by the ganglion cell circuit that also transports the spikes to the next processing stage; the rate of change is encoded in interevent intervals; panel (d) shows the response of an array of pixels to a natural scene (person moving in the field of view of the sensor). Events have been collected for some tens of milliseconds and are displayed as an image with ON (going brighter) and OFF (going darker) events drawn as white and black dots.

The pixel autonomously responds to relative changes in intensity at microsecond temporal resolution over six decades of illumination. These properties are a direct consequence of abandoning the frame principle and modeling three key properties of biological vision: the sparse, event-based output, the representation of relative luminance change (thus directly encoding scene reflectance change), and the rectification of positive and negative signals into separate output channels (ON/OFF).

The major consequence of the bioinspired approach and most distinctive feature with respect to standard imaging is that the control over the acquisition of the visual information is no longer being imposed to the sensor in the form of external timing signals such as shutter or frame clock, but the decision making is transferred to the single pixel that handles its own visual information individually and autonomously. Consequently, the sensor is "event driven" instead of clock driven; hence, like its biological model, the sensor *responds* to visual events happening in the scene it observes. The sensor output is an asynchronous stream of pixel address events [15, 16] that directly encode scene reflectance changes. The output data volume of such a self-timed, event-driven sensor depends essentially on the dynamic contents of the target scene as pixels that are not visually stimulated do not produce output. Due to the pixel-autonomous, asynchronous operation, the temporal resolution is not limited by an externally imposed frame rate. However, the asynchronous stream of events carries only change information and does not contain absolute intensity information; there are no conventional image data in the sense of gray levels. This style of visual data acquisition and processing yields a pure dynamic vision device which closely follows its paradigm, the transient pathway of the human retina.

Relative change events and gray-level image frames are two highly orthogonal representations of a visual scene. The former contains the information on local relative changes, hence encodes all dynamic contents, yet there is no information about absolute light levels or static parts of the scene. The latter is a snapshot that carries an absolute intensity map at a given point in time, however has no information about any motion; hence, if scene dynamics are to be captured, one needs to acquire many of those frames. In principle, there is no way to recreate DVS change events from image frames nor can gray-level images being recreated from DVS events.

2.3.4 Event-Driven Time-Domain Imaging: The ATIS

Besides limited temporal resolution, data redundancy is another major drawback of conventional frame-based image sensors. Each acquired frame carries the information from all pixels, regardless of whether or not this information has changed since the last frame had been acquired. This approach obviously results, depending on the dynamic contents of the scene, in a more or less high degree of redundancy in the recorded image data as pixel values from unchanged parts in the scene get recorded and transmitted over and over, even though they do not contain any (new) information. Acquisition, transmission, and processing of this unnecessarily inflated data volume waste power, bandwidth, and memory resources and eventually limit the performance of a vision system. The adverse effects of this data redundancy, common to all frame-based image acquisition techniques, would be tackled most effectively at the pixel level by simply not recording the redundant data in the first place.

Again, biology is leading the way to a more efficient style of image acquisition. ATIS is an image sensor that combines several functionalities of the biological "where" and "what" systems with multiple bioinspired approaches such as event-based time-domain imaging, temporal contrast dynamic vision, and asynchronous, event-based information encoding and data communication [20, 26]. The sensor is based on an array of fully autonomous pixels that combine a simplified magnocellular change detector circuit model and an exposure measurement device inspired by the sustained parvocellular pathway whereby the parvocell is driven by the response to relative changes in local illuminance of its corresponding magnocell.

The magno change detector individually and asynchronously initiates the measurement of a new exposure/gray scale value only if – and immediately after – a brightness change of a certain magnitude has been detected in the field of view of the respective pixel. This biology-inspired way of organizing the acquisition of visual information leads to an image sensor that does not use a shutter or frame clock to control the sampling of the image data, but in which each pixel autonomously defines the timing of its own sampling points in response to its visual input. If things change quickly, the pixel samples at a high rate; if nothing changes, the pixel stops acquiring redundant data and goes idle until things start to happen again in its field of view. In contrast to a conventional image sensor, the entire image acquisition process hence is not governed by a fixed external timing signal but by the visual signal to be sampled itself, leading to near ideal sampling of dynamic visual scenes.

The parvo exposure measurement circuit in each pixel encodes the instantaneous absolute pixel illuminance into the timing of asynchronous spike pulses, more precisely into interspike intervals comparable to a simple rate coding scheme (Figure 2.4). As a result, the ATIS pixel does not rely on external timing signals and autonomously requests access to an asynchronous and arbitrated AER output channel only when it has a new gray scale value to communicate. At the AER readout periphery, the pixel events are arbitrated, furnished with the pixel's array address by an address encoder and sent out on an asynchronous bit-parallel AER bus [15, 16].

The temporal redundancy suppression of the pixel-individual data-driven sampling operation ideally yields lossless focal-plane video compression with compression factors depending only on scene dynamics. Theoretically approaching infinity for static scenes, in practice, due to change detector background activity, the achievable compression factor is limited and reaches 1000 for

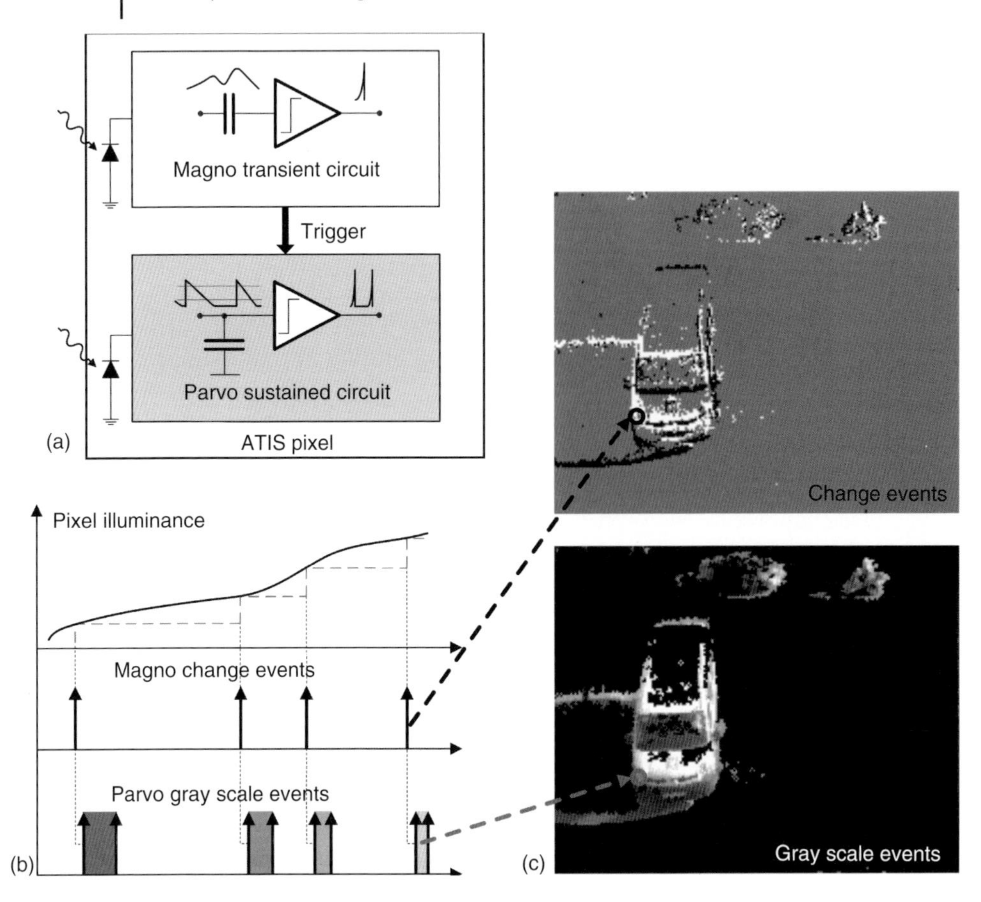

Figure 2.4 (a) Functional diagram of an ATIS pixel. (b) Arbitrary light stimulus and pixel response: two types of asynchronous "spike" events, encoding temporal change and sustained gray scale information, are generated and transmitted individually by each pixel in the imaging array. (c) Change events coded black (OFF) and white (ON) (top) and gray-level measurements at the respective pixel positions triggered by the change events (bottom).

bright static scenes. Typical dynamic scenes yield compression ratios between 50 and several hundreds. Figure 2.5 shows a typical traffic surveillance scene generating a 25–200 k events/s at 18 bit/event continuous-time video stream. The temporal resolution of the pixelwise recorded gray-level updates in this scene is about 1 ms, yielding 1000 frames per s equivalent video data. In addition, the time-domain encoding of the gray-level information results in exceptionally high DR beyond 120 dB and improved signal-to-noise ratio (SNR) of typically greater 50 dB [28].

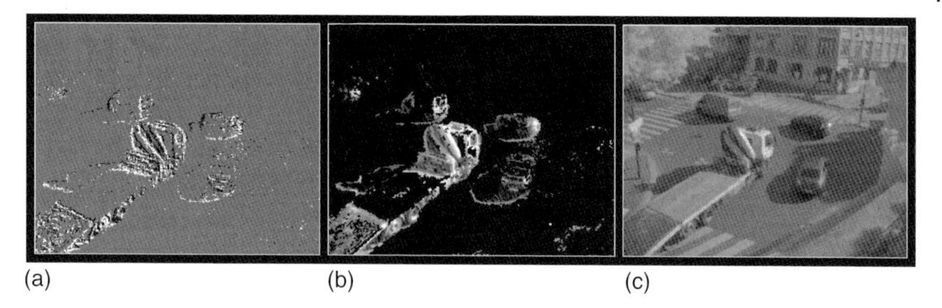

Figure 2.5 Instance of a traffic scene observed by an ATIS. (a) ON/OFF changes (shown white/black) leading to instantaneous pixel-individual sampling. (b) Associated gray scale data recorded by the currently active pixels displayed on black background – all black pixels do not sample/send information at that moment. (c) Same data with full background acquired earlier, showing a low-data-rate, high-temporal-resolution video stream. The average video compression factor is around 100 for this example scene; that is, only 1% of the data (with respect to a standard 30 frames per s image sensor) are acquired and transmitted, yet allowing for a near lossless recording of a video stream. The temporal resolution of the video data is about 1000 frames per s equivalent.

2.4 Conclusions and Outlook

Bioinspired vision technology is starting to outperform conventional, frame-based vision systems in many application fields and to establish new benchmarks in terms of redundancy suppression and data compression, DR, temporal resolution, and power efficiency. Demanding vision tasks such as real-time 3D mapping, complex multiobject tracking, or fast visual feedback loops for sensory–motor action, tasks that often pose severe, sometimes insurmountable, challenges to conventional artificial vision systems, are in reach using bioinspired vision sensing and processing techniques.

Fast sensorimotor action through visual feedback loops, based on the frame-free, event-driven style of biological vision, support, for example, autonomous robot navigation as well as micromanipulation and image-guided intervention in applications like scientific microscopy or robot-assisted surgery. Related developments are about to influence other fields such as human-machine systems involving, for example, air gesture recognition.

At the other end of the spectrum are biomedical applications like retina prosthetics and vision-assisted artificial limbs. For retinal implants, event-based and pulse-modulation vision chips that partially model human retina operation are naturally suitable to serve as the signal-generating front end. The sensors produce an output of pulse streams which can directly be used for evoking cell potentials. Furthermore, they can operate on very low voltages without degrading signal-to-noise ratio, which is an essential feature for implantable devices due to the need for low power dissipation, limiting heat generation, and extending battery lifetime. Finally, the intrinsic ultrahigh DR of this type of vision chips is very advantageous for the task of replacing biological photoreceptors. Recently, it has been shown that the computation of parallel filtering occurring in the mammalian retina can be reproduced based on data delivered by a neuromorphic sensor. With a simple linear–nonlinear model, it was possible to reconstruct the responses of the majority of ganglion cell types in the mammalian retina [29], such demonstrating the suitability of bioinspired vision sensor to serve as a transducer for retinal prosthetics.

The direct modeling of retinal (dys)functions and operational principles in integrated electronic circuits allows reproducing and studying retinal defects and diseases, potentially helping in devising novel ways of medical diagnosis and treatment. Research toward physical models of retinas and retinal defects in VLSI silicon, thus realizing artificial "patient's eyes," could facilitate large-scale experimental studies of particular retinal defects by partly replacing laborious and costly *in vivo* and *in vitro* studies, such as supporting design and construction of medical diagnosis and treatment devices and systems.

Finally, time-based imagers like ATIS deliver high DR, high-quality imaging, and video for scientific applications like astronomical imaging, fluorescence imaging, cell monitoring, or X-ray crystallography.

References

1. McCulloch, W. and Pitts, W. (1943) A logical calculus of the ideas immanent in nervous activity. *Bull. Math. Biol.*, **5**, 115–133.
2. Hebb, D. (1949) *The Organization of Behavior*, Wiley-VCH Verlag GmbH, New York.
3. Hodgkin, A. and Huxley, A. (1952) A quantitative description of membrane current and its application to conduction and excitation in nerve. *J. Physiol.*, **117**, 500Y544.
4. Mead, C. (1989) *Analog VLSI and Neural Systems*, Addison-Wesley.
5. Mead, C. (1990) Neuromorphic electronic systems. *Proc. IEEE*, **78**(10), 1629–1636.
6. Maher, M.A.C., Deweerth, S.P., Mahowald, M.A., and Mead, C.A. (1989) Implementing neural architectures using analog VLSI circuits. *Trans. Circuits Syst.*, **36**(5), 643–652.
7. Boahen, K.A. (2005) Neuromorphic microchips. *Sci. Am.*, **292**, 56–63.
8. Sarpeshkar, R. (2006) Brain power – borrowing from biology makes for low power computing – bionic ear. *IEEE Spectr.*, **43**(5), 24–29.
9. Indiveri, G. (2007) Synaptic plasticity and spike-based computation in VLSI networks of integrate-and-fire neurons. *Neural Inf. Process. - Lett. Rev.*, **11**(4-6), 135–146.
10. Furber, S. and Temple, S. (2007) Neural systems engineering. *J. R. Soc. Interface*, **2007**(4), 193–206.
11. Mead, C.A. and Mahowald, M.A. (1991) The silicon retina. *Sci. Am.*, **264**, 76–82.
12. Gollisch, T. (2009) Throwing a glance at the neural code: rapid information transmission in the visual system. *HFSP J.*, **3**(1), 36–46.
13. Masland, R. (2001) The fundamental plan of the retina. *Nat. Neurosci.*, **4**, 877–886.
14. Rodieck, R.W. (1998) The primate retina. *Comp. Primate Biol.*, **4**, 203–278.
15. Boahen, K. (2000) Point-to-point connectivity between neuromorphic chips using address events. *IEEE Trans. Circuits Syst. II*, **47**(5), 416–434.

16. Boahen, K. (2004) A burst-mode word-serial address-event link-I: transmitter design. *IEEE Trans. Circuits Syst. I*, **51**(7), 1269–1280.
17. Mahowald, M.A. (1992) VLSI analogs of neuronal visual processing: a synthesis of form and function. PhD Computation and Neural Systems, Caltech, Pasadena, CA.
18. Zaghloul, K.A. and Boahen, K. (2006) A silicon retina that reproduces signals in the optic nerve. *J. Neural Eng.*, **3**, 257–267.
19. Moini, A. (2000) *Vision Chips*, Kluwer Academic Publishers.
20. Posch, C., Matolin, D., and Wohlgenannt, R. (2011) A QVGA 143dB dynamic range frame-free PWM image sensor with lossless pixel-level video compression and time-domain CDS. *J. Solid-State Circuits*, **46**(1), 259–275.
21. Van Der Heijden, A.H.C. (1992) *Selective Attention in Vision*, Routledge, New York. ISBN: 0415061059.
22. Belbachir, A.N. *et al.* (2007) Real-time vision using a smart sensor system. 2007 IEEE International Symposium on Industrial Electronics, June 4–7, 2007.
23. Perez-Carrasco, J.A. *et al.* (2010) Fast vision through frameless event-based sensing and convolutional processing: application to texture recognition. *IEEE Trans. Neural Networks*, **21**(4), 609–620.
24. Serrano-Gotarredona, R. *et al.* (2009) CAVIAR: a 45k Neuron, 5M Synapse, 12G Connects/s AER hardware sensory–processing– learning–actuating system for high-speed visual object recognition and tracking. *IEEE Trans. Neural Networks*, **20**(9), 1417–1438.
25. Carneiro, J., Ieng, S.H., Posch, C., and Benosman, R. (2013) Asynchronous event-based 3D reconstruction from neuromorphic retinas. *Neural Networks*, **45**, 27–38.
26. Posch, C., Matolin, D., and Wohlgenannt, R. (2008) An asynchronous time-based image sensor. ISCAS 2008. IEEE International Symposium on Circuits and Systems, May 18–21 2008, pp. 2130–2133.
27. Lichtsteiner, P., Posch, C., and Delbruck, T. (2006) A 128 × 128 120dB 30mW asynchronous vision sensor that responds to relative intensity change. ISSCC, 2006, Digest of Technical Papers February 6–9, 2006, pp. 2060–2069.
28. Chen, D., Matolin, D., Bermak, A., and Posch, C. (2011) Pulse modulation imaging – review and performance analysis. *IEEE Trans. Biomed. Circuits Syst.*, **5**(1), 64–82.
29. Lorach, H. *et al.* (2012) Artificial retina: the multichannel processing of the mammalian retina achieved with a neuromorphic asynchronous light acquisition device. *J. Neural Eng.*, **9**, 066004.

3
Retinal Processing: From Biology to Models and Applications

David Alleysson and Nathalie Guyader

3.1
Introduction

To perceive the surrounding world, its beauty and subtlety, we constantly move our eyes, and, our visual system does an incredible job, processing the visual signal on different levels. Even though vision is the most frequently studied sense, we still do not fully understand how the neuronal circuitry of our visual system builds a coherent percept of the world. However, several models have been proposed to mimic some parts and some of the ways in which our visual system works.

Biologically inspired computer vision is not solely based on scientific results in the field of biology alone; various behavioral responses obtained through psychophysical experiments provide much needed data to propose models and to test their efficiency. Contrast sensitivity functions (CSF), which represent perceived luminance contrast as a function of spatial or temporal frequencies, were measured before the discovery of the neural network underlying them (see Section 3.3). Similarly, trichromacy, the fact that color vision lies in a tridimensional space, was described before the discovery of the three types of cones in the retina. In most cases, psychophysics precedes biology (See Chapter 5). To a certain extent, psychophysics remains biologically inspired, because everybody agrees that human visual behavior is "rooted" and emerges with the neural circuitry of the visual system. Often, it is simpler to find a model of human behavior instead of the biological structure and function of the neurons that support it.

However, the relationship between behavior and the underlying biology might be difficult to find, as is illustrated by the following three examples. The first one concerns the debate on random cone wiring to post-receptoral cells in color vision as discussed in Refs [1] and [2]. The second is about the fact that our visual world appears stable, whereas our eyes move continuously; and more especially, the neurons and mechanism involved in this processing are not known. The third concerns the neurophysiological explanation of the measured contrast sensitivity function. We still wonder whether the processing responsible for this perception occurs in the retina, in the transmission between retina and cortex, or in cortical areas.

Biologically Inspired Computer Vision: Fundamentals and Applications, First Edition.
Edited by Gabriel Cristóbal, Laurent Perrinet, and Matthias Keil.

Despite these examples, knowing exactly at which level a particular phenomenon is coded in our visual system is not essential to propose models of vision. The aim of models is to reproduce the functionalities of human vision to develop signal-processing algorithms that mimic the visual system functions. The limit to the imitation of nature by artificial elements is not a barrier to this if we are able to reproduce behavior. However, we must remain careful that a model, even one that works perfectly for behavior, is not necessarily an exact description of what happens in the visual system.

With the main objective being the description of biologically inspired computer vision, this chapter mixes results that come from psychophysics with results stemming from biology. We focus on the processes carried out at the retinal stage and we introduce some pathways between retinal biology and physiology, which are useful for computer vision and which result from psychophysical experiments and computational modeling. Because such pathways must encompass several different disciplines, we are not able to provide an exhaustive description and we choose to present some aspects. However, this introduction describes all the factors that should be taken together to understand the issue of computer vision.

The first part of the chapter is a short description of the anatomy and physiology of the retina. Useful links are given to the reader who wishes to obtain further information. The second part summarizes different types of retinal models. We present a panorama of these models and discuss their performance and usefulness both to explain vision and to explain their use in computer vision. The last part is dedicated to some examples of the utilization of models of vision in the context of digital color camera processing. This last part allows us to illustrate our purpose through real applications.

3.2 Anatomy and Physiology of the Retina

3.2.1 Overview of the Retina

Because vision is the most widely studied sense, a vast body of research in biology, physiology, medicine, neuroanatomy, psychophysics as well as computational modeling provide data to better understand this sense. This chapter goes through the basics of retinal functioning rather than providing a detailed description. The aim is to help a "vision researcher" to understand roughly how the retina works and how to model it for applications in computer vision. If one wants to know more about retina anatomy and physiology, he/she might want to read two free online books: (Webvision, The Organization of the Retina and Visual System by Helga Kolb and colleagues (*http://webvision.med.utah.edu/*) and Eye, brain and vision by David Hubel (*http://hubel.med.harvard.edu/book/bcontex.htm*).

The first detailed description of retinal anatomy was provided by Ramon y Cajal about a century ago [3]. Since then, with technological improvements, detailed

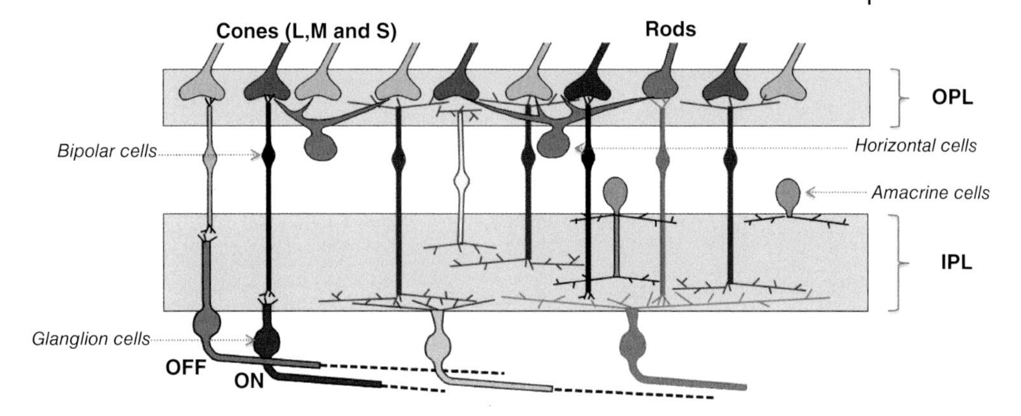

Figure 3.1 Schematic representation of the organization into columns of retinal cells. Light is transmitted through the retina from the ganglion cells to the photoreceptors. The neural signal is transmitted in the opposite direction from photoreceptors, their axons located at the outer plexiform layer (OPL), to ganglion cells, their dendritic field located in the inner plexiform layer (IPL), passing through bipolar cells (adapted from Ref. [4]).

knowledge about the organization of the retina and the visual system has been developed. A schematic representation of the retina is shown in Figure 3.1. This diagram is a simplified representation because the retina is much more complex and contains more types of cells. Basically, the retinal layers are organized into columns of cells, from the "input" cells, the photoreceptors, rods, and cones, to the "output" cells, the retinal ganglion cells, that have their synapses directly connected to the thalamus (lateral geniculate nucleus (LGN)), and from there to other areas, mainly the primary visual cortex, also called *V1*. In the retina, layers of cells are inversed compared to the transmission of light, with the photoreceptors located at the back of the retina. Two main layers of cells are generally described in the literature: (i) the outer plexiform layer (OPL), the connection between photoreceptors, horizontal cells, and bipolar cells and (ii) the inner plexiform layer (IPL), the connection between the bipolar, amacrine, and ganglion cells.

The photoreceptors convert the light, that is, the incoming photons, into an electrical signal through membrane potential; this corresponds to transduction, a phenomenon that exists for all the sensory cells. The electrical signal is transmitted to bipolar and ganglion cells. The electrical signal is integrated within ganglion cells and transformed into nerve spikes. Nerve spikes are a time-coded digital form of an electrical signal. This is used to transmit nervous system information over long distances, in this case through the optic nerve and into the visual cortical areas of the brain.

3.2.2
Photoreceptors

The retina contains an area without any photoreceptors where ganglion axons connect the retina through the LGN forming the optic nerve, to the primary visual

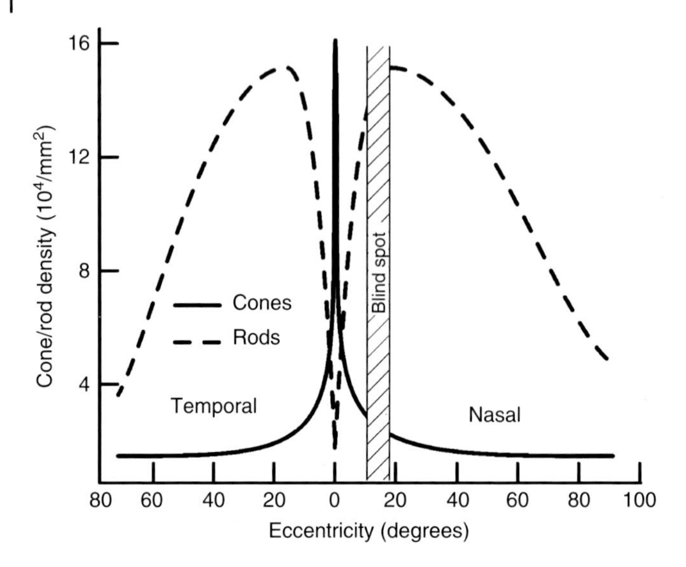

Figure 3.2 Number of photoreceptors (rods and cones) at the retina surface (in square millimeters) as a function of eccentricity (distance from the fovea center, located at 0). The blind spot corresponds to the optic nerve (an area without photoreceptor). Redrawn from Ref. [6]

cortex (back of the brain, occipital lobe). The center of the retina, called the *fovea*, is a particular area of the retina with the highest density of photoreceptors; it is located close to the center of the retina (the eye's optical axis) and corresponds to central vision. The area of the visual field that is gazed at is directly projected onto this central area. There are two main types of photoreceptors, which are named according to their physical shapes: rods and cones. The fovea only contains cones (Figure 3.2), the photoreceptors which are sensitive to color information. Three types of cones might be distinguished for their sensitivity to three different wavelengths: S-cones, sensitive to short wavelengths that respond to bluish color, M-cones, sensitive to medium wavelengths that correspond to greenish-yellowish color, and L-cones, sensitive to long wavelengths that correspond to reddish color. These photoreceptors are activated during day-vision (high light intensity). The number of cones drastically decreases with the eccentricity onto the retinal surface. On the periphery, the number of cones is very small compared to the number of rods. These latter photoreceptors are responsible for scotopic vision (vision under dim light). Not only are photoreceptors sensitive to light but a freshly discovered ganglion cell with light sensitivity through melanopsin expression is also responsible for circadian rhythm accommodation and pupil aperture [5].

The arrangement of photoreceptors on the retina surface has huge consequences on our visual perception. Our visual acuity is maximal for the region of the visual field gazed at (central vision), whereas it decreases rapidly for the surrounding areas (peripheral vision). It is important to note that the decrease in our visual acuity in peripheral vision is only partly due to retina photoreceptor distribution. In fact, other physiological and anatomical explanations might be given (see Section 3.2.3). Moreover, although we do not feel it in our everyday life, only our central vision is colored and peripheral vision is achromatic. Another particularity of cone arrangement is that the placing of L, M, and S cones is random in the retina (see Figure 3.5(b)) and differs from one individual to another [7].

Photoreceptors dynamically adapt their responses to various levels of illumination [8]. This property allows human vision to function and to be efficient for several decades of light intensity. In fact, we perceive the world and are able to recognize a scene on a sunny day as well as in twilight.

3.2.3
Outer and Inner Plexiform Layers (OPL and IPL)

At this stage, it is important to introduce the notion of *receptive field*. The receptive field of a cell is necessary to explain the functioning of retinal cells, and more generally, visual cells. This notion was introduced by Hartline for retinal ganglion cells [9]. According to Hartline's definition, the receptive field of a retinal ganglion cell corresponds to a restricted region of the visual field where light elicits response of the cell. This notion was extended to other retinal cells as well as other cortical or sensory cells. The receptive fields of visual cells correspond to a small area of the visual field. For example, the receptive field of a photoreceptor is a cone-shaped volume with the different directions in which light elicits a response from the photoreceptor [10]. However, most of the time, the receptive field of a visual cell is schematized in two dimensions corresponding to a visual stimulus located at a fixed distance from the participant.

The OPL corresponds to the connection between photoreceptors, bipolar cells, and horizontal cells. This connection is responsible for the center/surround receptive fields displayed by bipolar cells. Two types of bipolar cells can be distinguished: *on-center* and *off-center*. The receptive fields of retinal bipolar cells correspond to two concentric circles with different luminance contrast polarities. On-center cells respond when a light spot falls inside the central circle surrounded by a dark background and off-center cells respond to a dark stimulus surrounded by light [11]. The interaction between center and surround of the receptive fields causes contour enhancement, which can be easily implemented in image processing with a spatial high-pass filter.

The IPL corresponds to the connection between bipolar, ganglion, and amacrine cells. Ganglion cells still have center/surround receptive fields. Retinal ganglion cells vary significantly in terms of their size, connection, and response. As the number of photoreceptors decreases with retinal eccentricity, the size of the receptive fields of bipolar and ganglion cells increases. At the retinal center, there is the so-called one-to-one connection where one ganglion cell connects to one bipolar cell and to one photoreceptor ensuring maximal resolution around the optical axis [12]. However, at larger eccentricities, one bipolar cell connects up to several hundreds of photoreceptors.

There are essentially three types of anatomically and functionally distinct ganglion cells that project differentially to the LGN [13]. The most numerous type (around 80%) in the fovea and near fovea are the midget ganglion cells, so called because of their small size. They are the cells that connect one-to-one with photoreceptors in the fovea center. Their responses are sustained and slow, they are sensitive to high contrast, high spatial frequency as well as L-M color opponency. They project their axons to the parvocellular part of the LGN. In contrast, parasol ganglion cells are large and connect several bipolars and several photoreceptors even in the fovea. Their responses are transient and fast. They are known to be sensitive to low contrast and carry low spatial frequency, and are color blind. They also participate in night vision by integrating rod responses. They project their axons in the magnocellular pathway of the LGN. A third kind, called *bistratified ganglion cells* have dendrites which form two layers in the IPL. They are known to carry S-(L+M) color opponency and project in the koniocellular pathway of the LGN. Note that a new kind of ganglion cell containing melanopsin pigment that makes them directly sensitive to light has been recently discovered [5].

3.2.4 Summary

Most retinal ganglion cell axons are directly connected to the superior colliculi and the cortical visual areas passing through the lateral geniculate nuclei. The role of the retina is to preprocess the visual signal for the brain. Then the signal is successively processed by several cortical areas to bring a coherent percept of the world.

The main functions of the retina include the following:

- converting photons into neural signals that can be transmitted through neurons of our visual system;
- adapting its response to the mean luminance of the scene but also to the local luminance contrast;
- decomposing the input signal: low spatial frequencies are transmitted first to the cortex, immediately followed by high spatial frequencies and opponent color information.

In the next section, we present how models of retina simulate these different functions.

3.3 Models of Vision

3.3.1 Overview of the Retina Models

Models of the acquisition and preprocessing done by the retina mainly focus on modeling how retinal cells encode visual information. More specifically, these models address the following questions: are we able to reproduce the measured response of retinal cells using simple identified elements from biology? Or are we able to explain some sensation that we have when we perceive certain visual stimuli? These two questions are answered by two main types of model, the models that are based on the physiology of retinal cells and those that are based on the description of a stimulus.

A fundamental model was proposed by Hodgkin and Huxley [14]. It explains the dynamics of neural conduction, describing the conductance along the neuron's membrane with a differential equation. Mead and Mahowald [15] were the first to propose an equivalent electrical model of early visual processing and implement it in a silicon circuit. The advantage of defining an equivalent electrical model is double. First, an electrical model of analog signal and digital image processing algorithms might be translated into differential equations, offering a direct correspondence between retina functioning and signal/image processing. Such an approach allows different retinal cells to be simulated that have different connections to other cells (gap-junction, chemical synapse) with a slightly different electrical model. Finally, all these models are integrated into a network corresponding to the whole retina. Second, it is easy to infer the behavior of such a heterogeneous network of cells as long as the network follows certain rules [16].

In the following section, we briefly describe some models of the retina based on the Hodgkin and Huxley model of neural transmission and its equivalent electrical system.

Then, we introduce a different type of model based on information theory. Its goal is to preprocess the input signal into an efficient representation of the natural scene. Consequently, these models are based on so-called ecological vision because the preprocessing of the retina optimizes the representation of information coming from the visual scene.

We finish this part on retinal models by discussing a more general approach to neural modeling. Such models build a geometric space of neural activities and then infer the behavior of the network using the geometric rules stated at the neural scale. Neurogeometry, a term initially used by Jean Petitot, is a very persuasive example of such a category of model. Focusing on color vision, Riemannian geometry (e.g., it generalizes geometric spaces with a fixed metric by allowing a positionwise variation of their metrical properties) has been used extensively to explain color discrimination or differences from the LMS cone excitation space. We describe how these line-element models have tried to put human color discrimination capabilities in the context of Riemannian geometry.

3.3.2 Biological Models

3.3.2.1 Cellular and Molecular Models

A model at the cellular and molecular scale was proposed by van Hateren [17]. His model of the outer retina has been validated against the measured signal on the horizontal cells to wide field, spectrally white stimuli. Horizontal cells are known to change their sensitivities and control their bandwidth with background intensity. The response is nonlinear and the nonlinear function changes for background ranging from 1 to 1000 trolands.[1] All these properties are taken into account in the van Hateren model. He proposed a three-cascade feedback loop representing successively the calcium concentration of the cone outer pigment, the membrane voltage of the cone inner pigment, and the signal modulation of the horizontal cells to cone signals. His model shows that cone responses are the major factor regulating sensitivity in the retina. In a second version, he proposes to include pigment bleaching in the cone outer segment to better simulate cone response [18].

Van Hateren implements his model using an autoregressive moving-average filter that allows the processing to be modified in real time following the adaptation parameter (or the state at time t of the static nonlinearity part of the model) and fast calculus [17, 18]. He also manages to give all the constants involved in the model and implementation in Fortran and Matlab[2] which make the model very useful. The model is principally suited to simulating cone responses, but can be used in digital image processing such as high dynamic range processing [19, 20].

Even by focusing on cellular and molecular modeling, van Hateren falls on a more general kind of modeling that does not necessary follow biology closely. These general models use two kinds of building blocks, a linear (L) multicomponent filtering and a componentwise nonlinearity (N). The blocks are then considered in cascade, in what is called LNL or NLN models, to consider either the case of two linear operators L *sandwiched* the nonlinear N or vice versa. The NLN model (also called *Hammerstein – Wiener*) is less tractable than the LNL model [21, 22]. These models can be understood as a decomposition of the general nonlinear Volterra system [23, 24]. Because nonlinearity is componentwise, the only mixing between components (either spatial, spectral, or temporal) is through a linear function. These models have been found to be representative of the mechanism of gain control in the retina [25, 26] but may potentially be more interesting to model color vision [26 – 29].

Cellular and molecular network models of the retina are the most promising because they are on a perfect scale to fit electrophysiological measurement and the theory of the dynamic on heterogeneous neural network. However, they are difficult because they reach the limit of our knowledge about the optimization of complex systems.

3.3.2.2 **Network Models**

Models that simulate the way the retina works are numerous — from detailed models of a specific physiological phenomenon to models of the whole retina. In this chapter, we wanted to emphasize two interesting models that take into account the different aspects of retina function. These models naturally make

1) Troland is a unit of conventional retinal illuminance corrected for pupil size.
2) Implementation details are available at *https://sites.google.com/site/jhvanhateren/home/software-van-hateren-snippe-2007*

simplifications regarding the complexity of retinal cells but provide interesting results by replicating experiments either on particular response cells (e.g., the ganglion cells of cats) or on more global perceptual phenomena (contrast sensitivity function). An implementation of two of these models is provided through executable software that can be freely downloaded:

- Virtual retina from Wohrer and Kornprobst [30]: *http://www-sop.inria.fr/neuromathcomp/public/software/virtualretina/*
- The retina model originally proposed by Hérault and Durette [31] and improved and implemented by Benoit *et al.* [32] *http://docs.opencv.org/3.0-beta/modules/bioinspired/doc/retina/index.html*

The model proposed by Hérault [4] summarizes the main functions of the retina and it is compared to an image preprocessor. His simplified model of retinal circuits is based on the electrical properties of the cellular membrane that provides a linear spatiotemporal model of retinal cell processing. Beaudot *et al.* [33, 34] first proposed a basic electrical model equivalent to a series of cells including a membrane's leak current and connected by gap junctions. This model provides the fundamental equations of retina functions to finally model the retina circuitry through a spatiotemporal filter (see Figure 3.3 for its transfer function and Figure 3.4 for simulation on an image).

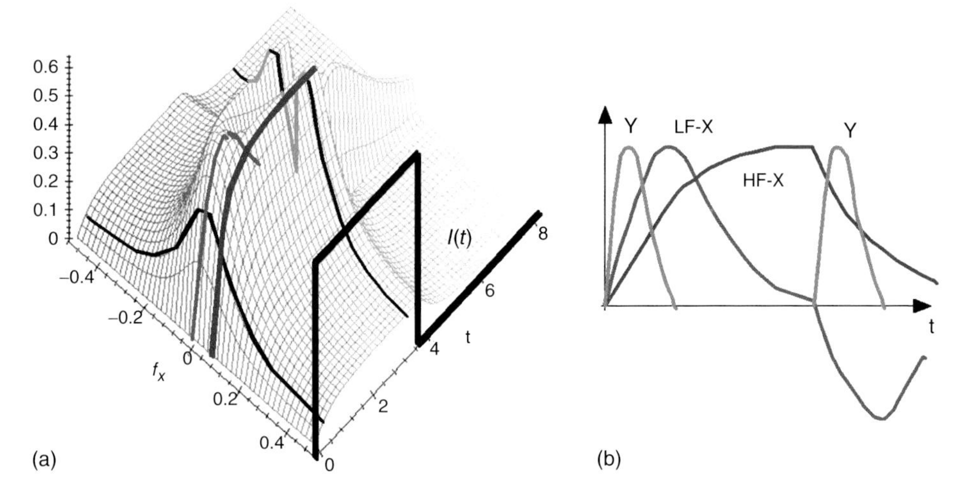

Figure 3.3 Simulation of the complexity of the spatiotemporal response of the retina in Hérault's model. (a) Temporal evolution of the modulus of the spatial transfer function (only in one dimension f_x) for a square input $I(t)$. When light is switched on, the spatial transfer function is low-pass at the beginning and becomes band-pass (black lines). Blue and red lines show the temporal profile for respectively low spatial frequency (magnocellular pathway) and high spatial frequency (parvocellular pathway). When the light is switched off, there is a rebound for low spatial frequency. (b) The corresponding temporal profiles for high spatial frequencies (HF-X, parvo) and low spatial frequencies (LF-X magno). The profile for Y cells is also shown (adapted with permission from Ref. [4]).

Figure 3.4 Simulation of Hérault's model on an image that is considered steady. Each image shows respectively the response of cones, horizontal cells, bipolar cells ON and OFF, midget cells (Parvo), parasol cells (Magno X), and cat's Y cells (Magno Y) in response to the original image.

It simulates the main function of the retina with a luminance adaptation as well as a luminance contrast mechanism that results in contrast enhancement. Finally, owing to the nonseparability of the spatial and temporal dimensions, the low spatial frequencies of the signal are transmitted first and later refined by the high spatial frequencies. This phenomenon reproduces the coarse to fine processing that occurs in the visual system [35]. The retina model has been integrated into a real-time C/C++ optimized program [32]. In addition, these authors proposed, with the retina model, a model of the cortical cells to show the advantages of bioinspired models to develop efficient modules for low-level image processing.

Wohrer and Kornprobst [30] also proposed a bioinspired model of the retina and proposed a simulation software, called *Virtual Retina*. The model is biologically inspired and offers a large-scale simulation (up to 100,000 neurons) in reasonable processing time. Their model includes a linear model of filtering in the OPL, a shunting feedback at the level of bipolar cells accounting for contrast gain control, and a spike generation process modeling ganglion cells. Their model reproduces several experimental measurements from single ganglion cells such as cat X and Y cells (like the data from Ref. [36]).

3.3.2.3 **Parallel and Descriptive Models**

Werblin and Roska [37] based their model on the identification of several different ways of processing by cells in the retina. The characteristics of the cells were investigated by electrophysiology. They considered the retina as a highly parallel processing system, providing several maps of information built by the network of different cells. The advantage of this approach is to provide images or image sequences of the different functions of the retina.

Similarly, Gollisch and Meister [38] have recently proposed the possibility that the retina could be embedded with high-level functions. They show elegantly how these functions could be implemented by the retinal network.

Another class of model has been called *computational models of vision* [39, 40]. They aim to model the scale-invariant processing of the visual system by integrating the spatiotemporal filtering of the retina into pyramidal image processing.

In conclusion to this part, no one knows to what degree of sophistication the retina is built, but what we do know is that almost 98% of the ganglion cells that have their dendrites connecting the retina to the LGN have been identified robustly. Nevertheless, their role in our everyday vision and perception has not yet been fully understood.

3.3.3 Information Models

Describing a scene viewed by a human observer requires consideration of every point of the scene as reflecting light. For a complex scene, this light field represents a huge amount of information. It is therefore not possible for a brain to encode every individual scene in that way to enable useful information to be retrieved. From a biological perspective, the retina has hundreds of millions of photoreceptors but the optic nerve that conducts information from the retina to the brain comprises only tens of millions of fibers. These arguments have led to the idea that a compression of information occurs in the visual system [41, 42].

From a computational point of view, this compression is similar to the compression of sound in mp3 or image in jpeg file format. A transform of the input data is done on a basis that reduces redundancy (the choice of the basis is given by statistical analysis of the expected input data). The representation of information onto this basis allows the dimension of the representation to be reduced.

From a biological point of view, Olshausen and Field [43, 44] found that learning a sparse, redundancy-reduced code from a set of natural images gives a basis that resembles the receptive fields of neurons in the visual system V1 (see Chapter 4). This work has been further extended [45, 46], specifically in the temporal [47, 48] and color domains with the correspondence between opponent color coding by neurons and statistical analysis of natural colored images [49–52]. A particular treatment of the theory focused on retinal processing [53–56] (see Chapter 14).

Ecology of vision, by stating that the visual system is adapted to the natural visual environment, gives a simple rule and efficient methods for testing and implementing the principle. Either principal component analysis (PCA), independent component analysis (ICA), or any sparsity methods applied to patches of natural color image sequences are used. However, there is no evidence that inside the retina or along the pathway from retina to cortex, there are neurons that implement such an analysis. Actually, information theory is a black box, describing the relationship between what enters the box and what exits the box well. But it does not give any rule of what is inside the box: in the case of vision, the processing of visual information by neurons.

In color vision, considering the sampling of color signals by the cone mosaic in the retina prevents the application of redundancy reduction [57, 58]. The idea is based on the fact that the statistical analysis performed in redundancy reduction implicitly supposes that the image is stationary in space (statistical analysis is done over patches of the image and should be coherent from one patch to another). But, because of the random arrangement of cones in the retina mosaic, even if the natural scene is stationary, the signal sampled is not.

Moreover, redundancy reduction models seek a unique model of information representation. However, because the cone mosaic differs from one individual to one another, it is likely that the image representation also changes from one individual to another [59]. It is thus possible that, on average, redundancy reduction offers a good model to explain the transformation from incoming light to neural representation into the cortex, but there is no evidence that it applies identically for each individual.

3.3.4
Geometry Models

Geometry models have been extensively developed for color vision and object location in 3D space. The reason is certainly that, as Riemann [60] argues, these two aspects of vision need a more general space than the Euclidean space. For him, they are the only two examples of a continuous manifold in three dimensions that cannot be well represented in Euclidean geometry. Apparently, this motivated his research on what is now called *Riemannian geometry*. The major distinction between Euclidean and Riemannian geometry is that, in the latter, the notion of distance or norm (length of a segment defined by two points) is local and depends on position in space. Conversely, in Euclidean geometry the norm is fixed all over space.

Soon after Riemann, Von Helmholtz [61] published the first Riemannian model for color discrimination. Before explaining this model in detail, let us consider the evolution of psychophysics at that time. It was shown by Weber that a variation of the sensation of weight depends on the weight considered. This law is denoted by $((\Delta I)/I) = K$, where K is a constant. Intuitively speaking, if you have a weight in your hand, let us say, I kilogram, then you feel a difference in weight if you add another weight such that the difference in weight ΔI exceeds a constant multiplying I. The same kind of law was found for light by Bouguer [62]. Formalizing the relationship between physical space and psychological space, Fechner [63] proposed that the law of sensation S be given by $S = k * \log(I)$, which is one possible solution of Weber's law because the derivative of Fechner's law gives $\Delta S = (k/I)\Delta I = k$.[3] It should be noticed that [65] proposes that the law of sensation is given by a power law, $S = k * I^a$, where a is real.[4] Because $\log(S) = \log(k) + a * \log(I)$ the relation is linear in log–log space whereas Fechner's law is linear in log space. This could be interpreted as adding a compressive law before reaching the space of sensation.

Helmholtz's idea was to provide a model of color discrimination based on Fechner's law embedded in a three-dimensional Riemannian space representing color vision space. The line element ds is therefore given by

$$ds^2 = (d\log R)^2 + (d\log G)^2 + (d\log B)^2 = \left(\frac{dR}{R}\right)^2 + \left(\frac{dG}{G}\right)^2 + \left(\frac{dB}{B}\right)^2 \quad (3.1)$$

where R, G, and B represent the excitation of putative color mechanism in the red, green, and blue parts of our color vision.[5]

It is, to some extent, a direct extension of Fechner's law in three-dimensional space as shown by Alleysson and Méary [66]. It is also a Riemannian space because the norm depends on the location of the space, given by the denominator values of R, G, and B. And, locally (i.e., for fixed R, G, and B values) the norm is the square norm. This class of discrimination model has been called a *line-element*. Many extensions of Helmholtz's model have been given (see, e.g., [26, 67–70]).

Despite these theoretical models, the prediction of color discrimination was not very accurate. MacAdam [71] proposes a direct measure of the color difference in CIExy space for a fixed luminance. His well known MacAdam's ellipses served as a basis measure for discrimination and the CIE proposed two nonlinear spaces CIE-LUV, CIE-Lab to take into account those discrimination curves. It has been shown that MacAdam ellipses could be modeled on a Riemannian space [72] although the corresponding curvatures become negative or positive following their position on the color space.

Recently, geometry has been used in a more general context than the color metric. It has been used to model the processing of assembly in neurons in the visual system in a field called *neurogeometry* [73, 74]. Using geometry rules based on the orientation function modeling the primary visual cortex V1, Jean Petitot was able to show the spatial link of neural activities that Kaniza's figure illusion elicits. The approach of neurogeometry is very elegant and fruitful and will probably be generalized to the whole brain. However, it is still challenging when considering color vision and the sampling of color information with retinal cone mosaic, which differs in arrangement and proportion from one individual to one another [66].

3) [64] states that in order to transform between Weber and Fechner, one would have to make the JNDs infinitesimally small which makes no sense in this context.
4) The exponent a depends on the type of experiment either protetic (measured) or metatic (sensation). In Reference [65] $0.3 \leq a \leq 2$.
5) Today we would probably use L, M, and S instead of R, G, and B to handle the LMS cone excitation space.

3.4
Application to Digital Photography

The main application of retina models is in the processing of raw images from digital cameras. In digital cameras, the image of the scene is captured by a single sensor covered with an array of color filters to allow for color acquisition. A particular filter is attached to each pixel of the sensor. The main color filter array (CFA) used today in the digital camera is called the *Bayer CFA* from the name of its inventor [75]. This filter array is made up of three colors (either RGB or CMY) and is a periodic replication of a subpattern of size 2×2 over the surface of the sensor. Consequently, one of the three colors is represented twice in the array (usually 1R, 2G, and 1B), see Figure 3.5(a). In comparison, Figure 3.5(b) represents a simulation of the random arrangement of L, M, and S-cones in the human retina.

At the output of the sensor, the image is digitalized into 8, 12, or more bits. However, the image is not a color image yet because (i) there is only a single color per pixel, (ii) color space is camera based, given by the spectral sensitivity of the camera, (iii) white may not be white following the tint of the light source, (iv) tones could appear unnatural because of nonlinearity in the processing of photometric data. This is the purpose of the digital processor inside the camera—to convert raw image coming from the sensor to a nicely viewed color image. Some of the papers and books that review this are Refs. [76–78].

Classically, the image processing pipeline is based on the following sequential operation: image acquisition by the sensor, some corrections on raw data (dead pixels, blooming, white pixels, noise reduction, etc.)demosaicing to retrieve three color information per pixel, white balancing, and tone mapping. In the following section, we will describe these operations and make references to the literature.

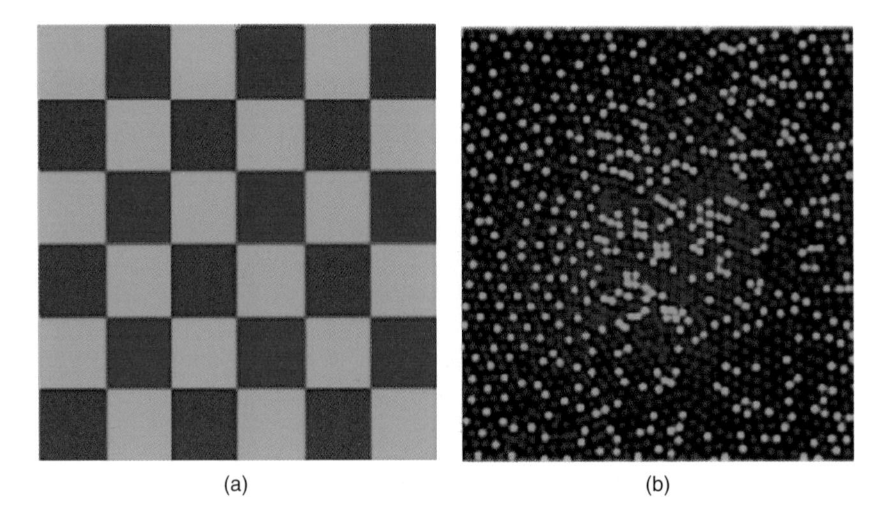

Figure 3.5 (a) Bayer color filter array (CFA) (b) Simulation of the arrangement of the L-cone, M-cone, and S-cone in the human retina (Redrawn from Ref. [7])

We are not being exhaustive in the description, but we focus on studies that use models of vision to state their methods directly or indirectly.

3.4.1 Color Demosaicing

Color demosaicing is an inverse problem [79] that estimates color intensity at various positions that are not acquired by the sensor. During the initial stages, demosaicing was investigated as an interpolation problem using a copy of neighboring pixels to fill in missing pixels or as a bilinear interpolation which was easily done by analog electronic elements such as delay, adder, and so on However, the emergence of digitalization of images at the output of the sensor has allowed more complex demosaicing.

The specificity of color image with inter-channel correlation has been taken into account and almost all methods propose interpolation of color difference instead of color channel directly [80, 81] to take advantage of the correlation between color channels. Another specificity of images that has been taken into account is the fact that images are two dimensional, made on flat surfaces as well as those with contours. By identifying the type of area present in a local neighborhood, it is possible to interpolate along edges rather than across them. This property of color image has been extensively used for demosaicing [80, 82]. Conjointly, some correction methods have been proposed [80–82] to improve the quality of the demosaiced image.

Another approach developed originally by Alleysson *et al.* [83, 84] and extended by Dubois [85] consists in modeling the expectation of the spatial frequency spectrum of an image acquired through a mosaic. In the case of the Bayer CFA of Figure 3.5, the spatial frequency spectrum shows nine zones of energy: one is in the center, corresponding to luminance (or achromatic information), whereas the remaining eight are on the edge of the spectrum, corresponding to chrominance (or color information at constant luminance) (See Figure 3.6 for an illustration. The demosaicing consists then in estimating (adaptively to image content or not) luminance and chrominance components, by selecting information corresponding either to luminance or chrominance in the spatial frequency spectrum.

This approach to demosaicing has been shown to be very fruitful. First, it is a linear method which can easily be optimized for a particular application or for any periodic CFA. Lu *et al.* [86] have shown the theoretical equivalence of frequency selection with alternating projection on convex sets (given by pixels and spatial frequency elements of the image) proposed by Gunturk *et al.* [81], whereas [87] showed the equivalence with a variational approach to dematricing. This model of information representation in a CFA image is also strongly related to the computational model of vision, but includes color. Nevertheless, the random nature of the cone mosaic in the human retina and the difference in arrangement between individuals makes the analogy difficult to establish formally. More generally, the role of eye movement in the reconstruction of color is still to be discovered and could be part of the demosaicing in human vision.

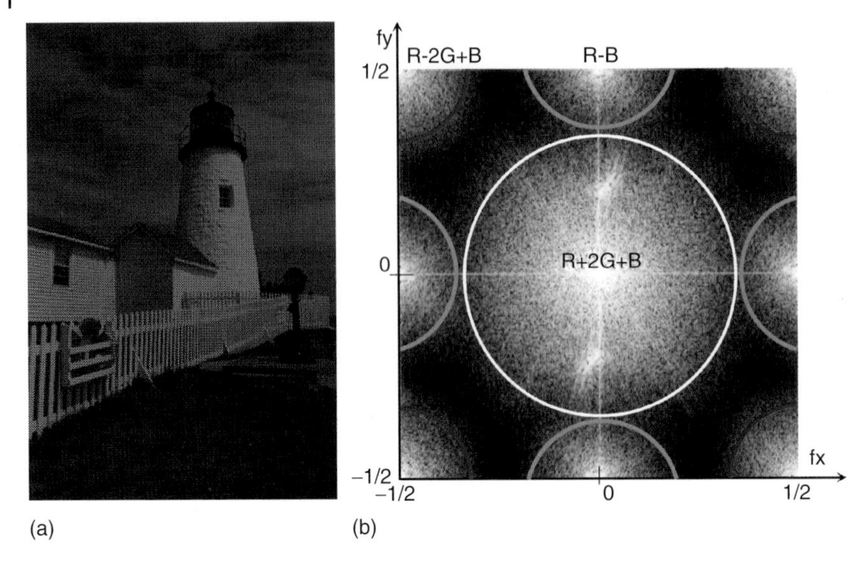

Figure 3.6 (a) Color image I_{CFA} acquired through the Bayer CFA. (b) The corresponding modulus of the spatial frequency spectrum $abs(fft2(I_{CFA}))$, showing spectrum of luminance (R+2G+B) information in the center and spectrum of chromatic information (R-B, R-2G+B) in the borders.

3.4.2
Color Constancy – Chromatic Adaptation – White Balance – Tone Mapping

The retina and visual system show a large extent of adaptation [8, 88], allowing human vision to be accurate even if the dynamic range of the scene is wide. Also, the perceived color of objects is very stable despite a large difference in their reflective light due to change in illumination. Think of the spectrum of a banana under daylight or indoor incandescent light, both of which are quite different, but the banana still appears yellow under both type of lighting. The so-called color constancy phenomenon is a consequence of chromatic adaptation that happens in the retina and visual system as illustrated by Land's retinex theory [89]. A large amount of literature has been produced on retinex theory and its application to image processing.

The principal effect on image quality of not implementing chromatic adaptation in a digital camera can be seen because white on objects or surfaces no longer appears white.

From a computational point of view, chromatic adaptation is often based on the von Kries model which stipulates that adaptation is a gain control in the responses of LMS cones. Roughly, the von Kries model of chromatic adaptation stipulates as follows: $\mathbf{y} = \mathbf{MDM}'\mathbf{x}$, where $\mathbf{x}$ is the vector containing the preadapted image, $\mathbf{y}$ is the vector containing the postadapted image, $\mathbf{M}$ is the 3×3 unitary matrix ($\mathbf{M}' = \mathbf{M}^{-1}$) for the transformation from image in the camera color space to image in the LMS color space and $\mathbf{D}$ is the adaptation transform given by

$$\mathbf{D} = \begin{bmatrix} L_2/L_1 & 0 & 0 \\ 0 & M_2/M_1 & 0 \\ 0 & 0 & S_2/S_1 \end{bmatrix} \tag{3.2}$$

where the index 1 represents adaptation under illuminant 1 and index 2 adpatation under illuminant 2. L, M, and S stand for the cone excitation space.

By studying the corresponding colors (colors that are perceived as identical under different light sources), many authors have tested for which LMS space the von Kries model is the most efficient [90–92]. Often, chromatic adaptation is limited to white balancing, and the so-called gray world assumption is made [93].

Tone mapping refers to the modification of the tone values in an image to match the dynamic range between sensor and display (i.e., 12 bits for cameras and 8 bits for display). This operation is particularly challenging when several dynamic ranges of the image are considered from acquisition to restitution. Generally, digital color sensors respond linearly to light intensity but result in a poor-looking image when displayed or printed. Display and printing have intrinsic nonlinearity and the process undergone in our visual system is also nonlinear. Tone mapping operators could be static and global, meaning that a unique nonlinear function is applied identically to all the pixels in the image. It could also be dynamic and local depending on the local content of the scene [89, 94]. Dynamic local tone mapping operators could, in principle, include all adaptive operations such as chromatic adaptation and white balance.

There are several examples of the use of retinal models for tone mapping [94–98].

3.5 Conclusion

In the past, computer vision was mainly concerned with the physical properties of light and how it can be sensed by artificial material. The final user, the human for example, when addressing photography, was not considered, and it was usual to think that all the problems of image rendering and interpreting were due to a poor image-acquisition system. This way of thinking was principally carried forth by the idea that the retina did not do much processing and that the brain was too complex to be modeled.

This view must change today. First, the retina processes the visual information sophisticatedly as explained in this chapter. And second, parts of neural circuitry of the visual system are well known and efficiently modeled. However, a few problems still remain owing to some misunderstandings. These misunderstandings lie in the knowledge that we have about the mechanisms involved when an observer decides the quality or interprets an image. All the efforts have been made in finding, in natural scenes, a universal model of our perception without considering that each individual has its own biological substrate.

Such an approach has been very fruitful because whatever the object of the visual system is, for adapting to the physical input world optimizing transduction or for compensating inter-individual differences, the mechanisms of the visual system are always interesting to be discovered. However, neglecting the specificity of each individual and how this specificity is compensated to enable us to share a common vision is certainly a major drawback of contemporary models of perception.

For computer color vision, we usually use the colorimetry reference observer CIE1924$V(\lambda)$ and CIE1931xy as the reference for color space perception and to develop all computer vision algorithms (reproduction on media, detection, identification, etc.). However, this reference observer for colorimetry is an average of the perception of several observers and hence is not suited to the perception of a particular individual and might in the same way not be suited to a particular application.

Color mosaic sampling, which differs from one individual to another, is a typical example of the compensation mechanism that occurs in our visual system to shape our perception [57–59]. This might be illustrated by the fact that in some circumstances, discrimination does not correspond to appearance [99, 100]. Discrimination is the ability of a system to quantify a difference between incoming signals. As such, it has a quantitative behavior. Appearance, on the other hand, is a qualitative measure because it is not necessarily quantifiable and because it says something directly and not through a few numbers. This is surprising because we usually design a boundary between categories with the locus of best discrimination. In many application systems, we measure the system's responses to a set of physical variables, optimized the processing and the representation for discrimination purpose between different categories of stimuli and we expect it will improve our decision in between multiple categories of actions.

Actually, the problem of ignoring compensations for inter-individual differences in color vision could be considered as noise. For all ecological models of vision, noise comes from the physical world and very few consider internal noise generated by neural processing. In color vision, we usually consider luminance and color to be subjected to the same independent, identically distributed noise. However, the fact that human cone mosaic is so different between observers will certainly generate a differential noise in luminance and chrominance among observers.

The problem of how individual biological differences in humans allow common vision remains to be solved. It is a challenge to explain how we share so much agreement in vision despite our very different biological material. To bring some explanations would certainly improve our knowledge of the functioning of the visual system and its underlying model. It would certainly provide an efficient way to reproduce natural vision in biologically inspired computer vision.

References

1. Solomon, S.G. and Lennie, P. (2007) The machinery of colour vision. *Nat. Rev. Neurosci.*, **8**, 276–286.
2. Buzás, P. *et al.* (2006) Specificity of M and L cone inputs to receptive fields in the parvocellular pathway: random

wiring with functional bias. *J. Neurosci.*, **26**, 11148–11161.
3. Piccolino, M., Strettoi, E., and Laurenzi, E. (1989) Santiago Ramon y Cajal, the retina and the neuron theory. *Doc. Ophthalmol.*, **71**, 123–141.
4. Hérault, J. (2010) *Vision: Images, Signals and Neural Networks-Models of Neural Processing in Visual Perception*, World Scientific Publishing.
5. Dacey, D.M. *et al.* (2005) Melanopsin-expressing ganglion cells in primate retina signal colour and irradiance and project to the LGN. *Nature*, **433**, 749–754.
6. Osterberg, G. (1935) *Topography of the Layer of Rods and Cones in the Human Retina*, Nyt Nordisk Forlag.
7. Roorda, A. and Williams, D.R. (1999) The arrangement of the three cone classes in the living human eye. *Nature*, **397**, 520–522.
8. Clark, D.A. *et al.* (2013) Dynamical adaptation in photoreceptors. *PLoS Comput. Biol.*, **9**, e1003289.
9. Hartline, H.K. (1938) The response of single optic nerve fibers of the vertebrate eye to illumination of the retina. *Am. J. Physiol.*, **121**, 400–415.
10. Alonso, J.-M. and Chen, Y. (2009) Receptive field. *Scholarpedia*, **4**, 5393.
11. Rodieck, R. (1998) *The First Steps in Seeing*, Sinauer Associates, Inc.
12. McMahon, M.J. *et al.* (2000) Fine structure of parvocellular receptive fields in the primate fovea revealed by laser interferometry. *J. Neurosci.*, **20**, 2043–2053.
13. Dacey, D.M. *et al.* (2003) Fireworks in the primate retina: in vitro photodynamics reveals diverse LGN-projecting ganglion cell types. *Neuron*, **37**, 15–27.
14. Hodgkin, A.L. and Huxley, A.F. (1952) A quantitative description of membrane current and its application to conduction and excitation in nerve. *J. Physiol.*, **117**, 500.
15. Mead, C.A. and Mahowald, M.A. (1988) A silicon model of early visual processing. *Neural Netw.*, **1**, 91–97.
16. Rojas, R. (1996) *Neural Networks: A Systematic Introduction*, Springer-Verlag.
17. van Hateren, H. (2005) A cellular and molecular model of response kinetics and adaptation in primate cones and horizontal cells. *J. Vis.*, **5**, 331–347.
18. Van Hateren, J. and Snippe, H. (2007) Simulating human cones from mid-mesopic up to high-photopic luminances. *J. Vis.*, 7.
19. Van Hateren, J. (2006) Encoding of high dynamic range video with a model of human cones. *ACM Trans. Graph. (TOG)*, **25**, 1380–1399.
20. Keil, M.S. and Vitrià, J. (2007) Pushing it to the limit: adaptation with dynamically switching gain control. *EURASIP J. Appl. Signal Process.*, **2007**, 117.
21. Crama, P. and Schoukens, J. (2004) Hammerstein–wiener system estimator initialization. *Automatica*, **40**, 1543–1550.
22. Billings, S. and Fakhouri, S. (1982) Identification of systems containing linear dynamic and static nonlinear elements. *Automatica*, **18**, 15–26.
23. Benardete, E.A. and Victor, J.D. (1994) An extension of the M-sequence technique for the analysis of multi-input nonlinear systems, in *Advanced Methods of Physiological System Modeling*, Springer-Verlag, pp. 87–110.
24. Volterra, V. (2005) Theory of Functionals and of Integral and Integro-differential Equations, *Dover Books on Mathematics Series*, Dover Publications.
25. Shapley, R. and Victor, J. (1979) Nonlinear spatial summation and the contrast gain control of cat retinal ganglion cells. *J. Physiol.*, **290**, 141–161.
26. Alleysson, D. and Hérault, J. (2001) Variability in color discrimination data explained by a generic model with nonlinear and adaptive processing. *Color Res. Appl.*, **26**, S225–S229.
27. Webster, M.A. and Mollon, J. (1991) Changes in colour appearance following post-receptoral adaptation. *Nature*, **349**, 235–238.
28. Zaidi, Q. and Shapiro, A.G. (1993) Adaptive orthogonalization of opponent-color signals. *Biol. Cybern.*, **69**, 415–428.

29. Chichilnisky, E. and Wandell, B.A. (1999) Trichromatic opponent color classification. *Vision Res.*, **39**, 3444–3458.
30. Wohrer, A. and Kornprobst, P. (2009) Virtual retina: a biological retina model and simulator, with contrast gain control. *J. Comput. Neurosci.*, **26**, 219–249.
31. Hérault, J. and Durette, B. (2007) Modeling visual perception for image processing, in *Computational and Ambient Intelligence*, Springer-Verlag, pp. 662–675.
32. Benoit, A. *et al.* (2010) Using human visual system modeling for bio-inspired low level image processing. *Comput. Vision Image Underst.*, **114**, 758–773.
33. Beaudot, W., Palagi, P., and Hérault, J. (1993) Realistic simulation tool for early visual processing including space, time and colour data, in *New Trends in Neural Computation*, Springer-Verlag, pp. 370–375.
34. Beaudot, W.H. (1996) Adaptive spatiotemporal filtering by a neuromorphic model of the vertebrate retina. Proceedings IEEE International Conference on Image Processing, Vol. 1, pp. 427–430.
35. Buser, P. and Imbert, M. (1987) *Neurophysiologie Fonctionnelle Tome 4: Vision*, Collection Méthodes, Hermann.
36. Enroth-Cugell, C. and Robson, J.G. (1966) The contrast sensitivity of retinal ganglion cells of the cat. *J. Physiol.*, **187**, 517–552.
37. Werblin, F.S. and Roska, B.M. (2004) Parallel visual processing: a tutorial of retinal function. *Int. J. Bifurcation Chaos*, **14**, 843–852.
38. Gollisch, T. and Meister, M. (2010) Eye smarter than scientists believed: neural computations in circuits of the retina. *Neuron*, **65**, 150–164.
39. Landy, M.S. and Movshon, J.A. (1991) *Computational Models of Visual Processing*, The MIT Press.
40. Heeger, D.J., Simoncelli, E.P., and Movshon, J.A. (1996) Computational models of visual processing. *Proc. Natl. Acad. Sci. U.S.A.*, **23**, 623–627.
41. Barlow, H.B. (2012) Possible principles underlying the transformation of sensory messages, in *Sensory Communication*, (ed. W.A. Rosenblith), The MIT Press, 2012. MIT Press Scholarship Online, 2013. DOI: *10.7551/mitpress/9780262518420.003.0013.*
42. Field, D.J. (1987) Relations between the statistics of natural images and the response properties of cortical cells. *J. Opt. Soc. Am. A*, **4**, 2379–2394.
43. Olshausen, B.A. and Field, D.J. (1996) Emergence of simple-cell receptive field properties by learning a sparse code for natural images. *Nature*, **381**, 607.
44. Olshausen, B.A. and Field, D.J. (2004) Sparse coding of sensory inputs. *Curr. Opin. Neurobiol.*, **14**, 481–487.
45. Bell, A.J. and Sejnowski, T.J. (1997) The "independent components" of natural scenes are edge filters. *Vision Res.*, **37**, 3327–3338.
46. Atick, J.J. (1992) Could information theory provide an ecological theory of sensory processing? *Netw. Comput. Neural Syst.*, **3**, 213–251.
47. van Hateren, J.H. and Ruderman, D.L. (1998) Independent component analysis of natural image sequences yields spatio-temporal filters similar to simple cells in primary visual cortex. *Proc. R. Soc. London, Ser. B: Biol. Sci.*, **265**, 2315–2320.
48. Dong, D.W. and Atick, J.J. (1995) Statistics of natural time-varying images. *Netw. Comput. Neural Syst.*, **6**, 345–358.
49. Buchsbaum, G. and Gottschalk, A. (1983) Trichromacy, opponent colours coding and optimum colour information transmission in the retina. *Proc. R. Soc. London, Ser. B.: Biol. Sci.*, **220**, 89–113.
50. Ruderman, D.L., Cronin, T.W., and Chiao, C.-C. (1998) Statistics of cone responses to natural images: implications for visual coding. *J. Opt. Soc. Am. A*, **15**, 2036–2045.
51. Tailor, D.R., Finkel, L.H., and Buchsbaum, G. (2000) Color-opponent receptive fields derived from independent component analysis of

natural images. *Vision Res.*, **40**, 2671–2676.
52. Lee, T.-W., Wachtler, T., and Sejnowski, T.J. (2002) Color opponency is an efficient representation of spectral properties in natural scenes. *Vision Res.*, **42**, 2095–2103.
53. Atick, J.J. and Redlich, A.N. (1992) What does the retina know about natural scenes? *Neural Comput.*, **4**, 196–210.
54. Atick, J.J. and Redlich, A.N. (1990) Towards a theory of early visual processing. *Neural Comput.*, **2**, 308–320.
55. Atick, J.J., Li, Z., and Redlich, A.N. (1992) Understanding retinal color coding from first principles. *Neural Comput.*, **4**, 559–572.
56. Doi, E. and Lewicki, M.S. (2007) A theory of retinal population coding. *Adv. Neural Inf. Process. Syst.*, **19**, 353.
57. Alleysson, D. and Süsstrunk, S. (2004) Spatio-chromatic ICA of a mosaiced color image, in *Independent Component Analysis and Blind Signal Separation*, Springer-Verlag, pp. 946–953.
58. Alleysson, D. and Süsstrunk, S. (2004) Spatio-chromatic PCA of a mosaiced color image. Conference on Colour in Graphics, Imaging, and Vision, Number 1 in 2004, Society for Imaging Science and Technology, pp. 311–314.
59. Alleysson, D. (2010) Spatially coherent colour image reconstruction from a trichromatic mosaic with random arrangement of chromatic samples. *Ophthalmic Physiol. Opt.*, **30**, 492–502.
60. Riemann, B. (1854) Ueber die hypothesen, welche der geometrie zu grunde liegen, *Habilitationsschrift, Abhandlungen der Königlichen Gesellschaft der Wissenschaften zu Göttingen*, vol. 13, *http://www.maths.tcd.ie/pub/HistMath/People/Riemann/Geom/Geom.pdf*; Translation English by Clifford, W. (1854) On the hypothesis which lie at the basis of Geometry. *Nature*, 1873, *http://192.43.228.178/classics/Riemann/WKCGeom.pdf*. Trad. Française par Houel J. Sur les hypothèses qui servent de fondement à la géométrie *http://ddata.over-blog.com/xxxyyy/0/31/89/29/Fusion-58/F58007.pdf*.
61. Von Helmholtz, H. (1866) *Handbuch der physiologischen Optik: mit 213 in den Text eingedruckten Holzschnitten und 11 Tafeln*, vol. 9, Voss.
62. Bouguer, P. (1729) *Essai d'optique, sur la gradation de la lumiere*, Claude Jombert.
63. Fechner, G.T. (1907) *Elemente der psychophysik*, vol. 2, Breitkopf & Härtel.
64. Wiener, N. (1921) A new theory of measurement: a study in the logic of mathematics. *Proc. London Math. Soc.*, **2**, 181–205.
65. Stevens, S.S. (1957) On the psychophysical law. *Psychol. Rev.*, **64**, 153.
66. Alleysson, D. and Méary, D. (2012) Neurogeometry of color vision. *J. Physiol. Paris*, **106**, 284–296.
67. Schrödinger, E. (1920) Grundlinien einer theorie der farbenmetrik im tagessehen. *Ann. Phys.*, **368**, 481–520.
68. Vos, J. and Walraven, P. (1972) An analytical description of the line element in the zone-fluctuation model of colour vision–I. Basic concepts. *Vision Res.*, **12**, 1327–1344.
69. Stiles, W. (1946) A modified helmholtz line-element in brightness-colour space. *Proc. Phys. Soc. London*, **58**, 41.
70. Koenderink, J., Van de Grind, W., and Bouman, M. (1972) Opponent color coding: a mechanistic model and a new metric for color space. *Kybernetik*, **10**, 78–98.
71. MacAdam, D.L. (1942) Visual sensitivities to color differences in daylight. *J. Opt. Soc. Am. A*, **32**, 247–273.
72. Kohei, T., Chao, J., and Lenz, R. (2010) On curvature of color spaces and its implications. Conference on Colour in Graphics, Imaging, and Vision, vol. 2010, Society for Imaging Science and Technology, pp. 393–398.
73. Petitot, J. (2003) The neurogeometry of pinwheels as a sub-riemannian contact structure. *J. Physiol. Paris*, **97**, 265–309.
74. Citti, G. and Sarti, A. (2010) Neuromathematics of vision. *AMC*, **10**, 12.
75. Bayer, B. (1976) Color imaging array. US Patent 3,971,065.

76. Ramanath, R. *et al.* (2005) Color image processing pipeline. *IEEE Signal Process. Mag.*, **22**, 34–43.
77. Li, X., Gunturk, B., and Zhang, L. (2008) Image demosaicing: a systematic survey, *Electronic Imaging 2008*, International Society for Optics and Photonics, pp. 68221J–68221J-15.
78. Lukac, R. (2012) *Single-Sensor Imaging: Methods and Applications for Digital Cameras*, CRC Press.
79. Ribes, A. and Schmitt, F. (2008) Linear inverse problems in imaging. *IEEE Signal Process. Mag.*, **25**, 84–99.
80. Hirakawa, K. and Parks, T.W. (2005) Adaptive homogeneity-directed demosaicing algorithm. *IEEE Trans. Image Process.*, **14**, 360–369.
81. Gunturk, B.K., Altunbasak, Y., and Mersereau, R.M. (2002) Color plane interpolation using alternating projections. *IEEE Trans. Image Process.*, **11**, 997–1013.
82. Kimmel, R. (1999) Demosaicing: image reconstruction from color CCD samples. *IEEE Trans. Image Process.*, **8**, 1221–1228.
83. Alleysson, D., Süsstrunk, S., and Hérault, J. (2002) Color demosaicing by estimating luminance and opponent chromatic signals in the fourier domain. *Color and Imaging Conference*, Number 1 in 2002, Society for Imaging Science and Technology, pp. 331–336.
84. Alleysson, D., Susstrunk, S., and Hérault, J. (2005) Linear demosaicing inspired by the human visual system. *IEEE Trans. Image Process.*, **14**, 439–449.
85. Dubois, E. (2005) Frequency-domain methods for demosaicking of bayer-sampled color images. *IEEE Signal Process Lett.*, **12**, 847.
86. Lu, Y.M., Karzand, M., and Vetterli, M. (2010) Demosaicking by alternating projections: theory and fast one-step implementation. *IEEE Trans. Image Process.*, **19**, 2085–2098.
87. Condat, L. (2009) A generic variational approach for demosaicking from an arbitrary color filter array. 2009 *16th IEEE International Conference on Image Processing (ICIP)*, IEEE, pp. 1625–1628.
88. Shapley, R. and Enroth-Cugell, C. (1984) Visual adaptation and retinal gain controls. *Prog. Retin. Res.*, **3**, 263–346.
89. Land, E.H. (1983) Recent advances in retinex theory and some implications for cortical computations: color vision and the natural image. *Proc. Natl. Acad. Sci. U.S.A.*, **80**, 5163.
90. Lam, K. (1985) Metamerism and colour constancy. PhD thesis. University of Bradford.
91. Finlayson, G.D. *et al.* (2012) Spectral sharpening by spherical sampling. *J. Opt. Soc. Am. A*, **29**, 1199–1210.
92. Finlayson, G.D. and Süsstrunk, S. (2000) Performance of a chromatic adaptation transform based on spectral sharpening. *Color and Imaging Conference*, vol. 2000, Society for Imaging Science and Technology, pp. 49–55.
93. Van De Weijer, J. and Gevers, T. (2005) Color constancy based on the grey-edge hypothesis. *IEEE International Conference on Image Processing*, 2005. ICIP 2005, vol. 2, IEEE, pp. II–722.
94. Meylan, L., Alleysson, D., and Süsstrunk, S. (2007) Model of retinal local adaptation for the tone mapping of color filter array images. *J. Opt. Soc. Am. A*, **24**, 2807–2816.
95. Mantiuk, R., Myszkowski, K., and Seidel, H.-P. (2006) A perceptual framework for contrast processing of high dynamic range images. *ACM Trans. Appl. Percept. (TAP)*, **3**, 286–308.
96. Ferwerda, J.A. *et al.* (1996) A model of visual adaptation for realistic image synthesis. Proceedings of the 23rd Annual Conference on Computer Graphics and Interactive Techniques, ACM, pp. 249–258.
97. Reinhard, E. and Devlin, K. (2005) Dynamic range reduction inspired by photoreceptor physiology. *IEEE Trans. Visual Comput. Graphics*, **11**, 13–24.
98. Larson, G.W., Rushmeier, H., and Piatko, C. (1997) A visibility matching tone reproduction operator for high

dynamic range scenes. *IEEE Trans. Visual Comput. Graphics*, **3**, 291–306.

99. Bachy, R. *et al.* (2012) Hue discrimination, unique hues and naming. *J. Opt. Soc. Am. A*, **29**, A60–A68.

100. Danilova, M. and Mollon, J. (2012) Foveal color perception: minimal thresholds at a boundary between perceptual categories. *Vision Res.*, **62**, 162–172.

4
Modeling Natural Image Statistics

Holly E. Gerhard, Lucas Theis, and Matthias Bethge

4.1
Introduction

Natural images possess complex statistical regularities induced by nonlinear interactions of objects (e.g., occlusions). Developing probabilistic models of these statistics offers a powerful means both to understanding biological vision and to designing successful computer vision applications. A long-standing hypothesis about biological sensory processing states that neural systems try to represent inputs as efficiently as possible by adapting to environmental regularities. With natural image models, it is possible to test specific predictions of this hypothesis and thereby reveal insights into biological mechanisms of sensory processing and learning. In computer vision, natural image models can be applied to a variety of problems from image restoration to higher-level classification tasks.

The chapter is divided into four major sections. First, we introduce some statistical qualities of natural images and discuss why it is interesting to model them. Second, we describe several models including the state of the art. Third, we discuss examples of how natural image models impact computer vision applications. And fourth, we discuss experimental examples of how biological systems are adapted to natural images.

4.2
Why Model Natural Images?

This chapter focuses on models of the *spatial structure* in natural images, that is, the content of static images as opposed to sequences of images. We will primarily focus on luminance as it carries a great deal of the structural variations in images and is a reasonable starting place for developing image models. In Figure 4.1, we analyze a single photographic image to illustrate some basic statistical properties of natural images. Photographic images tend to contain objects, an important cause for much of these properties. Because objects tend to have smoothly varying surfaces, nearby regions in images also tend to appear similar. As illustrated in

Biologically Inspired Computer Vision: Fundamentals and Applications, First Edition.
Edited by Gabriel Cristóbal, Laurent Perrinet, and Matthias Keil.

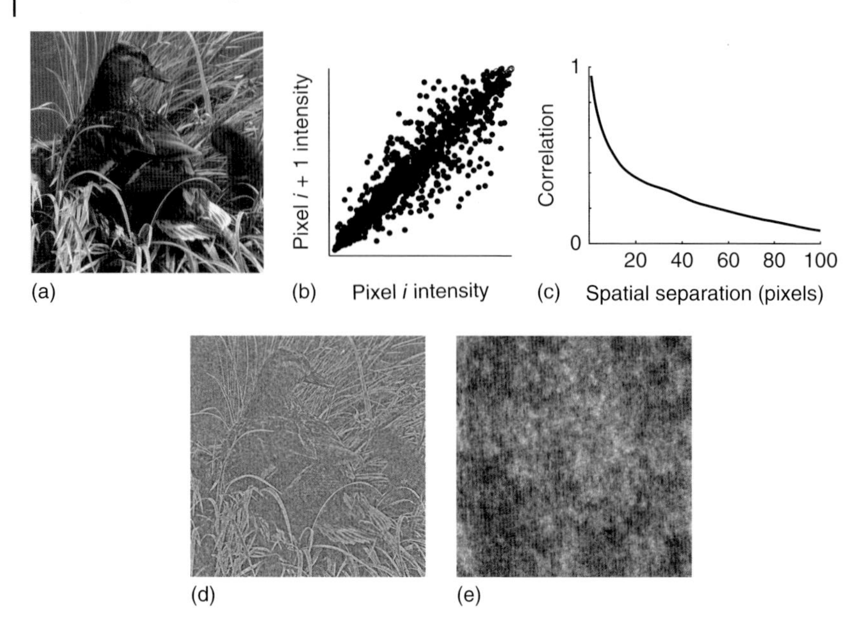

Figure 4.1 Natural images are highly structured. Here we show an analysis of a single image (a). The pairwise correlations in intensity between neighboring pixels are illustrated by the scatterplot ($\rho = 0.95$) in (b). Pairwise correlations extend well beyond neighboring pixels as shown by the autocorrelation function of the image in (c). We also show a whitened version of the image (d) and a phase scrambled version (e). Whitening removes pairwise correlations and preserves higher-order regularities, whereas Fourier phase scrambling has the opposite effect. Comparing the whitened and phase-scrambled images reveals that the *higher*-order regularities carry much of the perceptually meaningful structure. Second-order correlations can be modeled by a Gaussian distribution. The probabilistic models we will discuss are aimed at describing the higher-order statistical regularities and can be thought of as generalizations of the Gaussian distribution.

Figure 4.1, natural images therefore contain not only local pairwise correlations between pixel intensities (Figure 4.1b), but also long-range pairwise correlations (Figure 4.1c) and higher-order regularities as well (Figure 4.1d). The interested reader can also refer to Simoncelli and Olshausen [1] for a more detailed, yet accessible introduction to image measurements revealing higher-order regularities in natural images.

One can consider each photographic image as a point in a high-dimensional space where dimensions correspond to individual pixel values. If one were to analyze a large ensemble of photographic images and plot each in this space, it would become clear that they do not fill the space uniformly but instead represent only a small portion of the space of possible images, precisely because natural images contain statistical regularities. The goal of a probabilistic image model is to distribute probability mass throughout this space to best account for the true distribution's shape. The challenge lies in capturing the complex statistical properties of images, which, as we will describe in Section 4.3, requires sophisticated machine learning techniques.

In Section 4.3, we will discuss several important approaches to tackling this challenge, but first we ask: why is modeling natural images important, and what is it good for? We will argue that it not only informs our understanding of the physical world but also our understanding of biological systems and can lead to improved computer vision algorithms.

A primary reason probabilistic natural image models are so powerful is *prediction*. If one had access to the full distribution of natural images, one would, practical considerations aside, also have access to any conditional distribution and would be able to optimally predict the content of one image region given any other region. If the model also included time, one could additionally predict how an image will change over time. Being able to anticipate the structure of the external environment is clearly advantageous in a variety of scenarios. Intriguingly, many similarities have been found between representations used by the brain and the internal representations used by successful image models. Examples of such similarities will be discussed as we present various models in Section 4.3. In computer vision, natural image models have been directly applied to prediction problems such as denoising and filling-in (described in Section 4.4).

Historically, much of the inspiration for modeling environmental statistics stems from the efficient coding hypothesis put forward by Barlow [2] and Attneave [3], which states that biological systems try to represent information as efficiently as possible in a coding theoretic sense, that is, using as few bits as possible. Knowledge of the distribution of natural images can be used to construct a representation which is efficient in precisely this sense. We will present examples of experimental links made between natural image statistics and biological vision in Section 4.5.

Probabilistic image models have also been used to learn image representations for various classification tasks [4, 5]. Modeling natural images has great potential for enhancing object recognition performance as there is a deep relationship between objects and correlations of image features, and it also provides a principled route to exploiting unlabeled data (discussed in Section 4.4).

Before proceeding, we note that many kinds of natural scene statistics, for example, those related to color, depth, or object contours, are also active areas of research. The interested reader can consult [1, 6] for references to foundational work in those areas.

4.3
Natural Image Models

A wide spectrum of approaches to modeling the density of natural images has been proposed in the last two decades. Many have been designed to examine how biological systems adapt to environmental statistics, where the logic is to compare neural response properties to emergent aspects of the models after fitting to natural images. Similarities between the two are interpreted as indirect evidence that the neural representation is adapted to the image statistics captured by the model. This tradition stems from the efficient coding hypothesis [2, 3], which was originally formulated in terms of redundancy reduction. Intuitively, if an organism's nervous system has knowledge of the redundancies present in the sensory input, its neural representations can adapt to remove those redundancies and emphasize the interesting content of sensory signals.

Redundancy can be defined formally as the *multi-information* of a random vector **s**,

$$I[\mathbf{s}] = \sum_i H[s_i] - H[\mathbf{s}] \tag{4.1}$$

where H denotes (differential) entropy. The differential entropy in turn is defined as

$$H[\mathbf{s}] = -\int p(\mathbf{s}) \log\ p(\mathbf{s})\ d\mathbf{s} \tag{4.2}$$

where $p(\mathbf{s})$ denotes the probability density at **s**. Intuitively speaking, the entropy measures the spread of a distribution. If all of the variables s_i were independent of each other, the first term on the right-hand side of Eq. 4.1 would correspond to the entropy of the distribution over **s**, that is, $H[\mathbf{s}]$, the second term. The distribution of **s** would thus have zero multi-information, that is, no redundancies. This means that multi-information can be seen as measuring how much more concentrated the joint distribution is compared to a *factorial* (i.e., independent) version of it. This is visualized for two variables in Figure 4.2.

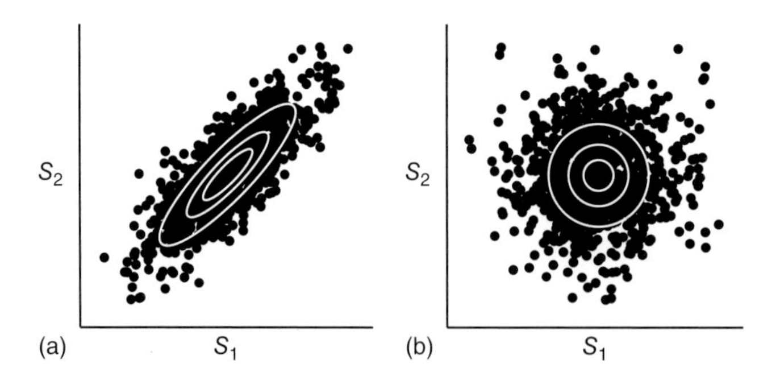

Figure 4.2 To illustrate how multi-information is measured as a proxy for how redundant a distribution is, we show an example joint distribution $p(\mathbf{s}) = p(s_1, s_2)$, visualized in (a) and its factorial form, $p(\mathbf{s}) = p(s_1)p(s_2)$, visualized in (b), that is, where the two variables are independent of each other. Multi-information is the difference between the joint distributions's entropy $H[\mathbf{s}]$ (corresponding to (a)) and the factorial form's entropy $\sum_i H[s_i]$ (corresponding to (b)). Intuitively speaking, multi-information measures the difference of the *spread* of the two distributions. The illustrated joint distribution therefore has a relatively high degree of multi-information, meaning that it is highly redundant.

The pixel values of natural images are highly redundant (e.g., Figure 4.1). To motivate the usefulness of redundancy reduction, imagine a world with white $N \times N$ images each containing just a single black disk positioned at a random location and a random diameter $d \in 1, \ldots, D\}$. It is clearly much more efficient to describe the image in terms of the object's position and diameter – which can be achieved with $2\log_2 N + \log_2 D$ bits – than to describe the binary value of all N^2 pixels independently, which would require N^2 bits. Finally, we note that knowledge of the full probability distribution of images can be used to compute a representation in the form of a transformation such that all components become independent [7]. In other words, a system with perfect knowledge of the input distribution could remove all redundancies from the input. However, the resulting representation is not unique, that is, for a given distribution, there are many transformations which lead to an independent representation. Reducing multi-information is thus not sufficient for deriving a representation, but it may nevertheless be used to guide and constrain representations.

An elegant early examination of the second-order correlations of natural images demonstrated the potential for efficient coding theory to explain how biological vision functions. Instead of working with correlations between pixels directly, it is often convenient to work with the power spectrum of natural images, which can be computed as the Fourier transform of the autocorrelation function. Starting from the observation that the power spectrum of natural images falls off approximately as $1/f^2$, Atick and Redlich [8] hypothesized that the goal of retinal processing is to remove this redundancy, that is, to decorrelate retinal input. They showed that previously measured contrast responses of monkey retinal ganglion cells are indeed consistent with the removal of the $1/f^2$ regularity from natural input. In addition, they derived decorrelating filters which they showed could predict human psychophysical measurements of contrast sensitivity with high accuracy. (See Chapter 5 for information about psychophysical measurements.)

One way to capture a distribution's pairwise correlations is to approximate it with a Gaussian distribution, where the covariance is set to be equal to the distribution's empirical covariance. In our review, we will discuss three main branches of approaches that extend from the Gaussian model (see Figure 4.3). Each branch adds additional modeling power by describing the higher-order correlations of natural images which are critically important for the structural or geometric shape-based content, for example, as illustrated in the whitened image in Figure 4.1(d), which highlights the higher-order correlations in the original photographic image. The Gaussian model, on the other hand, captures only as much content as is shown in the phase-scrambled image in Figure 4.1(e). Traversing down a branch in our diagram increases the degree of higher-order correlations captured by a model and hence its efficacy. On a technical note, each solid arrow points to a mathematically more general model which allows for even more types of regularities to be captured. (Dashed arrows indicate improved model efficacy but not increased generality.) For notational simplicity, we denote images by D-dimensional column vectors $\mathbf{x} \in \mathbb{R}^D$ where each dimension of $\mathbf{x}$ stores the intensity of a single pixel in the gray-scale image.

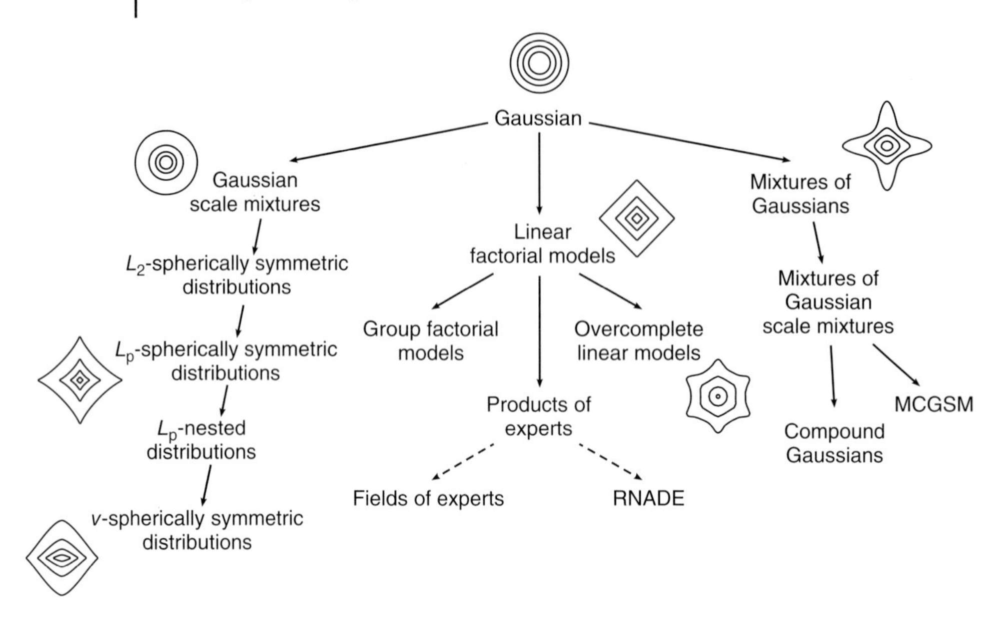

Figure 4.3 In this chapter, we review several important natural image models which we have organized into three branches of approaches, each extending the Gaussian distribution by certain higher-order regularities. Arrows with solid lines indicate a generalization of the model class.

We begin with *linear factorial* models. Conceptually simple yet highly influential, these models are an appropriate starting point for our review. Linear factorial models assume that an image of dimensionality D is generated by a random linear superposition of D basis functions:

$$\mathbf{x} = \mathbf{A}\mathbf{s} = \sum_{i}^{D} s_i \mathbf{a}_i, \tag{4.3}$$

where s_i is the weight applied to the ith basis function, $\mathbf{a}_i$, which has the same dimensionality as the images. An example set of basis functions is shown in Figure 4.4(a) for images of size 10×10 pixels. In a linear factorial model, the s_i, often referred to as the "sources" of the image, are assumed to be independently distributed:

$$p(\mathbf{s}) = \prod_{i=1}^{D} p_i(s_i) \tag{4.4}$$

meaning that an image's overall probability density factorizes into the sources' densities.

Under the assumptions of the linear factorial model, redundancies can be removed via a linear transformation $\mathbf{s} = \mathbf{W}\mathbf{x}$ with $\mathbf{W} = \mathbf{A}^{-1}$, which is often referred to as the *filter matrix*. In Figure 4.4(b), we visualize the corresponding filter matrix for the 10×10 pixel basis functions of Figure 4.4(a). The density of

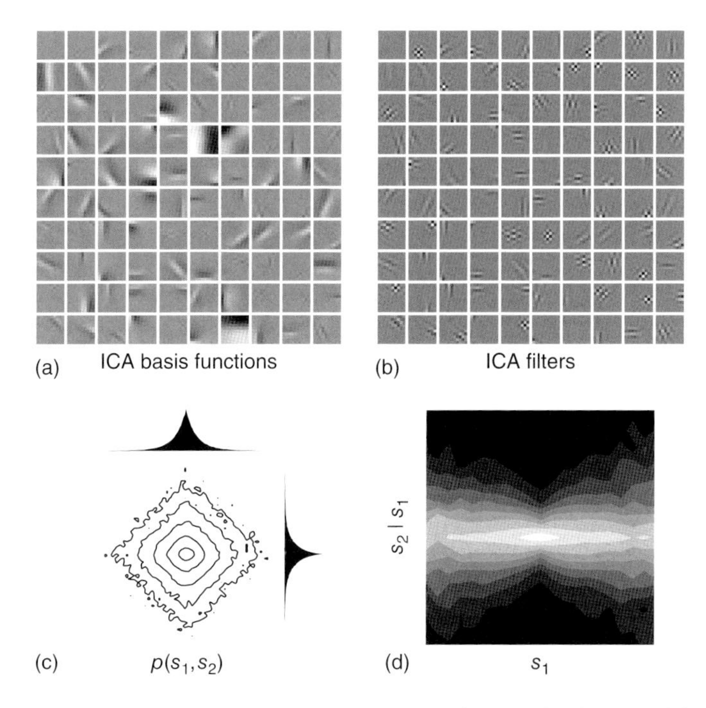

Figure 4.4 ICA filter responses are not independent for natural images. In panel a, we show a complete set of ICA basis functions ($\mathbf{A}$) trained on images 10×10 pixels in size ("complete" meaning that there are as many basis functions, 100, as there are dimensions in the data). Panel b visualizes the corresponding set of filters ($\mathbf{W} = \mathbf{A}^{-1}$). The filters are oriented, bandpass, and localized – prominent features shared by the receptive fields of simple cells in the primary visual cortex. In panels c and d, we examine the filter responses or "sources" for natural images. (c) shows the joint histogram of two filter responses, $p(s_1, s_2)$, where $s_i = \mathbf{w}_i^\top \mathbf{x}$. The joint distribution exhibits a diamond-shaped symmetry, which is well captured by an L_p-spherical symmetry with p close to 1. We also show the two marginal distributions, which are heavy-tailed, sparse distributions with a high probability of zero and an elevated probability of larger nonzero values, relative to the Gaussian distribution (i.e., filter responses are typically either very small or very large). (d) The higher-order dependence of the filter responses is shown by plotting the conditional distribution $p(s_2 \mid s_1)$ for each value of s_1. The "bow-tie" shape of this plot reveals that the variance of s_2 depends on the value of s_1.

an image is also given by

$$p(\mathbf{x}) = \prod_{i=1}^{D} p_i(\mathbf{w}_i^\top \mathbf{x})|\mathbf{W}| \tag{4.5}$$

where $\mathbf{w}_i^\top$ denotes the ith row vector and $|\mathbf{W}|$ the determinant of the filter matrix.

One has several choices in how to determine the transformation $\mathbf{W}$. Principal component analysis (PCA) represents one way of computing a filter matrix which removes pairwise correlations but ignores higher-order correlations (i.e., a "decorrelating" transformation). This would be enough if images and hence the sources

s were Gaussian distributed. However, it is well known that even random filters, which do not respond to flat images, lead to highly non-Gaussian, kurtotic (also commonly referred to as *sparse*) filter responses. Two example distributions of filter responses are shown in Figure 4.4(c). These marginal distributions exhibit heavy tails and high probabilities of near-zero values.

Unlike PCA, independent component analysis (ICA) tries to find the linear transformation which removes as much redundancy as possible by minimizing multi-information [9, 10]. In practice, this amounts to finding a set of filters with maximally sparse responses, an approach proposed by Olshausen and Field [11]. Equivalently, we can try to find the filter matrix with maximal likelihood under the linear factorial model. The resulting filters share three prominent features with the simple cells of the primary visual cortex: they are localized, oriented, and bandpass. ICA filters for 10×10 pixel images are shown in Figure 4.4(b). The emergence of these features after training on natural images suggests that the primary visual cortex may also be optimized according to similar rules for extracting statistically independent structure from natural images.

However, linear factorial models fail to achieve a truly independent representation of natural images. In the bow-tie plot of Figure 4.4(d) we demonstrate how the ICA sources for natural images still exhibit dependencies between each other, even though ICA is the best possible linear factorial model. Detailed analyses have shown that even when linear factorial models are optimized to capture higher-order correlations, as ICA is, they achieve only quite small improvements in modeling power compared to decorrelating transformations such as PCA [12, 13]. The physical reason for this failure is that image formation simply does not obey the rules of linear superposition but rather results from several nonlinear interactions such as occlusion. The assumptions made by the linear model are clearly too strong.

Before describing extensions of the linear factorial model, we wish first to describe a related family of models (the left branch of our diagram in Figure 4.3) that has been designed to exploit L_p-spherical symmetries exhibited by the natural image distribution. An example of such symmetry is shown in the joint histogram of natural image ICA sources in Figure 4.4(c) which exhibit an L_p-spherical symmetry with p close to 1, that is, a diamond-shaped symmetry. Not only do the joint responses of pairs of ICA filters exhibit this symmetry but joint wavelet coefficients and many kinds of oriented filter responses to natural images also exhibit L_p-spherical symmetry. An L_p-spherical distribution's density only depends on the L_p-norm, that is,

$$p(\mathbf{s}) = f(||\mathbf{s}||_p) = f\left(\left(\sum_{i=1}^{N} |s_i|^p\right)^{1/p}\right)$$

for some function f and $p > 0$. Note that the Euclidean norm corresponds to $p = 2$ and elliptical symmetry, a special case of L_p-spherical symmetry which can also describe a range of other symmetries when different values of p are used.

Gaussian scale mixtures (GSM) exploit this L_p-spherical symmetry and can be used to generate sparse, heavy-tailed distributions. The GSM specifies the density of an image $\mathbf{x}$ as

$$p(\mathbf{x}) = \int_{-\infty}^{\infty} p(z)\mathcal{N}(\mathbf{x}; 0, z\mathbf{C})\, dz \tag{4.6}$$

where z is the scale factor, $p(z)$ specifies a distribution over different scales, $\mathbf{C}$ determines the covariance structure, and $\mathcal{N}$ indicates the normal distribution. By mixing many Gaussian distributions with identical means and covariance structures yet different scale factors, one can generate distributions with very heavy tails. Wainwright and Simoncelli [14] introduced GSMs to the field of natural image statistics as a way of capturing the correlations between the wavelet coefficients of natural images, which are similar to the correlations shown in Figure 4.4(c) and (d), and Portilla and colleagues later successfully applied the model to denoising [15].

A generalization of the GSM is given by L_2-*elliptically symmetric* models (L2) [13, 16, 17], which only assume that the isodensity contours of the natural image distribution are elliptically symmetric. L_2-elliptically symmetric distributions can be defined in terms of a function of the Euclidean norm (i.e., the L_2 norm) after a whitening linear transformation $\mathbf{W}$, that is, one which removes pairwise correlations:

$$p(\mathbf{x}) \propto f(||\mathbf{Wx}||_2) \tag{4.7}$$

Importantly, the spherical symmetry assumption of L2 implies that it is invariant under arbitrary orthogonal transformations $\mathbf{Q}$ (as $||\mathbf{QWx}||_2 = ||\mathbf{Wx}||_2$), meaning that the particular filter shapes in $\mathbf{W}$ are unimportant as applying an orthogonal transformation destroys filter shape. Nonetheless, L2 outperforms ICA in fitting the distribution of natural images [13]. The redundancies captured by L2 can be removed by applying a nonlinear transformation after whitening [16, 17].

The L_p-*spherically symmetric* model (Lp), which replaces the L_2-norm with an L_p-norm in Equation 4.7, is even more general and allows for any shape of isoprobability contour in the class of L_p spheres. Sinz and Bethge have shown that the optimal p for natural image patches is approximately equal to 1.3 [17, 18], and they later also generalized the Lp model further to a class of models called the L_p-nested distributions [19]. All L_p-spherical models fall into a general class of models called ν-*spherical distributions* [19].

Models exploiting the L_p-spherical symmetry of natural images are intimately related to the contrast fluctuations in natural images. A common measure of local image contrast is root-mean-square contrast, which is also the standard deviation of the pixel intensities. Pixel standard deviation is directly proportional to $||\mathbf{x}||_2$ if the mean intensity of $\mathbf{x}$ has been removed. L_p-spherically symmetric models have thus been considered particularly apt for capturing contrast fluctuations and have also been linked with the physiological process of contrast gain control. Simoncelli and colleagues made this link by showing how a model of neural gain control, divisive normalization, can be used to remove correlations between filter responses [e.g., [14, 16, 17, 20]. Physiological measurements of population activity in the primary visual cortex demonstrate that such adaptive nonlinearities play an important role in the neural coding of natural images [21].

We now return to the middle branch of the diagram in Figure 4.3. One straightforward way to extend linear factorial models is to use an *overcomplete* basis set $\mathbf{A} \in \mathbb{R}^{D\times M}$ where $M > D$, that is, where there are more sources than dimensions in the data. The sparse coding algorithm proposed by Olshausen and Field [22] was the first to successfully learn an overcomplete basis resembling cortical representations. (See Chapter 14 for an overview of sparse models.) Subsequent analysis has shown that using overcomplete representations also yields a better fit to the distribution of natural images [23]. Sparse representations have additionally been highly influential in biological experiments of visual processing. In Section 4.5, we discuss examples of visual experiments examining whether sparse coding predicts neural activity.

Group factorial models represent another important extension in which the source variables in $\mathbf{s}$ are modeled as J independent groups of variables:

$$p(\mathbf{s}) = \prod_{j=1}^{J} p(\mathbf{s}_j) \tag{4.8}$$

The *independent feature subspace analysis* (ISA) model [24] is a group factorial model that assumes each group of source variables is spherically symmetric distributed. More specifically, ISA assumes that the coefficients $\mathbf{s}$ can be split into pairs, triplets, or m-tuples that are independent from each other while the coefficients within an m-tuple have a spherically symmetric distribution. The density of an image thus depends on the densities of each m-tuple's 2-norm:

$$p(\mathbf{x}) \propto \prod_{j=1}^{J} f_j\left(\sqrt{\sum_{i\in I_j} (\mathbf{w}_i^\top \mathbf{x})^2}\right) \tag{4.9}$$

where there are J independent m-tuples and I_j is the set of indices in the jth group. The model is strikingly analogous to models of complex cells in the primary visual cortex – the individual filters $\mathbf{w}_i$ can be thought of as simple cell-receptive fields so that the response of a complex cell (one m-tuple) can be identified with the sum of the squared responses to its input simple cells. When applied to natural images, the filter shapes of each m-tuple are similar in orientation, location, and frequency but vary in phase, consistent with complex cell response properties [24]. Extensions of ISA that allow both the size of the subspace, $|I_j|$, and the linear filters, $\mathbf{W}$, to be learned simultaneously perform even better at capturing natural image regularities [25].

A third important development in extending factorial models are *products of experts* (PoE) [26]. This class of models generalizes the linear factorial model to an overcomplete representation in a way which relaxes the assumption of statistically independent coefficients $\mathbf{s}$. A PoE defines the density of an image as the product of M functions of projections of the image (the "experts"),

$$p(\mathbf{x}) \propto \prod_{i=1}^{M} f_i(\mathbf{w}_i^\top \mathbf{x}) \tag{4.10}$$

For $M = D$, that is, when there are as many filters as dimensions in the data, and one-dimensional projections $\mathbf{w}_i \in \mathbb{R}^D$, that is, filters of the same size as the training images, PoE reduces to the linear factorial model as in Equation 4.5. The product of Student-t (PoT) distributions [27, 28] is a popular instance of the PoE in which the experts take the form

$$f_i(s_i) = (1 + s_i^2)^{-\alpha_i} \tag{4.11}$$

Using heavy-tailed PoT distributions as experts encourages the model to find sparsely distributed features. PoT distributions also allow for efficient model training even for highly overcomplete models. Other PoE models and extensions which show great promise are those built on the restricted Boltzmann machine (RBM), such as the mcRBM or mPoT of Ranzato and colleagues [5] or the masked RBM of Heess and colleagues [29]. Each of these models has been successfully applied in computer vision tasks, some of which we discuss in Section 4.4.

An extension of the PoE to images of arbitrary size is given by the *field of experts* (FoE) [30, 31]. Up till now, our discussion has focused on patch-based image models, that is, rather than modeling large images, the previously described models capture the statistics of small image patches up to 32×32 pixels in size. The patch-based approach is computationally convenient as the data contain far fewer dimensions. The FoE is able to model images of arbitrary size by applying the same experts to different parts of an image. If $\mathbf{x}_{(k)}$ is a neighborhood of $n \times n$ pixels centered around pixel k, then the FoE's density can be written as

$$p(\mathbf{x}) \propto \prod_{k} \prod_{i=1}^{M} f_i(\mathbf{w}_i^\top \mathbf{x}_{(k)}) \tag{4.12}$$

where the index k runs over all pixel locations in the image and therefore the model considers all overlapping $n \times n$ regions in the image. The key distinction from the PoE is the following: to train a PoE, one first gathers a large ensemble of small image patches and learns a set of experts assuming that the training samples are independent of each other. An FoE, on the other hand, is trained on all overlapping $n \times n$ regions in a large image and therefore explicitly captures the statistical dependencies existing between overlapping image regions. The same M experts are used for all regions, so that the number of parameters depends on n, not on the size of the entire image, and therefore remains tractable. FoE has been highly successful in various image restoration tasks and will be discussed further in Section 4.4.

We now turn our discussion to the final branch on the right of the diagram in Figure 4.3. *Mixtures of Gaussians* (MoG) are widely used in a variety of density estimation tasks and have been shown to perform very well as a model of natural image patches [e.g., [32, 33]]. Under MoG, image patches are modeled as a mixture of K Gaussian distributions each with its own mean $\boldsymbol{\mu}_k$ and covariance $\mathbf{C}_k$

$$p(\mathbf{x}) = \sum_{k=1}^{K} \pi_k \mathcal{N}(\mathbf{x}; \boldsymbol{\mu}_k, \mathbf{C}_k) \tag{4.13}$$

where π_k is the prior probability of the kth Gaussian. Modeling even small image patches with an MoG requires a large number of mixture components, K. The same performance can be achieved with far fewer components if Gaussians are replaced with Gaussian *scale* mixtures as in the *mixture of GSMs* (MoGSM),

$$p(\mathbf{x}) = \sum_{k=1}^{K} \pi_k \int_{-\infty}^{\infty} p_k(z) \mathcal{N}(\mathbf{x}; \boldsymbol{\mu}_k, z\mathbf{C}_k)\, dz \tag{4.14}$$

As presented earlier, a single GSM (Equation 4.6) can be used to model a distribution with narrow peaks and heavy tails. This allows GSMs to capture the strong contrast fluctuations of natural images, but not to generate more structured content (see an example of GSM samples in Figure 4.7). By allowing for a mixture of different GSMs each with its own covariance structure, one attains a better approximation of the natural image distribution. Guerrero-Colon and colleagues extended the GSM of wavelet coefficients [14, 15] to MoGSMs and showed improved denoising performance [34].

Another model proposed by Karklin and Lewicki [35] can be viewed as a continuous (or *compound*) mixture of an infinite number of Gaussian distributions and like ISA has been linked to complex cells of the primary visual cortex.

An extension of MoGSM to images of arbitrary size is achieved by the mixture of *conditional* Gaussian scale mixtures (MCGSM) [36]. Any distribution can be decomposed into a set of conditional distributions via the chain rule of probability theory,

$$p(\mathbf{x}) = \prod_{i,j} p(x_{ij} \mid \mathrm{Pa}_{ij}) \tag{4.15}$$

Rather than modeling the distribution of natural images directly, an MCGSM tries to capture the distribution of one pixel x_{ij} given a neighborhood of pixels Pa_{ij}. Such a neighborhood is illustrated in Figure 4.5. Assuming that the distribution of natural images is invariant under translations of the image, one can use the same conditional distribution for each pixel x_{ij}, so that only a single conditional distribution has to be learned. The translation-invariance assumption is analogous to the weight-sharing constraint used in convolutional neural networks, a regularization method that was critical to make discriminative learning of deep neural networks feasible. Theis *et al.* [36] derive a form for $p(x_{ij} \mid \mathrm{Pa}_{ij})$ by assuming that the joint distribution of $p(x_{ij}, \mathrm{Pa}_{ij})$ is given by an MoGSM. The MCGSM greatly improves modeling power over the MoGSM as the same amount of parameters used by the MoGSM to model the full distribution of image pixels is used to capture a single conditional distribution.

One of the main advances made by the MCGSM over patch-based models is its ability to capture long-range correlations. This is possible because the model is not restricted to learning the structure of independent small image regions, but like

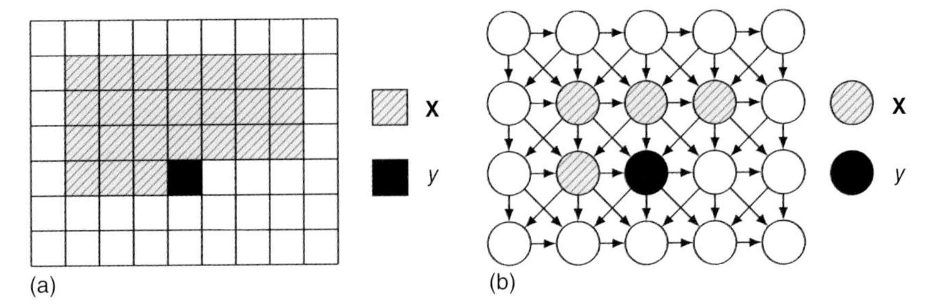

Figure 4.5 The MCGSM [36] models images by learning the distribution of one pixel, y, given a causal neighborhood of pixels $\mathbf{x}$ (a). A graphical model representation of the MCGSM is shown in (b), where for visualization purposes the neighborhood consists of only four pixels. The parents of a pixel are constrained to pixels which are above it or in the same row and left of it, which allows for efficient maximum likelihood learning and sampling.

the FoE model, it is trained using many overlapping neighborhoods. As a technical note, MCGSM has a practical advantage as well: unlike for FoEs, it is computationally tractable to evaluate the likelihood of an MCGSM. (We discuss such quantitative model comparisons in the following section.) The MCGSM is able to capture many kinds of perceptually relevant image structures, such as edges and texture features (see Figure 4.9 for some examples). By combining the MCGSM with a multiscale representation, the model's ability to capture correlations can be increased even further [36].

The same trick of turning a patch-based model into an image model has been applied to GSMs [37], MoG [38], and products of experts (RNADE) [39] to derive tractable models of natural images that scale well with image size.

4.3.1 Model Evaluation

In the previous section, we often alluded to how well the various models performed. An information-theoretic measure which quantifies how well a model distribution $q(\mathbf{x})$ agrees with a target distribution $p(\mathbf{x})$ is given by the Kullback–Leibler (KL) divergence,

$$D_{\mathrm{KL}}[p(\mathbf{x}) \,||\, q(\mathbf{x})] = -\int p(\mathbf{x})\log_2 q(\mathbf{x})\, d\mathbf{x} - H[p(\mathbf{x})] \tag{4.16}$$

It describes the average additional cost (in bits) incurred by encoding $\mathbf{x}$ using a coding scheme optimized for $q(\mathbf{x})$ when the actual distribution is $p(\mathbf{x})$. The KL divergence is always nonnegative, and it is zero if and only if p and q are identical. Unfortunately, evaluating the KL divergence requires knowledge of the distribution $p(\mathbf{x})$, which, in our case, is the distribution of natural images. However, using samples from $p(\mathbf{x})$, that is, a dataset of images, we can obtain an unbiased estimate of the first term, the negative log-likelihood or cross-entropy:

$$-\frac{1}{N}\sum_{i=1}^{N}\log\ q(\mathbf{x}_i) \tag{4.17}$$

The KL divergence is invariant under reparameterization of $\mathbf{x}$. This property is lost when only considering the cross-entropy term. In practice, this means that changes to the way images are represented and preprocessed, for example, differently scaled pixel values, affect the numbers we measure. One should thus only consider differences between the log-likelihoods of different models, which correspond to differences between the models' KL divergences (because the entropy term cancels out).

A more robust measure is obtained when we try to estimate the multi-information (Equation 4.1) by replacing the entropy with a cross-entropy,

$$\hat{I}[\mathbf{x}] = \sum_i H[x_i] + \frac{1}{N}\sum_{i=1}^{N}\log\ q(\mathbf{x}_i) \tag{4.18}$$

This measure is invariant under invertible pointwise nonlinearities and thus more meaningful on an absolute scale. In contrast to the KL divergence, it only requires knowledge of the marginal entropy of a pixel's intensity and this can be estimated relatively easily. Because the cross-entropy is always larger than the entropy, this yields an estimated lower bound on the true redundancy of natural images. It can thus be thought of as the amount of second- and higher-order correlations captured by a given model.

In Figure 4.6 we compare multi-information estimates for most of the models we reviewed. The parameters of all models were estimated using some form of maximum likelihood learning and a separate test set was used for evaluation. For patch-based models, we used 16×16 patches sampled uniformly from the dataset of van Hateren and van der Schaaf [10] which were subsequently log-transformed; that is, instead of working with linear pixel intensities, we model $\log\ \mathbf{x}$, as is common. By taking the logarithm, one tries to mimic the response properties of photoreceptors [40, 41]. We used 10,000 patches for evaluation. For some models, $\log\ q(\mathbf{x}_i)$ could not be evaluated analytically but had to be estimated (DBN [32], OICA, PoT [23]). For the MCGSM, we used a 7×4 neighborhood as in Figure 4.5(a) and 200,000 data points for evaluation. In addition, because the image intensities of the van Hateren dataset were discretized, we added independent uniform noise before log-transforming the pixels and evaluating the MCGSM. This ensures that the differential entropy and hence the cross-entropy is bounded below. Without adding noise, the model's likelihood might diverge (both on the training and the test sets). We only empirically observed this problem with the MCGSM, whose conditional structure allows it to pick up on the discretization.

Figure 4.6 shows that the stationarity assumption of the MCGSM allows it to capture much more correlations than its patch-based counterpart. This is the case despite the fact that the MCGSM has much fewer parameters. We used 32 components for the MoGSM but only 8 components for the MCGSM, and each component of the MCGSM has fewer parameters than one component of the MoGSM,

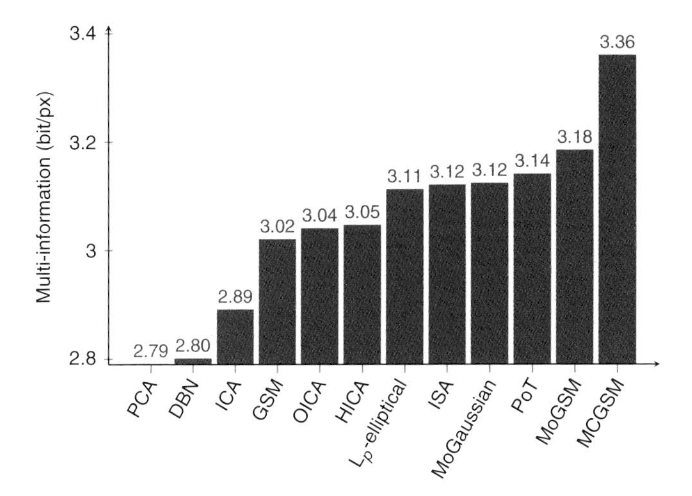

Figure 4.6 Redundancy reduction capabilities of various methods and models quantified in terms of estimated multi-information. PCA [12, 13] only takes second-order correlations into account and here serves as the baseline. ICA [12, 13] corresponds to the best linear transformation for removing second- and higher-order correlations. Overcomplete ICA (OICA) [23], ISA [25], and hierarchical ICA (HICA) [42] represent various extensions of ICA. Also included are estimates for deep belief networks (DBN), L_p-elliptical models [19], MoG (32 components), PoT, mixtures of GSMs, and MCGSMs.

because it depends on fewer pixels. The figure also shows that relatively simple models such as L_p-elliptical models can already capture much of the higher-order redundancies captured by more complex models such as MoG or PoT, which contain many more parameters and are more difficult to optimize.

Figure 4.7 shows samples generated by some of these models trained on image patches sampled from the van Hateren dataset [10]. It can be seen that the GSM improves over PCA by capturing the contrast fluctuations in natural images, and more sophisticated models introduce more and more structured content resembling branches. We note that the appearance of samples may change when a different dataset is used for training. For example, datasets used in computer vision more frequently contain urban scenes and handmade objects, leading to a stronger prevalence of high-contrast edges.

An important question is whether the differences in likelihood are perceptually meaningful. Recent psychophysical experiments in which human observers discriminated true natural images from model samples showed that even small changes in likelihood lead to appreciable perceptual differences and more generally that likelihood seems to have good predictive power about perceptual relevance [43]. Those experiments tested a wide range of patch-based models including a model capturing only second-order correlations, ICA, L2, Lp, and MoGSM. People performed very well on the task whenever images were 5×5 pixels in size or larger, indicating that the human visual system possesses far more detailed knowledge of the natural image distribution than any model tested.

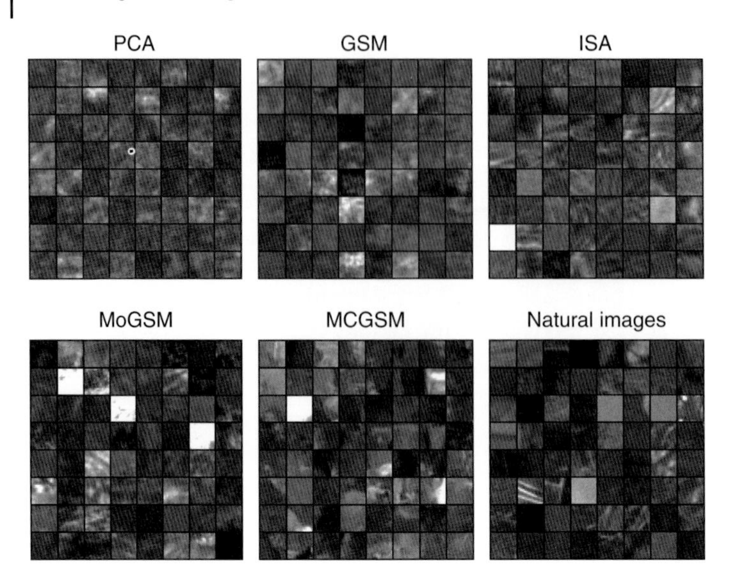

Figure 4.7 Image patches generated by various image models. To enhance perceptual visibility of the difference between the samples from the different models, all models were trained on natural images with a small amount of additive Gaussian noise.

Moreover, human performance depended on the model likelihood such that samples from lower likelihood models were easier to distinguish from natural images than samples from higher likelihood models were. (We will discuss this work in more detail in Section 4.5 when we discuss experimental links between biological vision and natural image models.)

Although likelihood provides an objective measure for model comparison that also shows good predictive power about perceptual relevance, it is not an absolute measure of model performance because the total amount of correlations present in natural images is unknown. Another difficulty is that it is not always straightforward or computationally tractable to evaluate model likelihood. Psychophysical discrimination measures can provide an absolute performance measure – either a model succeeds at fooling the human visual system, in which case, human observers will be at chance to discriminate its samples from real natural images, or not, but the technique developed in Ref. [43] requires the experimenter to generate model samples matched in joint probability to a set of natural images, which can be difficult for some models. Furthermore, it should be used in concert with likelihood estimation as it is trivial to construct a model which would bring human performance to chance but assigns a density of zero to almost all natural images (e.g., a model which assigns probability $1/N$ to each image in a training set of N images). Other methods for model evaluation have also been suggested, such as denoising performance [e.g., [34]], inpainting [e.g., [30, 31]], or measuring the similarity between model samples and true natural images [e.g., [44]].

4.4
Computer Vision Applications

Knowledge of natural image statistics is proving useful in a variety of computer vision applications. We will describe two important areas: image restoration and classification. We chose this focus because image models have an established and direct role in restoration and because contributions to classification applications show great promise for future impact as modeling techniques advance.

All image restoration tasks require a way of predicting the true content of an image from a corrupted version of the image. (Note that Chapter 16 also discusses image restoration.) A probabilistic image model tells us which images are a priori more likely and are thus clearly useful in such a setting. As an example, Figure 4.8 shows filling-in of missing pixels by computing the most likely pixel values under the distribution defined by an MCGSM.

For removing noise from an image, the common recipe starts with separately modeling the noise and the image distribution. One of the earlier examples is the GSM model of Portilla and colleagues [15]. In their approach, images are first transformed using a particular overcomplete wavelet transformation called the *steerable pyramid*, and dependencies between coefficients of the resulting representation are modeled using a GSM. In this model, noisy coefficients are assumed to result from the sum of a GSM and a Gaussian random variable. To denoise, the covariances of the GSM and the random Gaussian variable are estimated for each subband of the wavelet transformation and a Bayesian least squares estimator is used to predict the denoised coefficients. Guerrero-Colon and colleagues [34] extended the approach by using a finite MoGSM, a more sensitive image model that allows for variations in pixel covariance across different subregions of the image. It substantially outperforms a single GSM (in agreement with the models' likelihood ordering).

Another prominent example of modeling applied to image restoration is the work of Roth and Black [30, 31]. They used a FoE and showed that it performs well on both denoising and the related task of inpainting, in which unwanted parts of an image such as scratches are removed. Ranzato and colleagues [5]

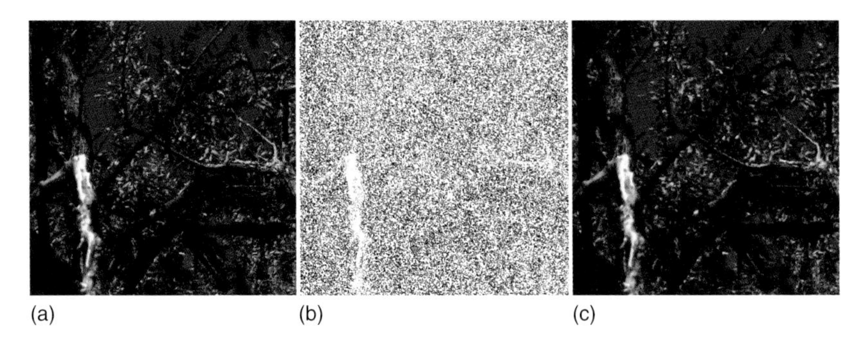

(a) (b) (c)

Figure 4.8 After removing 70% of the pixels from the image (a), the missing pixels from (b) were estimated by maximizing the density defined by an MCGSM (c).

have shown that PoE can be made to achieve competitive denoising performance by combining it with the nonlocal means algorithm [45]. Interestingly, the state-of-the-art denoising technique of Mairal and colleagues (LSSC) [46] relies on a combination of a sparse image model similar to that of Olshausen and Field [22] and a nonparametric heuristic exploiting the similarity of local regions across an image. Finally, Zoran and Weiss [33] achieved denoising results similar to LSSC using only large mixtures of Gaussians.

The success of natural image models at capturing perceptually meaningful image content is also highlighted in another application – texture modeling. Image models trained on various textures, that is, visually homogeneous images rather than scenes which typically contain many textures, can be used to develop useful image representations and to aid in discriminating textures and synthesizing new samples of a given texture. In Figure 4.9, we illustrate an example of natural textures and synthesized samples generated using the MCGSM model.

A more recent development is the application of natural image models to classification tasks. In this area, models have been used to design powerful image representations that often improve discriminability over methods relying on well-engineered features such as SIFT (See Chapter 15, section 15.4.1 for a description). Ranzato and colleagues have published a variety of recent results showing how natural image models can be brought to bear on high-level classification tasks, summarized in Ref. [5]. They applied a particular class of hierarchical image models based on the RBM that includes the mcRBM which is specialized to capture mean and covariance fluctuations across images and the mPoT, an extension of PoT. Their results show that the model-extracted representations

Figure 4.9 Computer vision applications of image models include the synthesis, discrimination, and computation of representations of textures. Here we illustrate the performance of one model in capturing the statistics of a variety of textures. The left image of each pair shows a random 256×256 pixel crop of a texture [47]. The right image of each pair shows a histogram-matched sample from the MCGSM trained on the texture. The samples provide a visual illustration of the kind of correlations the model can capture when applied to various visual textures.

lead to competitive performance on scene classification, object recognition, and recognition of facial expressions under occlusion.

A principled approach to improving classifiers using unlabeled data is to define a joint model of class labels k and images $\mathbf{x}$, $p(k, \mathbf{x})$, and to optimize the model's parameters with respect to the joint log-likelihood,

$$\sum_i \log\ p(k_i, \mathbf{x}_i) + \sum_j \log\ p(\mathbf{x}_j)$$

where the first sum is over labeled data and the second sum is over unlabeled data. Using a hierarchical extension of PoT, Ngiam and colleagues [4] showed that taking into account the joint density instead of performing purely discriminative training can have a regularizing effect and improve classification performance even in the absence of additional unlabeled data.

4.5
Biological Adaptations to Natural Images

Since the 1980s, several researchers have endeavored to directly measure how biological systems are adapted to the statistical regularities present in the environment. Although this goal remains highly challenging, several studies of biological vision have provided clear examples. In this section, we will focus on a small selection of some of the most compelling results. Our review will proceed from studies of early visual processing stages, such as those occurring in the eyes, to studies of the later processing stages occurring in subcortical and cortical sites of the mammalian visual pathway.

We start with one of the earliest experiments to connect biological vision with image statistics using efficient coding principles. In an elegant comparison of computational predictions and physiological measurements, Laughlin [48] illustrated how the contrast statistics of natural scenes are efficiently represented by the cells that code contrast in the fly's compound eye. Laughlin used a photodiode rig that replicated the imaging process of a fly photoreceptor to measure the contrast distribution over a variety of natural scenes. He then measured the physiological responses to contrast in the contrast-coding cells of the fly's compound eye. In the key comparison, the physiological response distribution was shown to be extremely well matched to the cumulative distribution of the measured contrast values, the optimal coding function for maximizing information transmission of a single input parameter using a single output parameter. This response distribution ensures that more frequent contrast values are represented with finer resolution in the space of possible physiological response states. This finding was one of the first to reveal a biological adaptation to natural visual input at the single-cell level.

In a recent analysis of primate retina (see Chapter 3 for an overview of retinal processing), Doi and colleagues [49] conducted a similarly elegant comparison of measured responses and efficient coding predictions. Their focus was on the retinal circuitry connecting cones, the photoreceptors active in daylight conditions, to complete populations of different classes of retinal ganglion cells, the cells relaying visual information from the eyes to the brain. They derived the optimal connectivity matrix of weights between cone inputs and retinal ganglion cells, that is, the one maximizing information transmission for natural images while also taking physiological constraints into account. The receptive field structures associated with the optimal connectivity matrix were remarkably similar to the measured ones. Doi and colleagues further showed that the information transmission of the measured ganglion cells was highly efficient relative to that of the optimal population, achieving 80% efficiency. This work beautifully illustrated that biological adaptations to natural images are also present in the connectivity structures linking visual neurons together.

In Section 4.3, we described earlier theoretical work by Atick and Redlich [8] which demonstrated that the responses of primate retinal ganglion cells are adapted for the $1/f^2$ power spectrum of natural images. Dan *et al.* [50] extended this work with physiological measurements in the cat lateral geniculate nucleus (LGN), a subcortical way station along the mammalian visual pathway where visual input is processed before being relayed to cortical areas. As with retinal ganglion cells, the receptive fields of LGN cells act as local spatial bandpass filters. Dan and colleagues measured LGN responses to natural movies and to white noise movies (free of both spatial and temporal correlations). The LGN responses to natural movies, but not to white noise stimuli, were flat in the frequency domain as predicted, which indicated the cells were adapted for removing temporal correlations in visual input.

Studies of the primary visual cortex in the mammalian visual pathway (V1) have revealed several forms of adaptation to natural images. Berkes and colleagues [51] measured neural activity in ferret V1 across the lifespan to identify developmental adaptations to natural images. Rather than testing predictions about efficient information transmission, they pursued an entirely different approach to measuring how neural processes adapt to the environment's statistical properties. Their motivation was to understand how organisms make inferences about the world despite ambiguous sensory input. They hypothesized that animals build an internal statistical model of the environment over the course of development and that this model should be reflected in the spontaneous activity of V1 neurons in the absence of visual stimulation. By comparing neural activations in the dark with activations evoked in response to natural movies and to non-natural control movies, they showed that as ferrets aged, their spontaneous activity grew more similar to the activity evoked by natural movies, an effect which was not present for activity evoked by non-natural movies. These measurements indicated that statistical properties of natural images can also influence complex neural dynamics in the absence of input, in a manner consistent with learning an internal model of the environment. (See Chapter 9 for a related discussion of perceptual inference using internal models of the visual world.)

Sparse coding is a mechanism proposed for learning in neural systems [11, 22] (introduced in our discussion of linear factorial models in Section 4.3 and discussed in detail in Chapter 14), and identifying evidence for sparse coding has been the target of several physiological experiments. Vinje and Gallant [52] first demonstrated that the responses of V1 neurons become more selective under natural viewing conditions, an effect causing an individual neuron's responses to be elicited less often (i.e., more sparsely) and causing pairs of neurons to become decorrelated. It is important to note that this effect did not depend on the spatiotemporal structure of the visual input: both grating patterns and natural images resulted in the same sparsity effects. The important factor was instead whether an individual neuron alone was stimulated or whether several neurons processing neighboring parts of the image were also stimulated simultaneously (as in natural viewing conditions) by an image having greater visual extent. A more recent study examined how the spatiotemporal correlations of natural input affect V1 population-level responses. Using a state-of-the-art, three-dimensional neural imaging technique, Froudarakis and colleagues [53] simultaneously measured the responses of up to 500 cells in mouse V1 while either natural movies or phase-scrambled versions were shown (containing the same second-order spatial-temporal correlations yet lacking the higher-order correlations of natural movies). The results confirmed that population activity during natural movies was sparser than during the phase-scrambled movies. The sparser representation allowed for better discrimination of the different natural movie frames shown, illustrating that sparsity is linked with improved readout of the neural signal. Importantly, Froudarakis and colleagues also showed that these results rested on a nonlinearity that rapidly adapts to the statistics of the visual input or a contrast gain control mechanism being active during natural but not unnatural stimulation. This work is one of the first to demonstrate the specific importance of higher-order natural scene regularities in shaping the complex population-level activity of V1 neurons.

Evidence for adaptations to the higher-order regularities of natural images can also be observed at the behavioral level using rigorous psychophysical methods. We previously developed a technique for evaluating natural image models using the human visual system ([43], introduced in Section 4.3.1). The experimental manipulation highlighted the local variations present in natural scenes. Subjects saw two sets of image patches arranged in tightly tiled square grids (Figure 4.10). They were told that one set was cut from real photographic images, whereas the other was a set of impostor images, and their task was to select the true set of photographic images. The impostors on each trial were model samples matched in joint-likelihood to the true images, such that a perfect model, that is, one capturing at least as much of the regularities as the human visual system represents, would force observers to be at chance on the discrimination task. We tested the efficacy of several natural image models in this way, from a model capturing only second-order correlations up to the patch-based model at the state of the art for capturing higher-order correlations (MoGSM). The results clearly showed that human observers were sensitive to more of the higher-order correlations present in the natural images than were captured by any of the models, which held true whenever the image patches were 5×5 pixels in size or greater.

The same psychophysical technique can also be used to test specific hypotheses about the image features that human observers use to discriminate model samples

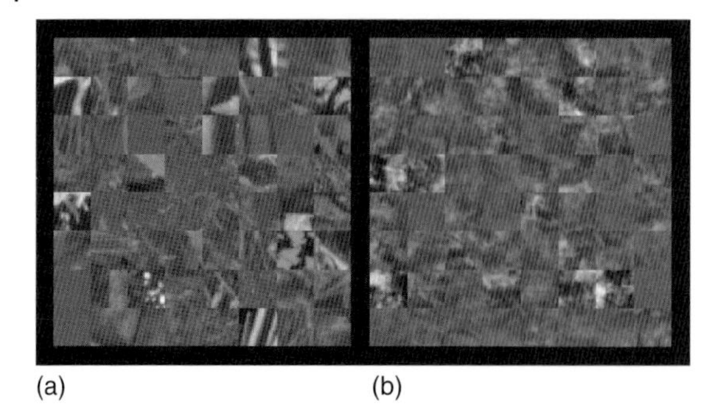

Figure 4.10 Illustration of a psychophysical stimulus pitting natural images against model samples following [43]. Observers viewed two sets of tightly tiled images. They were told that one set included patches from photographs of natural scenes, whereas the other contained impostors, and their task was to identify the true set ((a) here). The "imposters" here are samples from the L_p-spherically symmetric model. The model samples were always matched in joint probability to the natural images (under the particular model being tested, i.e., under Lp here). In this example, the patches are 20×20 pixels in size. As shown, the model samples fail to capture several prominent features of natural images, particularly their detailed geometric content.

from true natural images. We did so by performing a series of control experiments that were identical to the main experiment except that only select image features were made available in the stimuli. The results of the various feature-controlled experiments revealed three key model shortcomings: first, linear factorial models fail to capture the marginal distribution of natural images, second, L_p-spherical models fail to capture local contrast fluctuations across images, and third, all models fail to capture local shape statistics sufficiently. The third shortcoming was underscored by an experiment in which the image patches making up the stimuli were binarized, a manipulation emphasizing the shapes of the luminance contours present in the images, which, for example, relate to object silhouettes and shading patterns across three-dimensional objects. The human observers were able to discriminate natural images from model samples even with the impoverished binary stimuli, and the MoGSM trials were as easy as the second-order model trials. Taken together, the feature identification experiments suggested that increases in model likelihood were mostly driven by improvements in capturing local luminance and contrast statistics, rather than improvements in representing the detailed geometric content of natural scenes, an aspect to which human observers are highly sensitive.

In summary, a variety of neuroscience techniques have already revealed several ways biological visual systems are adapted for natural scenes: from the single-cell level, to connectivity and population dynamic effects, up to perceptual effects. Future improvements in natural image modeling ought to allow for exciting new possibilities for generating well-controlled experimental stimuli that could be used to probe further aspects of how visual neurons are adapted for natural scenes and also, importantly, to probe the learning mechanisms that support these adaptations. This in turn may well lead to new insights for designing improved machine and computer vision applications.

4.6 Conclusions

In summary, natural image models have been useful in the area of image restoration, show promise as a means to do unsupersived learning of representations for classification tasks, and have provided insights into the image representations used by biological vision systems. In Figure 4.11, we illustrate the types of image features captured by state-of-the-art natural image models by comparing natural images with model samples. Several meaningful aspects of natural images are clearly represented, such as texture and edge features, yet current state-of-the-art image models are still unable to represent more object-like structures. An important question for future research is thus how to achieve higher-level image representations.

It is important to note that many of the advances in density estimation performance are due to an improved ability to model contrast fluctuations. This is

Figure 4.11 (a) A crop of an example image from a natural image dataset collected by Tkačik *et al.* [54]. (b) A corresponding sample of an MCGSM trained on the dataset, illustrating the kind of correlations captured by a current state-of-the-art probabilistic model of natural images. As with textures (Figure 4.9), the nature of the samples depends highly on the training images. When trained on an urban scene (a crop of the scene is shown in (c); photograph by Matt Wiebe [55]), the samples express much more vertical and horizontal structure (d).

in large part due to the fact that model likelihood is very sensitive to the model's ability to capture luminance and contrast fluctuations. While objects themselves affect the pattern of contrast fluctuations apparent in an image, a lot of the variability in natural images is generated by changes in the illumination. It would be clearly advantageous for a model to be able to separate illumination effects from other variations in the scene's content. One way to focus more modeling power on other image content is to model contrast fluctuations separately, as is done by L_p-spherical models. An alternative approach would be to model other image representations than the pixel representation. For example, representations of image contours – e.g., binary black-and-white images showing only object silhouettes – already contain a great deal of relevant information for object recognition. Such representations are also much more stable under changes of lighting conditions than the pixel representation. Thus, modeling the statistics of black-and-white images would likely provide new insights for the development of useful image representations. A third approach could be to abandon likelihoods altogether and instead maximize other functions of the model's image distribution.

A complementary line of research has recently made it possible to learn high-level image representations using neural networks in a purely supervised manner [56, 57]. These techniques require huge amounts of labeled training data and even then can only be stopped from overfitting by using clever regularization techniques. An interesting question is whether the generalization performance of neural networks can be further improved using unsupervised learning, for example, by combining objectives from image modeling and classification in a semi-supervised setting as done by Ngiam and colleagues [4].

Although a few image models have been extended to video sequences, videos still provide a largely untapped source of information for unsupervised learning of image representations using probabilistic models. Obtaining large amounts of labels of high quality for videos is a more challenging endeavor than obtaining labels for images, so that we anticipate unsupervised learning to play a particularly important role for learning representations from videos.

To conclude, the field of image modeling is still in its formative stages, and the last couple of years have shown particularly promising bounds forward in terms of model performance, which is due partly to developments in machine learning techniques and partly to technological advances in computing. We expect this trend to continue in the coming years and the density estimation performance of image models to increase. As advances are made, one of the most exciting questions for future research will be how natural image statistics can be exploited to effectively improve visual inference in computer vision systems.

References

1. Simoncelli, E.P. and Olshausen, B.A. (2001) Natural image statistics and neural representation. *Annu. Rev. Neurosci.*, **24**, 1193–1216.
2. Barlow, H.B. (1959) Sensory mechanisms, the reduction of redundancy, and intelligence, in *The Mechanisation of Thought Processes*, Her

Majesty's Stationery Office, London, pp. 535–539.
3. Attneave, F. (1954) Some informational aspects of visual perception. *Psychol. Rev.*, **61** (3), 183–193.
4. Ngiam, J., Chen, Z., Koh, P.W., and Ng, A.Y. (2011) Learning deep energy models. *Proceedings of the 28th International Conference on Machine Learning*, pp. 1105–1112.
5. Ranzato, M., Mnih, V., Susskind, J.M., and Hinton, G.E. (2013) Modeling natural images using gated MRFs. *IEEE Trans. Pattern Anal. Mach. Intell.*, **35** (9), 2206–2222.
6. Geisler, W.S. (2008) Visual perception and the statistical properties of natural scenes. *Annu. Rev. Psychol.*, **59**, 167–192.
7. Chen, S.S. and Gopinath, R.A. (2000) Gaussianization, in *Advances in Neural Information Processing Systems*, MIT Press, vol. 13, pp. 423–429.
8. Atick, J. and Redlich, A. (1992) What does the retina know about natural scenes? *Neural Comput.*, **4**, 196–210.
9. Bell, A. and Sejnowski, T. (1997) The "Independent Components" of natural scenes are edge filters. *Vision Res.*, **37** (23), 3327–3338.
10. van Hateren, J.H. and van der Schaaf, A. (1998) Independent component filters of natural images compared with simple cells in primary visual cortex. *Proc. R. Soc. London, Ser. B*, **265** (1394), 359–366.
11. Olshausen, B.A. and Field, D.J. (1996) Emergence of simple-cell receptive field properties by learning a sparse code for natural images. *Nature*, **381**, 607–609.
12. Bethge, M. (2006) Factorial coding of natural images: how effective are linear models in removing higher-order dependencies? *J. Opt. Soc. Am. A*, **23** (6), 1253–1268.
13. Eichhorn, J., Sinz, F., and Bethge, M. (2009) Natural image coding in V1: how much use is orientation selectivity? *PLoS Comput. Biol.*, **5** (4), 1–16, doi: 10.1371/journal.pcbi.1000336.
14. Wainwright, M. and Simoncelli, E. (2000) Scale mixtures of Gaussians and the statistics of natural images, in *Advances in Neural Information Processing Systems*, MIT Press, vol. 12, pp. 855–861.
15. Portilla, J., Strela, V., Wainwright, M.J., and Simoncelli, E.P. (2003) Image denoising using scale mixtures of Gaussians in the wavelet domain. *IEEE Trans. Image Process.*, **12** (11), 1338–1351.
16. Lyu, S. and Simoncelli, E.P. (2009) Nonlinear extraction of independent components of natural images using radial Gaussianization. *Neural Comput.*, **21** (6), 1485–1519.
17. Sinz, F. and Bethge, M. (2009) The conjoint effect of divisive normalization and orientation selectivity on redundancy reduction, in *Advances in Neural Information Processing Systems*, MIT Press, vol. 21, pp. 1521–1528.
18. Sinz, F. and Bethge, M. (2008) The conjoint effect of divisive normalization and orientation selectivity on redundancy reduction, in *Advances in Neural Information Processing Systems*, MIT Press, vol. 21, pp. 1521–1528.
19. Sinz, F. and Bethge, M. (2010) L_p-nested symmetric distributions. *J. Mach. Learn. Res.*, **11**, 3409–3451.
20. Schwartz, O. and Simoncelli, E.P. (2001) Natural signal statistics and sensory gain control. *Nat. Neurosci.*, **4** (8), 819–825.
21. Heeger, D.J. (1992) Normalization of cell responses in cat striate cortex. *Visual Neurosci.*, **9**, 181–197.
22. Olshausen, B.A. and Field, D.J. (1997) Sparse coding with an overcomplete basis set: a strategy employed by V1. *Vision Res.*, **37**, 3311–3325.
23. Theis, L., Sohl-Dickstein, J., Bethge, M. (2012) Training sparse natural image models with a fast Gibbs sampler of an extended state space, in *Advances in Neural Information Processing Systems*, MIT Press, vol. 25, pp. 1133–1141.
24. Hyvärinen, A. and Hoyer, P. (1999) Emergence of topography and complex cell properties from natural images using extensions of ICA, in *Advances in Neural Information Processing Systems*, MIT Press, pp. 827–833.
25. Hyvärinen, A. and Köster, U. (2007) Complex cell pooling and the statistics of natural images. *Netw. Comput. Neural Syst.*, **18** (2), 81–100.

26. Hinton, G.E. (2002) Training products of experts by minimizing contrastive divergence. *Neural Comput.*, **14** (8), 1771–1800.
27. Welling, M., Hinton, G., and Osindero, S. (2003) Learning sparse topographic representations with products of student-t distributions, in *Advances in Neural Information Processing Systems*, MIT Press, vol. 15, pp. 1383–1390.
28. Teh, Y.W., Welling, M., Osindero, S., and Hinton, G.E. (2003) Energy-based models for sparse overcomplete representations. *J. Mach. Learn. Res.*, **4**, 1235–1260.
29. Heess, N., Le Roux, N., and Winn, J. (2011) Weakly supervised learning of foreground-background segmentation using masked RBMs, in *Artificial Neural Networks and Machine Learning*, Springer-Verlag, pp. 9–16.
30. Roth, S. and Black, M.J. (2005) Fields of experts: a framework for learning image priors. *IEEE Computer Society Conference on Computer Vision and Pattern Recognition*, 2005. CVPR 2005, pp. 860–867.
31. Roth, S. and Black, M.J. (2009) Fields of experts. *Int. J. Comput. Vision*, **82** (2), 205–229.
32. Theis, L., Gerwinn, S., Sinz, F., and Bethge, M. (2011) In all likelihood, deep belief is not enough. *J. Mach. Learn. Res.*, **12**, 3071–3096.
33. Zoran, D. and Weiss, Y. (2011) From learning models of natural image patches to whole image restoration. *2011 IEEE International Conference on Computer Vision (ICCV)*, pp. 479–486.
34. Guerrero-Colon, J., Simoncelli, E., and Portilla, J. (2008) Image denoising using mixtures of Gaussian scale mixtures. *15th IEEE International Conference on Image Processing, 2008. ICIP 2008*, pp. 565–568.
35. Karklin, Y. and Lewicki, M.S. (2009) Emergence of complex cell properties by learning to generalize in natural scenes. *Nature*, **457** (7225), 83–86, *http://www.ncbi.nlm.nih.gov/pubmed/19020501*, doi: 10.1038/nature07481.
36. Theis, L., Hosseini, R., and Bethge, M. (2012) Mixtures of conditional Gaussian scale mixtures applied to multiscale image representations. *PLoS ONE*, **7** (7), doi: 10.1371/journal.pone.0039857.
37. Hosseini, R., Sinz, F., and Bethge, M. (2010) Lower bounds on the redundancy of natural images. *Vision Res.*, **50** (22), 2213–2222.
38. Domke, J., Karapurkar, A., and Aloimonos, Y. (2008) Who killed the directed model? *IEEE Computer Society Conference on Computer Vision and Pattern Recognition*, doi: 10.1109/CVPR.2008.4587817.
39. Uria, B., Murray, I., and Larochelle, H. (2013) RNADE: the real-valued neural autoregressive density-estimator. *Advances in Neural Information Processing Systems*, MIT Press, vol. 26, pp. 2175–2183.
40. Naka, K.I. and Rushton, W.A. (1966) S-potentials from colour units in the retina of fish. *J. Physiol.*, **185**, 536–555.
41. Norman, R.A. and Werblin, F.S. (1974) Control of retinal sensitivity. I. Light and dark adaptation of vertebrate rods and cones. *J. Gen. Physiol.*, **63**, 37–61.
42. Hosseini, R., Sinz, F., and Bethge, M. (2010) New estimate for the redundancy of natural images, doi: 10.3389/conf.fncom.2010.51.00006.
43. Gerhard, H.E., Wichmann, F.A., and Bethge, M. (2013) How sensitive is the human visual system to the local statistics of natural images? *PLoS Comput. Biol.*, **9** (1), doi: doi/10.1371/journal.pcbi.1002873.
44. Heess, N., Williams, C.K.I., and Hinton, G.E. (2009) Learning generative texture models with extended fields-of-experts. British Machine Vision Conference.
45. Buades, A., Coll, B., and Morel, J. (2005) A non-local algorithm for image denoising. IEEE Conference on Computer Vision and Pattern Recognition, vol. 2, pp. 60–65.
46. Mairal, J., Bach, F., Ponce, J., Sapiro, G., and Zisserman, A. (2009) Non-local sparse models for image restoration. 2009 IEEE 12th International Conference on Computer Vision, pp. 2272–2279.
47. Brodatz, P. (1966) *Textures: A Photographic Album for Artists and Designers*, Dover Publications, New York.
48. Laughlin, S. (1981) A simple coding procedure enhances a neuron's information

capacity. *Z. Naturforsch. C*, **36** (910), 910–912.

49. Doi, E., Gauthier, J.L., Field, G.D., Shlens, J., Sher, A., Greschner, M., Machado, T.A., Jepson, L.H., Mathieson, K., Gunning, D.E., Litke, A.M., Paninski, L., Chichilnisky, E.J., and Simoncelli, E.P. (2012) Efficient coding of spatial information in the primate retina. *J. Neurosci.*, **32** (46), 16 256–16 264.
50. Dan, Y., Atick, J.J., and Reid, R.C. (1996) Efficient coding of natural scenes in the lateral geniculate nucleus: experimental test of a computational theory. *J. Neurosci.*, **16** (10), 3351–3362.
51. Berkes, P., Orbán, G., Lengyel, M., and Fiser, J. (2011) Spontaneous cortical activity reveals hallmarks of an optimal internal model of the environment. *Science*, **331** (6013), 83–87.
52. Vinje, W.E. and Gallant, J.L. (2000) Sparse coding and decorrelation in primary visual cortex during natural vision. *Science*, **287** (5456), 1273–1276.
53. Froudarakis, E., Berens, P., Ecker, A.S., Cotton, R.J., Sinz, F.H., Yatsenko, D., Saggau, P., Bethge, M., and Tolias, A.S. (2014) Population code in mouse V1 facilitates read-out of natural scenes through increased sparseness. *Nat. Neurosci.*, **17**, 851–857.
54. Tkačik, G., Garrigan, P., Ratliff, C., Milčinski, G., Klein, J.M., Seyfarth, L.H., Sterling, P., Brainard, D.H., and Balasubramanian, V. (2011) Natural images from the birthplace of the human eye. *PLoS ONE*, **6** (6), e20 409, doi: 10.1371/journal.pone.0020409.
55. Wiebe, M. (2014) Freedom Tower, *https://goo.gl/Egh9jx*.
56. Krizhevsky, A., Sutskever, I., and Hinton, G. (2012) Imagenet classification with deep convolutional neural networks, in *Advances in Neural Information Processing Systems*, MIT Press, vol. 25, pp. 1097–1105.
57. Zeiler, M.D. and Fergus, R. (2013) Visualizing and understanding convolutional networks, *arXiv.org*, **abs/1311.2901**.

5
Perceptual Psychophysics

C. Alejandro Parraga

5.1
Introduction

Since only a few decades ago, and in particular since the arrival of functional magnetic resonance techniques, there has been an explosion in our understanding of the functional brain structures and mechanisms that result in our conscious perception of the world. Although we are far from a complete understanding of cortical and extracortical processing, we know a fair amount of detail, perhaps enough to create computational models that replicate the workings of the best understood parts of the brain. We also know a fair amount about the environment and the evolutionary/neonatal constraints that shaped the workings of the perceptual machinery so that, even when the information reaching us is incomplete and subject to noise, our perception of the world is mostly stable: our brain filters out variability and attempts to recreate the environmental information based on millions of years of evolution and its own life experience. All this is done within the neural signal processing framework that characterizes our internal processes and drives our subjective responses. In this context, applying hard science methods to study how we humans reach subjective judgments has the appearance of an art or a craft, where each decision needs to be carefully weighted, experience is crucial, and small mistakes might render experimental results impossible to interpret.

5.1.1
What Is Psychophysics and Why Do We Need It?

Psychophysics is the branch of experimental psychology concerned with the relationship between a given stimulus and the sensation elicited by that stimulus. It engages problems by measuring the performance of observers in predetermined sensory (visual, auditory, olfactory, haptic, etc.) tasks. It has two remarkable roles, to describe and specify the sensory mechanisms underlying human performance and to test the hypotheses and models explaining these same mechanisms. Because of its nature, psychophysics inhabits two worlds: the world of physics where objects and interactions can be measured with ever-increasing

Biologically Inspired Computer Vision: Fundamentals and Applications, First Edition.
Edited by Gabriel Cristóbal, Laurent Perrinet, and Matthias Keil.

accuracy and the world of psychology where human performance is irregular and answers are determined by complex mechanisms and stochastic neural response processes. As a discipline, psychophysics was born from the need of psychologists to empirically study sensorial processes as a tool to understand more complex psychological processes. Its nineteenth-century origins are intertwined with those of experimental psychology (it occupies a very central position in that discipline), and its theoretical foundation dates from the publication of *Elemente der Psychophysik* by German physicist Gustav Fechner [1]. In this book, Fechner developed the theoretical and methodological background required to measure sensation and gave experimental psychology the tools necessary to begin its study of the mind. Subsequent progress such as the *theory of signal detection* has broadened the scope of psychophysics, enabling its outstanding contributions to such different areas as visual perception, language processing, memory, learning, social behavior, and so on.

Although it is generally possible to accurately measure the physical properties of a given stimulus, task performance measures are much more difficult to interpret and are usually expressed in terms of probabilities. To simplify the problem, the first psychophysical studies consisted in modifying a stimulus' strength until the subjects perceived it and responded behaviorally. These measurements are called "threshold measurements" and reveal the strength of a signal necessary to cause a determined performance. Different setups are needed to quantity performances when threshold measurements cannot be obtained or when the physical characteristic of interest are not properly defined.

Psychophysical results can be directly related to the output of computational models, since models predict observers' performance given a stimulus signal. For example, if a visual difference model predicts that picture A and picture B are indistinguishable, we can design a psychophysical experiment where the same images are presented to human observers in a setup where they have to distinguish them. Moreover, visual psychophysicists are not content with just asking observers whether they see any differences between the pictures, since such answers would be tainted by expectations and subjectivity. They will design a task that can only be performed if observers do distinguish between the pictures. The ability to design, execute, and interpret the results of such tests is the main skill of a psychophysicist.

5.2
Laboratory Methods

Psychophysical experiments are usually conducted in a laboratory where all stimuli are controlled and their properties quantified. All laboratory settings (chairs, temperature, illumination, etc.) should be carefully considered to make subjects comfortable while keeping them awake and fully attentive. Visual psychophysics setups are particularly complex since illumination should not interfere with the visual stimulus, and in most cases, light reflected from other objects and the walls should be minimized. All relevant dimensions like stimulus luminance, its distance to the subject, its subtended visual angle, the observer's head position, and so on, often need to be recorded and taken into account in the evaluation of the experiment.

5.2.1 Accuracy and Precision

Experimenters dealing with physical magnitudes need to know the notions behind estimating measurement uncertainty (measurement errors) and error propagation. Every measurement has an associated set of confidence limits, denoting its proximity to the true physical value (also called its *accuracy*). *Precision* of a measurement of a physical variable refers to its reproducibility: I can use a caliper to measure very accurately the size of a grain of sand, but for bigger samples (e.g., a handful from my local beach), this measure will vary greatly from one grain to the next. In this case, the largest error is linked to the statistical probability of obtaining a certain size when randomly picking grains of sand from the sample. This uncertainty is called the *absolute error* of the measure and in our case is defined as the *standard deviation* of all the grains. Suppose that we want to have an estimation of the size of the grains: to be precise, we have to measure all of them, calculate the mean $\overline{m}$ and the standard deviation s, and express the size as $\overline{m} \pm s$. The second term of this expression s is the absolute error (also called Δm) and denotes the confidence interval of our measure: the size of a randomly picked grain has a 68% probability to lie within the region $[\overline{m} - s, \overline{m} + s]$.

Measuring length provides a good example of how to deal with these concepts: suppose we want to measure the length of a table with a ruler whose smaller unit is a centimeter. After we set one end of the table against the zero of the ruler, it is likely that the other end will show up between any two marks so we would "round" our results to the closest mark. In this case, my confidence level is the size of the largest "rounding," which is half a centimeter. If the border is closest to the 91 cm mark, I can express the size of the table as 91 ± 0.5 cm (the real length of the table is somewhere between 90.5 and 91.5 cm).

It is a common convention in science to use significant figures to express accuracy (the last significant place in our 91 cm measure is the 1 cm unit). When the error is not explicitly stated, it is understood to be half of the last significant value (in our case, half a centimeter). Using this convention, our measure could be simply written as 91 cm, and adding more significant places, for example, writing 91.00 cm is confusing, since it also states that our precision was less than a hundredth of a centimeter.

If I am given a better rule (let's say with millimeter marks) and asked to measure the width of the table, I can apply the same criterion and obtain a second measurement, for example, 612 ± 1 mm. This time, I will adopt 1 mm as my uncertainty (my eyesight does not allow me to see anything smaller than that). In summary, I could safely state that my table is 91 ± 0.5 cm long and 61.2 ± 0.1 cm wide.

5.2.2
Error Propagation

Now, suppose we want to express the perimeter of the table. It is definitely the sum of all four sides ($91 + 91 + 61.2 + 61.2 = 304.4$ cm), but what about its uncertainty? In our case, it is easy; in the event of sum and subtraction of two independent measurements, we can estimate uncertainty using Eq. (5.1):

$$\begin{cases} L = a + b \\ L = a - b \end{cases} ; \quad \Delta L = \Delta a + \Delta b \tag{5.1}$$

where Δa and Δb represent the absolute error of each of the components of the sum. According to Eq. (5.1), the combined uncertainty is $0.5 + 0.5 + 0.1 + 0.1 = 1.2$ cm. The perimeter may be expressed as 304.4 ± 1.2 cm. However, if we follow our convention, by writing the last decimal (number 4) as the last significant digit, we imply that we can estimate millimeter units although our error is 12 mm! The correct approach is to round our last figure and express the perimeter as 304 ± 1.2 cm. Note that in this case, the error associated to the length predominates over the error associated to the width, making the millimeter ruler rather useless.

Now, let's calculate and write the area of the table. A simple product of length times width gives an area of 5569.2 cm^2; however, to calculate its uncertainty, we need to apply the following rule:

$$\begin{cases} A = a \cdot b \\ A = \frac{a}{b} \end{cases} ; \quad \frac{\Delta \mathrm{A}}{A} = \frac{\Delta a}{a} + \frac{\Delta b}{b} \tag{5.2}$$

where the terms $\Delta a/a$ and $\Delta b/b$ are called *relative uncertainties* of a and b. Equation (5.2) applies to independent measures and describes a method for calculating uncertainty in the cases of multiplication and division by operating with relative uncertainties. In our example, the absolute uncertainty of the table surface A is calculated by replacing in Eq. (5.2)

$$\Delta A = A \cdot \left(\frac{\Delta l}{l} + \frac{\Delta w}{w} \right) = 5569.2 \cdot \left(\frac{0.5}{91} + \frac{0.1}{61.2} \right) = 39.7 \text{ cm}^2 \tag{5.3}$$

Therefore, the error associated to A is approximately 40 cm^2, which makes the last three digits of 5569.2 of little value. In view of this, we can round them up and write $A = (55.7 \pm 0.4) \cdot 10^2 \text{cm}^2$, indicating our confidence limits.

In the general case, for example, when our final measure x is a function of many dependent variables (a, b, c, etc., also measured with their respective Δa, Δb, Δc, etc.), the absolute error Δx can be derived from the covariance matrix of the variables [2, 3]. In the case where fluctuations between a, b, c, and so on, are mostly independent and the number of observations is large, we can expect Eq. (5.4) to be a reasonable approximation:

$$\text{Given} : x = f(a, b, c, \ldots); |\Delta x| = \left|\frac{\partial f}{\partial a}\right| \cdot |\Delta a| + \left|\frac{\partial f}{\partial b}\right| \cdot |\Delta b| + \left|\frac{\partial f}{\partial c}\right| \cdot |\Delta c| + \cdots \tag{5.4}$$

Although Eq. (5.4) neglects the effects of correlations between the different variables, it is commonly used for calculating the effects of uncertainties on the final results.

5.3 Psychophysical Threshold Measurement

Once we know how to properly quantify all relevant physical magnitudes, we need to turn our attention to the problem of measuring sensation. The minimum amount of stimulus energy necessary to elicit a sensation is called *absolute threshold*. Given that our neural system is subject to noise, sensation measurements tend to fluctuate from a moment to the next, and several of them are necessary to obtain an accurate estimation by averaging. A second important concept in psychophysics is the *difference threshold* or the amount of change in the stimulus necessary to elicit a *just noticeable* increment in the sensation (just noticeable difference or *jnd*). For example, if I close my eyes and hold a bunch of paper clips in my hand, how many paper clips can you add until I notice a change in weight? Let's say two paper clips: their combined weight is the difference threshold.

5.3.1 Weber's Law

One of the first stimulus-sensation dependencies to be discovered was the relationship between difference threshold and the intensity level of a stimulus. Going back to the paper clips example, suppose that I hold 50 paper clips in my hand and we find out that the minimum number of paper clips necessary for me to notice a change in weight is 2. Is it going to be the same if I hold 150 paper clips in my hand? And if I hold 300? The answer was found by the German physiologist E.H. Weber in 1834 who determined that the amount of difference threshold is a linear function of the initial stimulus intensity [4]. In summary, if I hold 150 paper clips in my hand, I will need about 6 more to notice any change in weight. *Weber's law* can be expressed mathematically as

$$\Delta\omega = c\omega; \quad \text{or} \quad \frac{\Delta\omega}{\omega} = c \tag{5.5}$$

where ω is the initial intensity of the stimulus, $\Delta\omega$ is the amount of change required to produce a jnd, and c is a constant dependent of the sensory modality studied. The relationship in Eq. (5.5) holds for a wide range of stimulus intensities (except for very weak stimuli) and appears to be true for many types of signals: visual, auditive, tactile, olfactory, and so on. At very low levels of stimulation, the noise inherent to our neural system is of a magnitude similar to the stimulus necessary to produce a *jnd*, and a larger $\Delta\omega$ is needed for it to rise above the noise. Fechner [1] extended Weber's ideas by relating physical magnitudes of $\Delta\omega$ to their corresponding sensation *jnds*, assuming that all *jnds* where psychophysically equal. After a series of measurements, he came up with the following equation (known as *Fechner's law*):

$$S = k \log(\omega) \tag{5.6}$$

where S is the magnitude of the sensation experienced, ω is stimulus intensity, and k is a constant that also depends on sensory modality. Fechner's law can be obtained by writing Weber's law as a differential equation and integrating over $d\omega$, and both constants are related by $c = -k \log(\omega_0)$, where ω_0 is the stimulus threshold below which there is no perceived sensation.

Later studies have challenged the validity of Fechner's assumption on the uniformity of *jnd* increments for any level of stimulus sensation; however, his equation continues to be one of the most significant attempts to measure sensation in the history of science.

5.3.2
Sensitivity Functions

The equations stated previously belong to the small group of laws discovered in perceptual psychology, and their historical relevance cannot be underestimated. However, it is perhaps the measurement of *absolute thresholds* which is responsible for the most significant advances in our understanding of sensory mechanisms. Take the case of vision, where sensory thresholds are determined by properties of both the optics of the eye and the neural machinery of the visual system. Visual sensation starts at the retina, where photoreceptors (i.e., cones and rods) are excited by photons, sending signals that, in the case of rods, are spatially pooled together (*spatial summation*) and summed over time (*temporal summation*). In many aspects, their performance seems to match the limits imposed by the nature of light. For example, Hecht *et al.* [5] tested seven observers for several months and concluded that about 56–148 photons are necessary to elicit a sensation 60% of the time in absolute darkness and optimal conditions. After considering the optics of the eye [6], Hecht *et al.* estimated this to be equivalent to just 5–14 photons absorbed by a single rod photopigment. If a rod absorbed a smaller number of photons, the resulting signal could be confounded with signals due to spontaneous activity, and therefore, the output from many rods needs to be pooled together. The interaction between signal and noise in the visual system is studied with the tools of signal detection theory, and tasks such as visual discrimination and visual detection become inherently similar in the presence of noise: to detect a signal is the same as discriminating it from the noise.

5.4
Classic Psychophysics: Theory and Methods

Psychophysics has the dual role of providing us with quantitative descriptions of the limits, features, and capabilities of our senses and of offering ways of testing our hypothesis about the neural mechanisms behind these sensory capabilities. In the first role, it has contributed to the design of rooms and equipment optimized for human use such as auditoriums, concert halls, digitally compressed audiovisual imagery, ergonomically designed cabins and cockpits, LCD screens, and so on. In the second role of testing hypothetical neural mechanisms and models, psychophysicists must assume a biunivocal correspondence between the neural activity generated by a stimulus and its perception. This principle (which has been termed "*principle of nomination*" [7]) is what allows us to link psychophysical to biological data and computer models and is perhaps the single most relevant fact relating psychophysics and computer science. Computer models often attempt to match the performance of human observers in predetermined tasks, either by recreating the workings of neural processing ("mechanistic models") or by other means ("functional models"). In mechanistic models, the need for psychophysics and signal detection theory is clear: they provide the tools to infer the properties of neural mechanisms from observer's data. Functional models often rely in pattern inference theory to provide statistical descriptions of the scene structure, but there is always the final requirement of contrasting results against observer's performance. In the end, both approaches need to relate a measure of the observer's internal state (e.g., the perceived image as represented in the electronic machinery) to a physical quantity (e.g., the real image): this is what psychophysics does best though the principle of nomination.

5.4.1 Theory

To fully appreciate the power behind the principle of nomination, we should consider two different stimuli: if they produce identical neural response, their perceptual effects should be the identical too. The reflexive form of the principle states that if the sensory experience from two stimuli is identical, the neural responses should also be the same. This second form of the principle allows us to hypothesize and model the neural events that underlie our sensory experience by studying combinations of stimuli that elicit identical responses. It also allows us to combine findings from different disciplines such as neurophysiology, computational neuroscience, and psychophysics by linking physics to neural response and sensation. Furthermore, it is possible to classify psychophysical experiments according to their compliance with the reflexive principle of nomination [8] (i.e., by whether their stimuli are aimed at producing identical responses in the observer or not). One example of full compliance is the case of matching the surfaces of real objects to those produced by a computer monitor. Color monitors work by exploiting a property called "metamerism," that is, colors with different spectral power distributions are perceptually similar under certain conditions. Although the light produced by the three phosphors on the monitor screen has different properties from the light reflected by objects, monitors are routinely used to represent objects in psychophysical experiments and simulations. The "noncompliant" type of psychophysical experiments uses stimuli that are perceived as different except in a single perceptual dimension (e.g., chromaticity, brightness, speed, loudness, etc.). For example, experimenters may allow observers to manipulate the physical properties of one stimulus so that it matches the brightness of another, while they are still different in all other perceptual dimensions like chromaticity, texture, and so on. These observations are less rigorous than "fully compliant" observations, and their implications are limited (since there is no certainty that all neural responses are the same). However, they are still valid and provide the basis for many hypothesis-testing experiments.

Asking observers direct questions on whether or not they perceive a given stimulus while varying its intensity is the most basic psychophysical method for measuring thresholds. In many cases, the answers are likely to contain all the uncertainties common to experiments involving biological systems: when presented several times with the same stimulus, our "internal detector" will vary according to stochastic noise. For this reason, it is clear that thresholds need to be defined in terms of their statistical probability: in most cases a *threshold* is the stimulus level that is perceived in 50% of the trials. Figure 5.1 represents the results from a hypothetical threshold-measurement experiment where arrows in (a) denote the variability of "internal thresholds" in our neural system. The probability of the presence of these internal thresholds follows a normal (Gaussian) distribution, and the area under the Gaussian in (b) represents the percentage of positive answers for any given stimulus. The peak of the Gaussian corresponds to the stimulus that elicits 50% of positive responses in our observers: half the time observers will say "no" because noise will shift the internal threshold above that value and half the time they will say "yes" because noise will shift the internal threshold to the same value or below. The peak of the Gaussian meets our definition of threshold.

Fechner devised three different methods with their corresponding data analysis techniques for threshold measurement [1]. These methods, whose properties make them useful in different situations, are described in the following sections.

5.4.2
Method of Constant Stimuli

As its name indicates, the set of stimuli presented to the observer in the *method of constant stimuli* is always the same. The procedure starts by selecting stimuli, usually between 5 and 9 samples that range from the easily detectable to the almost impossible to detect. Their intensity (measured in physical units along some dimension) should be separated by equal steps. A simple experiment consists in presenting all the stimulus intensity steps to our observers in random order, repeating each presentation a number of times (e.g., 10 times) and noting the fraction of positive answers to the question "do you see the stimulus?" For large numbers of measurements, results tend to have a typical sigmoidal shape called *psychometric function* (see Figure 5.2) and are often best fit by a mathematical function called *ogive* which represents the area below a bell-shaped normal distribution curve.

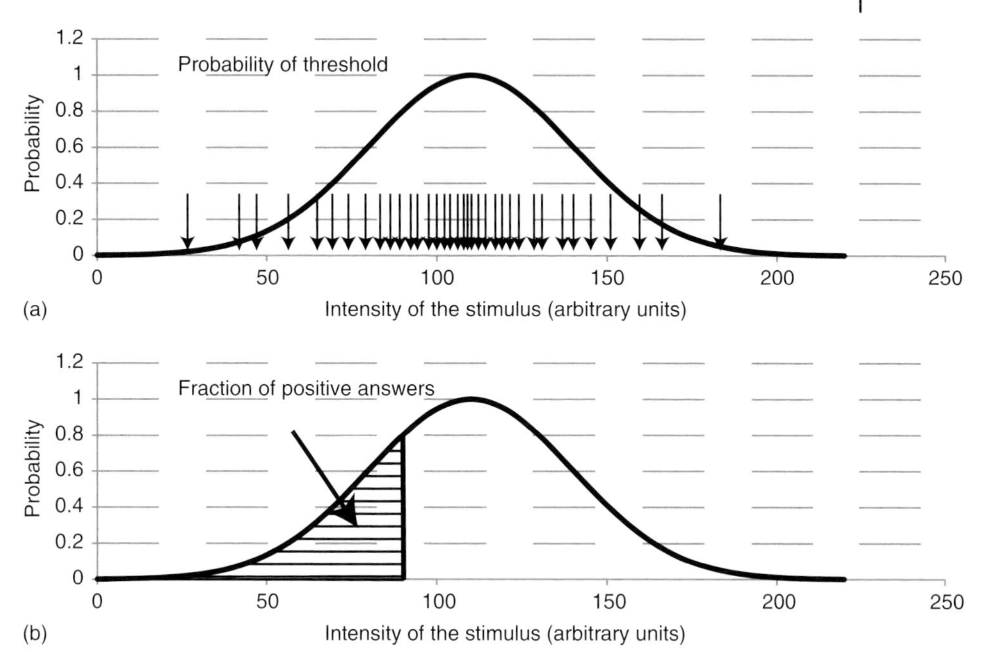

Figure 5.1 Gaussian distribution of perceived thresholds after a number of measures. This variability is the result of neural noise present in biological systems, experimental noise, and, in certain cases, quantal fluctuations of the stimulus. According to this, every time we look at the stimulus, our "internal" threshold is different and we respond accordingly. These internal thresholds follow a normal distribution with its peak in the most probable value as shown in (a). For any given stimulus intensity, the ratio of positive to negative answers to the question "do you see the stimulus?" will be determined by these internal threshold fluctuations and corresponds to the area under the Gaussian shown in (b).

Following our definition of threshold as the stimulus level that is detected in 50% of the trials, we just need to find the position of the curve where the ordinate y (fraction of positive answers) is equal to 0.5 and project to the abscissa x (stimulus intensity). Since the curve in Figure 5.2 is the integral of a normal distribution, the middle point of its "S-shape" (the threshold magnitude or where 50% of the answers are positive) corresponds to the peak of the Gaussian in Figure 5.1. Slimmer and taller bell-shaped Gaussians will result in steeper threshold transitions in Figure 5.2, and flattish Gaussians will result in smoother threshold transitions. The choice of unit measurement for the stimulus intensity is very important, since the abscissae should be measured in a linear scale. For example, monitors produce stimuli whose brightness is a nonlinear function of the gray-level values stored in memory (the so-called gamma function). For this reason, specifying the stimulus in terms of gray-level values is unlikely to produce curves like those of Figures 5.1 and 5.2.

Paradigms for measuring detection thresholds can be easily modified to measure *difference thresholds* between pairs of stimuli (i.e., observers are exposed to

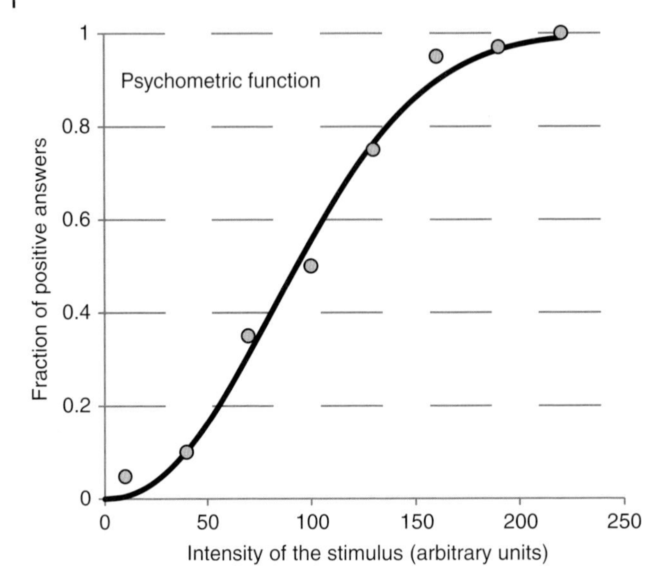

Figure 5.2 The psychometric function of a typical set of results using the method of constant stimuli. *y-Axis* values show the proportion of "yes" answers to the question "do you see the stimulus?" Each stimulus was presented an arbitrary number of times n (typically $n = 10$), and each gray circle is the average result of these presentations.

two stimuli and have to decide which produces the sensation of greatest magnitude). In this case, one of the stimuli is assigned the role of "reference" and the other the role of "test" stimulus. While the reference stimulus is kept constant, the test stimulus varies from trial to trial from values below those of the reference to values above in constant steps (usually 5–9). The test stimulus range needs to be chosen so that the stimulus of lowest magnitude is always judged smaller than the reference, the stimulus of largest magnitude is always judged larger, and the steps are equal. Reference and test stimuli are presented in pairs several times, with the test stimuli randomly chosen from the predetermined set, and observers report which of the two produces the greatest sensation. Ideally, both stimuli should activate the same set of receptors and neural mechanism at the same time. However, in practice, this is likely to be impossible, and a compromise needs to be reached by either presenting stimuli at the same time but in different physical positions or in the same position but at different times.

5.4.3 Method of Limits

This method is less precise but much faster to implement than other methods, making it perhaps the most popular technique for determining sensory thresholds. In the *method of limits*, it is generally indistinct whether the trials start from the top or the bottom of the stimulus staircase. A typical experiment begins

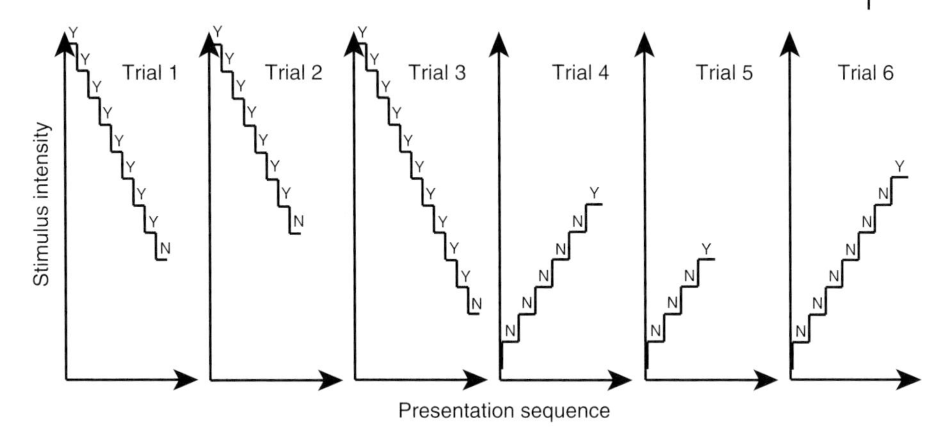

Figure 5.3 Fictional set of results for a typical method of limits experiment. There are six trials consisting of staircases (three ascending and three descending) of stimulus intensities. Observers answer "*yes*" (Y) or "*no*" (N) to indicate whether or not they have detected the stimulus. Individual trials start from one extreme (either from perfectly detectable or undetectable stimuli) and end as soon as observers alter their answers. In the example, no provision has been taken to counter habituation and expectation errors.

by presenting stimulus well below (or above) the observer's threshold, which is then sequentially increased (or decreased) in small amounts. The experiment ends when the observer reports the presence (or absence) of the expected sensation. The transition point is considered to be midway between the last two stimulus responses. The experiment is usually repeated using both ascending and descending stimulus series, and a threshold is obtained by averaging the responses of these (see Figure 5.3).

Although it is easy to implement, the results obtained by this method are influenced by two systematic errors, called *error of habituation* and *error of expectation*. The first affects experiments using descending stimuli and derives from the fact that observers may become used to give the same response for a period of time and continue to do so well after the threshold point has been reached. Conversely, in the case of ascending stimuli, observers may anticipate the arrival of the threshold by reporting its presence before they actually sense it. As a result, thresholds for ascending (or descending) stimuli will be misleadingly lower (or higher) than they actually are. Moreover, the magnitudes of these complementary errors are not necessarily the same, and it is not possible to average them out by repeating the experiment with alternating ascending (or descending) staircases. One possible improvement comes from varying the starting points of the staircases in each trial to stop observers "counting" steps or predicting the threshold from previous trials. Another improvement consists on alternating up and down staircases, switching direction as soon as the observer reports a threshold, and recording the reversing points. After a number of staircase reversals, the final threshold is obtained by averaging these reversing points. The precision of this method largely depends on the size of the steps considered.

5.4.3.1 Forced-Choice Methods

Although habituation and expectation biases in the method of limits can be reduced by training, it is impossible to be certain that observers are always honest and extraordinary results are always a consequence of extraordinary low/high thresholds and not of these errors. Furthermore, results are strongly influenced by observers' expectations and the probability of the test stimulus appearing: observers tend to learn whether the test stimulus is presented often and try to predict its appearance. In response to these criticisms, a series of methodological variations were developed where observers do not report directly whether they sense the stimulus but have to pick an item from several possible instead. For example, in a typical trial, the observer chooses an item from a set of stimuli, only one of which is different from the rest in some sensory dimension. In each trial, these items can be presented sequentially (temporal forced choice) or simultaneously (spatial forced choice), and the difference between the odd (test) stimulus and the others (references) is varied from one trial to the next following a staircase, making the task more difficult or easier. This method and its variations are termed "forced choice" because the observer always has to pick an option among others. When asked to select the odd item from a set of stimulus that are nearly the same, observers are unlikely to answer correctly most of the time if they cannot sense the test stimulus' difference. If this difference is below the perceptual threshold, the answer will be determined by chance, and if it is above, the answer will be above chance. Since this method relies on statistics, each step in the staircase procedure (each trial) needs to be repeated many times to rule out "lucky" answers. For example, in visual psychophysics, it is common to present observers with a choice of several closely looking images from which they have to pick one according to some instructions (the one where some signal is present, the odd one out, etc.). If the observer is presented with two choices whose difference is below threshold, about half the answers will be correct; however, if the difference is clearly visible, the test stimulus will be selected in all trials. At threshold, the test stimulus will be selected midway between 50 and 100% of the time, that is, 75%. In the case when observers are presented with two choices, the method is called 2AFC or 2-alternative forced choice.

Figure 5.4 shows an exemplary forced-choice version of the method of limits used to measure the threshold for discriminating between the photograph of an object and a slightly different (morphed) version of the same photograph [9]. The number of choices offered to the observer determines many of the characteristics of the method, such as the proportion of right answers obtained by chance and the threshold position, since results are always fitted by an ogive (see Figure 5.2). In all cases, the ogive's lowest y-value will correspond to the fraction of right answers obtained by randomly selecting stimuli and its highest y-value to the fraction of right answers obtained by clearly distinguishing the test from the reference stimulus. For example, in a 2AFC, the ogive's lowest value is always 0.5, and in theory, its highest value should be 1, but in practice, experimenters allow for mistakes made by observers accidentally making wrong choices. These mistakes (also called "finger errors") lower the chance of getting 100% correct responses even

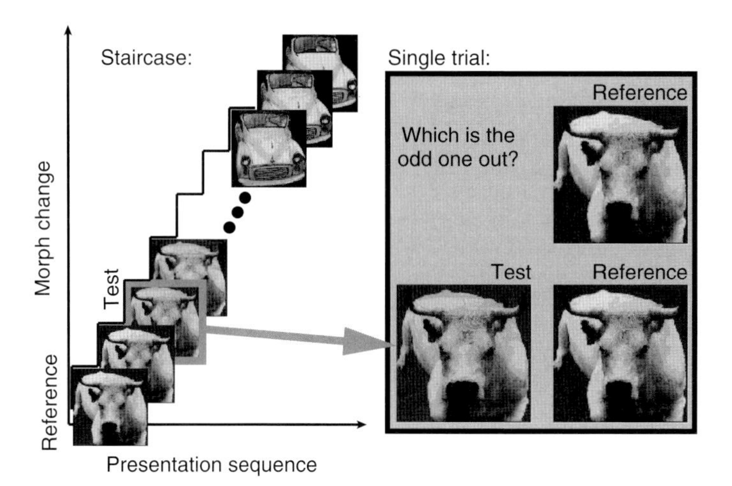

Figure 5.4 Example of a forced-choice version of the method of limits. The staircase consisted of a sequence of morphed pictures (a bull that morphs into a car in small steps). The experimental protocol searched for the point in the sequence where observers can just tell apart a modified picture from the original picture of the bull (discrimination threshold). To decide this, each step in the staircase originated a series of trials where observers had to select the corresponding test image (a morphed bull) from two references (the normal bull) presented in randomized order. By repeating each step a number of times, experimenters measured the morph change necessary for observers to tell the difference, without the inconvenience of habituation and expectation biases [9].

when the test stimulus is clearly sensed. To account for mistakes, the ogive's upper asymptote is usually set at 0.98 and the threshold to 0.74 (98 and 74% of correct responses, respectively).

Despite involving longer experimental hours and tiredness, forced-choice procedures are perhaps the best techniques to obtain perceptual thresholds free from response and observer expectation biases.

5.4.4
Method of Adjustments

The *method of adjustments* has much lower accuracy than the methods described earlier and tends to be selected in circumstances where there is an advantage in giving observers control over changes in the stimulus, making experiments more participative and arguably less tedious. In a single trial, the stimulus intensity is set far below (or above) the expected threshold, and the observers increase (or decrease) the intensity until the sensation is just perceptible (or just disappears). This method has the same shortcomings as the method of limits in terms of response and expectation biases and therefore requires randomized starting points and a large number of trials to reach acceptable accuracy levels, always combining ascending and descending runs and averaging the results to obtain the threshold. There are also issues regarding the use of discrete instead of continuously variable stimuli which makes more difficult to calculate the method's error. In summary, the method of adjustments is seldom used in rigorous laboratory contexts. It might have advantages when a "quick and dirty" measure of threshold is required or when the task is too long and tedious for observers to keep a reasonably high performance throughout the session.

5.4.5 Estimating Psychometric Function Parameters

Psychometric functions constructed from multiple-choice experiments differ from those constructed from yes–no experiments (see Figure 5.2) in that the position of the ogive is determined by the probability of a "chance" selection and "detection." For example, if the experiment contained four intervals and the observer cannot detect which interval contains the signal, results are likely to be correct in 25% of the trials, and this will determine the lower asymptote of the ogive.

The measurement of thresholds using multiple-choice paradigms with psychometric functions built from ascending and descending staircases could be long and tedious for observers. This is particularly true if the initial parameters of the procedure (step size, starting level, etc.) are not known in advance. Several algorithms have been proposed to optimize the collection of results, which reduces the number of experimental trials by aiming at the threshold magnitude. Two of the most popular are parameter estimation by sequential testing (PEST) [10] and QUEST [11], which consist of staircases that reverse their directions once the observer obtains results that are consistently different from the previous ones. They also start with a larger step size which is reduced as the experiment reaches threshold, effectively "homing" on the most likely value and reducing the number of trials. A comprehensive analysis of popular staircase algorithms alongside with some practical recommendations can be found in the work of Garcıa-Perez [12].

5.5 Signal Detection Theory

Signal detection theory was developed to understand and model the behavior of observers in detection tasks. It examines their actions by dissociating decision factors from sensory ones, incorporating ideas from the field of signal recovery in electronics which studies the separation of signal patterns from electronic noise [13–15]. In psychophysics, the signal is our stimulus and the noise refers to random patterns that may originate within both the stimulus itself and the neural activity in the observer's nervous system.

5.5.1 Signal and Noise

Much of the early work on signal detection theory studied the detection of targets on radar screens by military radar operators where important decisions had to be taken quickly and in the presence of uncertainty. For example, imagine a rather critical situation where a radar controller is examining a scan display, looking for evidence of enemy missiles. The controller's response is determined by intrinsic neural noise regardless whether there is an incoming missile or not. In this scenario, either there is an enemy missile (signal present) or not (signal absent) and the controller either sees it and activates countermeasures (fires the alarm) or not (does nothing). The possible outcomes are four: enemy missile detected and destroyed (hit), enemy missile incoming undetected (miss), no missile and no alarm (correct rejection), and no missile and alarm activation (false alarm). The first and third outcomes (hits and correct rejections) are desirable, while the others (misses and false alarms) are not. There is also a ranking of bad outcomes where a false alarm is much more preferable than a miss.

Figure 5.5 shows the probability distributions of the noise (broken lines) as a function of the observer's internal response and the same probability distribution when a signal is added (solid lines). The second is bigger in average, so the observer's task is to separate the stimulus containing signal from the pure noise by setting a criterion threshold: when his/her visual neural activity is larger than the criterion threshold, the "signal" is acknowledged and vice versa. This clearly defines the four sections that correspond to the possible outcomes (hits, misses, false alarms, and correct rejections) in Figure 5.5. On both hits and false alarms, our radar controller activates the countermeasures, and correspondingly, misses

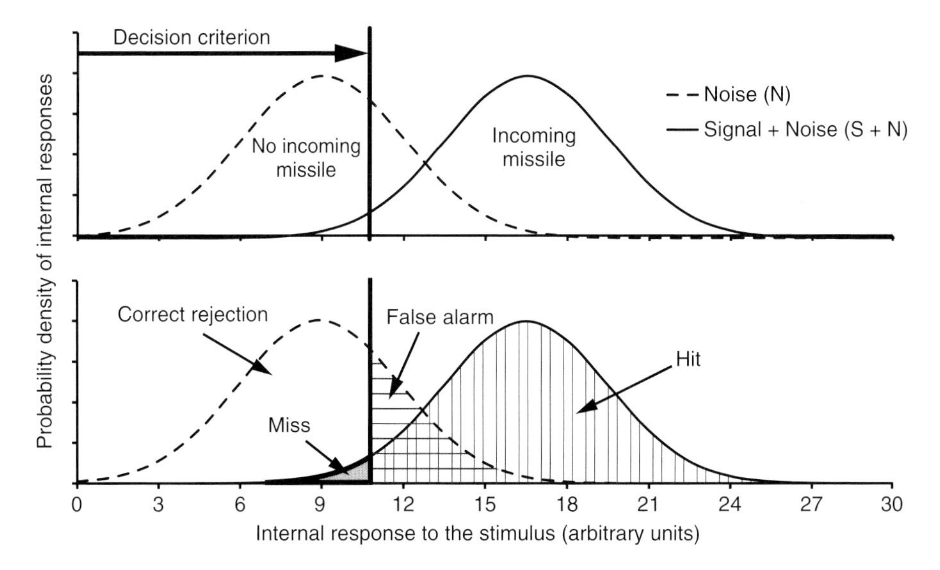

Figure 5.5 The hypothetical probability distribution of noise (N, in broken lines) and signal + noise ($S + N$, in solid lines) as a function of the observer's neural response. As expected, the $S + N$ curve generates more neural activity (e.g., spikes per second) in average than the N curve; however, there are sections where both curves overlap in x. The vertical line represents the observer's decision criterion, which could be anywhere along the *x-axis*. Depending on its position, four distinctive regions are determined: hit, miss, correct rejection, and false alarm.

occur when there is an incoming missile, but the neural activity it generates is insufficient to reach the decision criterion, as indicated in the figure.

In our example, the radar operator's job is to adjust the decision criterion according to the importance of each one of the possible outcomes. A low decision criterion means that the alarm will be set to almost everything, be it internal noise or not. The operator will never miss an incoming missile but similarly will have a high rate of "false alarms" (potentially leading to a waste of resources). On the contrary, setting a high criterion will lead to very few false alarms with the occasional undetected incoming missile.

5.5.2
The Receiver Operating Characteristic

In controlled lab conditions, our radar operator's task (judging the presence or not of a signal) would be treated as a yes–no procedure. In these procedures, observers are told in advance the different probabilities and costs associated to the different experiment outcomes (in the same way a radar operator knows the conditions and consequences of his actions). Such experiments typically consist of a large number of trials where observers must judge whether the signal was present or not.

Figure 5.6 shows how the results from a hypothetical yes–no experiment would look like if we consider the areas under the curves labeled "hit" and "false alarm" in Figure 5.5. If there is considerable overlap and the observer sets a low decision criterion, it is likely that both hits and false alarms will be high. As the decision criterion increases, both areas will decrease in a ratio that depends of their separation d and the overlap region (see inset in Figure 5.6). The curves in Figure 5.6 describing the possible relationships between hits and false alarm rates are called *receiver operating characteristic* (ROC) curves. They were first developed in Britain during World War II to facilitate the differentiation of enemy aircraft from noise (e.g., flocks of birds) by radar operators.

From the ROC curves that best fit experimental hit/false alarm data, one can obtain a measure of the strength of the signal that is independent of the observer's decision criterion:

$$d' = \frac{(\overline{SN} - \overline{N})}{\sigma_N} \tag{5.7}$$

where d' is called the observer's discriminability index, $\overline{SN}$ and $\overline{N}$ are the mean values of the $S + N$ and N distributions, respectively, and σ_N is the standard deviation of the N distribution.

The optimum decision criterion β_{opt} (i.e., the criterion that produces the largest winnings in the long run) can also be estimated from the probabilities of N and $S + N$ and the costs of the various decision outcomes using

$$\beta_{\text{opt}} = \left[\frac{P(N)}{P(SN)}\right] \cdot \left[\frac{\text{value of correct rejection} - \text{cost of false alarm}}{\text{value of hit} - \text{cost of miss}}\right] \tag{5.8}$$

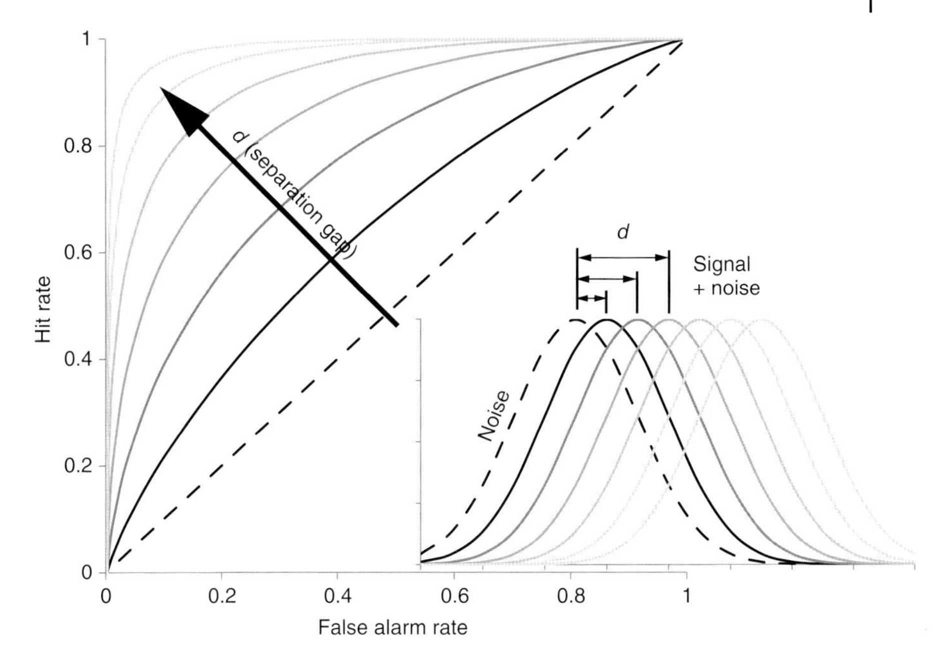

Figure 5.6 Receiver operating characteristic (ROC) curves describing the relationship between hits and false alarms for several decision criteria and signal strengths. As the observer lowers its decision criteria, both areas labeled "false alarm" and "hit" in Figure 5.5 increase in different proportions, depending of the proximity between the curves. The ROC family of curves describes all the options for our hypothetical radar controller regarding the separation between S and $S+N$ curves. When both curves are clearly separated, it is possible to obtain a nearly perfect hit rate without false alarms.

where costs are entered as negative numbers. Equation 5.8 has been found to reasonably represent the judgments made by actual observers [16].

ROC curves can be thought of as a plot of the fraction of true positives (also known in statistics as *sensitivity*) in the y-axis versus the fraction of false positives (also known as *specificity*) in the x-axis. Both measure the statistical performance of a binary classifier. While sensitivity measures the proportion of actual positives that are correctly identified (e.g., cancer patients correctly identified with the condition), specificity indicates the proportion of negatives correctly identified (e.g., healthy patients identified as not having the condition). Additionally, the area under the ROC curve is related to the accuracy of the method (how good is the classifier when discriminating between the two possible conditions). A perfect classifier has an area equal to one, while a classifier no better than chance has an area of 0.5. Because of its graphical nature, ROC analyzes are excellent tools for showing the interrelations between statistical results, and many algorithms/toolboxes have been developed [17].

Traditional psychophysical research methods have been criticized because of their inability to separate subjects' sensitivity from other response biases like the various costs of subjects' decision outcomes [13]. In this view, threshold measurements may be contaminated by arbitrary changes in the observer's decision criterion which may lead to faulty results. This explains, for example, the dependence of the method of constant stimuli's results with the fraction of trials that contain the signal: when the probability of a signal increases, observers lower their decision criterion, and the fraction of "yes" answers increases. As we have seen, signal detection theory provides a series of methods for computing d' and β which allows studying the effects of particular variables on each of them separately. In practice, it is possible to remove fluctuations in the decision criterion by using forced-choice experiments where many stimulus intervals are presented and the observer has to choose the one containing the signal. As seen before, forced-choice paradigms are not influenced by response biases, and d' measures can be obtained directly by measuring the proportion of correct responses [18]. The advantages of several-alternative procedures over the yes–no procedure reside in the fact that in every trial, observers are presented with more information (several intervals) whether in the yes–no procedure a single trial may contain the signal or not.

5.6 Psychophysical Scaling Methods

Early psychophysicists agreed that a complete mapping between physical stimulus and perceptual response could only be produced indirectly by measuring absolute and relative thresholds, with the *jnd* as the sensory magnitude unit. Their approach relied on two main assumptions: (i) that all *jnds* (for a given sensory task) are perceptually equal and (ii) it is always possible to control and measure the physical properties of the stimulus. Given these assumptions, it is possible to characterize the relationship between stimulus and response by measuring differential thresholds to ever-increasing stimulus and obtain the magnitude of a perceptual response simply by counting the number of *jnds* a stimulus is above absolute threshold. However, researchers working in the field of experimental aesthetics soon encountered situations that challenged these assumptions [19]. Indeed, attributes such as "pleasantness," "beauty," "naturalness," "repugnance," and so on are the product of higher cognitive cortical levels where processes like attention, memory, emotions, creativity, and language play a much more important role that is absent at perceptual, preconscious levels. To obtain a stimulus–response relationship in such complex cases, a different set of techniques were developed which either treat observers' judgments directly as sensory measurements or deal indirectly with stimuli whose relevant physical properties cannot be specified. This mapping of physical stimulus to perceptual (or most likely psychological) response resulted in the creation of many *psychophysical scales*: ranked collections of measurements of response to some property of the physical world.

5.6.1 Discrimination Scales

Although the methods described in previous sections are designed to measure discrimination responses to small stimulus changes, it is possible to extend them to construct psychophysical scales relating large changes in the physical stimulus to their corresponding psychological responses. This is a consequence of Weber's law (Eq. (5.5)), which states that $\Delta\omega$ (in physical units) is proportional to the magnitude ω, and Fechner's law (Eq. (5.6)), which assumes equal increments of the corresponding sensation *jnds* regardless of stimulus intensity. For example, if one obtains the signal change $\Delta\omega$ required to produce one *jnd* of sensation at many levels of signal intensity, it is possible construct a function over the whole range of the stimulus signal, relating the stimulus' physical units to psychological sensation. Panel (a) in Figure 5.7 shows an imaginary example of such construction. The y-axis represents the sensation (in number of *jnds*) elicited by the stimulus intensity (x-axis). As the stimulus intensity grows, larger increments ($\Delta\omega_2 > \Delta\omega_1$) are necessary to elicit the same sensation (1 *jnd*) until a saturation level is reached. Panel (b) shows a psychophysical scale obtained from adding the *jnds* in (a).

As we have seen before, Weber and Fechner's laws are related through their respective constants, meaning that the second is only valid if the first is correct.

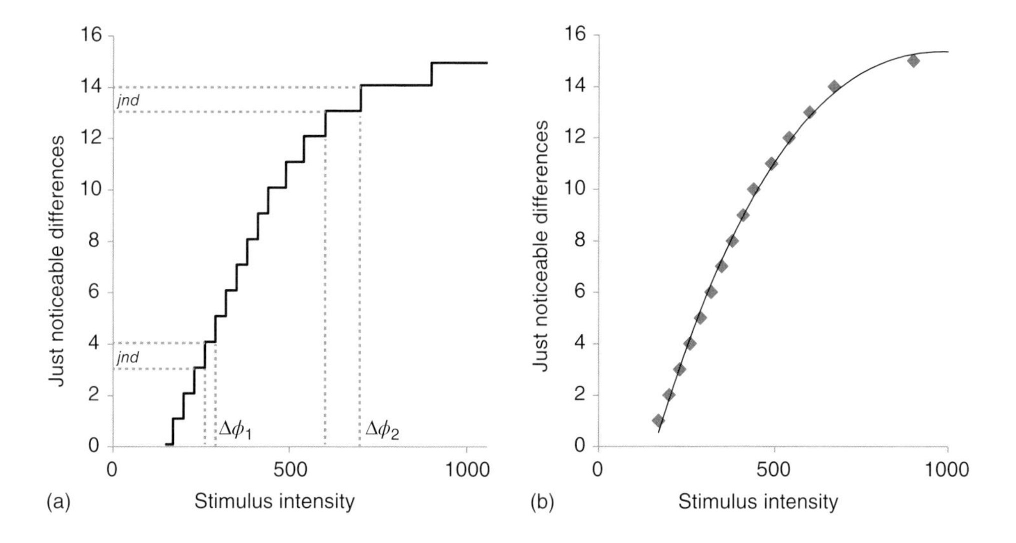

Figure 5.7 Hypothetical example of how to obtain a psychophysical sensation scale over a large stimulus range from measures of sensation discrimination at threshold (*jnd*). Panel (a) shows that to keep producing the same increment in sensation (equal steps in the *y-axis*), the stimulus intensity needs to be increased (unequal steps in the *x-axis*). Panel (b) represents the psychophysical scale obtained from the results in (a).

It has been shown [20] that the Weber fraction in Eq. (5.5) is not strictly constant over the whole range of stimulus intensities, which means that the data points of Figure 5.7 cannot be derived accurately from Eqs. 5.5 and 5.6 and have to be measured. After a hundred years of being accepted by the psychology and engineering community, today, Fechner's law is not considered to be an accurate description of the relationship between stimulus and sensation; however, his achievement still represents a good approximation and should not be underestimated.

5.6.2 Rating Scales

In some situations, it is not possible to produce psychophysical scales from indirect judgments such as detection or discrimination responses, and the only option is to use direct observer judgments to assign numerical values to stimulus attributes. For example, it might be necessary to psychophysically quantify the distortion produced by different compression levels in images or to arrange colors and textures in different categories (such as "red" or "blue" or "coarse" or "smooth"), translating physical attributes like hue into perceptual categories such as "magenta." In other cases, there is no quantifiable physical attribute to relate to, as in the task of ranking a series of objects according to a subjective judgment (e.g., beauty) and arranging their position in a numerical scale. In all these circumstances, it is convenient to produce a "*mapping*" that links physical to psychological attributes with steps matching the internal relationships of the psychological magnitudes considered.

5.6.2.1 Equipartition Scales

A common task in psychophysical research is to partition a psychological continuum into interval scales of equal *distances*. For example, two gray patches might be provided, a bright patch "A" and a dark patch "B," and observers may have to evaluate whether the lightness distance between pairs A and B is equal, more, or less than the distance between another pair of gray patches C and D. To systematize this task, one might want to relate the continuum of physical luminances to the perceived sensation distances between A – B and C – D by dividing this continuum in perceptually equal steps [21].

Figure 5.8 shows a hypothetical psychological sensation that varies as a nonlinear function of the physical stimulus in the range [0, 100]. In one partitioning method, observers are first exposed to fixed stimulus values (20, 40, 60, etc.) and then asked to manipulate them until all scale steps are perceptually equal. A second method starts by exposing observers to the full range of physical stimulus and asking them to find the perceptual middle point. The double arrows in Figure 5.8 show the equal intervals in which a given observer splits the sensation elicited by the physical signal in the interval [50, 100]. In our example, this perceptual middle point corresponds to physical stimulus of strength 80. In a second iteration of the experiment, the observer splits the range of sensations elicited by the physical stimulus of strength 80 and one extreme (i.e., strength = 100), subdividing the

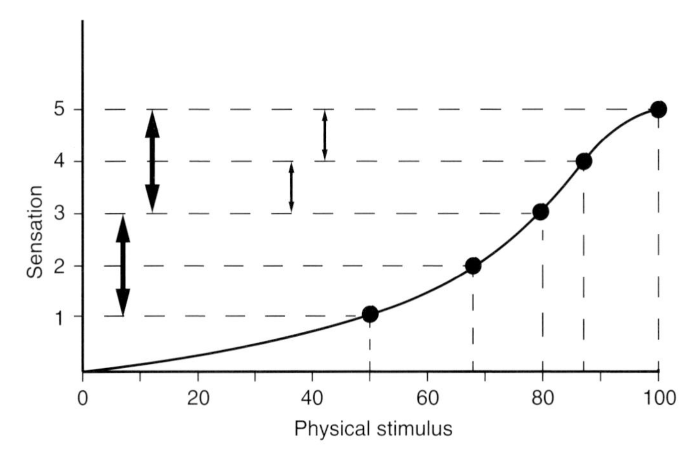

Figure 5.8 Hypothetical nonlinear relationship between a physical stimulus and its corresponding psychological sensation. The arrows show how a perceptual scale with uniform measurement units can be produced by partitioning the scale into smaller intervals.

scale in smaller evenly spaced intervals. After a few iterations, the psychological continuum is subdivided in equally spaced steps.

Another popular method of production of scales consists in adjusting stimuli so that their perceived intensity is a multiple of some original magnitude, extrapolating sensorial ratios rather than dividing a given interval. Since this task is also critically dependent on the observer's ability and training, results may vary, and there are some consistency tests available to evaluate their performance in such tasks [16].

5.6.2.2 **Paired Comparison Scales**

Paired comparison scales are preferred when the psychological sensation to quantify does not have any evident correspondence with a measurable physical attribute. As an example, consider again the experiment described in Figure 5.4, but now, suppose we are not interested in discriminations made between pictures *at threshold* but want to know the perceptual difference between pictures in the *suprathreshold* region (for instance, between the two extremes of the morphed sequence: the car and the bull). When image differences are very large (e.g., they contain different objects or are subject to extreme color, texture, or digital effect distortions), a problem arises since there is no obvious "morph metric" or any convenient physical measure to link to our perceptions [22]. A viable solution was proposed by Gustav Fechner, who was the first to systematically study this problem [19] whose theoretical foundations were established half a century later by Louis L. Thurstone as a *law of comparative judgment* [23]. Fechner's original aim was to quantify aesthetic preferences for objects, and to do that, he put forward the notion that the distance between two objects in some subjective pleasantness continuum was linked to the *proportion* of times an object was judged more pleasant than the other. By having subjects compare pairs of objects many times, he noticed that if object A is judged more pleasant than object B half the time, then both objects are equally pleasant. Correspondingly, if object C is consistently judged more pleasant than object A, then by the same measure, C is likely to be the most pleasant object. Thurstone generalized this concept into a measurement model which uses simple comparisons between pairs of stimuli (pairwise) to create a perceptual measurement scale. To do that, he linked p, the proportion of times a stimulus was judged greater than another, to the number of psychological scale units separating the two sensations. Similarly to the case of threshold detection (see Figure 5.1), the comparative judgment data dispersion in the psychological continuum was considered to be a product of the fluctuations of the neural system, and the psychological scale value was assumed to be the mean of these fluctuations (i.e., the mean of the frequency distribution). Since this frequency distribution is neural in origin and basically unknown, Thurstone devised an indirect method for calculating the psychological scale value from the proportion of positive comparison results p, following a series of careful considerations. In terms of signal detection theory, the psychological scale value that we want to obtain here is equivalent to the discriminability index (d') but calculated as the distance between two $S+N$ distributions instead of the classical S and $S+N$ (see Figure 5.6 and Eq. (5.7)).

Figure 5.9 shows the schematics of the statistical processes involved in a comparison between two different stimuli along an arbitrary psychological continuum. Panel (a) represents the distribution of a large number of judgments of stimulus A (solid curve) and stimulus B (broken lines curve). The downward-pointing arrows show the internal neural processes that determine judgment magnitudes corresponding to each stimulus (black arrows for stimulus A and gray broken arrows for stimulus B) in the psychological continuum. In a single pairwise comparison, a perceived value S_j for stimulus A is compared to the perceived value S_i for stimulus B, and distance $S_i - S_j$ is obtained. Because of the stochastic nature of neural processes, a tendency emerges after a number of comparisons with most judgments pairs concentrated around their mean values following the two Gaussian distributions of (a). In consequence, the average perceived distance between stimulus A and B is given by the difference between the means of the Gaussians ($\mu_B - \mu_A$).

The arrows in (a) show that in most cases, the value assigned to stimulus A will be larger than that of stimulus B, but in some cases, the opposite will occur (some broken gray arrows are located to the left of some continuous black arrows). The proportion of times where S_i is larger than S_j is determined by the proximity between the two Gaussians and their corresponding variances and can be calculated by integrating

$$P_{B-A}(u) = \iint_{-\infty}^{\infty} \frac{e^{-(x^2/(2\sigma_A^2))}}{\sigma_A\sqrt{2\pi}} \frac{e^{-(y^2/(2\sigma_B^2))}}{\sigma_B\sqrt{2\pi}} \delta((x-y)-u)\, dx\, dy \tag{5.9}$$

where P_{A-B} is the normal difference distribution, σ_A^2 and σ_B^2 are the variances of the Gaussians corresponding to stimulus A and B, and δ is a delta function [24].

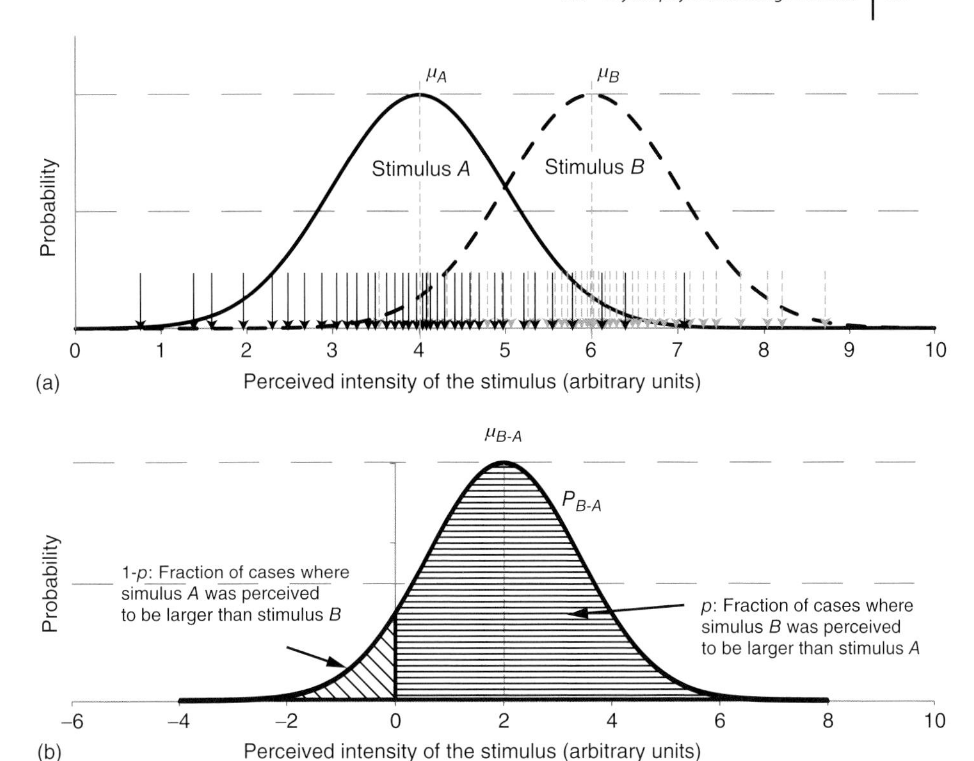

Figure 5.9 (a) Probability distribution of judgments for two different stimuli *A* and *B* in an arbitrary perceptual continuum. Their average perceptual distance is determined by $\mu_B - \mu_A$, the distance between the two means. (b) The probability distribution of the differences in perceptual judgments for two different stimuli, derived from (a). The mean $\mu_B - \mu_A$ of this new Gaussian is equal to the average perceptual distance between *A* and *B*, and its variance is also related to the variances of the distributions in (a).

The solution to Eq. (5.9) can be conveniently written as

$$P_{B-A}(u) = \frac{e^{((-[u-(\mu_B-\mu_A)]^2)/(2(\sigma_A^2+\sigma_B^2)))}}{\sqrt{2\pi(\sigma_A^2+\sigma_B^2)}} \tag{5.10}$$

which describes another normal distribution with mean value $\mu_{B-A} = \mu_B - \mu_A$ and variance $\sigma_{B-A}^2 = \sigma_A^2 + \sigma_B^2$. The relationship between this new Gaussian and the original pair is represented in Figure 5.9, where the difference between the means of the curves in (a) corresponds to the mean of the curve in (b) and the zero of the abscissa in (b) splits the shaded area under the curve according to the proportion of positive comparison results p. The case where the variables x and y in Eq. (5.9) are jointly normally distributed random variables is slightly more complex; however, the results of the integral are still normally distributed with a mean equal to $\mu_B - \mu_A$. The main change is that variances σ_A^2 and σ_B^2 are now correlated (and therefore not additive) which results in a slightly more complex value for the standard deviation of the normal difference distribution:

$$\sigma_{B-A} = \sqrt{\sigma_A^2 + \sigma_B^2 + 2\rho_{AB}\sigma_A\sigma_B} \tag{5.11}$$

In summary, the probability distribution of differences plotted in (b) provides the link between the psychological distance $\mu_B - \mu_A$ and the quantity p, which is easily computed by noting the proportion of times subjects report that the sensation produced by stimulus A is larger than that of stimulus B. To operate, we just need the following property of normal Gaussian distributions applied to $P_{B-A}(u)$:

$$z = \frac{\mu_{B-A}}{\sigma_{B-A}} \tag{5.12}$$

which defines the standard score z (also known as the z-score) for $P_{B-A}(u)$. z quantifies the number of standard deviations that occur between our zero and the mean value μ_{B-A} in (b). Traditionally, the calculation of z involved large lookup tables, but today, it can be obtained numerically, by inverting the cumulative distribution function, which describes the probability that a random variable will be found to have a value less than or equal to z. Once z is obtained from p, there are some remaining considerations about Eq. (5.11) before solving it for μ_{B-A} that were organized by Thurstone in five different cases [23]. Cases I and II consider the complete form of Eq. (5.11), Cases II and IV consider $\rho_{AB} = 0$ (no correlation between variances σ_A^2 and σ_B^2), and Case V considers the simplest solution, where there is no correlation and variances σ_A^2 and σ_B^2 are arbitrarily given a value of one, effectively choosing unity as *de facto* unit of measurement. Equation 5.13 shows the simplest and most common expression for Thurstone law of comparative judgment Case V:

$$\mu_B - \mu_A = z\sqrt{2} \tag{5.13}$$

Tables 5.1–5.3 illustrate how to produce a comparison scale from the statistical results of an imagined paired comparison experiment consisting of five stimuli labeled A, B, C, D, and E. These were compared against each other obtaining the results shown in Table 5.1, with rows showing the fraction of times they were perceived to be larger than columns. For example, the value [row A, col B] shows that $A > B$ in 48% of the trials. Table 5.2 shows the standard scores derived for the same values, and Table 5.3 shows the results of Eq. (5.13) applied to Table 5.2. Column 7 of Table 5.3 is just the average for each stimulus, and column 8 shows the definitive scale, where zero has been set to the lowest value.

Table 5.1 Hypothetical results of a pairwise comparison experiment.

	A	*B*	*C*	*D*	*E*
A	—	0.48	0.62	0.9	0.95
B	0.52	—	0.54	0.74	0.9
C	0.38	0.46	—	0.7	0.85
D	0.1	0.26	0.3	—	0.69
E	0.05	0.1	0.15	0.31	—

Table 5.2 Standard scores derived from the results in Table 5.1.

	A	*B*	*C*	*D*	*E*
A	—	−0.05	0.31	1.28	1.64
B	0.05	—	0.10	0.64	1.28
C	−0.31	−0.10	—	0.52	1.04
D	−1.28	−0.64	−0.52	—	0.50
E	−1.64	−1.28	−1.04	−0.50	—

Table 5.3 Perceptual distances calculated for each comparison and their averages.

	A	*B*	*C*	*D*	*E*	Average	Average (min)
A	—	−0.07	0.43	1.81	2.33	1.12	2.70
B	0.1	—	0.14	0.91	1.81	0.73	2.31
C	−0.4	−0.14	—	0.74	1.47	0.41	1.98
D	−1.8	−0.91	−0.74	—	0.70	−0.69	0.89
E	−2.3	−1.81	−1.47	−0.70	—	−1.58	0.00

5.7 Conclusions

Although great progress has been made in the century and a half since the birth of the discipline, many challenges remain. For example, there is still no set of fundamental laws or concepts that explain all psychophysical results, and to our days, different scaling methods may yield different results, with no certainty that these methods are not measuring different aspects of perceptual/cognitive processes. It might be the case that a single solution to all these problems might be developed by some kind of fundamental laws of psychophysics, but so far, there is no agreement on what these laws should be based on.

In this chapter, we have attempted to provide an overview of the most basic concepts necessary to both conduct and interpret psychophysical laboratory experiments. On a practical level, psychophysics experimentation is not to be taken lightly. Designing and implementing a task could be extremely uncertain, and many questions arise during the experimental design: How many trials? How long should the experiment take? What should I say to the observers? How should I reward my observers? All these questions can be answered from your theory, and if you consider that (i) you should try to make the measurement error small, (ii) you do not want your observers to get tired or to lose interest, (iii) you want to take the greatest advantage of your experimental hardware, and (iv) you do want your observers to understand exactly what they have to do. Besides, you do not want to waste time (and presumably money) in an experiment that is likely to yield noise as a result. To avoid mistakes, it is often recommended to run a pilot to see whether all these conditions are fulfilled and the final results are likely to answer your questions. It is also recommended that you remove yourself as much as possible from the final data collection to avoid introducing your own expectations and biases. As a general rule, assume that observers will always try to second-guess your intentions: only talk to your observers about the experiment *after* the experiment has finished to avoid giving them wrong cues and expectations.

As science progresses and communication among researchers from different disciplines increases exponentially, the boundaries between computer vision, computational neuroscience, and experimental psychology are blurring. Today, it is common to see computer vision laboratories that routinely collaborate with psychology departments establishing larger "vision research centers" or even resort to equip their own psychophysics laboratories. This partnership has not always been frictionless, especially because of the lack of a common language and background; however, the potential rewards are enormous. Today, psychophysics is behind much of the progress (through the use of ROC curves) in medical decision analysis, text retrieval and processing, the experimental underpinning of computational color vision, primary visual cortex modeling (both functional and developmental), computational models of visual attention, segmentation and categorization, image retrieval, and the list grows. Even larger, higher-level problems such as visual aesthetics are routinely explored applying statistical methods to "ground truth databases" where the performance of human observers has been recorded or/and analyzed. Indeed, ground truth databases are perhaps the single most conspicuous manifestation of the influence of psychophysical approaches and methods in computer science. It is in this role of users of psychophysical data that computer scientists most need to understand all measurement limitations inherent to each experiment to properly interpret how they may influence their own results.

References

1. Fechner, G.T. (1860) *Elements of Psychophysics*, 1966 edn, Holt, Rinehart and Winston, New York.
2. Bevington, P.R. and Robinson, D.K. (2003) *Data Reduction and Error Analysis for the Physical Sciences*, 3rd edn, McGraw-Hill, Boston; London.
3. Taylor, J.R. (1997) *An Introduction to Error Analysis: The Study of Uncertainties in Physical Measurements*, 2nd edn, University Science Books, Sausalito.
4. Weber, E.H. (1834) *De Pulsu, resorptione, auditu et tactu: Annotationes anatomicae et physiologicae*, C.F. Koehler.
5. Hecht, S., Shlaer, S., and Pirenne, M.H. (1941) Energy at the threshold of vision. *Science*, **93**, 585–587.
6. Ludvigh, E. and McCarthy, E.F. (1938) Absorption of visible light by the refractive media of the human eye. *Arch. Ophthalmol.*, **20**, 37–51.
7. Marks, L.E. (1978) *The Unity of the Senses: Interrelations Among the Modalities*, Academic Press, New York.
8. Brindley, G.S. (1970) *Physiology of the Retina and Visual Pathway*, 2nd edn, Edward Arnold, London.
9. Parraga, C.A., Troscianko, T., and Tolhurst, D.J. (2000) The human visual system is optimised for processing the

spatial information in natural visual images. *Curr. Biol.*, **10**, 35–38.
10. Taylor, M.M. and Creelman, C.D. (1967) PEST – Efficient estimates on probability functions. *J. Acoust. Soc. Am.*, **41**, 782.
11. Watson, A.B. and Pelli, D.G. (1983) QUEST: a Bayesian adaptive psychometric method. *Percept. Psychophys.*, **33**, 113–120.
12. Garcıa-Perez, M.A. (1998) Forced-choice staircases with fixed step sizes: asymptotic and small-sample properties. *Vision Res.*, **38**, 1861–1881.
13. Green, D.M. and Swets, J.A. (1966) *Signal Detection Theory and Psychophysics*, John Wiley & Sons, Inc., New York.
14. Macmillan, N.A. and Creelman, C.D. (2005) *Detection Theory: A User's Guide*, 2nd edn, Lawrence Erlbaum Associates, Mahwah.
15. Tanner, W.P. and Swets, J.A. (1954) A decision-making theory of visual detection. *Psychol. Rev.*, **61**, 401–409.
16. Gescheider, G.A. (1997) *Psychophysics: The Fundamentals*, 3rd edn, Lawrence Erlbaum Associates, Mahwah; London.
17. Stephan, C., Wesseling, S., Schink, T., and Jung, K. (2003) Comparison of eight computer programs for receiver-operating characteristic analysis. *Clin. Chem.*, **49**, 433–439.
18. Hacker, M.J. and Ratcliff, R. (1979) Revised table of D' for M-alternative forced choice. *Percept. Psychophys.*, **26**, 168–170.
19. Fechner, G.T. (1876) *Vorschule der Aesthetik*, 2. Aufl., Breitkopf & Härtel, Leipzig.
20. Murray, D.J. (1993) A perspective for viewing the history of psychophysics. *Behav. Brain Sci.*, **16**, 115–137.
21. Munsell, A.E.O., Sloan, L.L., and Godlove, I.H. (1933) Neutral value scales. I. Munsell neutral value scale. *J. Opt. Soc. Am.*, **23**, 394–411.
22. To, M.P., Lovell, P.G., Troscianko, T., and Tolhurst, D.J. (2010) Perception of suprathreshold naturalistic changes in colored natural images. *J. Vis.*, **10** (12), 11–22.
23. Thurstone, L. (1927) A law of comparative judgment. *Psychol. Rev.*, **34**, 273–286.
24. Weisstein, E.W. (2014) *Normal Difference Distribution, in MathWorld – A Wolfram Web Resource*, Wolfram Research, Inc.

Part II
Sensing

Biologically Inspired Computer Vision: Fundamentals and Applications, First Edition.
Edited by Gabriel Cristóbal, Laurent Perrinet, and Matthias Keil.

6
Bioinspired Optical Imaging

Mukul Sarkar

6.1
Visual Perception

Among the senses that help perceive the world around, vision plays a larger role for humans, as well as for many other animals. Vision provides veridical perception (from the Latin *veridicus*, truthful) about the environment and thus helps in successful survival without actual contact with the environment. Vision, however, would not have any meaning without the presence of light. Light is an electromagnetic radiation that travels in the form of waves. Visual perception is the ability to detect light and interpret it to acquire knowledge about the environment. Knowledge thus acquired helps in survival by properly interpreting objects inherent to the environment.

The early explanation of vision was provided by two major ancient Greek schools of thought. One believed in the "emission theory" championed by scholars such as Euclid and Ptolemy, according to which vision occurs when light rays emanate from the eyes and are intercepted by visual objects. The other school championed by scholars such as Aristotle and Galen believed in "intromission" where vision occurs by something entering the eyes representative of the object [1]. The Persian scholar Ibn al-Haytha, ("Alhazen") is credited for refining the intromission theory into the modern accepted theory of perception of vision [2]. In his most influential "Book of Optics," he defines vision to be due to the light from objects entering the eye [3, 4].

Perception is unavoidably selective; one cannot see all there is to see. Visual perception is not merely a translation of retinal stimuli. It is more a way to understand the environment and use the information about the environment to help in survival. Eyes are the first step to visual perception. The eyes can be classified broadly into single-aperture eyes of humans and the compound eyes of insects. The compound eyes of insects in comparison to single-aperture eyes have a much wider field of view, better capability to detect moving objects, higher sensitivity to light intensity, but much lower spatial resolution. Compound eyes are the most compact vision systems found in nature. The visual perception of eyes differs from that of the camera in that cameras can only take pictures and need not know the

Biologically Inspired Computer Vision: Fundamentals and Applications, First Edition.
Edited by Gabriel Cristóbal, Laurent Perrinet, and Matthias Keil.

objects whose pictures are being captured. When a conventional camera is used, only the picture information is captured, whereas eyes help in interpretation of the acquired image and thus help in knowing the environment. Insects, for example, have adapted their eyes to face environmental contingencies in a most efficient way in terms of utilization of resources and speed. Such systems are highly desirable in many machine vision applications. The principles of optical vision from natural systems can be used to build up a working artificial computer vision system for machine vision applications such as motion detection, object and gesture recognition, and so on

6.1.1 Natural Single-Aperture and Multiple-Aperture Eyes

Light is an important tool for visual perception. Most eyes experience light as having three features: color, brightness, and saturation. The color or hue of the light depends on its wavelength. The brightness of the light is related to the intensity of light the object emits or reflects. Intensity depends on the amplitude of the light wave. Saturation depends on the range of wavelengths in light. Figure 6.1 shows the light spectrum.

The visible spectrum is part of the light spectrum which includes the wavelength range from 400 nm to 700 nm. Wavelengths shorter than 400 nm are referred to as *ultraviolet light*, while wavelengths longer than 700 nm are referred to as *infrared radiations*.

Vision is perception of light. Light is perceived by means of optical systems, for example, eyes. Eyes can be simply defined as organs or visual systems for spatial vision. Eyes usually have millions of photoreceptors, which are specialized cells responding to light stimuli. The number of photoreceptors also defines the resolution of spatial vision. Spatial resolution defines how precisely objects in the environment can be differentiated. The higher the number of photoreceptors, the larger is the spatial resolution. This has been the inspiration behind human

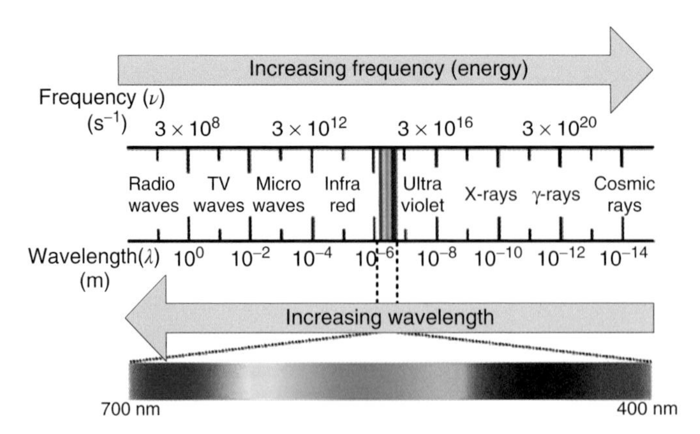

Figure 6.1 Light spectrum.

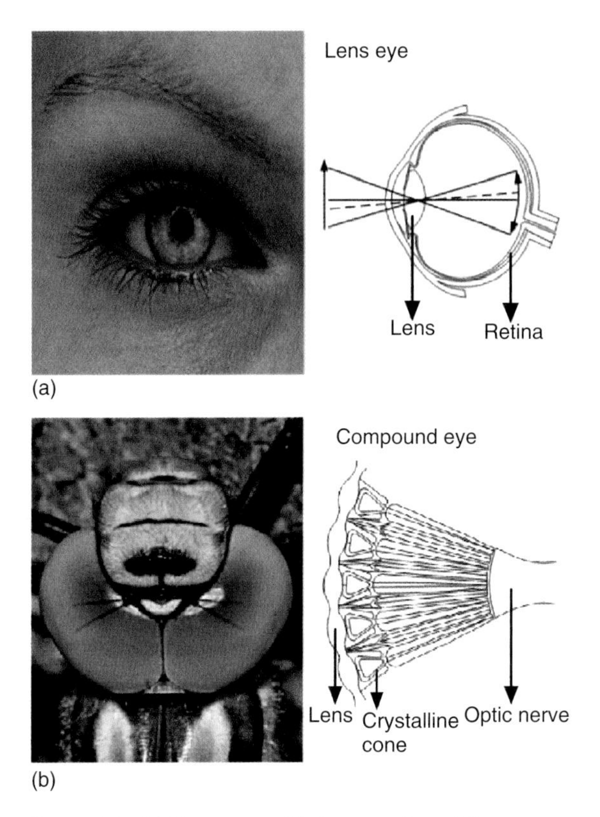

Figure 6.2 (a) Human eye, (b) Compound eye [6].

eyes, which have millions of photoreceptors. Higher spatial resolution can also be achieved by multiplying the visual system in its entirety [5] as observed in the eyes of insects. The insect's eyes are a compound of individual lenses, each with its own photoreceptor array. The single-aperture human eye and the compound eye of insects are shown in Figure 6.2.

6.1.1.1 **Human Eyes**

Human eyes are also called "camera-type eyes." The human eye focuses light on a light-sensitive membrane, the retina, just as a camera focuses light on a film. The human eye consists of two types of photoreceptors, rods and cones. On an average, there are 120 million rods and 6–7 million cones in the human eye. Rods are more sensitive to the intensity of light than are cones, but they have poor spectral (color) sensitivity, while cones provide spectral vision. The cones are divided into three colors – red, green, and blue – on the basis of their spectral sensitivity. The cones are responsible for high-resolution vision or visual acuity, whereas the rods are responsible for dark-adapted or scotopic vision. The sensitivity of the rods is very high and can be triggered by individual light photons [7]. The rods are thus better motion sensors.

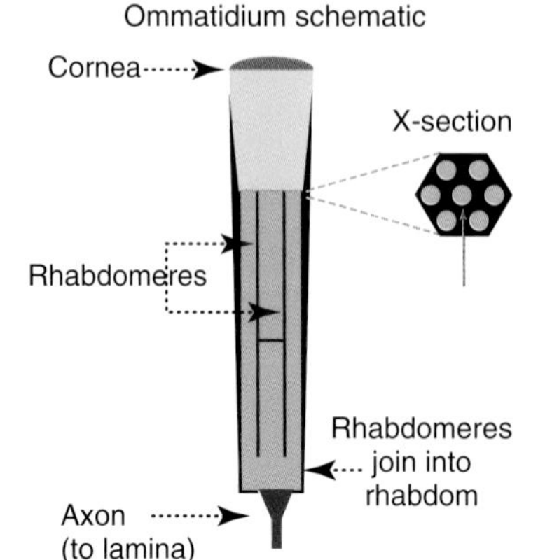

Figure 6.3 Ommatidia of a compound eye.

6.1.1.2 **Compound Eyes**

Most insects have compound eyes with multiple light-gathering units known as *ommatidia*. Each ommatidia as shown in Figure 6.3 consists of a lens that forms an image on the tip of the light-sensitive visual cell rhabdom, a transparent crystalline cone to guide the light and absorptive pigment between the ommatidium [8]. The optical part of the ommatidium includes the corneal lens and the crystalline cone, while the sensory part includes the retinula cells and the rhabdom. The light-sensitive parts of the visual cells are microvillis, an array of tubelike membranes where the pigment rhodopsin is located. Photons are captured by the rhabdome. The rhabdomeres are highly refractive materials; the light entering them are totally internally reflected and they are thus ideal for collecting available light with minimal losses. This visual information is transduced into electrical signals by the photoreceptor cells at the level of the retina.

Compound eyes can broadly be classified into apposition and superposition types [9]. In an apposition eye, each optical channel has its own receptor. The acceptance angle of light in each ommatidium is low, thus preventing the spreading of light among adjacent ommatidia. Apposition eyes are generally found in day-active insects such as ants, wasps, dragonflies, bees and cockroaches. In a superposition compound eye, light from a different adjacent channel is guided to a common receptor. Superposition eyes are typically more sensitive to light, and this type is found mainly in animals that live under low light conditions.

Insects and humans have very different types of eyes, each having its advantages and disadvantages. The following are the basic differences between the human eye and insect's eye:

6.1.1.3 **Resolution**

The compound eye is made up of multiple smaller optical structures called *ommatidia*. The higher the number of ommatidia or optical units in an eye, the higher is the spatial resolution (defined as the ability of the eye to resolve fine details). In the human eye, a single lens covers the entire retina with many sensory cells, wherein each sensory cell works as an optical unit or pixel. The density of photoreceptors is around 25 times higher in the human eyes compared to the compound eye. This allows the human eye to have a greater spatial resolution of the object. For insects, in order to reach a similar spatial resolution, the lenses need to be extraordinarily large. The small lenses of the compound eye are diffraction limited. The small optics are not capable of focusing light onto to a small enough point, thus the overall resolution is degraded.

Contrast is the ability to distinguish similar shades of the same color. The contrast is related to the intensity of the light levels; in dim light the contrast is low and in bright light the contrast is high. To improve the contrast in a dim environment, either a large aperture is needed to allow more light to reach the visual elements or the light needs to be integrated over a larger period of time. The eyes of insects are limited by the small aperture of each ommatidium in the compound eye as well as the time for which the light can be integrated, as the response time in a dynamic environment is quite low. Human eyes are also not adapted for night vision.

6.1.1.4 Visual Acuity

Acuity, in human vision, is defined as the reciprocal of the minimum resolvable angle measured in minutes of arc [10]. Visual acuity is also used to describe the smallest single object or the smaller details in the object that an eye can detect. Visual acuity defines the actual spatial resolution that the eyes can perceive. The spatial density of retinal cells in human eyes and ommatidia in compound eyes determine the number of visual elements in the final image and thus determine the visual acuity. Visual acuity is affected by the imperfections and optical limits of the lens system. The eyes of insects are limited by diffraction of light. A single point of light coming from an object appears as a blurred spot in the final image. The visual acuity of the compound eye is about one-hundredth that of the human eye.

6.1.1.5 Depth Perception

The binocular cues and monocular cues are used by human eyes to estimate the distance of the objects in the visual scene. Binocular cues require both eyes. There are two types of binocular cues: retinal disparity and convergence. Retinal disparity marks the difference between the two images captured by the two eyes. Retinal disparity increases as the eyes get closer to an object. The differences in the images captured by the two retina spaced apart in space are used to estimate the distance between the viewer and the object being viewed. Convergence is when eyes turn inward to cross on an object at close up. The eye muscles tense to turn the eye and this tense information is interpreted to estimate the distance of the object.

Monocular cues require only one eye. Monocular cues which help in estimating the distance of the objects are interposition, motion parallax, relative size and clarity, texture gradient, linear perspective, and light and shadow. In interposition, when one object is blocked by another object, the blocked object appears farther from the viewer. In motion parallax, when the viewer is moving with respect to a stationary object, the object tends to move in different directions and at different speeds on the basis of the location of the object in the visual field. A stationary object closer to the viewer appears to move backward with increasing speed compared to an object placed at a greater distance, which appears to move forward slowly. Distance can also be estimated by the relative size of the object. Farther objects make a smaller image on the retina and vice versa. The relative clarity with which the object forms an image on the retina also helps in estimating the object distance. If the image is clear and detailed, the object is near the viewer, while if the image is hazy then the object is far away. Further, smaller objects that are more thickly clustered appear farther away than objects that are spread out in space. Recently, it has been shown that jumping spiders (Hasarius adansoni) perceive depth by using image defocus [11].

The immobile compound eyes of the insects have fixed focus compared to human eyes. The compound eye has many little eyes looking in different directions as compared to human eyes which can swivel. The focus of the compound eyes cannot be changed dynamically and thus they are not able to determine the distance of the object by focusing the object on to the visual plane. However, insects are able to infer the ranges of the objects from the apparent motion of their images on their eyes. The apparent motion of objects, surfaces, and edges in a visual scene caused by the relative motion between an observer and the scene is known as *optic flow*. It usually contains information about self-motion and distance to potential obstacles and thus it is very useful for navigation [12]. The visual system can easily detect the motion of objects that move independently of the surroundings from the generated optical flow. The magnitude of the optical flow corresponds to angular velocity, which will be higher if the eye is moving in a straight line and the object is stationary and near. The range in the horizontal plane can be calculated as [13]

$$r = \frac{V}{\omega} \sin\theta \tag{6.1}$$

where, V is the speed of the eye, θ is the bearing of the object, ω is the apparent angular velocity $(d\omega/dt)$ and r is the range of the object.

6.1.1.6 Color Vision

The color or hue of light depends on its wavelength. Color vision in humans occurs because of the presence of three types of cones in the retina, which responds to light of three different wavelengths, corresponding to red, green, or blue (see Chapter 2). The combinational activation of the cones is responsible for the perception of colors. The ability to see colors is rare in insects and most of them only see light and dark. However, a few insects, such as bees, can see more colors than humans [14]. The spectral response of the housefly shows spectral sensitivities to at least six wavelengths from the ultraviolet (300 nm) to the red (600 nm) part of the spectrum [15, 16]. Compared to humans, insects have a higher sensitivity to shorter wavelengths. The color vision of the human eye is limited to wavelengths in the visible spectrum in Figure 6.1, while many insects are able to perceive higher frequency (ultraviolet) than the visible spectrum. Color sensitivity of the compound eyes in the ultraviolet spectrum plays an important role in foraging, navigation, and mate selection in both flying and terrestrial invertebrate animals.

6.1.1.7 Speed of Imaging and Motion Detection

For insects to survive, they need to track changes in the environment effectively, thus motion vision is important. For insects, motion vision provides cues for tasks such as maintaining flight trajectory, avoiding colliding with approaching obstacles in the flight path, and identifying and detecting targets such as the motion of a potential prey. The housefly, for example, is able to navigate in an unconstrained environment while avoiding any obstacles even in a densely populated and dynamic environment [17, 18].

The compound eyes of insects are excellent in detecting motion. The compound eye uses an array of elementary motion detectors (EMDs) as smart, passive ranging sensors [19]. As the object moves in the visual field, the individual receptor channels (ommatidia) are progressively turned on and off owing to the changes in the light intensity received. This creates a "flicker effect". The rate at which the ommatidia are turned on and off is known as the *flicker frequency*. The flicker effect allows insects to detect moving objects in their visual field. The high flicker frequency of the compound eye allows for much faster processing of images than the single-chambered eye. The reaction time reported for honeybees to an object suddenly appearing in the visual field is 0.01 s, while that for humans is 0.05 s. In addition, the division of the scene into multiple sub-images allows parallel image processing and nonsequential readout of the selected sub-images. This facilitates high-speed object tracking while reducing image-processing bottlenecks.

6.1.1.8 Polarization Vision

The three basic characteristics of light are intensity, wavelength, and polarization. Light is a transverse electromagnetic wave, thus an infinite number of planes for vibrations exist. If the light is composed of electric field vectors oriented in all directions, the light is unpolarized, while if the electric field vector is oriented in a specific direction, the light is said to be *linearly* polarized. The direction of the electric field oscillation is defined as the polarization direction.

Polarization provides a more general description of light than either the intensity or the color alone, and can therefore provide richer sets of descriptive physical constraints for the interpretation of the imaged scene. The single-aperture human eye is very mildly sensitive to the polarization state of the light, whereas the compound eye of the insect is able to perceive polarization patterns in the incident light wave. The microvillar organization of the insect's photoreceptors makes its visual systems inherently sensitive to the plane of polarization of the light. The visual pigment rhodopsin absorbs a maximum of linearly polarized light when the electric field vibrates in the direction of their dipole axis. In human eyes, the axes of the rhodopsin molecules are randomly oriented; however, in insects' eyes, they are specifically aligned as shown in the Figure 6.4. The rhodposin molecules

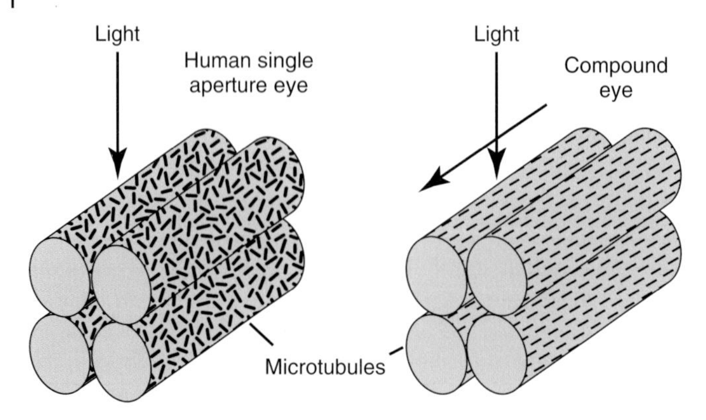

Figure 6.4 Alignment of visual molecules.

are aligned preferentially parallel to the axes of the microvilli tubes in the ommatidia. Thus each visual cell is maximally sensitive to polarized light parallel to its microvilli orientation.

The sensitivity to polarization has been shown to be useful in a number of visual functions including increase in contrast, orientation, navigation, prey detection, predator avoidance, and intraspecies signaling. Aquatic animals, such as the cuttlefish, are able to change the light polarization by changing their skin properties; this is used for communication and to distinguish objects under water [20]. Ctenophore plankton is completely transparent in unpolarized light but can be easily detected when placed between cross polarizers. The electric field vector patterns in the sky are used by insects such as ants and bees as a reference for compass information in navigation.

In summary, the compound eyes of insects have properties that are ideally suited for machine vision applications. A vision system can be greatly augmented by adapting different sensor models with different capabilities. The polarization vision of insects has an impact on the image formation and thus significantly affects the image features used for decision making. These polarization features can be used with conventional camera technology for image formation, leading to numerous biologically inspired applications. Polarization vision is already being explored in the fields of robotic orientation and navigation, automatic target detection and recognition, scene segmentation, camouflage breaking and communication, and underwater vision. Light rays get reflected when they strike a reflecting surface, and the reflected light is polarized depending on the incident light and the nature of the reflecting surface. This polarized component can be used to detect the nature of the surface, for example, to discriminate between a metal and a dielectric surface [21, 22].

The application of polarization vision in detecting transparent objects for object recognition and detection is presented in Section 6.2. Flying insects have extraordinary visual capabilities. Their ability to detect fast motion in the visual scene and avoid collision using low-level image processing and little computational power makes their visual processing interesting for real time motion/collision detection in machine vision applications. The ability of the compound eyes to detect fast-moving objects is explored in machine vision applications in Section 6.3.

6.2 Polarization Vision - Object Differentiation/Recognition

In machine vision applications, object differentiation or recognition in the visual scene is very important. To determine the shape of the objects, usually an active or passive light technique is used. Active light techniques include laser range scanning, coded structured light systems, digital holography, or time-of-flight scanners, whereas passive techniques are mainly stereovision, photogrammetry, or shape from techniques such as shading, optical flow, motion, focus, and so on. Active range scanning techniques actively control the light in the scene as compared to passive sensing and thus are more reliable in detecting patterns of light and features of the scene [23].

An important machine vision problem for object recognition is to recognize transparent or specular objects and opaque objects in the visual scene. In applications such as environment monitoring, security surveillance, or autonomous robotic navigation using visual sensors, the recognition of a transparent object such as glass from opaque objects is particularly important. If the transparent objects in the environment are not detected, the autonomous agents would fail in tasks such as manipulation or recognition of objects. A collision, for example, would be inevitable if the robot tried to manipulate a known object but is incapable of recognizing a transparent object located in between.

Transparent objects are common in any environment; they allow seeing the content directly owing to their transparency. The perception of transparent surface is a difficult vision problem due to the absence of body and surface reflections. Further, the multiple reflections inside the transparent objects make them difficult to detect. When a transparent object is present in an environment or scene, the light through the scene is characterized by three basic processes – specular reflection at an object's surface, specular transmission at the surface of a transparent object, and linear propagation within an object's interior and through empty space [24]. The light reflected from a transparent object is mostly specular, while in case of opaque objects, the reflections are mostly diffuse [21, 22]. Specular reflection is the reflection of light from a surface in which light from a single incoming direction is reflected into a single outgoing direction. It obeys the laws of reflection. Diffuse reflection is the reflection of light at many angles rather than a single angle as in the case of specular reflection. It is diffuse reflection that gives color to an object. For normal camera systems, transparent objects are almost invisible except for specularities. This makes their detection and shape estimation difficult.

Despite the prevalence of transparent objects in human environments, the problem of transparent object recognition has received relatively little attention. The available vision algorithms to detect or model transparent surfaces can be classified into two groups. The first group of algorithms tries to enhance the surface-related properties to define the surface's shape. The shape-from-distortion algorithms tries to reconstruct the 3D shape of the transparent surface by analyzing the patterns that are distorted by specular reflection [25–27]. The shape-from-specularity approaches uses the highlights created at points on the surface owing to specular reflection of the incident light for the reconstruction of the surface points [28, 29]. The structure-from-motion algorithms uses the apparent motion of the features in the image plane of a moving camera to reconstruct the transparent surface [30, 31]. The second group of algorithms tries to synthesize a realistic image of transparent objects without using a 3D shape information. These algorithms try to capture multiple-view points to create novel views using interpolation [24]. Many algorithms also determine the surface shape by separating overlapping images of reflected and transmitted images from the transparent surface. The specular stereo algorithm relies on a two-camera configuration or a moving object observer to reconstruct the transparent objects [28, 32].

The appearance-based algorithms for object and category recognition are not appropriate for transparent objects where the appearance changes dramatically, depending on the background and illumination conditions. Detection of transparent objects using the background scene contents is also popular. The method captures a local edge energy distribution characteristic of the refraction using the latent factor model. Machine-learning-based methods, for instance, the additive model of patch formation using latent Dirichlet allocation (LDA) is quite efficient, but it is computationally intensive and results vary with illumination conditions [33]. Other methods based on absorption of particular light wavelengths by transparent objects are computationally less expensive as compared to the former methods, but require sophisticated and expensive equipment such as time-of-flight cameras, LIDAR sensors, and so on [34]. These methods try to use reflection of infrared radiation from transparent objects. The methods are computationally intensive and not suited for real-time applications.

In nature, organisms are often required to differentiate between opaque and transparent objects for their survival. For example, flying insects need to detect transparent objects to avoid collision, while aquatic animals need it for navigation under water. The visual predators need to detect transparent aquatic animals for their survival. The transparent prey uses camouflage in which the animal reflects or produces partially polarized light to reduce the dark area of the body and matches the background illumination, thus making it visually transparent. The polarization sensitivity of the predator's eyes helps in overcoming the radiance matching effects and thus allows predators to detect transparent prey, break camouflage, and increase the detection range of the target. As discussed in Section 6.1, the photoreceptors of in the compound eye of the insects react differentially to partially polarized light [35].

The reflection from a transparent object is mostly specular, while reflection from an opaque object is mostly diffused. Because of specular reflection from transparent objects, the reflected light is linearly polarized, while the light reflected from opaque objects is not polarized. Hence, analysis of light reflected from different objects using a polarizing filter can be used to classify objects into being transparent or opaque. Polarization-based algorithms are often used to describe or enhance surface-related properties of transparent objects. These algorithms are gaining popularity due to their effectiveness and simplicity in detecting and measuring specular or surface reflection. Polarization-based algorithms, however, suffer from surface reflection and ambiguity problems. Further polarization-based analysis is difficult when transmission, rather than specular reflection, dominates image formation [24].

In this section, a bioinspired algorithm using polarization of reflected light to detect transparent objects is presented. The bioinspired algorithm can be used in many applications of machine vision where objects can be classified in real time. This method is computationally inexpensive and only a polarizing filter together with a camera is required.

6.2.1 Polarization of Light

An electromagnetic wave is said to be *linearly* polarized if the oscillation of its electric and magnetic vectors are confined to specific directions. If the electric and magnetic vectors are not defined in any direction, the electromagnetic wave is said to be *unpolarized*. Sunlight is considered to be unpolarized by nature. Polarization can also be approximated as intensity in one direction, thus unpolarized light has equal magnitude in all directions. When this light is passed through a polarizing material, specularly reflected from any surface, or scattered by particles (e.g., Rayleigh scattering by air molecules), it gets partially polarized. Figure 6.5 shows the polarization of light by reflection and transmission. The incident ray makes an

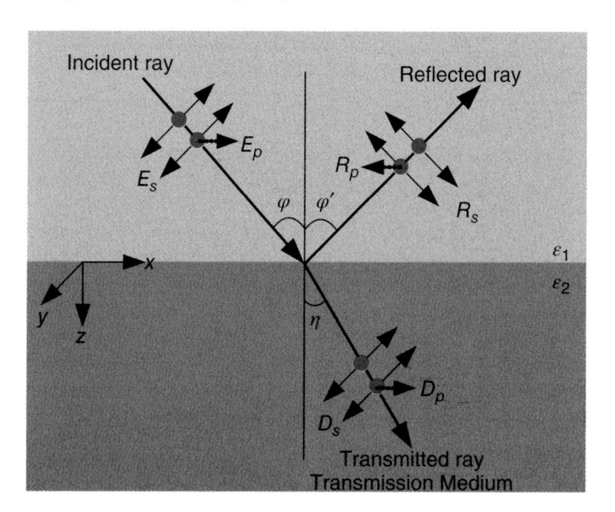

Figure 6.5 Polarization of light by reflection and transmission.

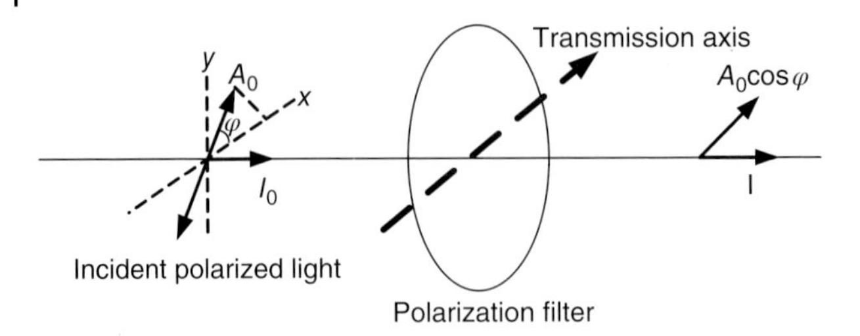

Figure 6.6 Transmission of a polarized beam of light through a linear polarizer.

angle of incidence of φ with the plane of incidence xz. The incident ray has two polarization components, parallel (E_p) and perpendicular (E_s). When the incident ray reaches the boundary, it is divided into a reflected and a refracted (transmitted) wave component. The reflected and transmitted rays also consist of two polarization wave components, parallel (R_p, D_p) and perpendicular (R_s, D_s) respectively. The reflected and transmitted light are polarized on the basis of the nature of the surface and the incident angle.

The transmission of polarized light through a linear polarization filter is shown in Figure 6.6. The intensity of the transmitted partially polarized light wave through a linear polarization filter is given by Malus' law. According to Malus' law, when a perfect polarizer is placed in a path of a polarized beam of light, the transmitted intensity, I, through the linear polarizer is given by

$$I = I_0 \cos^2 \varphi \tag{6.2}$$

where I_0 is the incident intensity and φ is the angle between the light's initial polarization direction and the transmission axis of the linear polarizer. The intensity of the transmitted partially polarized light can also be expressed as

$$I = \frac{I_{max} - I_{min}}{2} \cos(2\varphi) + \frac{I_{max} + I_{min}}{2} \tag{6.3}$$

where I_{min} and I_{max} represent the minimum and maximum intensity transmitted through the linear polarization filter as shown in Figure 6.7.

The values I_{max} and I_{min} are obtained for different angles depending on the type of polarizer used. For a 0° polarizer, I_{max} and I_{min} would be equal to I_{0° and I_{90° respectively because the horizontally polarized component would pass

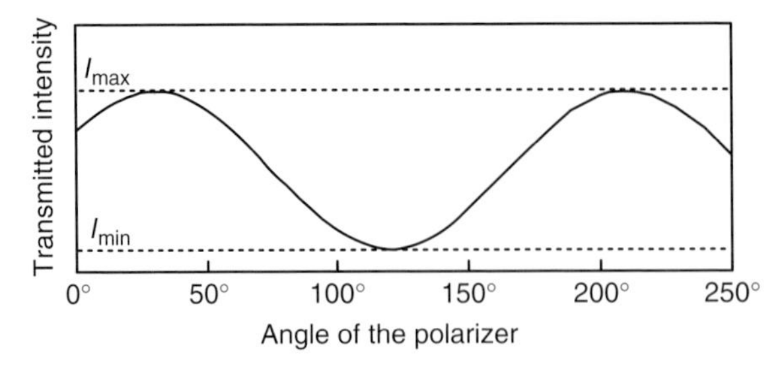

Figure 6.7 Transmitted light intensity.

completely, while the vertical component would be blocked completely. Similarly, for a 90° polarizer, I_{max} and I_{min} would be equal to I_{90}° and I_{0}°, respectively. The total intensity of the light and the degree of polarization (the proportion of polarized components in the partially polarized light) can be expressed as

$$I = I_{max} + I_{min} \tag{6.4}$$

$$\text{DOP} = \frac{I_{max} - I_{min}}{I_{max} + I_{min}} \tag{6.5}$$

Partially polarized light is composed of a dominating perfectly polarized component and traces of unpolarized components. The degree of polarization (DOP) measures the degree of the linearly polarized light component in a beam of partially polarized light. If the reflected light is unpolarized, DOP is 0, while if the reflected light is polarized, the DOP is 1. Completely linearly polarized light is observed when the incident angle and the reflecting angle are at the Brewster angle [36]. At the Brewster angle, the light reflected from the surface is completely linearly polarized. The degree of polarization depends on angle of incidence and refractive index of the material. It is given by the equation

$$\text{DOP} = \frac{2\sin^2\varphi\sqrt{n^2 - \sin^2\varphi - n^2\sin^2\varphi + \sin^4\varphi}}{n^2 - \sin^2\varphi - n^2\sin^2\varphi + 2\sin^4\varphi} \tag{6.6}$$

where φ is the angle of incidence and n is the refractive index of the material.

The description of the state of polarization of light is provided by Stokes' polarization parameters. Stokes' parameters were introduced in 1852 by Sir George Gabriel Stokes. The polarization ellipse is only valid at a given instant of time, a time average (Eq. (6.7)) of the ellipse gives the Stokes' parameters of light.

$$E_{0x}{}^2 + E_{0y}{}^{22} - E_{0x}{}^2 - E_{0y}{}^{22} - (2E_{0x}E_{0y}\cos\delta)^2 = (2E_{0x}E_{0y}\sin\delta)^2 \tag{6.7}$$

where E_{0x} and E_{0y} are the amplitudes of the light wave and δ is the phase difference between the E_{0x} and E_{0y} waves.

Equation (6.7) can be expressed in terms of the Stokes' parameters S_0, S_1, S_2 and S_3

$$S_0{}^2 = S_1{}^2 + S_2{}^2 + S_3{}^2 \tag{6.8}$$

where

$$\begin{aligned} S_0 &= I = E_{0x}{}^2 + E_{0y}{}^2 \\ S_1 &= Q = E_{0x}{}^2 - E_{0y}{}^2 \\ S_2 &= U = 2E_{0x}E_{0y}\cos\delta \\ S_3 &= V = 2E_{0x}E_{0y}\sin\delta \end{aligned} \tag{6.9}$$

The first Stokes' parameter S_0 represents the total intensity, the parameters S_1 and S_2 represent the linear polarization state, and S_3, the circular part [37]. The Stokes' parameters allow for determination of the degree of polarization directly from the measured intensity values. The Stokes' degree of polarization is given by equation

$$\mathrm{DOP} = \frac{\sqrt{S_1{}^2 + S_2{}^2 + S_3{}^2}}{S_0} \tag{6.10}$$

If the light wave is assumed to be linearly polarized, the degree of polarization is often referred to as *degree of linear polarization (ρ)* and is expressed as

$$\rho = \frac{S_1}{S_0} \tag{6.11}$$

Besides the degree of polarization and Stokes' degree of polarization, the polarization Fresnel ratio (PFR) is also often used to characterize linear polarized light. The PFR was introduced by Wolff [21] and is found to be a very useful tool to distinguish between the two linearly polarized states, horizontal and vertical, respectively. Using the Fresnel equations, the PFR can be written as

$$PFR = \frac{F_{\perp}}{F_{\parallel}} \tag{6.12}$$

where $F_{\perp}$ and $F_{\parallel}$ refer to Fresnel perpendicular and parallel reflection coefficients respectively [21]. The Fresnel coefficients are derived from the Fresnel reflectance model. This model describes the behavior of electromagnetic waves when reflected from a surface. It gives the reflection coefficients parallel and perpendicular to the plane of incidence in terms of refractive index, angle of incidence, reflection, and refraction.

PFR can be expressed in terms of maximum and minimum light intensity as is DOP, by using Fresnel reflectance model. The Fresnel reflectance model implies the following two equations [38]:

$$I_{\max} = \frac{1}{2}I_d + \frac{F_{\perp}}{(F_{\perp} + F_{\parallel})} \tag{6.13}$$

$$I_{\min} = \frac{1}{2}I_d + \frac{F_{\parallel}}{(F_{\perp} + F_{\parallel})} \tag{6.14}$$

where, $F_{\perp}$ and $F_{\parallel}$ are same as in Eq. (6.12) and I_d is the diffuse component of reflection. From Eqs (6.12) and (6.13) [37],

$$\frac{F_{\perp}}{F_{\parallel}} = \frac{I_{\max} - \frac{1}{2}I_d}{I_{\min} - \frac{1}{2}I_d} \tag{6.15}$$

Since, in specular highlight regions, $I_d \ll I_s$ (I_s refers to the specular component of reflected light), Eq. (6.15) could be approximated to [38]

$$\frac{F_{\perp}}{F_{\parallel}} = \frac{I_{\max}}{I_{\min}} \tag{6.16}$$

If a polarizer with a transmission axis at 0° is used, then

$$I_{\max} = I_{0°} \tag{6.17}$$

$$I_{\min} = I_{90°} \tag{6.18}$$

Hence from Eqs. (6.16), (6.17), and (6.18) we have

$$PFR = \frac{I_{0°}}{I_{90°}} \tag{6.19}$$

From Eqs (6.5), (6.17), and (6.18), degree of polarization can be calculated as

$$\rho = \frac{I_{0°} - I_{90°}}{I_{0°} + I_{90°}} \tag{6.20}$$

To calculate the degree of polarization in terms of transmitted intensity using Stokes' parameters, Eq. (6.10) can be further simplified as shown in Eq. (6.21). The Stokes' parameters S_2 in Eq. (6.10) can be neglected for perfectly linear polarized light. The degree of polarization obtained after ignoring S_2 is called the *degree of linear polarization* (*DOLP*).

$$p = \frac{S_1}{S_0} \tag{6.21}$$

The Stokes' parameters in terms of the transmitted intensity and the transmission axis of the linear polarizer can be written as

$$S_0 = I_{0°}{}^2 + I_{90°}{}^2 \tag{6.22}$$

$$S_1 = I_{0°}{}^2 - I_{90°}{}^2 \tag{6.23}$$

Therefore, from Eqs (6.21), (6.22), and (6.23), Stokes' degree of polarization (p) can be expressed as

$$p = \frac{I_{0°}^2 - I_{90°}^2}{I_{0°}^2 + I_{90°}^2} \tag{6.24}$$

The degree of polarization and the angle of polarization parameters are related to the geometry of the optical element. Thus these parameters help to determine the shape of the transparent objects.

6.2.2 Polarization Imaging

Polarization imaging offers an alternative way of visualizing and distinguishing transparent objects from their environment. Saito *et al.* presented a method which used partial polarization to highlight surface points [39]. If the light is polarized by reflection, the polarization is minimal in the plane of reflection, that is, in the plane containing the incident light ray, the surface normal, and the viewing ray. Chen *et al.* [40] uses polarization and phase shifting of light to target surface scanning of transparent objects. Phase shifting enables the separation of specular components from diffuse components [41]. The polarization components of the specular and diffuse components are then used to detect transparent surfaces. Two orthogonally polarized images are used to remove multiple-scattering from the object's surface, thus enabling the recovery of the object's shape.

The detection of shape of single-colored objects is difficult from captured images from a single view. A polarization camera, however, makes it easier to detect differences in the surface orientation of single-colored objects. The polarized state of light from a subject differs depending on the surface orientation of that subject. This is because the unpolarized light upon reflection from the object becomes partially linearly polarized, depending on the surface normal and the refractive index of the medium. These differences can be captured as polarization images with a polarization camera. The constraints on the surface normal can be determined using the Fresnel reflectance model [21]. Figure 6.5 shows the specular reflection of an unpolarized light wave on a surface.

The reflection from a transparent object is mostly specular while reflection from an opaque object is mostly diffused. Because of the specular reflection from transparent objects, the reflected light is partially polarized, while the light reflected from an opaque object is unpolarized. Hence, analysis of light reflected from different objects using a polarizing filter can be used to classify objects into being transparent or opaque. This method of classification can be used in real time in machine vision applications. This method is computationally inexpensive and only a linear polarizing filter together with a camera is required.

In this section, an efficient way to discriminate between transparent and opaque objects using degree of polarization of reflected light is presented. The algorithm works efficiently when the polarization state of the light reflected from the objects is preserved during image capture and analysis.

6.2.2.1 Detection of Transparent and Opaque Objects

A polarization camera is a generalization of the conventional intensity camera and it allows acquisition of polarization information from an object. The polarization camera consists of a normal intensity camera and a polarization filter. Rotation of the polarization filter in front of the camera allows the polarized information to be recorded. A polarization filter has the property either to transmit light vibrating in one direction producing linearly polarized light or to transmit light with a sense of rotation producing circularly polarized light. Basically, a polarization filter translates polarization information into intensity difference information. This is important as humans are able to perceive intensity information but not polarization information. Thus a polarization sensor has to sense the intensity of light, the polarized component of the incident light, and the angle of the linearly polarized component. State-of-the-art polarization image sensors are based either on a standard CMOS/CCD camera coupled with an external polarization filter controlled externally, or on integrated polarization filters fabricated on top of the pixel array. The latter can measure polarization information in real time and increase the speed of sensing polarization components of the reflected light. Embedded wire grid polarizers on the imaging sensors for real time polarization measurements are described in Refs [6, 42, 43].

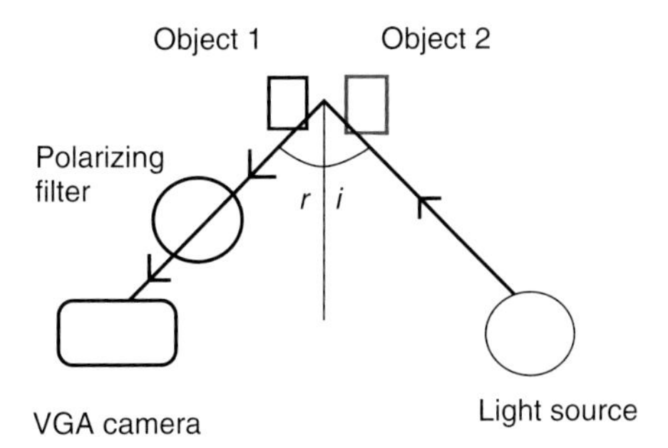

Figure 6.8 Experimental setup.

Theoretically, transparent objects reflect light specularly. So if a transparent object is observed through a polarization filter and the transmission axis of the filter is rotated, a sinusoidal variation in the intensity of light which is reflected from the transparent object should be observed. The experimental setup to detect reflected polarized light is shown in Figure 6.8 [44]. In the figure, '*i*' and '*r*' denote the angle of incidence and angle of reflection, respectively. The light source used was a halogen bulb.

Objects 1 and 2 in the figure correspond to a transparent and an opaque object respectively. The opaque object was chosen such that the specular reflection component from it was small as compared to its diffuse reflection component. The polarizing filter used was an absorption-based polarizer. Absorption-based polarizers are made of materials whose crystals preferentially absorb light polarized in a specific direction. This makes them good linear polarizers. The camera used was a VGA webcam from Logitech. Owing to the camera sensitivity, noise, and disturbance in the polarization state of light due to air molecules, the polarization phenomenon could not be observed if the objects were farther than 25 cm away from the camera. This maximum distance can be improved if imaging conditions are brighter and the camera is more sensitive to lower light intensities. The light source used was a 12 V halogen bulb. Owing to natural illumination, the polarization state of the light reflected from the objects got distorted. So, to capture the lighting effects on polarized images, the images were taken in the dark with almost zero natural illumination. From Eq. (6.9), it is clear that three images with different polarizer orientations are required to determine the components of the polarization imaging. The three orientations of the polarizer filters are often selected to be 0°, 45°, and 90°.

The light reflected from a transparent object is polarized; therefore, on rotating the transmission axis of the polarizing filter, the intensity of the light reflected from transparent object should vary according to Malus' law. Since the angle of incidence at the measured pixel is small (measurement taken at the specular highlight region), it could be assumed that the intensity is directly proportional to square of the cosine of the angle. Using the experimental setup of Figure 6.8, the variations in intensity on changing the transmission axis of the linear polarizer is shown in

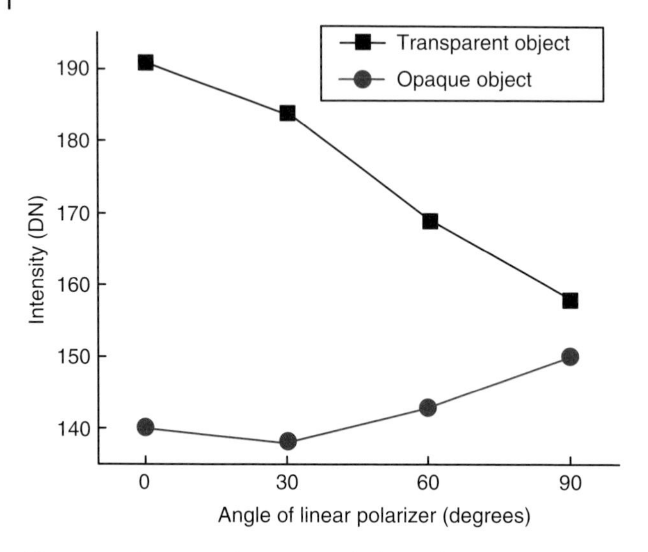

Figure 6.9 Variation of intensity with the angle of the linear polarizer.

the Figure 6.9 [44]. The transmission axis of the polarization filter was varied in steps of 30°.

The intensity variation plot for the transparent object, as expected, has a negative slope in accordance with Malus' law. For the opaque object, no particular trend was expected. But the intensity plot had a positive slope, indicating that the intensity increased with the angle of the transmission axis of the polarizing filter. However, the intensity variation of the object pixels due to the polarization of light in the case of the opaque object was not much higher as compared to the intensity variation obtained in the case of the transparent object.

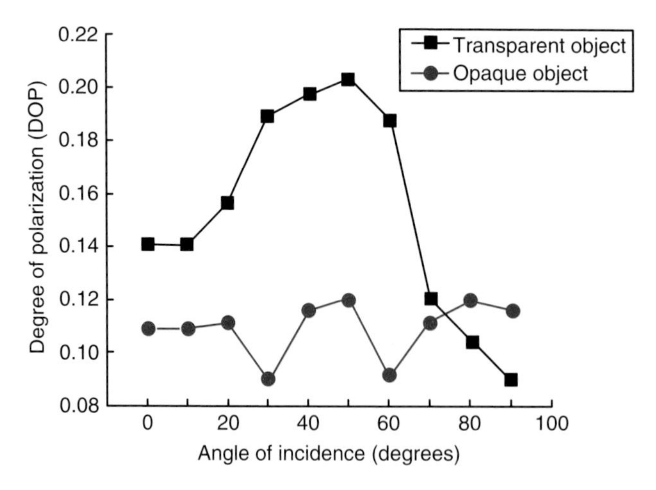

Figure 6.10 Variation in the degree of polarization with angle of incidence.

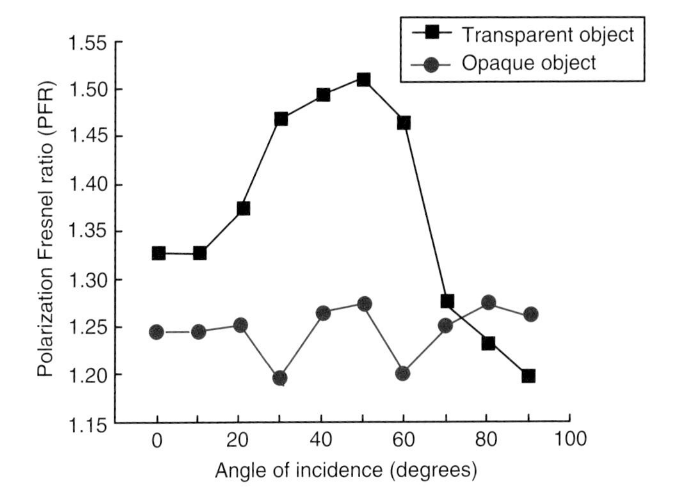

Figure 6.11 Variation of PFR with angle of incidence.

As discussed in Section 6.2.1, the DOP is a measure of the polarized component of light. To measure the DOP, it is assumed that the imaged object has uniform surface texture and refractive index throughout. The degree of polarization then depends only on the angle of incidence. The degree of polarization is expected to be a maximum at Brewster's angle. The variations in the degree of polarization with varying angles of incidence are shown in Figure 6.10 [44].

It can be seen from the figure that as the transmission axis of the linear polarizer is increased, the degree of polarization for the transparent object increases to a maximum value and thereafter decreases. The peak is observed around 50°, which is approximately equal to Brewster's angle for air to glass transition – where the degree of polarization should be theoretically maximum and equal to unity. For opaque objects, the variations observed were random and quite small as compared to that for transparent objects. The magnitude of degree of polarization for opaque objects was always quite lower than that for transparent objects.

The variations in the PFR and Stokes' degree of polarization with varying angles of incidence are shown in Figures 6.11 and 6.12, respectively.

The profiles for Stokes' degree of polarization and polarization Fresnel ratio (Tables 6.1 and 6.2)were similar to the degree of polarization even though their values were different from each other. This similar trend of increase in the PFR and Stokes' degree of polarization with angle of incidence was observed for transparent objects, while for opaque objects, there was not much variation.

The polarization state for the diffuse and specular components of the reflection also depend on the reflecting surface, and the measurement of polarization state of the reflected light serves as an indicator for the type of material surface. In metals, the specular component of reflection is greater than the diffuse component of reflection, while in plastics, the diffuse component dominates the specular component of reflection. This varies the degree of polarization of the reflected light,

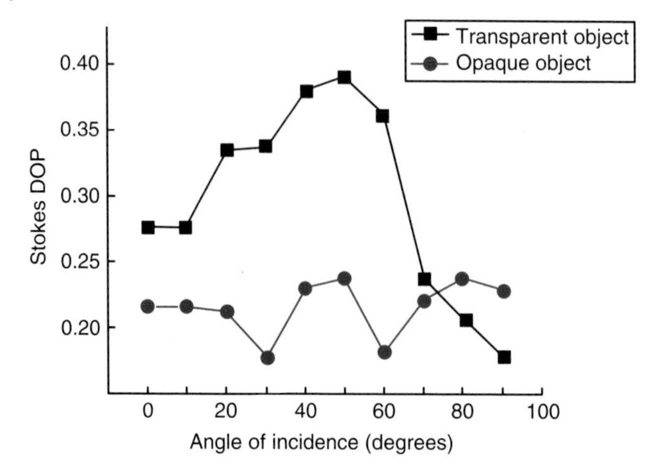

Figure 6.12 Variation of Stoke's degree of polarization with varying angle of incidence.

Table 6.1 Degree of polarization.

Angle of incidence (degrees)	DOP (transparent object)	DOP (opaque object)
0	0.1464	0.08904
10	0.1464	0.08904
20	0.1526	0.1089
30	0.1911	0.08658
40	0.1951	0.1093
50	0.1818	0.08695
60	0.1549	0.0945
70	0.09499	0.1015
80	0.08466	0.09875
90	0.05780	0.05862

Table 6.2 Polarization Fresnel ratio.

Angle of incidence (degrees)	DOP (transparent object)	DOP (opaque object)
0	1.3430	1.1955
10	1.3430	1.1955
20	1.3602	1.2444
30	1.4725	1.1896
40	1.4848	1.2454
50	1.4444	1.1905
60	1.3666	1.2087
70	1.2099	1.2259
80	1.1850	1.2191
90	1.1227	1.1245

Table 6.3 Stokes' degree of polarization.

Angle of incidence (degrees)	DOP (transparent object)	DOP (opaque object)
0	0.2866	0.1767
10	0.2866	0.1767
20	0.3233	0.2143
30	0.3687	0.1718
40	0.3759	0.2159
50	0.352	0.1726
60	0.3023	0.1873
70	0.1884	0.2009
80	0.1683	0.1955
90	0.1154	0.1168

which is shown to be able to differentiate among various reflecting surfaces on the basis of intensity variations [22] (Table 6.3).

6.2.2.2 Shape Detection Using Polarized Light

The detection of shape of a single-colored objects is difficult from captured images from a single view. A polarization camera, however, makes it easier to detect differences in the surface orientation of single-colored objects. The polarized state of light from a subject differs depending on the surface orientation of that subject. This is because the unpolarized light upon reflection from the object becomes partially linearly polarized, depending on the surface normal and the refractive index of the medium. These differences can be captured as polarization images with a polarization camera. The constraints on the surface normal can be determined using the Fresnel reflectance model [21] as described in Section 6.2. The surface normals are computed using the polarized images and then the 3D surface of the object is obtained by integrating the surface normals [45]. Miyazaki *et al.* [36] uses two polarization images taken from two different directions: one polarization image is taken before the object is rotated and the other polarization image is taken after rotating the object through a small angle. The degree of polarization of identical points on the object surface is compared between the two images. The correspondence between the two images is then used to estimate the shape of the transparent object.

Polarization of light is also used to recover the surface shape of transparent objects [36]. The two optically important directions of linear polarization after reflection from the surface are the directions perpendicular and parallel to the surface plane of incidence. After reflection from the surface, the direction of the polarization perpendicular and parallel to the plane of incidence do not change; however the phase and the intensity of the light changes depending on the direction of polarization and the angle of incidence. The degree of polarization of light reflected from the object's surface, depends on the reflection angle which, in turn, depends on the object's surface normal. Thus by measuring the degree

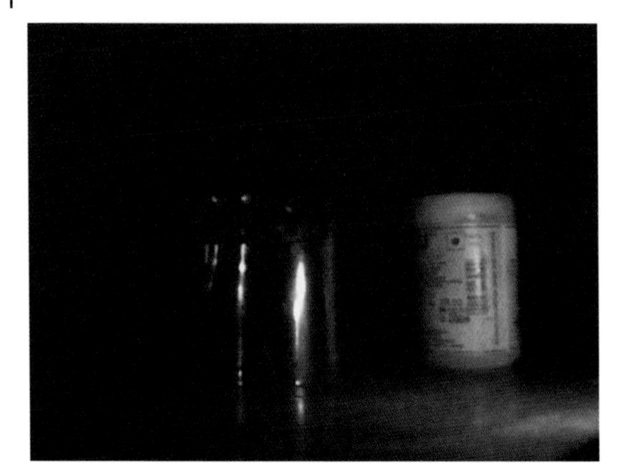

Figure 6.13 Original test image.

of polarization, the surface normal of the object can be calculated. The potential applications of the shape determination of a transparent object includes object recognition for single-colored objects in a production line. The polarization camera also makes it easier to detect the presence of foreign material.

As observed in the previous subsections, the magnitude of degree of polarization (also Stokes' DOP and PFR) is significantly higher for image pixels corresponding to transparent objects as compared to the rest of the image pixels. This has been used to segment the transparent object from the background. Figure 6.13 shows the original image containing a transparent and an opaque object illuminated by a small halogen light source.

DOP values were obtained for every image pixel by taking two images (angle of rotation of polarizer 0° and 90°) and using Eq. (6.20). The matrix containing DOP values was converted into a grayscale image which is shown in Figure 6.14. The image can be converted into a binary image by taking a low threshold value and this image can be analyzed in a similar way thereafter. It can be seen that transparent object pixels are more "white" than the remaining image pixels, including the background and that opaque objects are "black." Using a suitable threshold and some image-processing operations such as region segmentation, morphological operations (dilation, erosion, closing, and opening), and edge detection, the approximate shape of the object could be determined even in a dark imaging environment as shown in Figure 6.15.

The method can be used to determine the shape of transparent objects. Since measures of polarization are easy to calculate and need less equipment, this is a computationally inexpensive and economical method. To guide autonomous agents, a system to capture images at required angles of rotation is desirable. A motorized polarizer should be installed in front of the camera. For taking images at a specific angle of rotation, a control circuit for synchronizing the camera and the polarizer is also required. Images acquired by this system should then be analyzed by the processor responsible for image-processing operations for the agent.

Figure 6.14 Image formed by degree of polarization matrix.

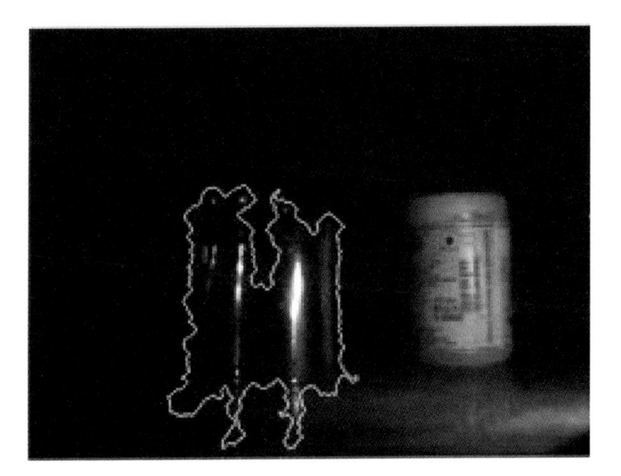

Figure 6.15 Classified transparent object after applying image-processing functions on the polarization matrix image.

With minor operations, the location and shape of the transparent object can be detected. The processor can thus control the navigation of the autonomous agent appropriately.

6.3 High-Speed Motion Detection

As discussed in Section 6.1.1, flying insects are able to detect fast motion in the visual scene and avoid collision using low-level image processing and little computational power. This makes their visual processing interesting for real-time motion/collision detection in machine vision applications. Real-time detection of targets with minimal energy consumption for surveillance using wireless sensor nodes is becoming popular. The sensor nodes consist of multiple autonomous functional units that cooperatively monitor the desired property.

A camera as a sensor node in the wireless sensor network would be desirable as it can be used in many applications such as surveillance, environment monitoring, smart homes, virtual reality, and medicine. As charge coupled devices (CCDs) and complementary metal oxide semiconductor (CMOS) images/video cameras become smaller, less expensive, and more power efficient, the cameras in wireless sensor nodes will become more powerful and widespread in use. The CMOS image sensor is more suited for application in the wireless sensor node to CCD owing to its low power operation and possibility of incorporating processing power within the pixel. The in-pixel processing of sensed information helps in reduction of the signal information being sent out and thus reduces the amount of data transmitted over sensor nodes. A wireless sensor node with an integrated image sensor would have the following requirements: (i) low power consumption (in microwatts), (ii) high dynamic range (>100 dB), (iii) digital interface with few or no extra components, (iv) data compression at sensor level, (v) low cost, and (vi) fast readout.

The power consumption of the vision sensors used in wireless sensor nodes should be minimal. Therefore, computationally intensive vision algorithms cannot be used as they are typically power hungry. Conventional vision sensors such as CMOS or CCD image sensors are not suitable for wireless node applications as they sample the visual field at tens to thousands of times per second, generating kilobytes or megabytes of data for each frame. The resulting information must be heavily processed to derive any result, resulting in complex systems and increased power consumption. Furthermore, these vision sensors are not suitable for all visual environments as most of the available vision sensors cannot handle low-contrast environments. The dynamic range of the available vision sensors is limited and thus cannot image the entire available light range. The dynamic range of a sensor expresses the ability to capture dark as well as bright sections within an image while maintaining the highest fidelity. The readout speed of the vision sensors in wireless sensor nodes also needs to be very high to be able to detect any changes in the visual scene. For fast readout of images, high-speed readout circuits are needed, which are power hungry. Further, as processing algorithms become more and more demanding, processing capabilities reach a bottleneck in realizing efficient vision sensor nodes.

Most of the available vision sensors are modeled on the basis of the single-aperture human eye. As discussed in Section 6.1.1, the compound eye is one of the most compact vision systems found in nature. This visual system allows an insect to fly with a limited intelligence and brain-processing power. In comparison to single-aperture eyes, compound eyes have a much wider field of view *(FOV)*, better capability to detect moving objects, higher sensitivity to light intensity, but much lower spatial resolution. Thus this type of vision system is ideal for wireless sensor nodes. A vision sensor for wireless sensor networks that performs local image capture and analysis has been presented [46]. The vision sensor was shown to be useful in object detection and gesture recognition. The wireless sensor node proposed by Kim *et al.* [47] was operated in rolling-shutter mode. In rolling-shutter mode, imaging a moving object introduces motion artifacts. The power consumption of 1 mW by the image sensor was also on a higher side with the output buffer alone consuming about 72% of the power [47].

In this section, some bioinspired properties that can be implemented in a wireless sensor node are discussed. Motion detection using temporal differentiation is used. The temporal differentiation in binary domain replicates the *"flickering effect"* of the insect eye when the object is in motion. The binary motion detection can be very fast and energy efficient because of considerable reduction in the output bandwidth of information.

6.3.1 Temporal Differentiation

Temporal differentiation helps to attain high dynamic range, data compression at the sensor level, and fast readout. The active differential imaging process involves getting the image of a scene at two different instances of time and subtracting the two frames as shown in Figure 6.16(a). The two images are separately captured and stored in the memory. The subtraction of the two images can either be done in analog or digital domain. In temporal differentiation, the stationary objects with constant illumination will be erased by the imaging system. If there is a motion

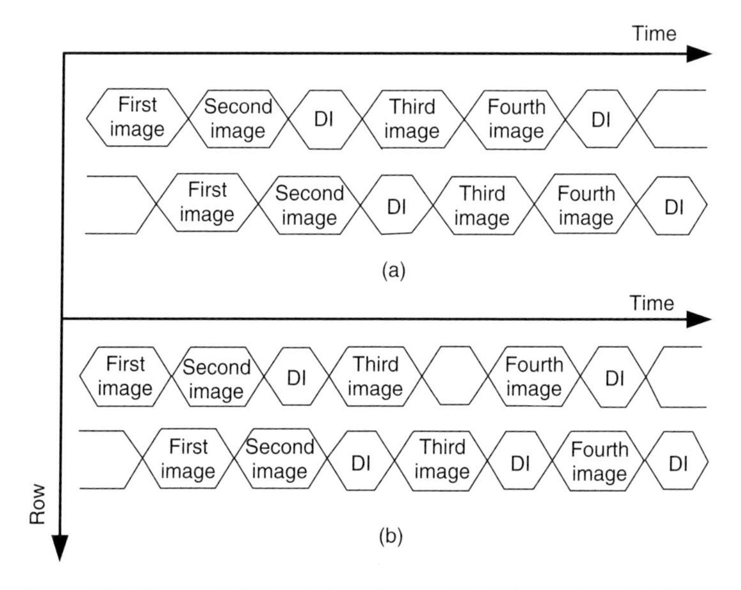

Figure 6.16 Temporal differentiation schemes (DI – difference image). (a) Differential imaging and (b) visual tracking or saccadic imaging.

between the two frames, then the motion information will be recorded. This serves in imaging high light environments. These algorithms are widely used in industrial and scientific fields for motion detection, object tracking, or recognition and so on.

A variant of temporal differentiation is the visual tracking mode or saccadic imaging used by insects as shown in Figure 6.16(b). Here the background information is stored and the current image frame is subtracted from the background image to determine change in state or motion in the visual field. These processes are very apt for fast tracking. In this model, the background information is always maintained. The temporal difference image enhances the background subtraction capability and thus is very useful in motion detection. Some of the applications of this invariance to ambivalent light are human face recognition, pseudo 3D sensing, and so on.

The use of digital subtractor would need a fast readout circuit together with a frame buffer. These systems are highly complex and they consume a lot of power. A differential image sensor with high common mode rejection ratio (CMRR) was proposed by Innocent and Meynants [48], where both the images are stored in the analog domain and a difference signal is obtained during readout. Common-mode rejection ratio of a device is the measure of the tendency of the device to reject a common input signal. A high CMRR is desired for applications where the signal of interest is a small voltage fluctuation superimposed on a large voltage offset or when the required information is contained in the voltage difference of two signals.

A high-CMRR differential CMOS imager sensor, capable of producing both analog and digital differential images using two in-pixel analog and digital memories has been designed. The details of the CMOS image sensor are presented elsewhere [6]. The subtraction is done pixel by pixel and the analog image signal is transferred to an external 12-bit ADC, to get the digital pixel data. The process is repeated for each image line and thus a differential image is obtained without the requirement for an off-chip computation [6]. Pixel-level processing helps in decreasing the data rate from the sensor, compared to frame-level processing. This often results in compact, high-speed, and low-power solutions. Focal plane computations are also free from temporal aliasing, which is usually a problem in motion detection algorithms.

However, processing analog/digital information takes computational power and resources. The compound eyes of the insects are known to process motion information in binary format aided by the *"flickering effect"* in the eyes when the object moves from one eye unit to another. Figure 6.17 shows the horizontal motion detection using binary optical flow information. Two consecutive frames of a light source moving over the image sensor are shown in Figure 6.17. The first image shows the light source at its initial position and the second image shows it after a slight movement. The two images look very similar, as only a very small motion was introduced. The histograms of the two images are also shown in Figure 6.17. The subtraction of the two images results in a difference image, and the histogram

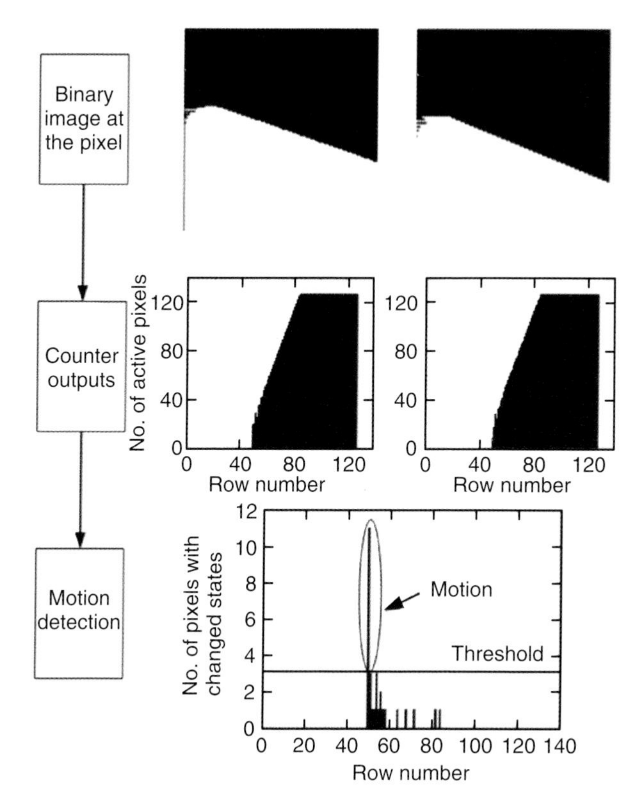

Figure 6.17 Horizontal motion detection using spatially integrated binary optical flow.

of the pixels which changed states are shown at the bottom of the figure. By selecting a proper threshold, accurate detection of motion can be done [49, 50]. This can detect motion very fast and also as only binary information are produced; pixel-level computations are simple and can be easily implemented. The power consumption of the system will also be very low. Thus such algorithms are best for being adopted in the visual sensor node.

Temporal differentiation can also be used to enhance the dynamic range of the system. It is also used to compress pixel information as shown in Figure 6.18. Figure 6.18(a) and (b) shows temporal differential images and Figure 6.18(c) and (d) shows histogram equalization (HE) of the temporal images. Figure 6.18(a) is brighter than (b) as the integration time difference between the temporal samples is larger. 6.18(b) has fewer pixels near saturation, thus able to image higher illuminations, increasing the dynamic range. After HE, the two images have near equal distribution of the contrast. This can be used as an on-chip image compression tool where only the brighter pixels can be scanned for readout, thus increasing the readout speed. This will increase the output bandwidth and also help with power consumption as not all the pixels will be read.

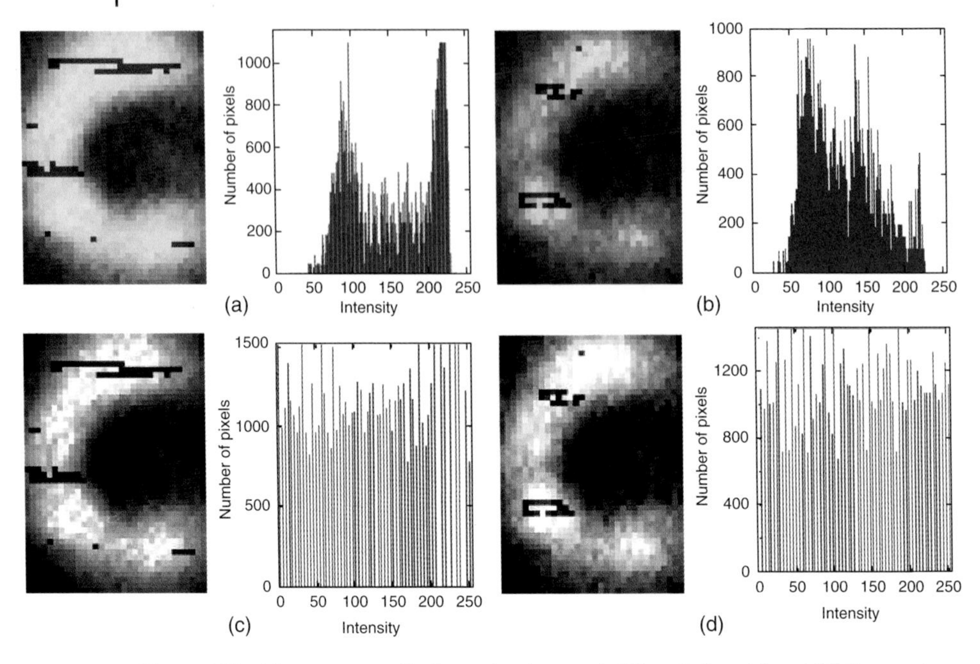

Figure 6.18 Histogram equalization using temporal differentiation. (a) and (b) Temporal images with the first frame captured at 10 ms and the second frame 20 ms, respectively; (c) and (d) Histogram equalization of the temporal images.

6.4
Conclusion

In this chapter, algorithms inspired by the compound eyes of insects are presented, that could be modeled to enhance the visual perception of standard camera systems. Separation of transparent and opaque objects in a visual scene is important in many machine vision applications. The incident light on a surface is reflected and this reflected light is found to be polarized. The polarization of the reflected light depends on the surface properties of the object. Using DOP, PFR, and Stokes' degree of linear polarization, transparent and opaque objects can be identified in a visual scene with relative ease. For most incident angles, the DOP values for transparent objects are quite higher when compared to those for opaque objects. These measurements could be used to differentiate among transparent and opaque objects by using an appropriate threshold in real time. Further, the polarization matrix can also be used to recover the shape of the transparent object as well. This method is computationally inexpensive and relatively easy to implement in real-time vision systems. The polarization sensing ability in the CMOS image sensor for object differentiation can be useful for many applications such as environment monitoring and surveillance.

Algorithms inspired by the compound eyes of insects for high-speed motion detection for the visual sensor nodes in wireless networks are also presented.

A CMOS image sensor operating in temporal differential mode which spatially integrates 1D binary information can be very useful for fast motion detection. The binary information can be generated in-pixel, thus reducing the computational overload making them low-power systems. Temporal differentiation can also be used to compress information.

References

1. Visual perception, *http://en.wikipedia.org/wiki/Visual_perception* (accessed 4 May 2015).
2. Sabra, A.I. (1989) *The Optics of Ibn al-Haytham*, Warburg Institute, London, ISBN: 0854810722.
3. Sabra, A.I. (2003) Ibn al-Haytham, brief life of an Arab mathematician: died circa 1040. Harvard Magazine, pp. 54–55.
4. Sabra, A.I. (1966) Ibn al-Haytham's criticisms of Ptolemy's Optics. *J. Hist. Philol.*, **4**, 145–149.
5. Nilsson, D.E. (2009) The evolution of eyes and visually guided behaviour. *Philos. Trans. R. Soc. London, Ser. B: Biol. Sci.*, **364** (1531), 2833–2847.
6. Sarkar, M. (2011) A biologically inspired CMOS image sensor. PhD dissertation, ISBN: 978909025980-2.
7. Rieke, F. and Baylor, D.A. (1993) Single-photon detection by rod cells of the retina. *Rev. Mod. Phys.*, **70** (3), 1027.
8. Land, M.F. and Fernald, R.D. (1992) The evolution of eyes. *Annu. Rev. Neurosci.*, **15**, 1–29.
9. *http://watchingtheworldwakeup.blogspot.ae/2009/11/amazing-housefly-part-2-coolest-eye.html.*
10. Land, M.F. and Nilsson, D.E. (2012) *Animal Eyes*, 2nd edn, Oxford University Press, Oxford, ISBN: 978-0199581146.
11. Nagata, T., Koyanagi, M., Tsukamoto, H., Saeki, S., Isono, K., Shichida, Y., Tokunaga, F., kinoshita, M., Arikawa, K., and Terakita, A. (2012) Depth perception from image defocus in a jumping spider. *Science*, **335** (6067), 469–4071.
12. Srinivasan, M.V., Zhang, S.W., Chahl, J.S., Barth, E., and Venkatesh, S. (2000) How honeybees make grazing landings on flat surfaces. *Biol. Cybern.*, **83**, 171–183.
13. Srinivasan, M.V. (1992) Distance perception in insects. *Curr. Dir. Psychol.*, **1** (1), 22–26.
14. Goldsmith, T.H. (2006) What birds see. *Sci. Am.*, **295**, 68–75.
15. Franceschini, N. (1985) Early processing of colour and motion in a mosaic visual system. *Neurosci. Res.*, (Suppl. 2), S17–S49.
16. Hardie, R.C. (1986) The photoreceptor array of the dipteran retina. *Trends Neurosci.*, **9**, 419–423.
17. Nachtigall, W. and Wilson, D.M. (1967) Neuro-muscular control of dipteran flight. *J. Exp. Biol.*, **47**, 77–97.
18. Heide, G. (1983) Neural mechanisms of flight control in Diptera, in *Biona-Report 2* (ed. W. Nachtigall), pp. 35–52.
19. Reichardt, W. (1987) Evaluation of optical motion information by movement detectors. *J. Comp. Physiol. A*, **161** (4), 533–547.
20. Mäthger, L.M., Shashar, N., and Hanlon, R.T. (2009) Do cephalopods communicate using polarized light reflections from their skin? *J. Exp. Biol.*, **212** (14), 2133–2140.
21. Wolff, L.B. (1990) Polarization-based material classification from specular reflection. *IEEE Trans. Pattern Anal. Mach. Intell.*, **12** (11), 1059–1071.
22. Sarkar, M., San Segundo Bello, D., Van Hoof, C., and Theuwissen, A. (2011) Integrated polarization analyzing CMOS image sensor for material classification. *IEEE Sens. J.*, **11** (8), 1692–1703.
23. Ihrke, I., Kutulakos, K.N., Lensch, H.P.A., Magnor, M., and Heidrich, W. (2010) Transparent and specular object reconstruction. *Comput. Graph. Forum*, **29** (8), 2400–2426.
24. Kutulakos, K.N. and Steger, E. (2005) A theory of refractive and specular 3D

shape by Light-Path Triangulation. IEEE ICCV, pp. 1448–1455.

25. Schultz, H. (1994) Retrieving shape information from multiple images of a specular surface. *IEEE Trans. Pattern Anal. Mach. Intell.*, **16** (2), 195–201.
26. Halstead, M.A., Barsky, B.A., Klein, S.A., and Mandell, R.B. (1996) Reconstructing curved surfaces from specular reflection patterns using spline surface fitting of normals. Proceedings of ACM SIGGRAPH, pp. 335–342.
27. Bonfort, T. and Sturm, P. (2003) Voxel carving for specular surfaces. Proceedings of IEEE International Conference on Computer Vision (ICCV), pp. 591–596.
28. Blake, A. and Brelstaff, G. (1988) Geometry from specularity. Proceedings of IEEE International Conference on Computer Vision (ICCV), pp. 297–302.
29. Blake, A. (1985) Specular stereo. Proceedings of International Joint Conference on Artificial Intelligence (IJCAI), pp. 973–976.
30. Oren, M. and Nayar, S.K. (1996) A theory of specular surface geometry. *Int. J. Comput. Vision*, **24** (2), 105–124.
31. Hartley, R. and Zisserman, A. (2000) *Multiple View Geometry*, Cambridge University Press, Cambridge.
32. Nayar, S.K., Sanderson, A.C., and Simon, D. (1990) Specular surface inspection using structured highlight and Gaussian images. *IEEE Trans. Rob. Autom.*, **6** (2), 208–218.
33. Fritz, M., Black, M., Bradski, G., Karayev, S., and Darrell, T. (2009) An additive latent feature model for transparent object recognition. Neural Information Processing Systems.
34. Klank, U., Carton, D., and Beetz, M. (2011) Transparent object detection and reconstruction on a mobile platform. IEEE International Conference on Robotics and Automation, p. 5971.
35. Nilsson, D.E. and Warrant, E. (1999) Visual discrimination: seeing the third quality of light. *Curr. Biol.*, **9**, R535–R537.
36. Miyazaki, D., Kagesawa, M., and Ikeuchi, K. (2004) Transparent surface modeling from a pair of polarization images. *IEEE Trans. Pattern Anal. Mach. Intell.*, **26** (1), 73–82.
37. Ferraton, M., Stolz, C., and Meriaudeau, F. (2008) Surface reconstruction of transparent objects by polarization imaging. IEEE International Conference on Signal Image Technology and Internet based Systems, pp. 474–479.
38. Al-Qasimi, A., Korotkova, O., James, D., and Wolf, E. (2007) Definitions of the degree of polarization of a light beam. *Opt. Lett.*, **32**, 1015–1016.
39. Saito, M., Sato, Y., Ikeuchi, K., and Kashiwagi, H. (1999) Measurement of surface orientations of transparent objects using polarization in highlight. Proceedings of IEEE Conference on Computer Vision and Pattern Recognition (CVPR), vol. 1, pp. 381–386.
40. Chen, T., Lensch, H.P.A., Fichs, C., and Seidel, H.P. (2007) Polarization and phase-shifting for 3d scanning of translucent objects. Proceedings of IEEE Computer Society Conference on Computer Vision and Pattern Recognition (CVPR), pp. 1–8.
41. Nayar, S.K., Krishnan, G., Grossberg, M.D., and Raskar, R. (2006) Fast separation of direct and global components of a scene using high frequency illumination. Proceedings of ACM SIGGRAPH, pp. 935–944.
42. Tokuda, T., Yamada, H., Sasagawa, K., and Ohta, J. (2009) Polarization-analyzing CMOS image sensor with monolithically embedded polarizer for microchemistry systems. *IEEE Trans. Biomed. Circuits Syst.*, **3** (5), 259–266.
43. Sarkar, M., San Segundo Bello, D., Van Hoof, C., and Theuwissen, A. (2010) Integrated polarization-analyzing CMOS image sensor. Proceedings of IEEE International Symposium on In Circuits and Systems (ISCAS), pp. 621–624.
44. Mahendru, A. and Sarkar, M. (2012) Bio-inspired object classification using polarization imaging. International Conference on Sensing Technology (ICST), pp. 207–212.
45. Morel, O., Meriaudeau, F., Stolz, C., and Gorria, P. (2005) Polarization imaging applied to 3D reconstruction of specular metallic surfaces. Electronic Imaging, International Society for Optics and Photonics, pp. 178–186.

46. Rahimi, M., Baer, R., and Iroezi, O.I. (2005) Cyclops: in situ imagesensing and interpretation in wireless sensor networks. Proceedings of the 3rd International Conference on Embedded Networked Sensor Systems.
47. Kim, D., Fu, Z., Park, J.H., and Culurciello, E. (2009) A 1-mW temporal difference AER sensor for wireless sensor networks. *IEEE Trans. Electron. Devices*, **56** (11), 2586–2593.
48. Innocent, M. and Meynants, G. (2005) Differential image sensor with high common mode rejection. Proceedings of European Solid-Sate Circuits Conference, pp. 483–486.
49. Sarkar, M., San Segundo, D., Van Hoof, C., and Theuwissen, A.J.P. (2011) A biologically inspired CMOS image sensor for polarization and fast motion detection. Proceedings IEEE Sensors Conference, pp. 825–828.
50. Sarkar, M., San Segundo, D., Van Hoof, C., and Theuwissen, A.J.P. (2013) Biologically inspired CMOS image sensor for fast Motion and polarization detection. *IEEE Sens. J.*, **13** (3), 1065–1073.

7
Biomimetic Vision Systems

Reinhard Voelkel

7.1
Introduction

Biomimetics, combined from the Greek words *bios*, meaning life, and *mimesis*, meaning to imitate, is the imitation of the models, systems, and elements of nature for the purpose of solving complex human or technical problems [1]. How could nature inspire an engineer designing next generations of camera and computer vision systems? This depends on the application and restrictions for space and resources. If size, costs, computing power, and energy consumption do not matter, the task is relatively simple. The engineer will combine leading edge components and build a bulky expensive system providing high speed and high resolution. When the vision system needs to be very small, very fast or has to operate with minimum resources, the task gets more challenging. Then, nature is an excellent teacher. The smaller a creature is, the more expensive vision becomes for it and the more carefully the animal vision system needs to be adapted to size, brain, metabolism, and behavior [2]. Nature has done a great job in optimizing animal vision systems to the very limits.

The oldest known animal vision systems date back to the Cambrian period, about 530 million years ago. In the Cambrian period, also referred to as the "age of the trilobites", life on earth made a huge jump forward and a rich fauna of animals evolved. This phenomenon is usually known as the Cambrian explosion. A very likely scenario is that the appearance of the first eyes was the trigger for this radical evolutionary event. Visually guided predation could have caused soft-bodied wormlike creatures of the Precambrian period to evolve to much larger, essentially modern types of animals. Vision might have created a tremendous selection pressure on all species to either quickly evolve vision themselves or at least evolve protective measures such as body armor, high mobility, or other sensors. Eyes might have been the decisive key enabling technology triggering the radical evolution of all species. From simple ocelli to large vertebrate eyes, a large variety of animal eyes evolved within this short period of only 5 million years [2]. Astonishingly, most of the intermediate solutions in eye evolution are still used today. In technology, a user will always opt for the highly developed end

Biologically Inspired Computer Vision: Fundamentals and Applications, First Edition.
Edited by Gabriel Cristóbal, Laurent Perrinet, and Matthias Keil.

product. In nature, only larger animals opted for high-resolution large field-of-view (FOV) single-aperture eyes. Typically, the smaller an animal is, the more primitive the animal's eye system is. This is related to the "scaling law of optics," stating that scaling down of a lens system drastically reduces its performance [3]. The scaling law of optics explains why it is so difficult to design miniaturized vision systems.

Understanding the scaling law of optics is mandatory for all engineers designing vision systems. Especially, as scaling down of optics follows different rules than scaling down in electronics. In electronics, scaling down of components and electric wiring is a very powerful strategy to improve the performance. Computers have evolved from room-filling and low-performance to highly miniaturized, lightweight, and very powerful devices. Over the last 50 years, the semiconductor industry has made incredible progress by making electronics smaller and smaller, down to the very limits, the atomic level. Following this success story, it would be logical to suppose that scaling down of optics would allow a miniaturized camera with very high resolution to be built. However, optics does not scale like electronics. The miniaturization of a vision system always means a significant loss of image quality and fidelity [4]. Miniaturizing the pixels of an image sensor below the wavelength is useless. Scaling down optics and building perfect miniaturized computer vision systems require different strategies. Nature has struggled with the scaling law of optics for 530 million years and, for example, replaced single-aperture eye by cluster or compound eyes. This is the main focus of this chapter. How natural concepts for miniaturization could be imitated for building computer vision systems with perfect adaption to small size, special tasks, and specific applications.

7.2 Scaling Laws in Optics

7.2.1 Optical Properties of Imaging Lens Systems

The basic parameters of a lens system are the lens pupil diameter D, the focal length f, and the size XY of the image sensing unit, as shown schematically in Figure 7.1. The power of a lens is given by its f-number $N = (f/D) = (1/(2\sin(u))) = (1/(2NA))$, defined as the ratio of focal length f to lens pupil diameter D and proportional to the reciprocal of the numerical aperture $NA = \sin(u)$. The smallest feature a perfect lens with a circular aperture can make is the diffraction-limited spot, referred to as the Airy disk. The size of the Airy disk (diameter to the first zero) is approximately $\delta x = 2.44 \cdot \lambda(f/D)$, thus related to the lens diameter D, the focal length f, and the wavelength λ only. For the photographic lens shown in Figure 7.2(a) with f-number $f/2.8$, the size of the diffraction-limited spot is $\delta x \approx 4$ µm using green wavelength where our eyes and most image sensors see most brightly.

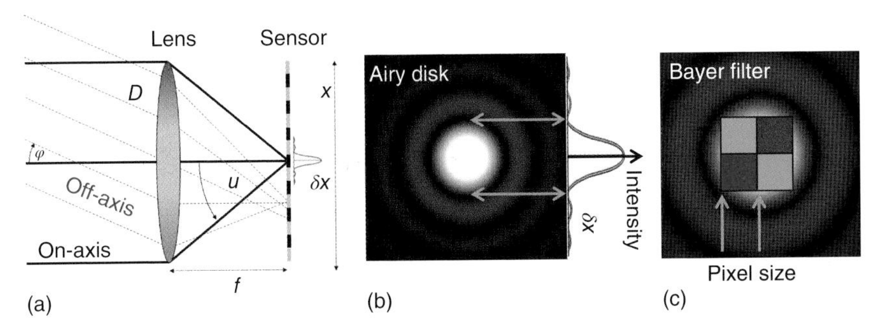

Figure 7.1 (a) Scheme of a lens with lens pupil diameter *D*, focal length *f*, and angle u for the maximum cone of rays contributing to the focal spot δx (Airy disk); (b) intensity distribution Airy disk free of aberrations; and (c) 2 × 2 pixels of image sensor with Bayer color filter covering the size of the Airy disk.

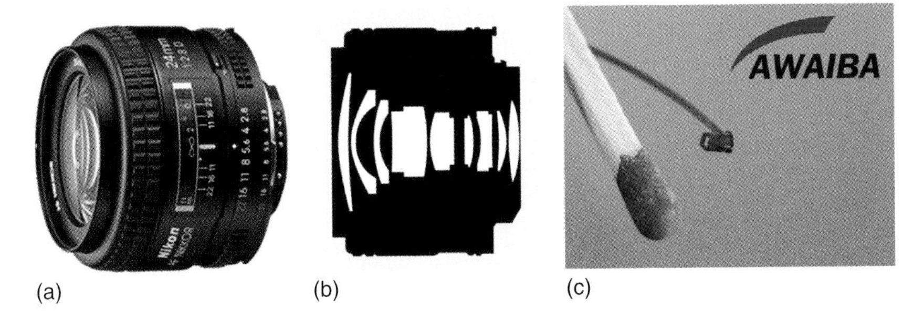

Figure 7.2 (a) Photographic lens (Nikon) with focal length $f = 24$ mm and *f*-number of *f*/2.8, often written as 1 : 2.8; (b) schematic view of the optical design of the photographic lens shown (a); and (c) a pair of miniaturized *NanEye* cameras (Awaiba) with *f*/2.7, a focal length $f = 0.66$ mm, and a lens diameter of $D = 0.244$ mm.

The diffraction limit is the fundamental limit of resolution for lens systems. In practice, the performance of a lens system is further reduced by aberrations, like distortion, spherical aberration, coma, astigmatism, and defocus. Unfortunately, these aberrations scale with the second, third, or even fourth power of $\sin(u)$. A "fast lens" system with a small *f*-number of *f*/1.4 requires very elaborate and expensive aberration corrections. Aberration correction is achieved by multiple lenses of different forms and glass material, as shown in Figure 7.2(a) and (b). For perfectly corrected lens systems, the image disturbance due to aberration is smaller than the size of the diffraction-limited spot for the full image field. The lens is referred to as *diffraction limited*.

For the sake of simplicity, it is assumed that each diffraction-limited spot, that is, each Airy disk, is captured by a matrix of 2 × 2 pixels of the image sensor, as shown in Figure 7.1(c). For square image pixels, the hypotenuse of each pixel will then be $\delta x/2$. For color detection, each pixel of the 2 × 2 matrix is equipped with a color filter. In practice, a 2 × 2 Bayer filter, consisting of two green, one red, and one blue filters, is applied. The usage of two green elements mimics the physiology

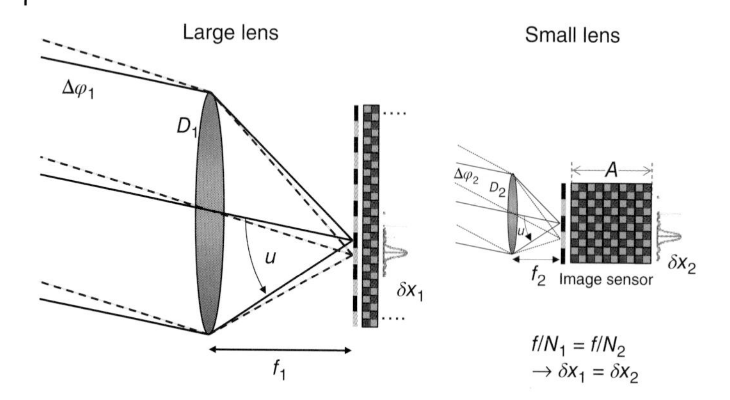

Figure 7.3 Schematic illustration of the scaling law of optics. Lenses with an equal *f*-number have an equal best focused diffraction-limited spot of δx size (Airy pattern). Thus, the smallest resolved feature in the image is independent from the lens scaling for different lenses with an equal *f*-number. However, the number of resolvable pixels in the image plane, also referred to as space–bandwidth product (SW), is drastically reduced for a small lens, as the size of the image scales with the lens size.

of the human eye and the wavelength sensitivity curve of image sensors where the green light detection offers highest brightness.

7.2.2 Space Bandwidth Product (SW)

The smaller a lens is, the easier it is to achieve diffraction-limited performance. This will be illustrated for the examples of the two different camera systems shown in Figure 7.2: (a) a large photographic camera lens and (c) a miniaturized camera for stereoendoscopes. Both lenses have an equal FOV, equal *f*-numbers of *f*/2.8, and an equal diffraction-limited image spot size around 4 μm for green light. In this case, the number of diffraction-limited pixels of the photographic lens is about 54 million pixels (6000 × 9000) for an image field of 24 mm × 36 mm. The sharpness of the image is significantly higher than the spatial resolution of today's top-level digital photographic cameras providing 36 million pixels with approximately 5 μm size. The photographic lens consists of 9 lens groups with a total of 13 lenses, shown schematically in Figure 7.2(b). For the miniaturized NanEye camera, shown in Figure 7.2(c), the number of recorded image pixels is 62 thousand pixels (250 × 250). The NanEye camera consists of a simple plano-convex lens and a pupil. Both lenses have the same diffraction-limited resolution. The miniaturized lens is much easier to manufacture, but the number of image pixels is drastically reduced. Scaling down of a vision systems impacts the transmitted information about the object space [3, 5]. This relationship is referred to as the *space–bandwidth product* (SW) and shown schematically in Figure 7.3.

The scaling law of optics is a fundamental restriction in designing cameras for consumer applications like smartphones or tablets. Due to industrial design constraints, the overall package of a mobile phone camera should not be thicker than 5 mm, leading to focal length $f < 4$ mm. This constraint limits the lens diameter to $D \approx 1.4$ mm for $f/2.8$. Typically, the corresponding image sensors are around 4 mm × 3 mm with a pixel size from 1.4 to 1.75 μm. This example illustrates why it is impossible to build mobile phone cameras with more than 10 megapixels and fit them into a package of less than 5 mm height. Customers' demands for higher and higher resolution of mobile phone cameras require new strategies for image capturing, for example, using camera arrays. In Section 7.3.4, will be explained how nature has perfectly mastered this size–resolution restriction for jumping spiders.

7.3
The Evolution of Vision Systems

The very basic unit of every vision system is the light-sensing unit, the sensor that converts the incident light to electrical signals. Whereas plants and bacteria have evolved a variety of different light-harvesting and light-sensing molecules, all animals share the same absorber, the opsin protein. Thus, it is very likely that the last common ancestor of all animals evolved a light-sensing opsin long before eyes appeared in the Cambrian period [2]. The simplest form of an eye is a single or multiple photoreceptor cells, referred to as an *eye spot*, which can only sense the presence of ambient light intensities. The animals can distinguish light from dark. This is sufficient for animals to adapt to the circadian rhythm of day and night. The technical equivalent to eye spots is the ambient light sensor in consumer products like smartphones and tablets. Their only purpose is to augment or dim the brightness of the display for optimum visibility with minimum energy consumption. The next step in the evolution of animal eyes is a *pit eye*, where the photoreceptor cells are caved into a cup shape. The cup of the pit eye shields the light-sensitive cells from exposure in all directions except for the cup opening direction. An enhancement of the light direction detection is the *pinhole camera* eye, a pit eye, where the edges of the cup go together to form a small opening. The size of the pinhole defines the angular resolution, an array of photoreceptors form the retina. Small pinholes correspond to a high resolution, but low light passes through the pinhole. To protect the inner chamber of the pinhole camera from contamination and parasitic infestation, a transparent cell closing the pinhole evolved. These transparent cells then evolved to a lens able to collect light more efficiently.

Figure 7.4 shows schematically the evolution of eyes from (a) a simple photoreceptor to (b) a pit or pinhole eye, (c) a high-resolution focused camera-type eye, and (d) a reflector eye for low-light environment [2].

Interestingly, all intermediate solutions are still represented in animal species living today. This is surprising. Why did selection pressure not force all animals to develop a perfect high-resolution eye? Regarding the costs for making eyes, the required brain power for vision, and the scaling laws for optics relating resolution

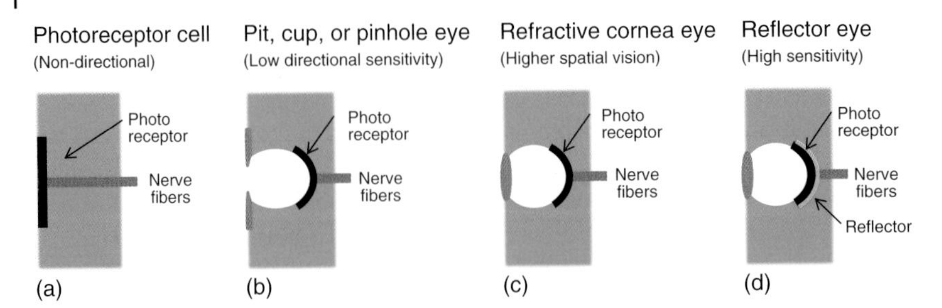

Figure 7.4 Eye evolution from a simple one-cell photoreceptor to a high-resolution focused camera-type eye. It is assumed that the whole process of eye evolution was accomplished in less than 400 000 generations. (Modified from Land [2].)

with eye size, it is well understandable that not every creature on this planet can have high-resolution eyes. The smaller a creature is, the more expensive is its vision. The smaller a creature is, the more carefully the eye needs to be adapted to size, brain, metabolism, and behavior. Thus, eye evolution stopped at different stages for different types of animals. However, all intermediate solutions are also fully operational "end products." They are fully operational vision systems which help the specific animal to improve its fitness for evolution and to survive. Still today, eyes are the most important sensory organ for many animals, providing detailed information about the environment both close up and far away [2].

7.3.1 Single-Aperture Eyes

The human eye is a single-aperture image capturing system similar to photographic or electronic camera systems, as shown schematically in Figure 7.5. The eye consists of a flexible lens for focusing, a variable pupil (iris) for fast sensitivity adaptation, and the retina, the image detector. A nice definition of the human eye is given by Hofmann [6]: "The human eye is a special version of a pin-hole camera, where Fresnel diffraction is compensated by the introduction of a focusing system." The diameter of the pinhole or iris is variable from 1.8 to 8 mm for bright light and dark viewing. The lens of the human eye is spherically overcorrected by graded index effects and aspherical surface profiles. The refractive index of the lens is higher in the central part of the lens; the curvature becomes weaker toward the margin. This correction offsets the undercorrected spherical aberration of the outer surface of the cornea. Contraction or relaxation of muscles changes the focal length [7].

The retina contains nerve fibers, light-sensitive rod and cone cells, and a pigment layer. There are about 6 million cones, about 120 million rods, and only about 1 million nerve fibers. The cones of the fovea (center of sharp vision) are 1 – 1.5 μm in diameter and about 2 – 2.5 μm apart. The rods are about 2 μm in diameter. In the outer portions of the retina, the sensitive cells are more widely spaced and are multiply connected to nerve fibers (several hundred to a fiber). The field of vision of

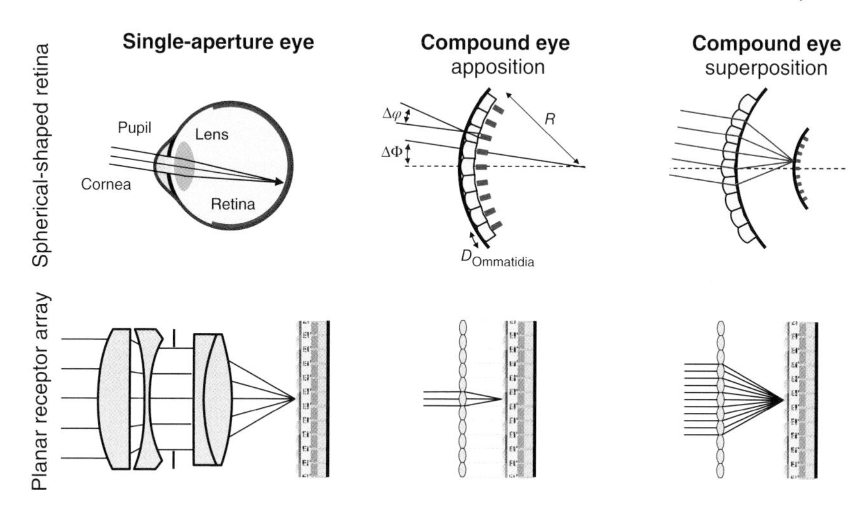

Figure 7.5 Different types of natural eye sensors and their artificial counterpart. Artificial eye sensors are based on planar components due to manufacturing restrictions.

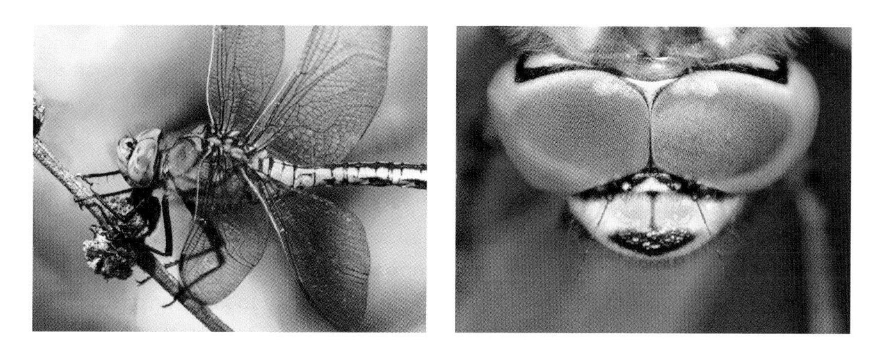

Figure 7.6 Compound eyes of dragonflies (*Anisoptera*).

an eye approximates an ellipse about 150° high by about 210° wide. The angular resolution or acuity $\Delta\Phi$ is around 0.6–1 min of arc for the fovea [7].

7.3.2 Compound Eyes

Compound eyes are multiaperture optical sensors of insects and crustaceans and generally divided into two main classes: *apposition* compound eyes and *superposition* compound eyes, as shown schematically in Figure 7.5. An apposition eye consists of an array of lenses and photoreceptors each of the lenses focusing light from a small solid angle of object space onto a single photoreceptor. Each lens–photoreceptor system is referred to as *ommatidia* (Greek: ommation meaning *little eye*). Apposition eyes have some hundreds up to tens of thousands of these ommatidia packed in nonuniform hexagonal arrays [8] (Figure 7.6).

The superposition compound eye has primarily evolved on nocturnal insects and deepwater crustaceans. The light from multiple facets combines on the surface of the photoreceptor layer to form a single erect image of the object. Compared to apposition eyes, the superposition eye is more light sensitive. Nature has evolved two principal optical methods to create superposition images: gradient refractive index and reflective ommatidia [2]. Some insects use a combination of both types of compound eyes. Variable pigments switch between apposition (daylight) and superposition (night) or change the number of recruited facets making up the superposition image [9]. A very interesting modification is the neural superposition eye of a housefly, where each ommatidium has seven detector pixels or rhabdomeres. Signals of different adjacent ommatidia interact within the neurons of the fly's brain system. This allows fast directionally selective motion detection, which enables the housefly to maneuver perfectly in three-dimensional space [7, 10].

7.3.3 The Array Optics Concept

Array optics seems to be the appropriate solution for miniaturized vision systems in nature. Compound *array* eyes allow small invertebrates to obtain sufficient visual information about their environment in a small eye volume with a large FOV, at the cost of comparatively low spatial resolution. These ingenious animal eyes have always inspired optical engineers. But array optics is not only a powerful design concept for miniaturization, also larger scale array optics, like camera arrays for light fields and computational photography, and even much larger array optics, like the Very Large Telescope (VLT) array for ground-based astronomy are using array configurations. The four 8.2 m diameter main mirrors, shown in Figure 7.7, work together to form a giant interferometer, the ESO's Very Large Telescope Interferometer (VLTI).

Using four distant mirrors with coherent superposition should have allowed the astronomers to achieve superresolution, about 25 times higher than the resolution of the individual telescopes – in theory. In practice, it turned out to be very difficult to realize. Due to the long path and more than 30 optical elements for light guiding, less than 5% of the collected light reached the detection unit

Figure 7.7 Very large telescope (VLT) Array, 4 × 8.2 m Unit Telescopes, Cerro Paranal, Chile (ESO).

of the interferometer. The stabilization of the light path with a precision of fractions of the wavelength turned out to be almost impossible. After more than 10 years of delay to the original schedule, a first successful attempt to interfere the light from all four telescopes was accomplished in 2012, now achieving the equivalent resolution of a 130 m diameter mirror. The problems to form one integral superposition image are quite similar in the micro- and macroworld: the subimages need to fit, it is difficult to transport them to a central unit, and the superposition itself is often not trivial.

What then, is now the basic concept of array optics? A bulky and expensive one-aperture system is replaced by a multitude of smaller, flatter, or cheaper elements. Typically, these elements are arranged in a regular matrix, an array. If the arrangement is not regular, it is often referred to as a cluster. The actual scale of *small* does not matter much. Small could be a few micrometers and small could be the 8.2 m diameter astronomic mirror. When the performance of a one-channel system cannot be further optimized, the task is split up and distributed to a multitude of smaller units.

In a more general sense, this statement applies to all parallel systems. A similar strategy is also used in computers. Single-processor performance rose steadily from some 10 MHz in 1986 to 4 GHz in 2004 when it abruptly leveled off. Since then, the processor speeds of desktop and laptop computers range from 2 to 3 GHz, with a tendency toward multicore processors of 2, 4, 8, and sometimes even 16 cores. However, using a cluster of 16 processors does not lead to 16 times more computational power. The coordination and combining of parallel information typically reduce the achievable overall system performance. Parallel systems need a very careful design, architecture, and algorithms to be efficient. However, if a parallel system is set up properly and applied to appropriate tasks, this system could achieve superpower or superresolution, much greater than would be expected from the arithmetical sum of all individual component.

In conclusion: array optics is a beautiful concept, which allows optical designers to overcome fundamental limits of "normal" optics and to design smaller, flatter, lighter, or cheaper optical systems. However, there are some fundamental restrictions, which need to be understood before opting for array optics.

When should an optical designer opt for an array optics, a biomimetic solution? It is not easy to give a good answer. Probably, the best answer is: "Only when normal optics doesn't allow him to solve a problem." When a classical one-aperture system is not small enough or sensitive enough, when standard single-aperture optics does not provide enough resolution or the space–bandwidth product (SW) is too low. More strictly speaking, it could be said: "never use array optics – except if you really have to." Array optics should be used with much care and there are good reasons for this statement.

7.3.4 Jumping Spiders: Perfect Eyes for Small Animals

Looking at the eye position, two main categories could be distinguished: predators and prey. Predators tend to have eyes on the front of their head, good for focusing

Figure 7.8 (a) Forward-directed eyes of a cat allowing stereoscopic view and direct distance measurement; (b) prey animals like the mouse have their eyes on the side allowing round vision; and (c) a male jumping spider (*Pseudeuophrys lanigera*). These jumping spiders live inside the houses in Europe and have excellent vision required to sight, stalk, and jump on prey, typically a smaller insects like a booklice.

on a prey animal, tracking its movements and coordinating the attack. Forward-directed eyes offer a larger binocular overlap improving stereoscopic view and distance measurement. Prey animals tend to have their eyes on the side of their head, good for all round vision, allowing an attack from the side or behind to be detected. For larger animals, like a cat or a mouse, the sizes of the eye and the brain are large enough to see and detect prey or a dangerous predator. Things get more difficult if animals are very small. The main outcome of the previous section was that small animals could have sharp eyes, but their eyes will only transmit a low number of pixels, leading to very rudimentary information about the environment if a large FOV or panoramic view is required.

If now an animal is both, a predator, which requires sharp eyes and also prey, which needs to detect a potential attack from behind, things get more complicated. But evolution has developed a perfect solution for this task, the very impressive vision system of jumping spiders, shown in Figure 7.8(c). Jumping spiders (Salticidae), a family of arthropods with a body length ranging from 1 to 20 mm, have evolved amazing eyes, combining both excellent vision for hunting and a near 360° view of the world. Jumping spiders have four pairs of eyes. Two pairs are directed anteriorly to view objects in front of the spider. The visual fields of the eight eyes of a jumping spider are shown schematically in Figure 7.9 [2, 11].

The posterior lateral eyes (PLE) are wide-angle motion detectors which sense motion from the side and behind. The posterior median eyes (PME) are very low resolution also detecting motion. The anterior lateral eyes (ALE) are the most complex of the secondary eyes, have the best visual acuity, and provide even stereoscopic vision. They are triggered by motion and are able to distinguish some details of the surroundings. The most impressive part of the jumping spider's vision system is the anterior medial eyes (AME), the primary eyes. They are built like a telescopic tube with a corneal lens in the front and a second lens in the back that focus images onto a four-layered retina [11]. The retina of the primary eyes is a narrow boomerang-shaped strip oriented vertically and has four different kinds of receptor cells. The first two layers closest to the surface contain ultraviolet-sensitive pigments, while the two deepest contain green-sensitive pigments.

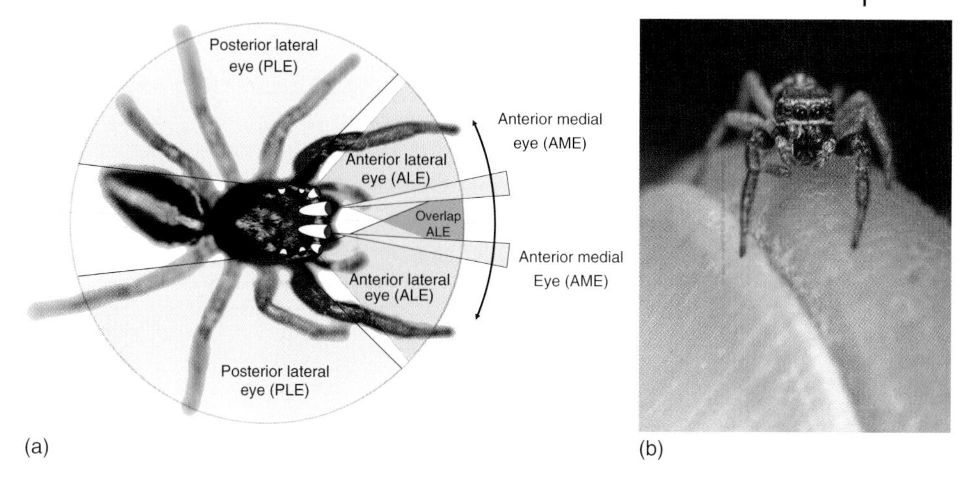

Figure 7.9 (a) Scheme of the visual fields of the four eye pairs of a jumping spider and (b) a male jumping spider (*Pseudeuophrys lanigera*) sitting on a fingertip.

The incoming green light is only focused on the deepest layer, while the other one receives defocused or fuzzy images. By measuring the amount of defocus from the fuzzy layer, the jumping spider is able to calculate the distance to the objects in front of them. The primary telescope eyes have a very high resolution, but a narrow FOV, typically 2–5° (horizontally). Unlike vertebrate eyes, the lens of the primary eye remains still. Only the retina moves. To compensate the narrow FOV, the retina of the primary (AME) eye is moved by six muscles around three axes allowing different areas to be explored, as shown (arrow) in Figure 7.9. The jumping spider observes the surroundings with its anterior lateral "wide-angle" eyes and then redirects the high-resolution anterior medial "telescope" eyes to the point of interest. Jumping spiders are diurnal hunters that stalk their prey in much the same way that a cat stalks a bird [11].

7.3.5 Nocturnal Spiders: Night Vision

Nocturnal spiders, like *Deinopis*, an ogre-faced or net-casting spider, have evolved very sensitive night vision (NV) eyes [2, 11]. For this, the 1.3 mm (diameter) lenses of the PME have a very small *f*-number of *f*/0.58. The lens is made of two components of different refractive indices for aberration correction. The photoreceptors (rhabdom) in the PME of *Deinopis* are around 20 μm wide, 10 times larger than in the AME of jumping spiders. A reflector layer (tapetum lucidum) behind the retina improves the light sensitivity by collecting twice-, direct-, and back-reflected light. Compared to the jumping spiders, the net-casting spiders trade resolution in better sensitivity, a perfect solution for net-casting spiders, who – as the name already indicates – catch flying insects by holding a square net between the two pairs of forelegs. As their eyes are very sensitive to sunlight, they hide in a dark spot during daytime [2, 11].

Jumping spiders and net-casting spiders, using a cluster of different single-aperture eyes, are a perfect example for miniaturization of a vision system. Dividing the vision system into different subsystems allows the tiny and light-weight spider to have three completely different vision systems: panoramic motion detection, wide angle, and telescope vision. The movable retina with a small number of pixels avoids overloading the brain with image information. Jumping spiders are able to detect prey, mates, and enemies. The movable retina allows a jumping spider track prey motionless before jumping [2, 11]. After discussing biologically vision systems in much detail, the following section focuses on the manufacturing of miniaturized optics.

7.4 Manufacturing of Optics for Miniaturized Vision Systems

7.4.1 Optics Industry

Optics manufacturing started some thousand years ago, when our ancestors polished rock crystal to manufacture loupes or burning glasses for fire making. Reading stones, a primitive plano-convex lens made by cutting a glass sphere in half, appeared in the eleventh century. The invention of spectacles or eyeglasses in the thirteenth century was the start of a first optics industry in the north of Italy. Hans Lippershey, a German-Dutch spectacle maker, filed the first patent application for a telescope on the 25th of September 1608. Building better telescopes was not only a challenge for astronomers but also very important for military purposes. Discovering and observing an enemy from a larger distance are a huge advantage for jumping spiders – and humans. Antonie van Leeuwenhoek (1632–1723) manufactured miniaturized ball lenses by melting small rods of soda lime glass in a hot flame. These high-quality glass spheres improved the resolution of his microscope viewers beyond current limits. He was the first to observe and report the existence of single-cell microorganisms, the very beginning of microbiology. Robert Hooke (1635–1703), another pioneer of microscopy, published his famous book "Micrographia," a collection of microscope observations, in 1665 [12].

Pioneers like Carl Zeiss, Otto Schott, and Ernst Abbe significantly improved lens manufacturing and microscopy in the nineteenth century. The invention of photography further accelerated the growth of the optics industry. The optics industry was mainly dealing with large bulky lenses, but their grinding and polishing techniques were also applicable to the manufacture of miniaturized lenses for microscopes and endoscopes. Today, the optics industry provides a large variety of high-class optical elements from millimeter-scale lenses to large-scale astronomic mirrors with a surface and profile accuracy in the atomic range.

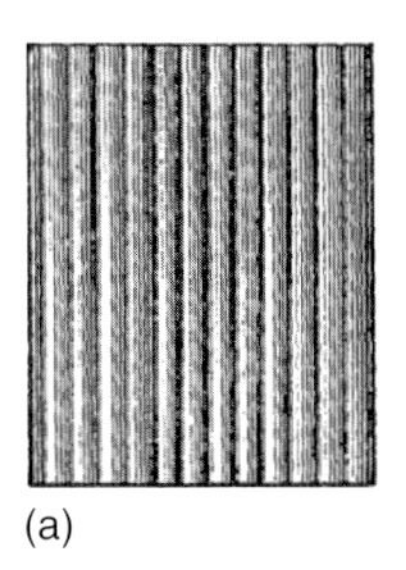
(a)

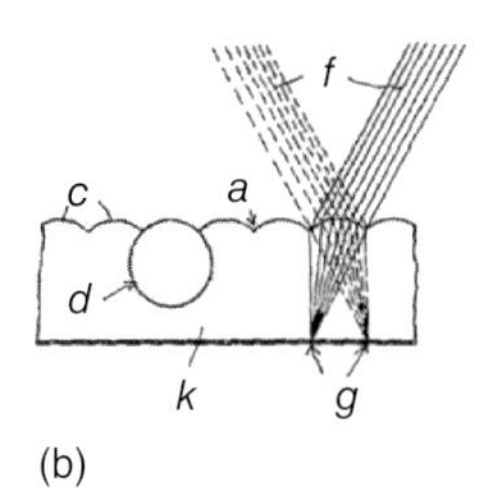

(b)

Figure 7.10 Schematic drawing (a) top view and (b) enlarged cross section of a cylindrical microlens array on a photographic layer for integral photography as proposed by Walter Hess in 1912 [13].

7.4.2
Planar Array Optics for Stereoscopic Vision

The development of planar array optics, also referred to as lenticular lens, lens, or microlens arrays, is much connected with photo- and cinematography. In 1908, Gabriel Lippmann invented the "integral photography," an autostereoscopic method to display 3D images for observation with the naked eye by using a microlens array. Integral photography uses an array of small microlenses to record multiple subimages of a scene in a photographic layer. Each microlens acts like a miniaturized camera recording an individual subimage. Observing the developed photo plate through a similar lens array, the superimposed subimages form an autostereoscopic integral image, a 3D image.

Walter Hess, a Swiss ophthalmologist, learned about Lippmann's invention and founded a company to manufacture 3D photographs based on integral photography in 1914. His approach, shown schematically in Figure 7.10, was the use of cylindrical microlens arrays made of celluloid [13]. The company did not succeed and Hess continued to work in neurophysiology. It is not known how much stereoscopic vision triggered his career as neurophysiologist, but in 1949, he was awarded the Nobel Prize for mapping the different areas of the brain involved in the control of internal organs. Still today, where 3D cinemas are booming, stereoscopic vision and its influence on human brains are not fully understood. Watching 3D films is quite a challenge for the brain, and frequently, people get sick and have to leave the cinema. The problem for the brain is the mismatch of focus and convergence. The eyes are focusing on the projection screen or display, while the eyeballs follow the 3D action and converge in another plane. Lippmann's invention of integral photography is also used for plenoptic cameras.

7.4.3
The Lack of a Suitable Fabrication Technology Hinders Innovation

Over the last 100 years, many researchers published and patented inventions, where planar array optics was the decisive key element. Often, they were brilliant ideas, but only few of them could be realized and even fewer were a commercial success. For the early lens array applications, the insurmountable entrance barrier was the availability of suitable microoptical elements for reasonable costs. At that time, microlens arrays and diffractive gratings were usually engraved or polished, for example, on a lathe. This piece-by-piece fabrication was very time-consuming and expensive, and the arrays were not very uniform. Later, glass molding, casting, and pressing were used, for example, to manufacture fly's eye condensers for slide and film projectors. Often, the quality of these microlens arrays remained poor. Surface roughness, defects, lens profile accuracy, and nonuniformities in the array constrained their field of application to illumination tasks. For more sophisticated applications, requiring, for example, a stack of two or three microoptics layers, the lateral mismatch (grid imperfections) and the array-to-array alignment were problematic. Using microlens arrays was considered to be an exotic idea usually not leading to success. The situation changed with the rapid progress of microstructuring technology in the second part of the last century, when wafer-based manufacturing was promoted by the semiconductor industry.

7.4.4 Semiconductor Industry Promotes Wafer-Based Manufacturing

In the semiconductor industry, the term "wafer" appeared in the 1950s to describe a thin round slice of semiconductor material, typically germanium or silicon. The breakthrough for planar wafer manufacturing in semiconductor industry came with the invention of the "planar process" by Jean Hoerni in 1957. The next decisive invention was to connect these transistors and form integrated circuits (IC) or microchips. Robert Noyce, cofounder of Fairchild and later Intel, patented his planar IC in 1959, and Fairchild started IC manufacturing in 1960. Hundreds, thousands, later millions, and today billions of electronics components could now be manufactured in parallel. Replacing discrete piece-by-piece electronics fabrication by Fairchild's planar process revolutionized the semiconductor industry. This was the start of the IC explosion in what is now called Silicon Valley, which created tens of thousands of new workplaces within only a few years. Semiconductor industry had a major impact on both essential parts of a miniaturized vision system, the lens, and the image sensors. Regarding biomimetic vision systems, the dominance of semiconductor technology is also one of the major restrictions for system designers. Wafers are 2D – planar. Manufacturing of microoptics and image sensors is restricted to two dimensions. Natural animal vision systems consist of 3D components like a graded index lens and a half-sphere-shaped retina. It is difficult to transfer 3D ideas to a 2D technical world.

7.4.5 Image Sensors

The photographic process, based on silver halide crystals, was the dominant image recording technology from its invention in the 1820s until the appearance of electronically recorded TV images a century later. The progression of image sensors from video camera tubes to charge-coupled devices (CCD) and active pixel sensors (CMOS) is inextricably linked to the success of wafer-based manufacturing in semiconductor industry. An image sensor is, in simple terms, an array of tiny solar cells, each of which transforms the incident light into an electronic signal. In CCD sensors, the charge is transported across the chip, read out at one side, and converted into a digital value. In CMOS devices, the electronic signal is amplified by transistors, located beside the receptor, and the signal is transported by electric wires. Until recently, CCD sensors were credited to have less visual noise and distortion and were preferred for high-quality image capturing. CMOS (Complementary metal-oxide semiconductor) sensors have now taken the lead. CMOS sensors are significantly cheaper than CCD, they are faster, and backside-illuminated CMOS sensors make them more sensitive for low-light situations. Another advantage of CMOS image sensors is that they are manufactured with standard microprocessor technology allowing reuse of obsolescent semiconductor fabs for their production.

In a front-illuminated image sensor, a matrix of transistors (active pixel sensors) and metal wiring is placed in front of the photosensitive layer. This maze of wires and transistors blocks a part of the incident light, as shown schematically in Figure 7.11(a). In backside-illuminated image sensors, the transistor matrix and wiring is moved behind the photosensitive layer, as shown in Figure 7.11(b). A 100% fill-factor image sensor with improved low-light sensitivity is obtained. For the manufacturing of backside-illuminated sensors, the wafer with the CMOS image sensor is flipped and the silicon is thinned from the backside, until it is thin enough to let the light shine through.

Interestingly, nature also evolved similar 100% fill-factor retinas for the eyes of cephalopods, active marine predators like the nautilus and the octopus. In the

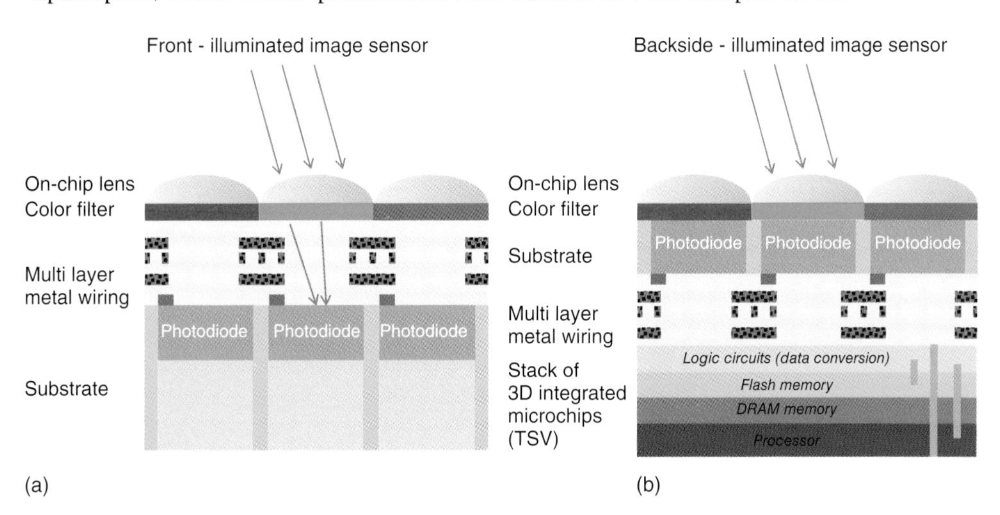

Figure 7.11 (a) Front-illuminated image sensor and (b) backside-illuminated image sensor bonded to a stack of a 3D integrated chips with through-silicon via (TSV) connections.

human retina, the nerve fibers are before the retina, blocking light and creating a blind spot where the fibers pass through the retina. In cephalopod eyes, the nerve fibers route behind the retina layer, very similar to the backside illumination image sensor.

Beside the higher sensitivity, the backside illumination CMOS sensor technology has another huge advantage compared to the CCD sensor. As shown in Figure 7.11(b), the image sensor could be bonded to a stack of other thin electronic layers (processors, memory, etc.). Through-silicon via (TSV) connections allow signals and data to be transferred vertically through the different electronic layers. Wafer-thinning, wafer-level stacking, or wafer-level packaging (WLP) technologies are referred to as 3D IC manufacturing. The new 3D IC trend in microelectronics is likely to have a grand impact on future generations of CMOS image sensors. Integrated 3D CMOS image sensors, using highly parallel data treatment, transport, and on-sensor storage, will provide ultrahigh-speed, low-noise, high-dynamic range (HDR), and hyperspectral image sensors in a small package at low costs.

7.4.6
Wafer-Based Manufacturing of Optics

The manufacturing of miniaturized optics by using classical optics manufacturing technologies, like grinding and polishing, is possible but very cumbersome and expensive. The rise of the wafer-based semiconductor industry also had a considerable impact on microoptics manufacturing [14]. In the mid-1960s, Adolf W. Lohmann made the first step by combining computers, plotters, and photolithography to manufacture computer-generated holograms (CGH). In the 1970s, dry plasma etching technology was introduced for the transfer of microoptical structures from structured photoresist into glass wafers. In the mid-1980s, Zoran Popovic proposed a microlens fabrication technology based on microstructuring of the photoresist by photolithography and a subsequent resist melting process, as shown in Figure 7.13.

Planar optics, the integration of different optical functions on one planar substrate, was one of the most promising future technologies in the 1980s. Optics, later relabeled as photonics, hoped to repeat the semiconductor hype in generating infinite growth and wealth of a *planar optics* industry, similar to the IC explosion in Silicon Valley. The first companies manufacturing microoptics on wafer scale appeared in the 1990s. However, neither the optical computer nor the later *photonics* hype really took off. Planar optics could not redo what semiconductor industry had done. The fundamental difference is that electronics allows complete devices to be built – including the input sensor, CPU, memory and output, a display, sound, or movement. Optics and microoptical elements are typically only parts of a larger mechanical or electronic device used to redirect, shape, or switch the light. But planar microoptics is a decisive key enabling technology for miniaturization of optical systems, like vision systems.

7.4.7
Manufacturing of Microlens Arrays on Wafer Level

Microlens arrays appeared already in the Cambrium. Trilobites, shown in Figure 7.12(a), a fossil group of marine arthropods, had complex compound eyes with microlenses made of calcite. Still today, similar compound eyes are found in many small creatures. Refractive microlens arrays seem to be an appropriate solution for miniaturized vision systems in nature.

How to manufacture refractive microlenses and microlens arrays for biologically inspired vision sensors? Many technologies have been developed to generate arrays of small lenses. The most accurate, and therefore widely used in industry, is the melting resist technology [14], shown schematically in Figure 7.13. The melting resist process is also used for on-chip microlenses on the CMOS image sensor pixel, as shown in Figure 7.11. Here, the microlenses are not used for imaging, they are used for fill-factor enhancement, funneling the incident light onto the photosensitive layer.

For the melting resist process, a thick layer of positive photosensitive resist is spin coated on a glass wafer and exposed with ultraviolet light in a mask aligner. After wet-chemical development and drying, the resist structures are melted in an oven or on a hot plate. The melting procedure itself is quite simple [14]. Above the softening temperature, the edges of the resist structure start melting. Above the glass transition temperature, the amorphous resist polymer changes into a glass state system. The surface tension tries to minimize the

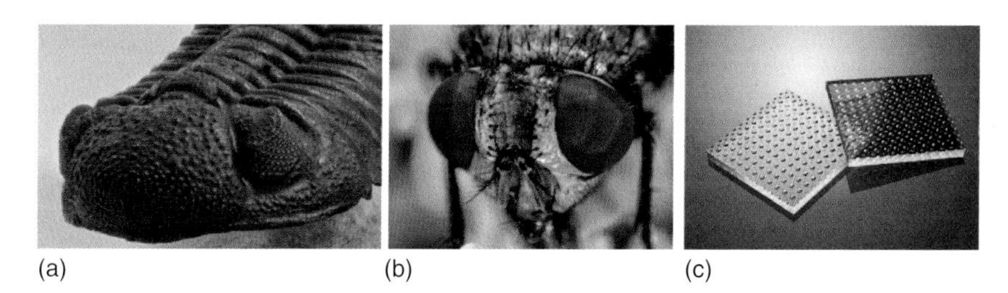

Figure 7.12 (a) Trilobite eye – microlens arrays made of calcite, (b) compound apposition eyes of a fly, and (c) microlens array manufactured by wafer-based technology in quartz glass (fused silica) [14].

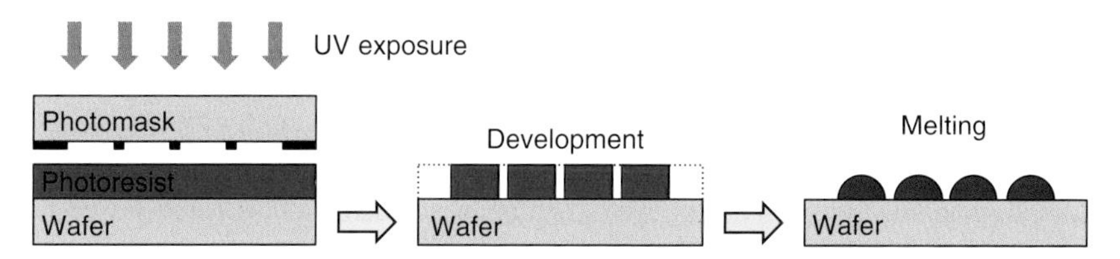

Figure 7.13 Melting photoresist technique for manufacturing refractive microlenses on wafer level.

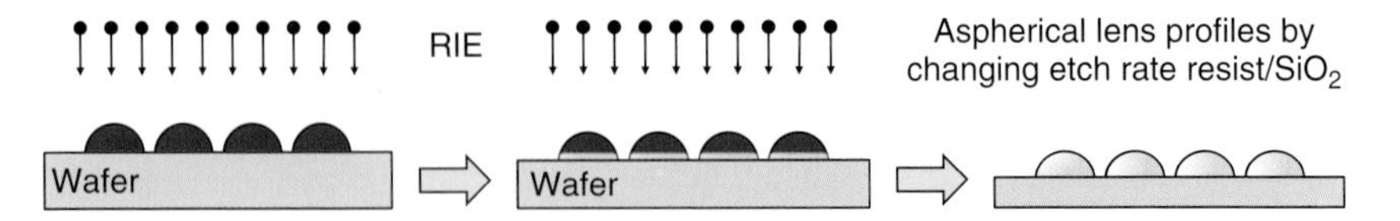

Figure 7.14 Transfer of resist microlenses into the wafer bulk material by reactive ion etching (RIE).

surface area by rearranging the liquid masses inside the drop. Ideally, the resist melts completely, the masses are freely transported, and surface tension forms a spherical microlens. In practice, the lens melting process needs careful process optimization and precise control of all process parameters to obtain good lens-to-lens uniformity within one wafer and from wafer to wafer. Repeatability and uniformity of melted resist lenses are key factors for the following etch process.

In the next step, the microoptical components are transferred from the photoresist state into wafer bulk material by reactive ion etching (RIE) as shown in Figure 7.14. The etching process removes atoms from the resist and the wafer surface at different etch rates. Surface areas covered by resist structures are protected until the covering resist layer is removed.

Melted resist microlenses are usually very close to a spherical lens profile with a conic constant around $k \approx 0$ after melting. The transfer of the melted resist lens by RIE allows the lens profile to be changed. This is done by varying the mixture of the etch gases and oxygen during the etch process. If the etch rate for resist is higher than for the wafer bulk material, the resulting lens profile will be flatter than the resist lens profile. A continuous change of all etch parameters allows to obtain aspherical lens profiles.

7.4.8
Diffractive Optical Elements and Subwavelength Structures for Antireflection Coatings

Diffractive optical elements (DOEs) and gratings are manufactured on wafer level by using subsequent photolithography and sputter or plasma etching steps, as shown in Figure 7.15 for a four-level DOE [15]. A very critical process step for the manufacturing of multilevel DOEs is the proper alignment during photolithography and the precise transfer of the resist structures into the wafer bulk material.

Photolithography in wafer stepper or e-beam direct writing allows manufacturing highly efficient DOEs with minimum feature sizes smaller than the wavelength of the light. For most optical applications, it does not make sense to manufacture structures with subwavelength feature size. The light is not able to resolve the structures; the light *sees* a composite material of which the optical properties are between those of air and those of the base material. Nonetheless these subwavelength structures are becoming very popular as new type of antireflection (AR) coating recently [16].

When light passes through a glass surface, a part of the light – typically some 4–5% – is reflected at the surface. Photographic objectives, as shown in Figure 7.2(a), typically consist of some 10 individual lens or doublet elements

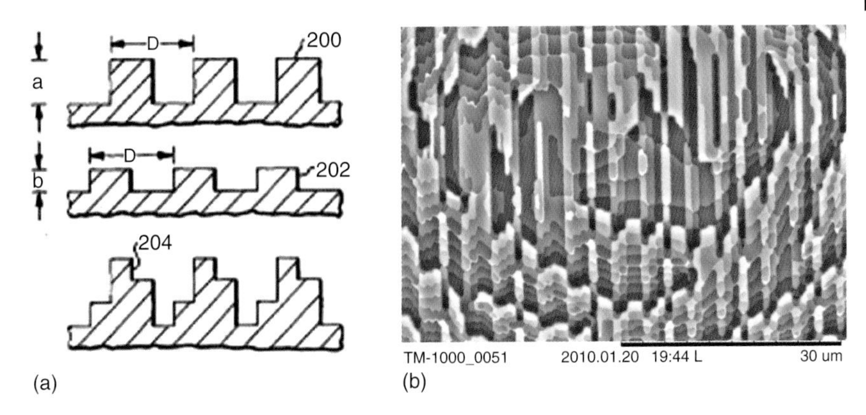

Figure 7.15 (a) Multilevel diffractive optical elements (DOE) manufactured with a planar process using subsequent photolithography and dry etching [15]. (b) Eight-level diffractive optical element in fused silica glass material.

comprising some 20 glass to air (or other glass material) surfaces. At each surface, some light is reflected. This light continues its path in the objective generating stray light, ghosting, and flare images. Antireflective coatings, typically thin-film structures with alternating layers of high and low refractive index, were introduced in photographic lenses in the 1970s. They significantly reduce the loss of light in lens systems and improve the contrast and imaging performance of lenses for vision systems. Unfortunately, the choice of suitable thin-film materials is very limited, and their refractive indices are close to glass.

Recently, major camera lens manufactures introduced subwavelength AR coatings, often referred to as nanoparticle, nanocrystal, or subwavelength structure coatings (SWC), for their high-end lenses. Subwavelength AR coatings have some major advantages. Subwavelength AR coatings are, for example, made of wedge-shaped nanostructures causing a continuously changing refractive index or spongy layers with variable cavity density within their volume. This allows material with a very low refractive index close to air to be manufactured. Instead of multiple layers of different material, now a continuous change of the index from glass to air is applied. This significantly reduces the losses and reflections at the lens surfaces. Today, subwavelength coatings are mainly applied to wide-angle lenses where the curve of the lens surfaces is very large.

Not surprisingly, nature had already invented subwavelength AR coatings for nocturnal insects. The corneas of some moth eyes are covered with conical protuberances of subwavelength size, providing a graded transition of refractive index between air and the cornea [17] – very similar to the novel nanocoatings for high-end photographic lenses. For these nocturnal insects, the purpose of the AR coating is twofold. Low loss of light at the cornea's surface is certainly of some importance. Much more important is the lowest possible reflection from the cornea's surface, because shining and glowing eyes could make the insect visible to a potential predator [2].

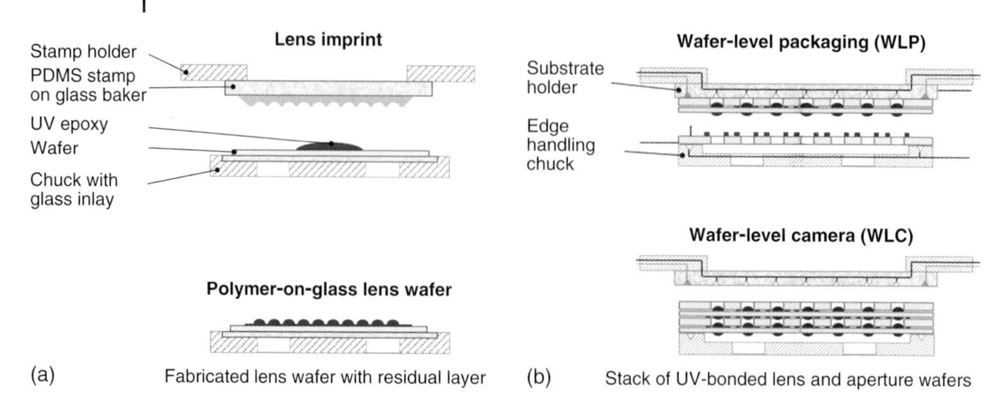

Figure 7.16 UV imprint lithography and wafer-level packaging (WLP): (a) A PDMS stamp is embossed on a wafer with a drop of a photosensitive polymer. The resist is formed by the PDMS pattern and exposed by UV light for curing. After hardening of the resist, the PDMS stamp and lens wafer are separated; (b) lens and aperture wafers are mounted in a mask aligner to obtain a stack of wafer-level cameras (WLC).

7.4.9 Microlens Imprint and Replication Processes

The wafer-based technologies using lithography and plasma etching are well suited to obtain high-quality microoptical components on wafer scale. However, these manufacturing technologies are too expensive for high production volumes. For consumer applications, the average sales price (ASP) of a miniaturized camera module is typically in the range of US$ 1–20. For mass production of wafer-level camera (WLC) lenses, the preferred technology is UV imprint lithography, using a mask aligner and photosensitive polymers [4, 18]. A microlens wafer is used as a master, copied to a PDMS stamp and then replicated into polymer on glass.

Figure 7.16 shows the basic process of UV imprint lithography in a mask aligner. A stamp is embossed in the liquid polymer, and exposure with ultraviolet light provides a quick curing of the polymer. UV imprint allows to manufacture full lens wafers in one step and to achieve high uniformity and precise lateral placement of the microlenses within the array [19]. Process improvements like prior dispensation of microdroplets on the wafer or in the stamp, as well as enhanced imprint methods, like surface conformable imprint lithography (SCIL), are applied to improve the uniformity and yield. The microlens wafers are then mounted by WLP to wafer-level optics modules [20].

7.4.10 Wafer-Level Stacking or Packaging (WLP)

Kenichi Iga and his colleagues attracted attention in the early 1980s with their idea of *stacked planar optics* [21]. In the early 1990s, the first systems requiring

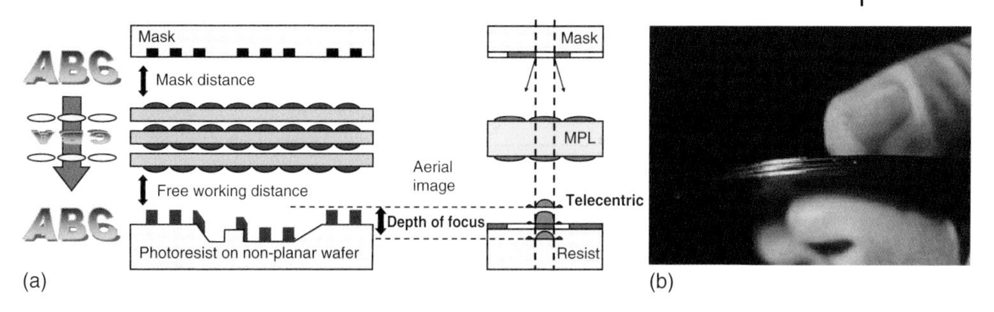

Figure 7.17 (a) Schemes of a microlens projection lithography (MPL) system for 1 : 1 projection of a photomask onto a wafer in a mask aligner and (b) photograph of an early prototype system, a stack of three microlens wafers.

a stack of microlens wafers were built – for example, a 1 : 1 microlens projection lithography (MPL) system for mask aligners, projecting a photomask onto a wafer located at a large distance by using some 300 000 individual lens channels [22] (Figure 7.17).

The most critical process step for such wafer-level projection systems is the proper alignment of multiple wafers. Thickness variation of the wafers, bending and warp, lateral displacement, and bonding or gluing of multiple wafers could be quite cumbersome. WLP of microoptics became very popular some years ago, when industry tried to implement the WLC technology in mass production of cameras for consumer products like smartphones and tablets [4].

7.4.11 Wafer-Level Camera (WLC)

Today's mobile phone cameras consist of some 10–20 different components such as plastic- or glass-molded lenses, pupils, baffles, actuators, lens holders, barrel, filters, and the image sensor. These components are manufactured and assembled piece by piece. The WLC approach is rather simple: All components are manufactured on wafer level, the lens wafers are mounted together with the CMOS image sensor wafer, and the wafer stack is diced into some thousands of individual camera modules, as shown in Figure 7.16(b). The complete mobile phone camera, including the optics, is manufactured and packaged on wafer level using standard semiconductor technology [4].

WLC for low-cost applications such as mobile phones and disposable endoscopes had gained much popularity in industry some years ago. However, most companies failed to succeed with WLC cameras on the market. The major problem of WLC was the overall yield. A defect in one of the multiple layers makes the camera unusable leading to low yield for complete wafer stacks. Another source of problems is the restriction of the optical design. In WLC, all lenses have to be located on the front or the backside of a wafer. This is a severe handicap for the optical designer. Lenses have to be separated by the thickness of the supporting wafer. For example, it is not possible to manufacture meniscus-shaped lenses.

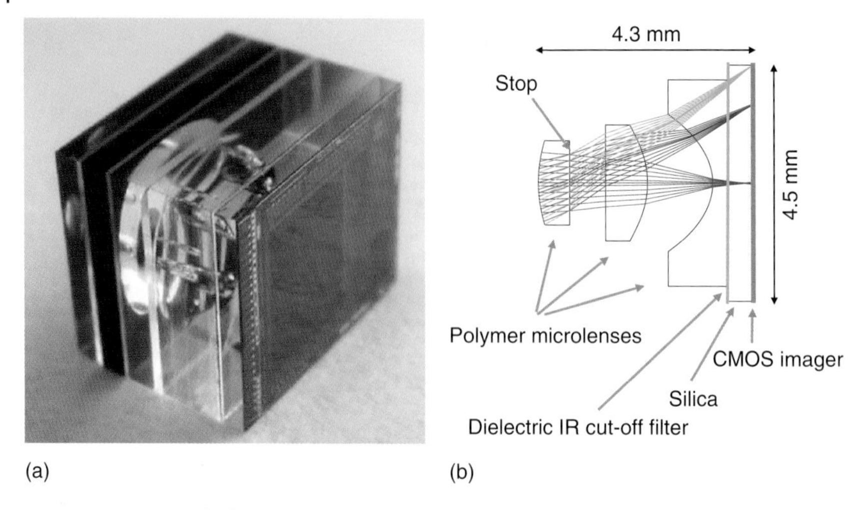

Figure 7.18 (a) Wafer-level camera (WLC) built within European Research Project WALORI in 2005 [23]. Backside illumination (BSI) through thinned CMOS image sensor, 5.6 μm pixel VGA; (b) optical design of the WLC shown (a).

The first wafer level cameras (WLC) such as the one shown in Figure 7.18, combining microoptics wafers made from polycarbonate and thinned CMOS imagers (backside illumination) were developed in 2002 [23].

The next step is wafer-level integration (WLI), combining wafer-level microoptics with optoelectronic components such as photodiodes, LEDs, or lasers arranged and mounted on wafer level. The WLI approach gained much success recently for manufacturing high-volume components integrated in the latest generations of smartphones. Several thousands of lenses or other microoptical elements can be fabricated simultaneously on a single 200 mm glass wafer. The process chain includes lithography for apertures, double-sided and aligned replication of lenses, stacking of wafers, automated optical testing, and wafer dicing. Providing miniaturized optoelectronic modules for smartphones is the key technology for future biomimetic compound vision systems.

7.5
Examples for Biomimetic Compound Vision Systems

Biomimetic compound vision systems have been investigated by numerous research teams within the last 25 years [2, 10, 24]. Artificial compound eyes have been implemented and tested for many different applications. Robots for flying and swimming have been equipped with lightweight biomimetic vision systems, enabling them to maneuver successfully in 3D space. Novel approaches to miniaturize and to improve vision systems for next generations of robots, drones, and – even more important – for automotive and consumer applications are on their way. It is impossible to cite or reference all these very interesting approaches. In the following section, a small selection of biomimetic compound vision systems is presented as examples.

7.5.1 Ultraflat Cameras

Taking pictures whenever and wherever has become standard nowadays. Miniaturized cameras, integrated in consumer devices, like smartphones, tablets, and laptops, create a permanent availability of a camera. This was not always the case.

In the beginning of photography, the photographic plates had to be developed right after the exposure. To take pictures outside their studio, the early photographers needed to be equipped with a mobile darkroom, typically a tent or a van. In 1913, Oskar Barnack built a first prototype of a Leica camera, working with standard cinema 35 mm film. His Leica camera set the standard with 24 mm × 36 mm image size on negative film and became the first permanently available consumer camera manufactured in large volume in the 1930s. The Minox, a miniaturized camera with 8 × 11 mm image size, was the first camera that could be carried around in a trouser pocket. Its small size and macrofocusing ability made it also very popular for espionage. Compact digital still cameras entered the consumer market in the mid-1990s, gradually replacing photographic film cameras. Camera phones first appeared on the market in 2002. Today, wearable cameras integrated in clothes or eyeglasses are on the rise.

An earlier approach in the pre-camera phone area was the credit card camera [24]. Besides taking pictures, a credit card camera could enhance the card security by providing embedded face or iris detection. However, the 0.76 mm normalized thickness of a credit card would require an ultraflat camera design. Obviously, a single-aperture camera system would not work out; a compound *array* camera design could be the only option.

In a first approach, mimicking an apposition compound eye, a single microlens array was used. Each microlens captures a different demagnified subimage as indicated in Figure 7.5. All subimages are recorded on a thin CMOS image sensor. Figure 7.19 shows (a) schematically the principle of the single microlens array imaging system and (b) the obtained subimages for an array of microlenses with lens pupil diameter $D = 135\,\mu m$, focal length $f = 320\,\mu m$, $f/2.4$, and $3\,\mu m$ diffraction-limited spot size. In this approach, the individual images are superposed electronically to form a complete image.

In a more complex approach, mimicking a superposition compound eye, a stack of three microlens arrays, as shown schematically in Figure 7.20(a), was used. The lenses of the first array image demagnify and invert subimages of the object into the plane of the second lens array. The lenses of the third array image these inverted subimages to the image sensor plane. The lenses of the second array serve as field lenses, imaging the pupil of the first array to the corresponding entrance pupils of the third lens array. As shown in Figure 7.20(a), size, focal length, and distances in the system are optimized to obtain correct superposition of the individual subimages in the plane of the image sensor.

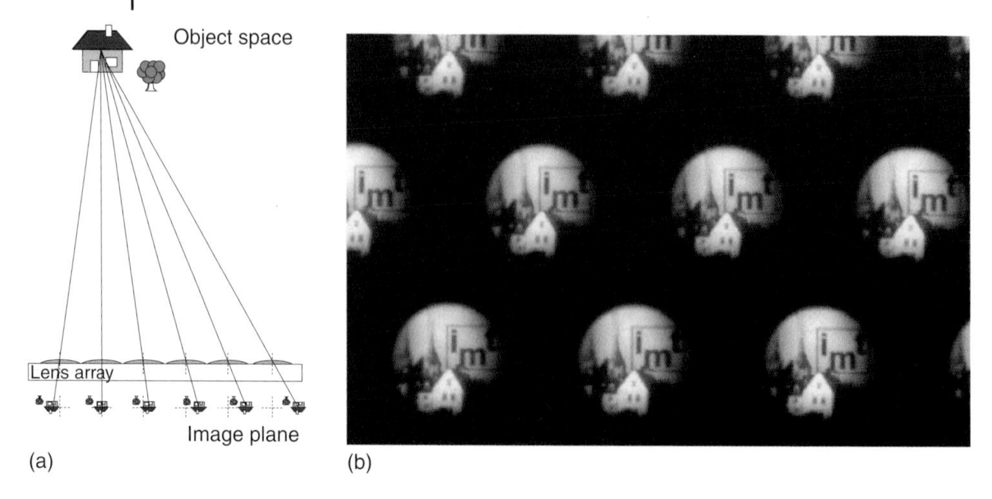

Figure 7.19 (a) Microlens array camera imaging individual subimages from an object to an image plane and (b) recorded subimages for *f*/2.4 array camera.

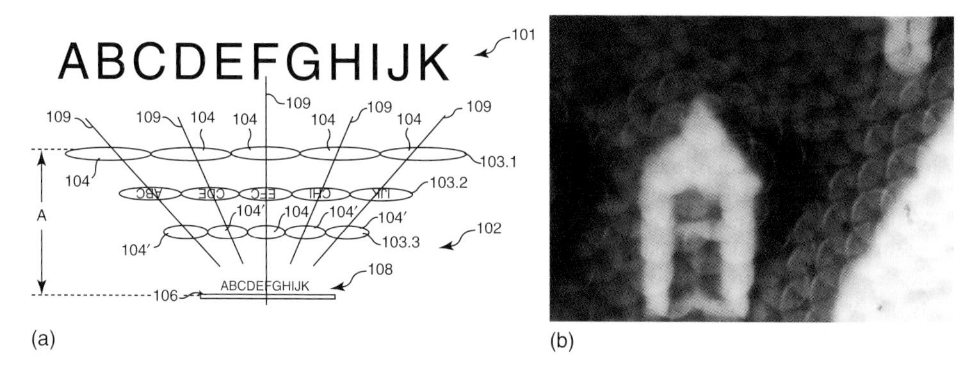

Figure 7.20 (a) Ultraflat camera system consisting of three layers of microlens arrays and (b) superposition of individual subimages obtained from a 1 : 1 multichannel imaging system as shown in Figure 7.16, imaging the same object also shown in Figure 7.18.

Integrating a camera in a credit card is obsolete since camera phones have taken over the consumer market. However, both the apposition and the superposition version of the proposed ultraflat cameras are still investigated today [25, 26].

Future applications are, for example, large-object field microscopes, shown in Figure 7.21. If it becomes possible to build ultrathin and flexible cameras, a large range of possible products comes into sight.

7.5.2 Biologically Inspired Vision Systems for Smartphone Cameras

The eight-eye vision system of jumping spiders, shown in Figure 7.9, seems to be the optimum solution nature evolved for small animals: 360° panoramic motion

(a)

(b)

Figure 7.21 (a) Electronic cluster eye (eCLEY) with a FOV of 52° × 44° and VGA resolution, an apposition compound design with electronic superposition and (b) an ultrathin array microscope with integrated illumination, a superposition compound vision design. Both devices are built at Fraunhofer IOF in Germany. Both photos compare the biomimetic vision system (right in the photos) to the standard optics, a bulky objective, and a standard microscope [24, 25].

detection, a wide-angle viewing capability to identify prey or an enemy, and a high-resolution vision with a range finder to track a target to jump on. These three different vision systems are integrated in a small lightweight spider body and operated by a tiny brain.

What then, is the perfect solution for future miniaturized cameras to be integrated in future consumer products? Making a wish list, a future miniaturized vision system for consumer products should have:

- High resolution, preferably in the range of 50 megapixels and more
- HDR, from very low light to the bright sunlight at the beach
- High color depth
- Fast autofocus (AF) and zoom
- Stereoscopic view and detailed depth map
- High speed with 1000 frames per second
- NV and thermal imaging

It is – from the current point of view – not possible to integrate all these functions in a single vision module. Additionally, the height restriction of some 4–5 mm correlates with an achievable resolution of some 5–8 megapixels due to the limited space–bandwidth product (SBP), as discussed in Section 7.2.2. The most promising approach is to mimic the eyes of the jumping spider and to split vision to a cluster of single-aperture cameras. One interesting biologically inspired approach is the 4 × 4 PiCam cluster camera discussed in more details in the next section.

7.5.3 PiCam Cluster Camera

An interesting approach in this direction is the 4 × 4 PiCam developed by Pelican Imaging [27] and shown in Figure 7.22. All 16 cameras capture an image at the same time using 1000 × 750 pixels leading to a synthesized image of 8 megapixels

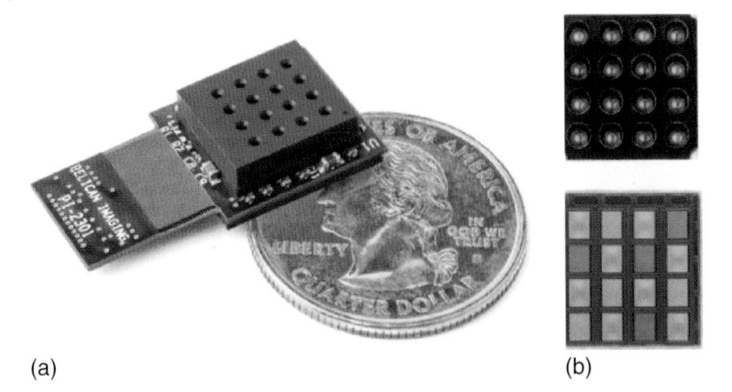

Figure 7.22 (a) PiCam module consisting of 4 × 4 individual cameras with 0.75 megapixel resolution and (b) lens and sensor array with RGB color filters (Pelican Imaging).

after date processing. A clever trick of the PiCam cluster approach is to move the RGB color filters from the sensor pixel plane to the lens level. Three types of *f*/3.1 lenses for red, green, and blue, a 56° (diagonal) FOV and a focal length of 2 mm are used. The total height of the camera module is less than 3.5 mm. As shown in Figure 7.22(b), the Bayer filter pattern with typically two green, one red, and one blue is applied to the camera lens level. Two green, one red, and one blue cameras form 2 × 2 subarrays.

Using cameras with a smaller chromatic range significantly relaxes the optical design for the lenses. Reduced chromatic aberrations lead to significantly better images with fewer lenses. Preferably, the image plane is adapted to the wavelength range for the individual cameras. Avoiding a Bayer filter pattern on each image sensor pixel reduces the height of the pixel stack. As shown in Figure 7.11, a thinner pixel stack enables the pixel to accept a wider cone of incident light, thus improving the efficiency of the light capturing and reducing crosstalk and blur [27].

The hyperfocal distance, defined as the closest distance a lens can focus while keeping objects at infinity acceptably sharp, scales with the lens diameter. For the PiCam, the hyperfocal distance is 40 cm. Thus, the lens images all objects from 20 cm to infinity sharp, if the fixed focus is set to the hyperfocal distance. The parallax detection and superresolution provided by the 4 × 4 arrangement allows a postcapture refocus, similar to a light-field camera. The absence of an AF unit is not only a cost factor; it also avoids the typical AF delay in image capturing. Today, the majority of AF camera modules are based on voice coil motors (VCM) to move one or more lenses along the optical axis. A major disadvantage is that the VCM AF does not provide fixed positions. A typical AF procedure is to move the focus through all positions, capture, and analyze the resulting images. By repeating this, the best focus for the scene is obtained. A typical camera can have as many as 10–20 steps between near and far extreme focus and easily take >1 s to acquire a sharp image. This *focus first and capture later* procedure is quite annoying, especially for capturing quickly moving objects or scenes. The cluster camera approach of the PiCam allows a *capture first and refocus later approach* – a clear advantage for the user [27].

Although the cluster array concept, followed by the 4 × 4 PiCam, has very clear advantages, it remains challenging. Similar to the jumping spiders, a cluster camera requires a lot of computational power and intelligent software algorithms to compose the high-resolution images. Computation is already a problem for jumping spiders. For some jumping spiders, the brain takes up to 80% of the space in the body. To achieve superresolution, a cluster camera needs special processor hardware for ultrafast and parallel image processing. Similar to the jumping spider, cluster vision is expensive with regard to energy respectively battery consumption. The next example for biomimetic compound vision systems is related to fast movement detection, similar to the compound apposition eyes of *Drosophila* and dragonflies.

7.5.4 Panoramic Motion Camera for Flying Robots

Engineering a vision system for small flying robots is a very challenging task [28]. The vision system should be small, lightweight, and provide distortion-free wide FOV. For flight control and collision avoidance, the systems should be optimized for fast motion perception, meaning a low resolution to enable ultrafast data treatment using low computational power. Flying insects perfectly meet all of these requirements. Compound eyes, consisting of a mosaic of small ommatidia, offer large FOV of up to 360°. The even more impressive feature of these insect eyes is the high temporal resolution. Human vision is sensitive to temporal frequencies up to 20 Hz. Some compound eyes respond to temporal frequencies as high as 200–300 Hz [28]. This explains why it is so difficult to catch a fly. For a 10 times faster motion detection system, the movement of our hand appears in slow motion and the fly has all time to move away [2, 28].

Flying arthropods like bees, houseflies, and dragonflies have been extensively studied by biologists for decades. The compound vision and neuronal system is very well understood today. The sensory and nervous systems of flies have been mimicked as neuromorphic sensors. However, despite a deep knowledge of the vision system of flying insects, it remains very difficult to reproduce artificial systems that can display the agility of a simple housefly.

A curved artificial compound eye has been investigated recently the CurvACE research project [29], shown in Figure 7.23.

The major challenge was to manufacture a miniature artificial compound eye on a curved surface, as shown schematically in Figure 7.23. Similar to a compound apposition eye, shown in Figure 7.5 (center), each ommatidia or lens channel consisted of one microlens and one corresponding image pixel on the CMOS image sensor. To reduce crosstalk, two low-reflective opaque metal layers with apertures (pinholes) were placed in between lens and pixel. The technical approach was WLP of the two optical layers (microlenses and apertures) and the CMOS image sensor chip on a flexible printed circuit board

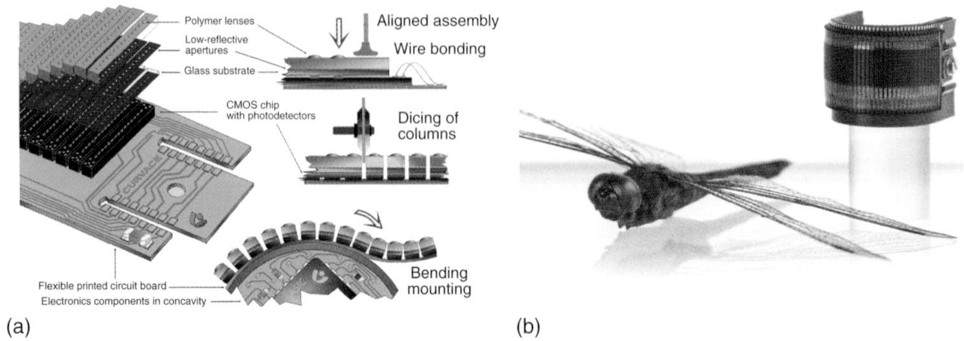

Figure 7.23 (A) Scheme of the manufacturing of a curved artificial compound eye from the research project CurvACE [29]. Polymer-on-glass lenses, apertures, photodetectors (CMOS), and interconnections (PCB) were manufactured on a flexible layer, then diced, and mounted on a rigid semicylindrical substrate; (B) image of the CurvACE prototype with a dragonfly.

(PCB) containing the electrical interconnections. In a next step, the assembled stack was diced into columns down to the flexible interconnection layer, which remained intact. Finally, the ommatidial array was curved along the bendable direction and attached to a rigid semicylindrical substrate with a radius of curvature of 6.4 mm. Two rigid circuit boards containing two microcontrollers, one three-axis accelerometer, and one three-axis rate gyroscope were inserted into the rigid substrate concavity and soldered to the sides of the ommatidia through dedicated pads. The prototype of a curved artificial compound eye contained 630 lens channels with an interommatidial angle of around 4° providing a FOV of 180° in the curved direction. These values correspond exactly to the values of a *Drosophila melanogaster* compound eye, but with the only difference that the fly's eye is 10 times smaller in scale [29]. The curved artificial compound eye, shown in Figure 7.23, is a lightweight, energy-efficient, miniature vision sensor with panoramic view. The restriction to a horizontal only bending of the image sensor enables a robot to navigate in flat environments only – still a long way to go to manufacture artificial compound sensors with excellence comparable to nature – but it is a first step and it opens up new avenues for vision sensors with alternative morphologies and FOVs of up to 360° in small packages. The recent trend to develop wearable electronics and cameras will certainly help to quickly improve the CurvACE concept to achieve more elaborate bioinspired vision sensors.

7.5.5 Conclusion

Mobile phone cameras are a flagship feature of a flagship device. Customers will always opt for the *better* or more prestigious camera. The scaling law of optics and the related space–bandwidth product (SBW) make it difficult to manufacture

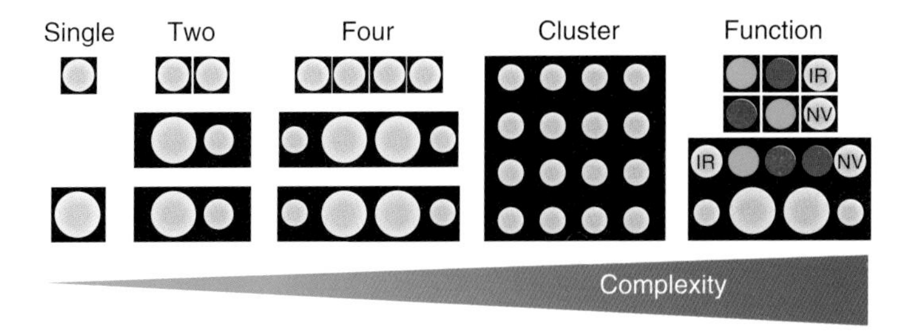

Figure 7.24 Different possible arrangements of camera modules for integration in smartphones or tablets.

a single-aperture camera with more than 10 megapixels fitting in a package of less than 5 mm overall thickness. Thus, the demand for higher and higher image quality of mobile phone cameras requires new strategies for image capturing sensor systems. The jumping spider's eye, a 4×2 cluster eye, is probably the best high-resolution vision system within a small package that nature developed in 530 million years. Is it possible to derive something similar in the technical world?

The pure cluster camera approach, for example, the 4×4 PiCam, is expensive in terms of manufacturing costs, expensive in terms of image computation hardware, and expensive in terms of energy consummation per picture taken. Another important issue for the acceptance of cluster cameras – and more generally for all computational image capturing technologies – is their inability to take pictures in critical environmental light conditions. Pure cluster cameras with computational image capturing have severe problems producing acceptable image quality in difficult lighting situations like in bright sunlight at the beach, in a room, in a nightclub, or at a concert. Scrambled or distorted images with artifacts are not acceptable for customers.

A biomimetic approach, for example, the combination of jumping spider's eye with *Deinopis* NV eyes, might be a better match than pure cluster cameras. The idea is to combine two cameras with different FOV: a wide-angle and a telescopic camera. Figure 7.24 shows different possible arrangements of camera modules for integration in smartphones or tablets.

Zooming from large field to small field could be achieved by combining both images. A dual camera concept also allows gives high resolution in the center of an image. Like the jumping spider, combine these two lenses with other lenses for specific tasks, for example, integrating thermal imaging (IR) in a mobile phone, as shown in Figure 7.25, or infrared front-facing cameras to detect the eyeball position to provide dynamic 3D perspective.

Stereoscopic vision, range finder, NV, and thermal vision (IR) will be available for consumer products soon and will immediately become standard for cars' onboard camera systems. Detecting a deer running on collision course or a fallen tree blocking the road in the dark forest will save many lives worldwide. For mobile phone cameras, it is more difficult to predict what will be the future camera systems convincing customers to purchase next generations of mobile

Figure 7.25 Portrait of the author performing a self-portrait using a FLIR ONE™, the first thermal imager designed for mobile phones and presented by FLIR Systems in 2014 [30].

phones and tablets. But biologically inspired cluster eye concepts will play a decisive role, and the rapidly advancing CMOS image sensor technology will help to revolutionize miniaturized vision systems in the near future.

Image sensor manufacturers recently announced that they are working on curved image sensors with more than 20 megapixels. If this technology becomes mature, it might trigger a true revolution in imaging. For an imaging system, a flat object normal to the optical axis will be imaged to a curved image plane, the Petzval field curvature. Upto now, photographic plates, films, CCD and CMOS image sensors, and all available image capturing devices have been planar. Correcting an objective for field curvature (Petzval) is very difficult for lens designers, especially for large object and image fields. A curved image sensor would allow to be simplified the lens design significantly. Using a curved image sensor allows to design high-speed lenses with very large image fields. For example, using the optical specifications of a *Deinopis* nocturnal eye, that is, 1.3 mm lens diameter, *f*/0.6, and a curved image sensor, this single-aperture lens could provide more than >10 megapixels within package of 5 mm size. If industry could also provide full-frame curved image sensors, all photographic cameras and lenses might change completely in design, size, and resolution. For now, just a dream – but imaginable.

References

1. Wikipedia Biomimetics, *http://en.wikipedia.org/wiki/Biomimetics* (accessed 15 April 2015).
2. Land, M.F. and Nilsson, D.-E. (2012) *Animal Eyes*, 2nd edn, Oxford University Press, ISBN: 978-0-19-958113-9.

3. Lohmann Adolf, W. (1989) Scaling laws for lens systems. *Appl. Opt.*, **28** (23), 4996–4998.
4. Voelkel, R. and Zoberbier, R. (2009) Inside wafer-level cameras. *Semicond. Int.*, 28–32.
5. Ozaktas, H.M., Urey, H., and Lohmann, A.W. (1994) Scaling of diffractive and refractive lenses for optical computing and interconnections. *Appl. Opt.*, **33** (17), 3782–3789.
6. Hofmann, C. (1980) *Die optische Abbildung*, Akademische Verlagsgesellschaft Geest & Portig K.-G., Leipzig.
7. Voelkel, R. (1999) Natural optical design concepts for highly miniaturized camera systems. *Proc. SPIE*, Design and Engineering of Optical Systems II, **3737**, 548. doi: 10.1117/12.360049.
8. Sanders, J.S. and Halford, C.E. (1995) Design and analysis of apposition compound eye optical sensors. *Opt. Eng.*, **34**, 222–235.
9. Snyder, A.W. (1977) Acuity of compound eyes: physical limitations and design. *J. Comp. Physiol. A*, **116**, 161–182.
10. Franceschini, N., Riehle, A., and Le Nestour, A. (1989) in *Facets of Vision* (eds D.G. Stavenga and R. Hardie), Springer-Verlag, Berlin, pp. 360–390.
11. Wikipedia. *http://en.wikipedia.org/wiki/Jumping_spider*, (accessed June 2014).
12. Hooke, R. (1665) *Micrographia*, J. Martyn and J. Allestry, Royal Society, London.
13. Hess, W.R. (1912) Stereoscopic pictures. US Patent 1128979.
14. Voelkel, R. (2012) Wafer-scale micro-optics fabrication. *Adv. Opt. Technol.*, **1**, 135–150. doi: 10.1515/aot-2012-0013 (Review article).
15. Gale, M.T., Lehmann, H.W., and Widmer, R.W. (1977) Color diffractive subtractive filter master recording comprising a plurality of superposed two-level relief patterns on the surface of a substrate. US Patent 4155627.
16. Lohmueller, T., Brunner, R., and Spatz, J.P. (2010) Improved properties of optical surfaces by following the example of the moth eye, in *Biomimetics Learning from Nature* (ed M. Amitava), InTech. ISBN: 978-953-307-025-4.
17. Clapham, P.B. and Hutley, M.C. (1973) Reduction of lens reflection by the 'moth eye' principle. *Nature*, **244**, 281–282. doi: 10.1038/244281a0
18. Rudmann H, Rossi M (2004) Design and fabrication technologies for ultraviolet replicated micro-optics. *Opt. Eng.*; **43**(11), 2575–2582.
19. Schmitt, H. *et al.* (2010) Full wafer microlens replication by UV imprint lithography. *Microelectron. Eng.*, **87** (5-8), 1074–1076.
20. Voelkel, R., Duparre, J., Wippermann, F., Dannberg, P., Braeuer, A., Zoberbier, R., Hansen, S., and Suess, R. (2008) Technology trends of microlens imprint lithography and wafer level cameras (WLC). 14th Micro-Optics Conference (MOC, 08), Brussels, Belgium, Technical Digest, September 25–27, 2008, pp. 312–315.
21. Iga, K., Kokubun, Y., and Oikawa, M. (1984) *Fundamentals of Microoptics: Distributed-Index, Microlens, and Stacked Planar Optics*, Academic Press Inc., ISBN-10: 0123703603.
22. Voelkel, R., Herzig, H.P., Nussbaum, Ph., Singer, W., Weible, K.J., and Daendliker, R., and Hugle, W.B. 1995 Microlens lithography: a new fabrication method for very large displays. Asia Display'95, pp. 713–716.
23. EU-IST-2001-35366, Project WALORI (2002-2005) WAfer Level Optic Solution for Compact CMOS Imager. Partners: Fraunhofer IOF, CEA LETI, ATMEL, IMT Neuchâtel, Fresnel Optics, SUSS MicroOptics.
24. Voelkel, R. and Wallstab, S. (1999) Flat image acquisition system. Offenlegungsschrift DE 199 17 890 A1, WO 00/64146, Apr. 20 1999.
25. Wippermann, F. and Brückner, A. (2012) Ultra-thin wafer-level cameras. *SPIE Newsroom*. doi: 10.1117/2.1201208.004430
26. Brückner A., Leitel R., Oberdörster A., Dannberg P., Wippermann F., Bräuer A. (2011) Multi-aperture optics for wafer-level cameras, *J. Micro/Nanolithog., MEMS MOEMS*; **10**(4), 043010-043010-10, doi: 0.1117/1.3659144.

27. Venkataraman, K., Lelescu, D., Duparré, J., McMahon, A., Molina, G., Chatterjee, P., and Mullis, R. (2013) PiCam: an ultra-thin high performance monolithic camera array. 6th ACM Siggraph Conference and Exhibition on Computer Graphics and Interactive Techniques in Asia, published in ACM Transactions on Graphics (TOG), 32(6).

28. Zufferey, J.-C. (2008) *Bio-inspired Flying Robots*, CRC Press. ISBN 978-1-4200-6684-5.

29. Floreano, D., Pericet-Camara, R., Viollet, S., Ruffier, F., Brückner, A., Leitel, R., Buss, W., Menouni, M., Expert, F., Juston, R., Dobrzynski, M.K., L'Eplattenier, G., Recktenwald, F., Mallot, H.A., and Franceschini, N. (2013) Miniature curved artificial compound eyes. *Proc. Natl. Acad. Sci. U.S.A.*, **110** (23), 9267–9272. doi: 10.1073/pnas.1219068110

30. FLIR Systems Inc. *http://flir.com/flirone/* (accessed 15 April 2015)

8
Plenoptic Cameras

Fernando Pérez Nava, Alejandro Pérez Nava, Manuel Rodríguez Valido, and Eduardo Magdaleno Castellò

8.1
Introduction

The perceptual system provides to biological and artificial agents information about the environment in which they inhabit. This information is obtained by capturing one or more of the types of energy that surrounds them. By far, the most informative is light, electromagnetic radiation in the visible spectrum. The sun fills the earth with photons that are reflected and absorbed by the environment and finally perceived by visual sensors. This initiates several complex information processing tasks that supply the agents with an understanding of their environment. There is an enormous interest to comprehend and replicate the biological visual perceptual system in artificial agents. The first step of this process is deciding how to capture light. To design visual sensors in artificial agents, we are not constrained by evolutionary or biological reasons and we may question if it is appropriate to replicate the human visual system. Although at first sight it seems reasonable that sensors resembling our eyes are more appropriate for the tasks that we are interested for computers to automate, other eye designs present in animals or not explored by the evolution may provide better performance for some tasks.

Eye design in animals began with simple photosensitive cells to capture light intensity. From these remote beginnings, eye evolution has been guided toward increasing the information that can be extracted from the environment [1]. When photosensitive cells were placed on a curved surface, directional selectivity was incorporated to the recorded light intensity. The curvature of this inward depression increased and finally formed a chamber, with an inlet similar to a pinhole camera. Pinhole cameras face a trade-off between low light sensitivity (increasing with bigger holes) and resolution (increasing with smaller holes). The solution both in photography and animal eye evolution is the development of a lens. The resulting eye is the design found in vertebrates. The main advantages of single-aperture eyes are high sensitivity and resolution, while its drawbacks are the small size of the field of view and its large volume and weight [2]. For small invertebrates,

Biologically Inspired Computer Vision: Fundamentals and Applications, First Edition.
Edited by Gabriel Cristóbal, Laurent Perrinet, and Matthias Keil.

eyes are very expensive in weight and brain processing needs, so the compound eye was developed as a different solution toward achieving more information. Rather than curving the photoreceptive surface inward, evolution has led to a surface curved outward in these creatures. This outward curvature results in a directional activation of specific photosensitive cells that is similar to the single-chamber eye, except that this configuration does not allow for the presence of a single lens. Instead, each portion of the compound eye is usually equipped with its own lens forming an elemental eye. Each elemental eye captures light from a certain angle of incidence contributing to a spot in the final image. The final image of the environment consists of the sum of the light captured by all the elemental eyes, resulting in a mosaic like structure of the world. Inspired by their biological counterpart, artificial compound eye imaging systems have received a lot of attention in recent years. The main advantages of compound eye imaging systems over the classical camera imaging design are thinness, lightness, and wide field of view. Several designs of compound eye imaging systems have been proposed over the last years [2].

In this chapter, we will present the plenoptic camera and study its properties. As pointed out in Ref. [3], from the biological perspective, the optical design of the plenoptic camera can be thought of as taking a human eye (a camera) and replacing its retina with a compound eye (a combination of a microlens and a photosensor array). This allows the plenoptic camera to measure the 4D light field composed by the radiance and direction of all the light rays in a scene. Conventional 2D images are obtained by 2D projections of the 4D light field. The fundamental ideas behind the use of plenoptic cameras can be traced back to the beginning of the previous century. These ideas have been recently implemented in the field of computational photography, and now, there are several commercial plenoptic cameras available [4, 5].

In this chapter, we will expose the theory behind plenoptic cameras and their practical applications. In Section 8.2, we will introduce the light field concept and the devices that are used to capture it. Then, in Section 8.3, we will give a detailed description of the standard plenoptic camera, describing its design, properties, and how it samples the 4D light field. In Section 8.4, we will show several applications of the plenoptic camera that extend the capabilities of current imaging sensors like perspective shift, refocusing the image after the shot, recovery of 3D information, or extending the depth of field of the image. Plenoptic camera also has several drawbacks like a reduction in spatial resolution compared with a conventional camera. We will explain in Section 8.4 how to increase the resolution of the plenoptic camera using superresolution techniques. In Section 8.5, we will introduce the generalized plenoptic camera as an optical method to increase the spatial resolution of the plenoptic camera and explore its properties. Plenoptic image processing requires high processing power, so in Section 8.6, we will review some implementations in high-performance computing architectures. Finally, in Section 8.7, we summarize the main outcomes of this chapter.

8.2 Light Field Representation of the Plenoptic Function

8.2.1 The Plenoptic Function

Every eye captures a subset of the space of the available visual information. The content in this subset and the capability to extract the necessary information determine how well the organism can perform a given task. To extract and process the visual information for a specific task, we need to know what is the information that we need and the camera design and image representation that optimally facilitate processing such visual information [6]. Therefore, our first step must be to describe the visual information available to an agent at any point in space and time. As pointed out in Ref. [7], complete visual information can be described in terms of a high-dimensional plenoptic function (plenoptic comes from the Latin *plenus,* meaning full, and *opticus,* meaning related to seeing or vision). This function is a ray-based model for radiance. Rays are determined in terms of geometric spatial and directional parameters, time, and color spectrum. Since the plenoptic function concept is defined for geometric optics, it is restricted to incoherent light and objects larger than the wavelength of light. It can be described as the 7D function:

$$P(x, y, z, \theta, \phi, \lambda, t) \tag{8.1}$$

where (x, y, z) are the spatial 3D coordinates, (θ, ϕ) denote any direction given by spherical coordinates, λ is a wavelength, and t stands for time. The plenoptic function is also fundamental for image-based rendering (IBR) in computer graphics that studies the representation of scenes from images. The dimensionality of the plenoptic function is higher than is required for IBR, and in Ref. [8] the dimensionality was reduced to a 4D subset, the light field, making the storage requirements associated to the representation of the plenoptic function more tractable. This dimensionality reduction is obtained through several simplifications. Since IBR uses several color images of the scene, the wavelength dimension is described through three color channels, that is, red, green, and blue channels. Each channel represents the integration of the plenoptic function over a certain wavelength range. As the air is transparent and the radiances along a light ray through empty space remain constant, it is not necessary to record the radiances of a light ray on different positions along its path since they are all identical. Finally, it is considered that the scene is static; thus, the time dimension is eliminated. The plenoptic function is therefore reduced to a set of 4D functions:

$$L(x, y, \theta, \phi) \tag{8.2}$$

Note that we have one 4D function for each color plane making the plenoptic function more manageable.

8.2.2
Light Field Parameterizations

The light field can be described through several parameterizations. The most important are the spherical–Cartesian parameterization, the sphere parameterization, and the two-plane parameterization. The spherical–Cartesian parameterization $L(x, y, \theta, \phi)$, also known as the one-plane parameterization, describes a ray's position using the coordinates (x, y) of its point of intersection with a plane and its direction using two angles (θ, ϕ), for a total of four parameters. The sphere parameterization $L(\theta, \varphi, \phi, \psi)$ describes a ray through its intersection with a sphere. The 4D light field can also be described with the two-plane parameterization as $L(x, y, u, v)$, where the parameters (x, y) and (u, v) represent the intersection of the light ray with two parallel reference planes with axes denoted as x, y and u, v at one unit of distance [8]. The point on the x, y plane represents position, and the point on the u, v plane represents direction, so the terms spatial and angular reference planes are used to describe each plane. The coordinates of the intersection point with the u, v plane may be defined as relative coordinates with respect to the intersection on the x, y plane or as absolute coordinates with respect to the origin of the u, v plane. Note that all previous parameterizations represent essentially the same information, the 4D light field, and conversions between them is possible except where rays run parallel to the reference planes in the two-plane parameterization. In Figure 8.1, we show the different parameterizations in 3D space. In order to make the notation and graphical representations of the 4D light field simpler, we will use through the chapter a 2D version of the 4D light field and 2D versions of the 4D parameterizations. The generalization to the 4D case is straightforward. All parameterizations are shown in Figure 8.1.

The light field can also be described in the frequency domain by means of its Fourier transform. Using the two-plane parameterization, we have

$$\hat{L}(\xi, \eta) = \int L(x, u) e^{-2\pi j(x\xi + u\eta)} dx du \tag{8.3}$$

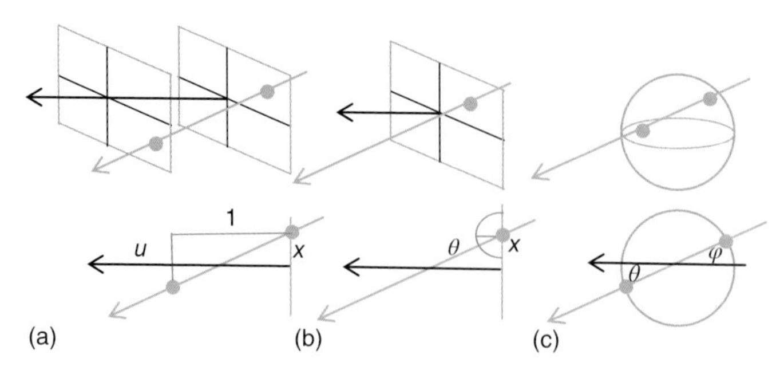

Figure 8.1 Light field parameterizations. (a) Two-plane parameterization, (b) spherical–Cartesian parameterization, and (c) sphere parameterization.

where (ξ, η) are the spatial frequency variables and $\hat{L}$ is the Fourier transform of the light field.

8.2.3 Light Field Reparameterization

When light traverses free space, the radiance along the light ray is unchanged, but the light field representation depends on the position of the reference planes. Different reference planes will be fixed to the optical elements of the plenoptic camera to study how rays change inside the camera body. Thus, the reference planes can be moved to arbitrary distance d or be coincident with the plane of a lens with focal length f.

If we displace the reference planes by d units, we can relate the original L_0 and new representation L_1 through a linear transform defined by a matrix $\boldsymbol{T}_d$. Similarly, we can describe the light field when light rays pass through a converging lens with focal length f aligning the spatial reference plane with the lens. In this case, we also obtain a linear transform defined by a matrix $\boldsymbol{R}_f$ considering the thin lens model in Gaussian optics [9].

The matrix description of the transforms is shown below in the spatial and frequency domain using the two-plane parameterization with relative spatial coordinates (x, u) and relative frequency coordinates (ξ, η) [10]:

$$L_1((x,u)^{\mathrm{T}}) = L_0(\boldsymbol{T}_d(x,u)^{\mathrm{T}}),\ \boldsymbol{T}_d = \begin{pmatrix} 1 & -d \\ 0 & 1 \end{pmatrix}, \hat{L}_1((\xi,\eta)^{\mathrm{T}}) = \hat{L}_0(\boldsymbol{T}_d^{-\mathrm{T}}(\xi,\eta)^{\mathrm{T}}) \quad (8.4)$$

$$L_1((x,u)^{\mathbf{T}}) = L_0(\boldsymbol{R}_f(x,u)^{\mathrm{T}}),\ \boldsymbol{R}_f = \begin{pmatrix} 1 & 0 \\ 1/f & 1 \end{pmatrix}, \hat{L}_1((\xi,\eta)^{\mathbf{T}}) = \hat{L}_0(\boldsymbol{R}_f^{-\mathrm{T}}(\xi,\eta)^{\mathrm{T}}) \quad (8.5)$$

Finally, it may be desirable to describe the light field by two reference planes separated by a distance B in absolute coordinates instead of the relative coordinates with unit distance separation. This is useful to parameterize the light field inside a camera where the spatial reference plane is aligned with the image sensor and the directional plane is aligned with the main lens. In this case, we can relate the original L_0 and new representation L_1 through a linear transform defined by a matrix $\boldsymbol{S_B}$:

$$L_1((x,u)^{\mathrm{T}}) = \frac{1}{B} L_0(\boldsymbol{S}_B(x,u)^{\mathrm{T}}),\ \boldsymbol{S}_B = \begin{pmatrix} 1 & 0 \\ -\frac{1}{B} & \frac{1}{B} \end{pmatrix}, \hat{L}_1((\xi,\eta)^{\mathbf{T}}) = B\hat{L}_0(\boldsymbol{S}_B^{-\mathrm{T}}(\xi,\eta)^{\mathrm{T}}) \quad (8.6)$$

where the scaling factor is the Jacobian of the transform. Since the transformations described above are linear, they can be combined to model more complex transforms simply by multiplying their associated matrices. Inverting the transform, it is also possible to describe the original light field L_0 in terms of the new representation L_1.

8.2.4 Image Formation

An image sensor integrates all the incoming radiances into a sensed irradiance value $I(x)$:

$$I(x) = \int L(x, u)du \tag{8.7}$$

Therefore, a conventional image is obtained integrating the light field with relative spatial coordinates along the directional u-axis. In Eq. (8.7), there is also an optical vignetting term. In the paraxial approximation [9], this term can be ignored. Geometrically, to obtain an image, we project the 2D light field into the spatial reference plane. In general case, for an n-dimensional signal, integration along $n-k$ dimensions generates a k-dimensional image. For a 4D light field, we integrate along the 2D directional variables to obtain a 2D image. According to the slice theorem [11], projecting a signal in the spatial domain is equivalent to extracting the spectrum of a signal in the Fourier domain. This process is known as a slicing operation and can be described as

$$\hat{I}(\xi) = \hat{L}(\xi, 0) \tag{8.8}$$

The advantage of this formulation is that integration in the spatial domain is replaced by taking the slice $\eta = 0$ in the frequency domain that is a simpler operation.

8.2.5 Light Field Sampling

Light field recording devices necessarily capture a subset of the light field in terms of discrete samples. The discretization comes from two sources: the optical configuration of the device and the discrete pixels in the imaging sensor. The combined effect of these sources is to impose a specific geometry on the overall sampling of the light field. Understanding the geometry of the captured light field is essential to compare the different camera designs. In this section, we will explore different solutions to sample the light field and explore its properties.

8.2.5.1 Pinhole and Thin Lens Camera

The sampling geometry of an ideal pinhole camera is depicted in 2D in Figure 8.2. While this model treats pixels as having a finite spatial extent, the aperture is modeled as being infinitesimally small. The integration volume corresponding to each pixel of size Δp is a segment in the x-axis, taking on a value of zero everywhere except for $u = 0$. The sampling geometry of the thin lens model with an aperture of size D is shown in Figure 8.3. Now, the integration volume for each pixel is a rectangle in the x, u plane, but as we can see from the sampling geometry, very low resolution is obtained in the directional reference plane u, so several approaches have been proposed to improve the directional resolution of the light field.

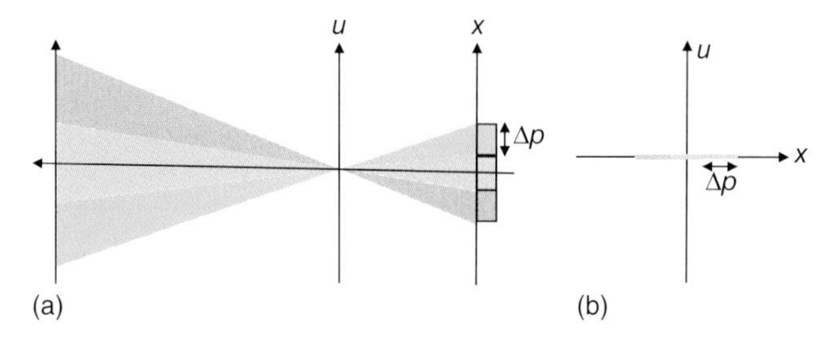

Figure 8.2 Pinhole camera. (a) Optical model and (b) light field representation.

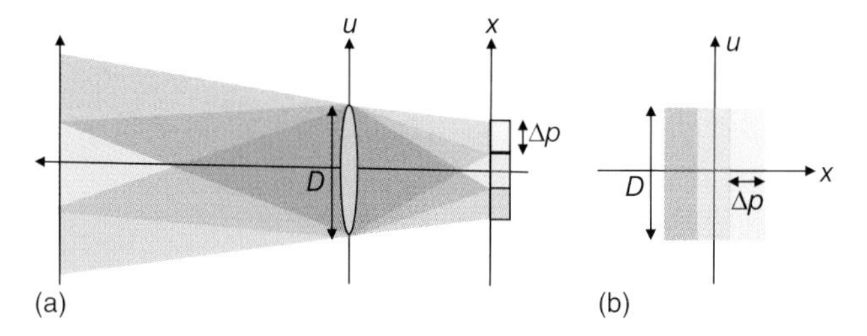

Figure 8.3 Thin lens camera. (a) Optical model and (b) light field representation. D is the lens aperture and Δp is the pixel size.

8.2.5.2 Multiple Devices

A straightforward approach to capture more information from the light field is the use of multiple devices. This is a solution also adopted in the biological domain that can be replicated with the use of an array of conventional 2D cameras distributed on a plane [12, 13]. If we adopt the pinhole approximation, sampling the light field from a set of images corresponds to inserting 2D slices into the 4D light field representation. Considering the image from each camera as a 2D slice of the 4D light field, an estimate of the light field is obtained from concatenating the captured slices. Camera arrays need geometric and color calibration and are expensive and complex to maintain, so there has been an effort to produce more manageable devices. In Figure 8.4, we show an example of a compact camera array composed of nine OV2640 cameras. The separation between them is 35 mm, the whole board is governed by a field programmable gate array (FPGA), and data transmission is performed through an Ethernet 1000 MB under UDP/IP. More compact arrays for mobile phones have also been developed [14].

8.2.5.3 Temporal Multiplexing

Camera arrays cannot provide sufficient light field resolution in the directional dimension for several applications. This is a natural result of the camera size that physically limits the cameras from being located close to each other. To handle this limitation, the temporal multiplexing approach uses a single camera to capture multiple images from the camera array by moving its location [8, 15]. The position of the camera can be controlled precisely using a camera gantry or estimated

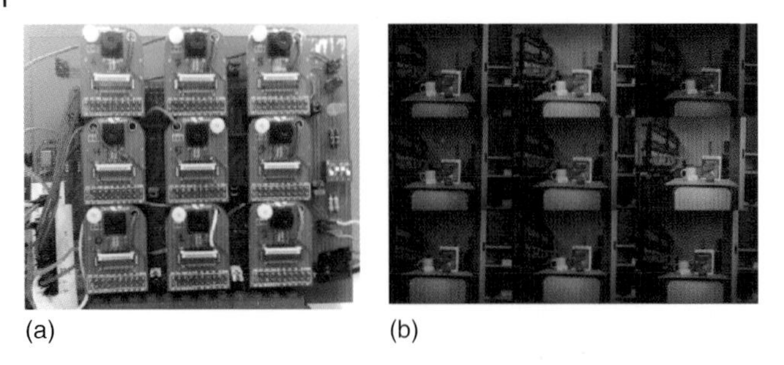

(a) (b)

Figure 8.4 (a) A compact camera array. (b) Raw images from the array. Note the need for geometric and color calibration.

from the captured images by structure from motion (SFM) algorithms. The disadvantages of the time-sequential capture include the difficulty of capturing dynamic environments and videos, its slowness because the position of the camera or the object has to be altered before each exposure, and the necessity of either gantry or the SFM algorithms to work in controlled environments.

8.2.5.4 Frequency Multiplexing

Temporal multiplexing solves some of the problems associated with camera arrays, but it has to be applied to static scenes, so other means of multiplexing the 4D light field into a 2D image are required to overcome this limitation. In Ref. [16], the frequency multiplexing approach is introduced as an alternative method to capture the light field using a single device. The frequency multiplexing method encodes different slices of the plenoptic function in different frequency bands and is implemented by placing nonrefractive light-attenuating masks in front of the image sensor. These masks provide frequency-domain multiplexing of the 4D Fourier transform of the light field into the Fourier transform of the 2D sensor image. The main limitation of these methods is that the attenuating mask limits the light entering in the camera.

8.2.5.5 Spatial Multiplexing

Spatial multiplexing produces an array of elemental images within the image captured with a single camera. This method began in the previous century with the works of Ives and Lippmann on integral imaging [17, 18]. Spatial multiplexing allows capturing dynamic scenes and videos but transfers spatial sampling resolution in favor of directional sampling resolution. One implementation of a spatial multiplexing system uses an array of microlenses placed near the image sensor. This device is called a plenoptic camera and will be described in Section 8.3. Examples of such implementations can be found in [3, 19–22].

8.2.5.6 Simulation

Light fields can also be obtained by rendering images from 3D models [23, 24]. By digitally moving the virtual cameras, this approach offers a low-cost acquisition method and easy access to several properties of the objects like their geometry

Figure 8.5 Simulated images from a light field.

or textures. It is also possible to simulate imperfections of the optical device or the sensor's noise behavior. Figure 8.5 shows an example of a simulated light field.

8.2.6 Light Field Sampling Analysis

In the previous section, we reviewed several devices designed to sample the light field. If enough samples are taken from them, we expect to reconstruct the complete continuous light field. However, the light field is a 4D structure, so storage requirements are important and it is necessary to keep the samples to the minimum. The light field sampling problem is difficult because the sampling rate is determined by several factors, including the scene geometry, the texture on the scene surface, or the reflection properties of the scene surface. In Ref. [25], it was proposed to perform the light field sampling analysis applying the Fourier transform to the light field and then sampling it based on its spectrum for a camera array. Assuming constant depth, a Lambertian surface, and no occlusions, the light field in the spatial domain is composed of constant slope lines, and the Fourier spectrum of the light field is a slanted line. By considering all depths or equivalently all slanted lines, the maximum distance Δu between the cameras was found to be

$$\Delta u = \frac{1}{Bfh_d}, \; h_d = \frac{1}{z_{\min}} - \frac{1}{z_{\max}} \tag{8.9}$$

where B is the highest spatial frequency of the light field, $z_{\min}$ and $z_{\max}$ are the depth bounds of the scene geometry, and f is the focal length of the cameras. A more detailed analysis can be found in Ref. [26], where non-Lambertian and nonuniform sampling are considered. Sampling analysis of the plenoptic camera will be detailed in Section 8.4.1.

8.2.7 Light Field Visualization

In order to understand the light field properties, it is useful to have some way of visualizing its structure. Since it is not possible to display a 4D signal in a 2D image, the 4D light field will be represented joining multiple 2D slices from the 4D signal to obtain a 2D multiplexed image. The directional information for a light field with relative spatial coordinates (x, y, u, v) can be visualized extracting a u,

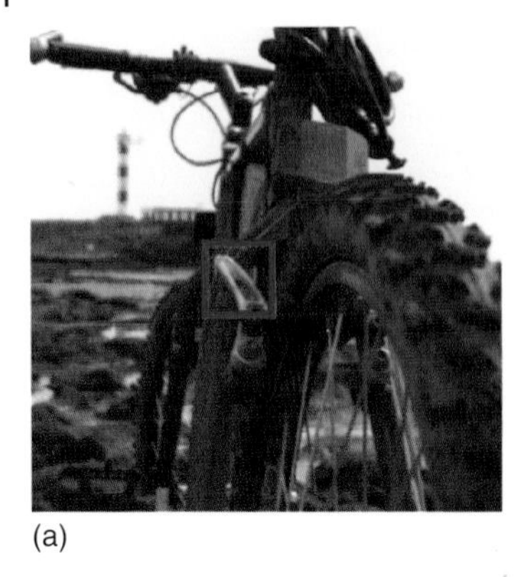
(a)

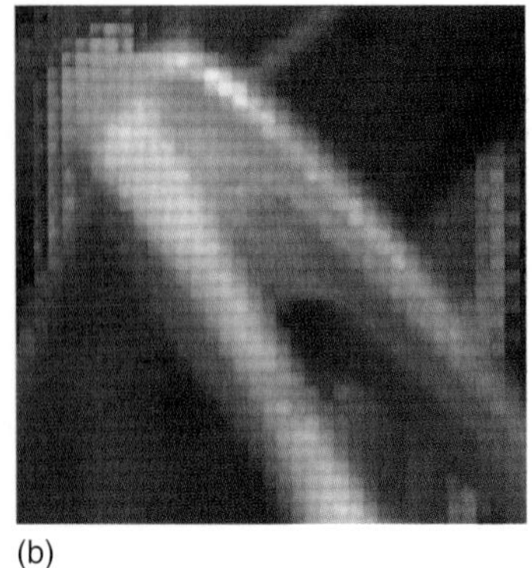
(b)

Figure 8.6 Fixing the spatial variables (*x*, *y*) and varying the directional variables (*u*, *v*) give the directional distribution of the light field. On (a), we see the 4D light field as a 2D multiplexed image. The image in (b) is a detail from (a). Each small square in (b) is the (*u*, *v*) slice for a fixed (*x*, *y*).

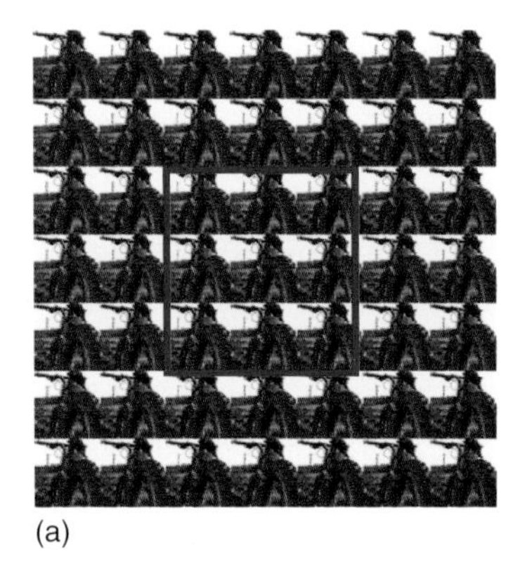
(a)

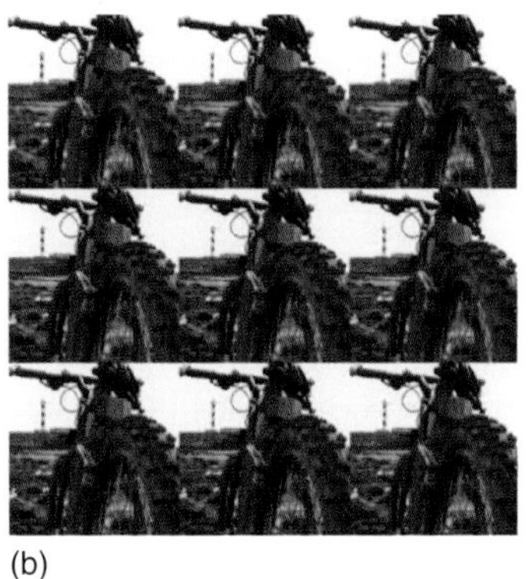
(b)

Figure 8.7 Fixing the directional variables (*u*, *v*) and varying the spatial variables (*x*, *y*) can be interpreted as taking images from a camera array. On (a), we see the 4D light field as another 2D multiplexed image. Image (b) is a detail from (a). Each image in (b) is the (*x*, *y*) slice for a fixed (*u*, *v*).

v slice of the light field fixing the spatial variables x and y. In Figure 8.6, we see all the u, v slices spatially multiplexed. Another alternative shown in Figure 8.7 is to fix a direction u, v and let the spatial variables x and y vary. These images can be interpreted as images from a camera array. Finally, in Figure 8.8, the pair of variables x and u (or y and v) are fixed and the other pair vary. The resulting images are called epipolar images of the light field. These epipolar images can be used to estimate the depth of the objects using the slope of the lines that compose them [27]. In a plenoptic camera, a line with zero slope indicates that the object is placed on the focus plane, while nonzero absolute values for slope indicate that the object is at a greater distance from the focus plane.

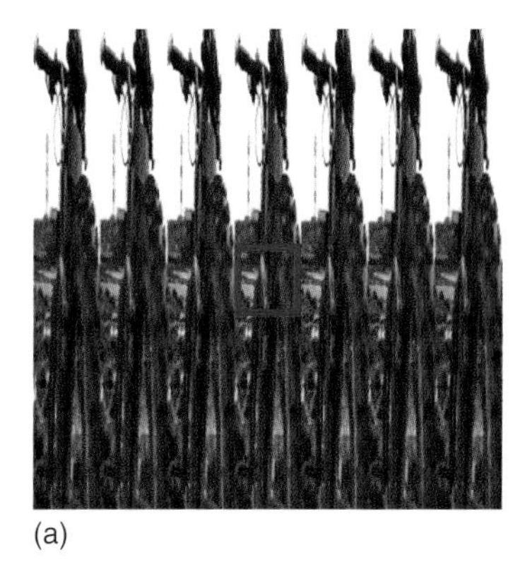

(a)

(b)

Figure 8.8 Fixing one-directional variable (*u* or *v*) and one spatial variable (*x* or *y*) generates another 2D multiplexed image. This is the epipolar image that gives information about the depth of the elements in the scene. On (b), we see a detail from (a). Note the different orientations corresponding to different depths in the scene.

8.3 The Plenoptic Camera

A plenoptic camera captures the 4D light field by placing a microlens array between the lens of the camera and the image sensor to measure the radiance and direction of all the light rays in a scene. The fundamental ideas behind the use of plenoptic cameras can be traced back to the beginning of the previous century with the works of Lippmann and Ives on integral photography [17, 18]. A problem of the Lippmann's method is the presence of pseudoscopic images (inverted parallax) when the object is formed at reconstruction. Different methods have been proposed for avoiding pseudoscopic images in real-time applications [28, 29]. One of the first plenoptic cameras based on the principles of integral photography was proposed in the computer vision field by Adelson and Wang [19] to infer depth from a single image. Other similar systems were built by Okano *et al.* [30] and Naemura *et al.* [31] using graded-index microlens arrays. Georgeiv *et al.* [32] have also used an integral camera composed of an array of lenses and prisms in front of the lens of an ordinary camera. Another integral imaging system is the Shack–Hartmann sensor [33] that is used to measure aberrations in optical systems. The plenoptic camera field experimented a great impulse when Ng [20] presented a portable handheld plenoptic camera. Its basic optical configuration comprises a photographic main lens, an $N_x \times N_x$ microlens array, and a photosensor array with $N_u \times N_u$ pixels behind each microlens. Figure 8.9 illustrates the layout of these components. Parameters D and d are the main lens and microlens apertures, respectively. Parameter B is the distance from the microlens x plane to the main lens u plane, and parameter b is the distance from the microlens plane to the sensor plane. Δp is the pixel size. The main lens may be translated along its optical axis to focus at a desired plane at distance A.

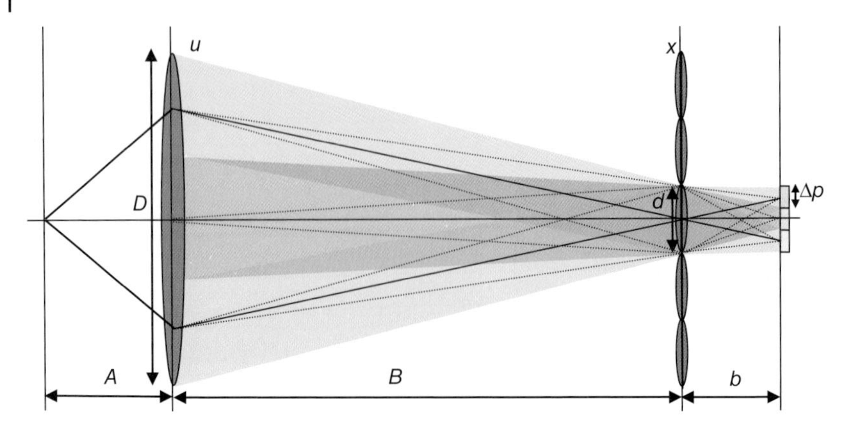

Figure 8.9 Optical configuration of the plenoptic camera (not shown at scale).

As shown in Figure 8.9, rays of light from a single point on space converge on a single location in the plane of the microlens array. Then, the microlens at that location separates these light rays based on direction, creating a focused image of the main lens plane.

To maximize the directional resolution, the microlenses are focused on the plane of the main lens. This may seem to require a dynamic change of the separation between the photosensor plane and microlens array since the main lens moves during focusing and zooming. However, the microlenses are vanishingly small compared to the main lens, so independently of its zoom or focus settings, the main lens is fixed at the optical infinity of the microlenses by placing the photosensor plane at the microlenses' focal depth [19].

To describe the light rays inside the plenoptic camera, an initial light field L_0 is defined aligning a directional reference plane with the image sensor and a spatial reference plane with the microlens plane. We can describe in absolute coordinates the light field inside the camera L_1 after light rays pass through the microlens s of length d and focal length f aligning the same spatial reference plane with the microlens plane and a directional reference plane with the main lens plane. The relation between L_0 and L_1 can be described using the transforms in Section 8.2.3 as

$$L_1((x,u)^{\mathrm{T}}) = \left(\frac{b}{B}\right) L_0(\boldsymbol{P}\boldsymbol{S}_b^{-1}\boldsymbol{T}_b\boldsymbol{R}_f\boldsymbol{S}_B(x-ds, u-ds)^{\mathrm{T}} + (ds, ds)^{\mathrm{T}}) \tag{8.10}$$

where $\boldsymbol{P}$ is a permutation matrix that interchanges the components of a vector since rays in L_0 and L_1 point in a different direction.

Equation (8.10) can be explained as follows. We take a ray in light field L_1 defined from a point x on the microlens plane to a point u on the main lens plane and denote it by (x, u), and we want to obtain the corresponding ray (x, u') defined from the microlens plane to the sensor plane as seen in Figure 8.10. To obtain (x, u'), we change the optical axis from the main lens to the microlens s of length d obtaining the ray $(x-ds, u-ds)$ with coordinates relative to the new optical axis. As shown in Figure 8.10, the distance between the microlens plane and the main

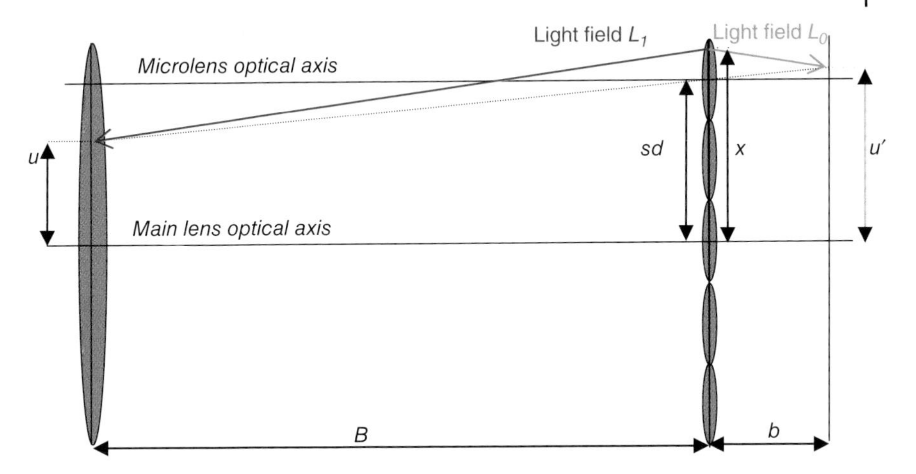

Figure 8.10 Ray transform in the plenoptic camera.

lens plane is B, so we use $\boldsymbol{S}_B$ to rewrite $(x-ds, u-ds)$ using a unit distance reference plane from the microlens plane in the main lens direction. Ray $\boldsymbol{S}_B(x-ds, u-ds)^{\mathrm{T}}$ passes through the microlens s with focal length f that changes its direction, so we use $\boldsymbol{R}_f$ to obtain the transformed ray $\boldsymbol{R}_f\boldsymbol{S}_B(x-ds, u-ds)^{\mathrm{T}}$ defined from the microlens plane to the unit distance plane. To obtain the corresponding ray in L_0, we have to move both reference planes b units using $\boldsymbol{T}_b$ to obtain the ray defined by $\boldsymbol{T}_b\boldsymbol{R}_f\boldsymbol{S}_B(x-ds, u-ds)^{\mathrm{T}}$. Now, we separate the unit distance reference plane to distance b obtaining the ray $\boldsymbol{S}_b^{-1}\boldsymbol{T}_b\boldsymbol{R}_f\boldsymbol{S}_B(x-ds, u-ds)^{\mathrm{T}}$ from the sensor plane to the microlens plane. Since we are interested that rays in L_0 point from the microlens plane to the sensor plane, we permute their coordinates using P. The final step is to move again the optical axis to the main lens axis obtaining the ray $(x, u')^{\mathrm{T}} = \boldsymbol{P}\boldsymbol{S}_b^{-1}\boldsymbol{T}_b\boldsymbol{R}_f\boldsymbol{S}_B(x-ds, u-ds)^{\mathrm{T}} + (ds, ds)^{\mathrm{T}}$ in L_0. The relation between the radiance of rays (x, u) in L_1 and (x, u') in L_0 is defined in Eq. (8.10).

Computing the matrix products, Eq. (8.10) can be rewritten as

$$L_1(x, u) = \left(\frac{b}{B}\right) L_0 \left(x, ds\left(1 + \frac{b}{B}\right) - \left(\frac{b}{B}\right) u\right) \tag{8.11}$$

In order to use as many photosensor pixels as possible, it is desirable to choose the relative sizes of the main lens and microlens apertures so that the images under the microlenses are as large as possible without overlapping. Using Eq. (8.11), we see that the image of the main lens aperture in relative coordinates to microlens s is approximately $-(b/B)u$ since $(1 + b/B) \approx 1$. To use the maximum extent of the image sensor, we must have the f-number condition $d/D = b/B$ since u varies within the aperture values $|u| \leq D/2$. Note that the f-number condition is defined in terms of the image-side f-number, which is the diameter of the lens divided by the separation between the corresponding planes. Using Eq. (8.11), we can represent the sampling geometry of each pixel in L_1 as shown in Figure 8.11. Note the difference between the sampling pattern of Figures 8.3 and 8.10. In a plenoptic camera, the directional coordinate u is sampled, while in a conventional camera,

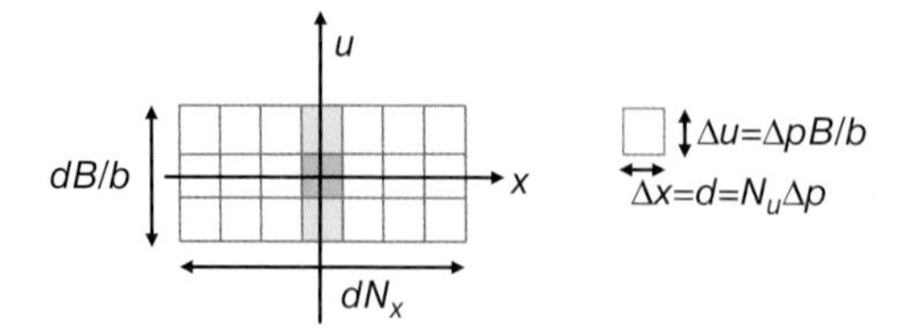

Figure 8.11 Sampling geometry of the plenoptic camera in L_1. Each vertical column corresponds to a microlens. Each square corresponds to a pixel of the image sensor.

the corresponding samples are integrated. Another result that can be seen from Figure 8.11 is that each row through the x-axis corresponds to an image taken from an aperture of size D/N_u. This implies that the depth of field of such images is extended N_u times relative to a conventional photograph.

The plenoptic model of this section represents a simplified representation of reality since it does not account for lens distortion or unwanted rotations between the microlens array and the photosensor plane. Therefore, a prior step to use this model in practical applications is to calibrate the plenoptic camera [34]. Images recorded from several plenoptic cameras can be obtained from [35–38].

8.4 Applications of the Plenoptic Camera

8.4.1 Refocusing

In conventional photography, the focus of the camera must be fixed before the shot is taken to ensure that selected elements in the scene are in focus. This is a result of all the light rays that reach an individual pixel being integrated as seen in Figure 8.3. With the plenoptic camera, the light field is sampled directionally, and the device provides the user with the ability to refocus images after the moment of exposure. If L_1 is a light field from the microlens plane to the main lens plane, the image is refocused by virtually moving the microlens plane to a distance $B\alpha$ from the main lens plane obtaining a light field L_r and applying the image formation operator in Eq. (8.7) as seen in Figure 8.12(b). For a 2D light field, we have

$$L_r(\alpha x, u) = L_1(\boldsymbol{S}_B^{-1}\boldsymbol{T}_{(1-\alpha)B}\boldsymbol{S}_{B\alpha}(\alpha x, u)^{\mathrm{T}}) = L_1((1 - 1/\alpha\,)u + x, u) \tag{8.12}$$

Then, applying the image formation operator, we obtain

$$I_{\alpha B}(\alpha x) = \frac{1}{\alpha B}\int L_r(\alpha x, u)du = \frac{1}{\alpha B}\int L_1\left(\left(1 - \frac{1}{\alpha}\right)u + x, u\right)du \tag{8.13}$$

From Eq. (8.13), refocusing to a virtual plane is equivalent to integration along lines with slope $(1 - 1/\alpha)$. It can be interpreted as shifting for each u the

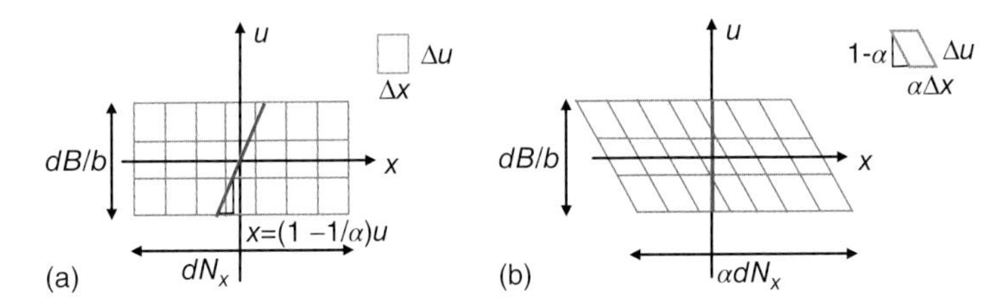

Figure 8.12 (a) Light field L_1 inside the camera. Refocusing is equivalent to integration along lines with x–u slope $(1 - 1/\alpha)$. (b) Equivalent light field L_r with x–u slope $(1 - \alpha)$ after virtually moving the focusing plane.

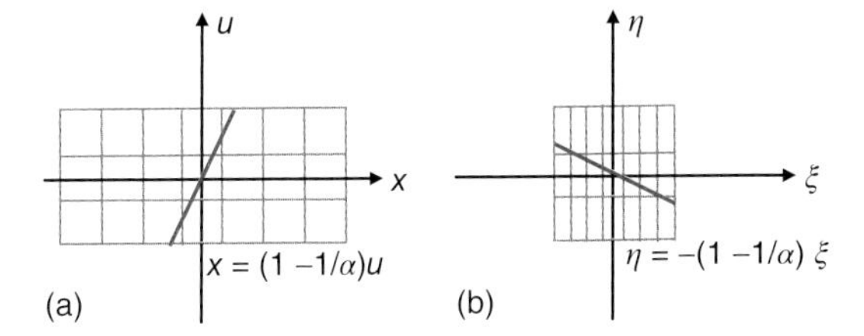

Figure 8.13 Fourier slice theorem. (a) Spatial integration. (b) Fourier slicing.

subaperture image $L_1(x\,,u)$ with an amount of $(1 - 1/\alpha\,)u$ and adding all the shifted subaperture images. If we assume that a plenoptic camera is composed of $N_x \times N_x$ microlenses and each microlens generates an $N_u \times N_u$ image with $\mathrm{O}(N_x) = \mathrm{O}(N_u)$, then the full sensor resolution is $\mathrm{O}((N_x\ N_u)^2) = \mathrm{O}(N^4)$. In this case, the shift and add approach needs $\mathrm{O}(N^4)$ operations to generate a single photograph refocused on a determined distance.

A significant improvement to this performance was obtained in Ref. [20] with the Fourier slice photography technique that reformulates the problem in the frequency domain. This technique decreased the computational cost to $\mathrm{O}(N^4\log(N))$ to generate a set of photographs with size $\mathrm{O}(N^2)$, giving $\mathrm{O}(N^2\log(N))$ operations to compute a $\mathrm{O}(N^2)$ refocused image. This technique is based on the slicing operation in Eq. (8.8). Then, using Eq. (8.13), we obtain Eq. (8.14) indicating that a refocused photograph is a slanted slice in the Fourier transform of the light field. This can be seen graphically in Figure 8.13:

$$\hat{L}_1\left(\left(1 - \frac{1}{\alpha}\right)u + x\,,u\right)(\xi,\eta) = \hat{L}_1\left(\xi, -\left(1 - \frac{1}{\alpha}\right)\xi + \eta\right)$$
$$\hat{I}_{\alpha B}(\alpha x)(\xi) = \frac{1}{(\alpha B)}\hat{L}_1\left(\xi, -\left(1 - \frac{1}{\alpha}\right)\xi\right) \tag{8.14}$$

Then, in order to obtain a photograph, we have to compute first the Fourier transform of the light field $\hat{L}_1$. After that, we have to evaluate the slice $\hat{L}_1(\xi, -(1 - 1/\alpha)\xi)$, and finally, we have to compute the inverse Fourier transform of the slice. A discrete version of the Fourier slice theorem can be found in Ref. [39]. An example of the refocusing technique is shown in Figure 8.14. To characterize the performance of the refocusing process, it is necessary to obtain the range of depths for which we can compute an exact refocused photograph.

Figure 8.14 Image refocusing at several distances using the discrete focal stack transform [39].

If we assume a band-limited light field, the refocusing range is [20]

$$\left|1 - \frac{1}{\alpha}\right| \leq \frac{\Delta x}{\Delta u} \tag{8.15}$$

In this slope interval, the reconstruction of the photograph from the samples of the light field is exact.

8.4.2 Perspective Shift

In addition to the refocusing capability of the plenoptic camera, it is also possible to view the scene from different perspectives, allowing the viewer a better angular knowledge of the world. As explained in Section 8.3, the pixels under each microlens collect light rays from a certain direction. Thus, considering the same pixel under each microlens, it is possible to compose a picture of the scene viewed from that direction. Perspective images are shown in Figure 8.7.

8.4.3 Depth Estimation

Plenoptic cameras can also provide 3D information about the scene indicating the depth of all pixels in the image [39, 40]. In fact, the plenoptic camera as a depth-sensing device is equivalent to a set of convergent cameras [41]. Based on this 3D information, it is also possible to obtain extended depth of field images in which all the elements in the image are in focus. In Figure 8.15, we show the basic geometry of depth estimation.

It can be seen in the figure above that a Lambertian point at distance Z in the scene is transformed through the lens to a point at distance αB from the main lens. The set of rays emanating from this point is simply a line in 2D plenoptic space with equation $x = (1 - 1/\alpha)u$. Then, we can obtain the depth of the point estimating the slope of the line in the epipolar image that was presented in Section 8.2.7. This approach is widely employed due to its simplicity [19, 42]. For example, we may write the light field L for all points (x_i, u_i) in a neighborhood $N(x, u)$ of a point

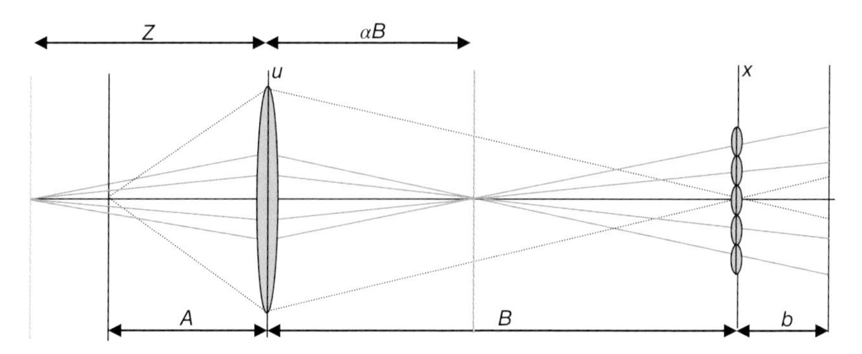

Figure 8.15 Geometry of depth estimation with a plenoptic camera.

(x, u) as $L((x_i, u_i) + \mathbf{v}) \approx L((x_i, u_i)) + \nabla L(x_i, u_i)^{\mathrm{T}}\mathbf{v}$. Then, noting that in the direction $\mathbf{v} = ((1 - 1/\alpha), 1)$ the light field is constant due to the Lambertian assumption, we obtain the set of equations $e(x_i, u_i) = L_x(x_i, u_i)(1 - 1/\alpha) + L_u(x_i, u_i) = 0$. Then, we can define a variance measure $VM(x, u, \alpha)$ as the sum of the $e(x_i, u_i)^2$ in the neighborhood $N(x, u)$, and we can estimate the unknown slope through minimizing $VM(x, u, \alpha)$ as [19]

$$\left(1 - \frac{1}{\alpha}\right) = -\frac{\sum_{(x_i,u_i)\in N(x,u)} L_x(x_i, u_i)L_u(x_i, u_i)}{\sum_{(x_i,u_i)\in N(x,u)} L_x(x_i, u_i)^2} \tag{8.16}$$

$$VM(x, u, \alpha) = \sum_{(x_i,u_i)\in N(x,u)} \left(L_x\left(x_i, u_i\right)\left(1 - \frac{1}{\alpha}\right) + L_u(x_i, u_i)\right)^2 \tag{8.17}$$

Note that the estimation is unstable when the denominator is close to zero. This occurs in regions of the image with insufficient detail as is common in all passive depth recovery methods. Another alternative is to use a depth from focus approach. If we take a photograph of the Lambertian point at distance αB from the main lens with the sensor placed at a virtual distance of γB from the main lens, we obtain a constant element with radius

$$\mathrm{FM}(x, \alpha) = D\left|\left(1 - \frac{1}{\alpha}\right) - \left(1 - \frac{1}{\gamma}\right)\right| \tag{8.18}$$

Then, we can obtain refocused images for several values of γ and select the sharpest one (the image with the smallest radius). There are several focus measures FM that can be employed which are basically contrast detectors within a small spatial window in the image. Unfortunately, the selection of the sharpest point is also unstable when the image has insufficient detail and some regularization method has to be employed to guide the solution. In Ref. [39], the Markov random field approach is used to obtain the optimal depth minimizing an energy function. Other regularization approaches include variational techniques [42]. Some examples of depth estimation are shown in Figure 8.16. In order to

Figure 8.16 Depth estimation and extended depth of field images.

use the plenoptic camera as a depth recovery device, it is interesting to study its depth discrimination capabilities. This is defined as the smallest relative distance between two on-axis points, one at the focusing distance Z of the plenoptic camera and the other at Z' which would give a slope connecting the central pixel of the central microlens with a border pixel of an adjacent microlens. The relative distance can be approximated as [43]

$$\frac{(Z' - Z)}{Z} \approx \pm dk\left(\frac{\theta}{S}\right)^2 Z \tag{8.19}$$

where d is the microlens size, k is the f-number of the camera, θ is the field of view, and S is the sensor size. Therefore, to increase the depth discrimination capabilities, it is necessary to select small values of the microlens size, f-number, field of view, and focusing distance and to select high values of the sensor size.

8.4.4 Extended Depth of Field Images

Another feature of plenoptic cameras is their capability to obtain extended depth of field images also known as all-in-focus images [3, 39]. The approach in light field photography is to capture a light field, obtain the refocused images at all depths, and select for each pixel the appropriate refocused images according to the estimated depth. Extended depth of field images is obtained in classical photography reducing the size of the lens aperture. An advantage of the plenoptic extended depth of field photograph is the use of light coming through a larger lens aperture allowing images with a higher signal to noise ratio. Examples of extended depth of field images are shown in Figure 8.16.

8.4.5 Superresolution

One of the major drawbacks of plenoptic cameras is that capturing a certain amount of directional resolution requires a proportional reduction in spatial resolution. If the number of microlenses is $N_x \times N_x$ and the pixels behind each microlens are $N_u \times N_u$, the refocused images have a resolution of $N_x \times N_x$ and sensor resolution is reduced with a factor of $(N_u)^2$. Two approaches to increase the resolution can be found in plenoptic cameras. In this section, we will present some software-based methods, while in Section 8.5, hardware-based methods

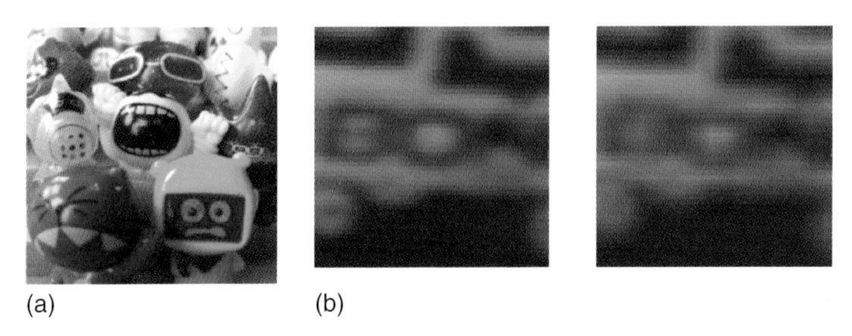

(a) (b)

Figure 8.17 (a) Superresolution photographs from a plenoptic camera. (b) Comparison of superresolution (left) against bilinear interpolation (right).

will be shown. Software-based superresolution methods discard the optimal sampling assumption of the light field of Section 8.4.1 and try to recover the light field using the aliased recorded plenoptic image. In Ref. [44], it is shown how to build superresolution refocused photographs that have a common size of approximately one-fourth of the full sensor resolution. It recovers the superresolution photograph, assuming that the scene is Lambertian and composed of several planes of constant depth. Since depth in the light field is unknown, it is estimated by the defocus method in Section 8.4.3. In Ref. [45], the superresolution problem was formulated in a variational Bayesian framework to include blur effects reporting a 7× magnification factor. In Ref. [46], the Fourier slice technique of [39] that was developed for photographs with standard resolution is extended to obtain superresolved images. Depth is also estimated by the defocus method in Section 8.4.3. In Figure 8.17, some results of the Fourier slice superresolution technique are shown.

8.5 Generalizations of the Plenoptic Camera

The previous section described a plenoptic camera where the microlenses are focused on the main lens. This is the case that provides a maximal directional resolution. The major drawback of the plenoptic camera is that capturing a certain amount of directional resolution requires a proportional reduction in the spatial resolution of the final photographs. In Ref. [3], a generalized plenoptic camera is introduced that provides a simple way to vary this trade-off, by simply reducing the separation between the microlenses and photosensor. This generalized plenoptic camera is also known as the focused plenoptic camera or plenoptic 2.0 camera [21]. To study the generalized plenoptic camera, β is defined to be the separation measured as a fraction of the depth that causes the microlenses to be focused on the main lens. For example, the standard plenoptic camera corresponds to $\beta = 1$, and the configuration where the microlenses are pushed against the photosensor corresponds to $\beta = 0$. A geometrical configuration of the generalized plenoptic camera is shown in Figure 8.18.

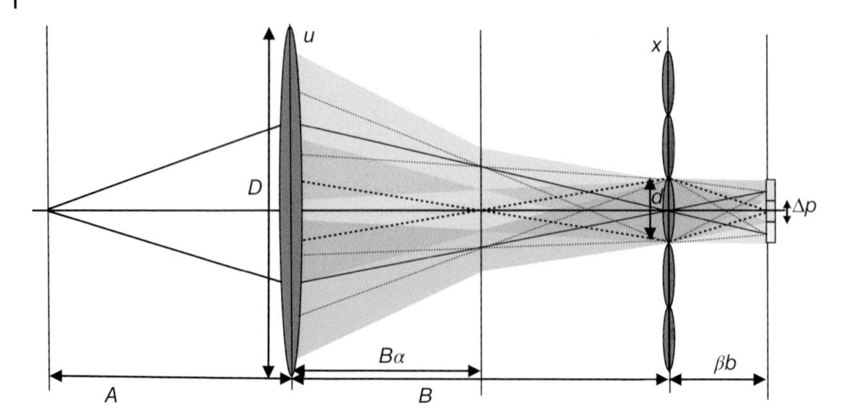

Figure 8.18 Geometry of a generalized plenoptic camera. Microlenses are now placed at distance βb from the sensor and focused on a plane at distance $B\alpha$ from the main lens. Standard plenoptic camera corresponds to $\beta = 1$, $\alpha = 0$.

As in Section 8.3, to describe the light rays inside the camera, a light field L_0 is defined aligning a directional reference plane with the image sensor and a spatial reference plane with the microlens plane. We also describe the light field inside the camera L_1 after light rays pass through the microlens s of length d aligning a spatial reference plane with the microlens plane and a directional reference plane with the main lens plane. The relation between L_0 and L_1 can be described as

$$L_1((x,u)^{\mathrm{T}}) = \left(\frac{b\beta}{B}\right) L_0(\boldsymbol{P}\boldsymbol{S}_{b\beta}^{-1}\boldsymbol{T}_{b\beta}\boldsymbol{R}_f\boldsymbol{S}_B(x-ds, u-ds)^{\mathrm{T}} + (ds, ds)^{\mathrm{T}}) \quad (8.20)$$

That can be explicitly written as

$$L_1(x,u) = \left(\frac{b\beta}{B}\right) L_0\left(x, ds\left(1+\frac{b\beta}{B}\right) + (x-ds)(1-\beta) - \left(\frac{b\beta}{B}\right)u\right) \quad (8.21)$$

Using Eq. (8.21), we can represent the sampling geometry of each pixel as shown in Figure 8.19(a). Note the difference between the sampling pattern of the standard plenoptic camera in Figure 8.11 and the generalized plenoptic camera. As the β value changes from 1, the light field sampling pattern shears within each microlens. This generates an increase in spatial resolution and a decrease in the directional resolution. If the resolution of the sensor is $N_p \times N_p$ and there are $N_u \times N_u$ pixels behind each of the $N_x \times N_x$ microlenses, the refocused images have a resolution of $N_r \times N_r$, where

$$N_r = \max((1 - \beta)\, N_p,\ N_x) \quad (8.22)$$

and the increment in resolution is $((1-\beta)\, N_u)^2$ that is $(1-\beta)^2$ times the number of pixels in each microlens [3, 47]. Note that the change of pixel size in the x-axis is reflected in Figures 8.11 and 8.19 varying from d to $\Delta p/(1-\beta)$ leading to the increment in resolution in Eq. (8.22).

In order to exploit the resolution capabilities of the generalized plenoptic camera, the focus of its main lens has to be done differently than in a conventional

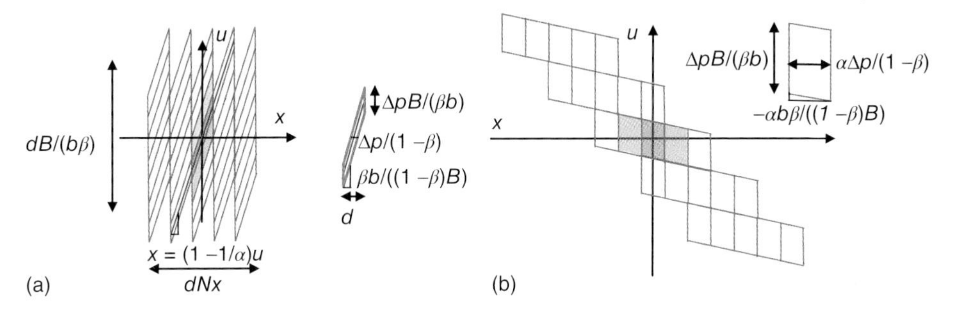

Figure 8.19 Sampling geometry of the generalized plenoptic camera. (a) Sampling in L_1. Each vertical column corresponds to a microlens. Each sheared square corresponds to a pixel of the image sensor. Pixel slope in the x–u plane is $\beta b/((1-\beta)B)$. (b) Sampling in L_r where the microlens plane is at distance $B\alpha$ from the main lens and α verifies Eq. (8.23).

or plenoptic camera [3]. In conventional and standard plenoptic cameras, best results are obtained by optically focusing on the subject of interest. On the other side, as shown in Figure 8.19(a), the highest final image resolution occurs when the refocusing plane verifies:

$$1 - \frac{1}{\alpha} = \frac{\beta b}{((1-\beta)B)} \tag{8.23}$$

or equivalently $\alpha = \beta b/(B(1-\beta))$. The sampling pattern for a light field Lr where the microlens plane is at distance $B\alpha$ from the main lens plane and α verifies Eq. 8.23 is shown in Figure 8.19(b). If we want to obtain the maximum resolution at depth B, we should optically focus the main lens by positioning it at depth B/α for an optimal focusing distance of

$$B_{\text{opt}} = B - \frac{b\beta}{(1-\beta)} \tag{8.24}$$

Refocused photographs from a generalized plenoptic camera are shown in Figure 8.20. Other generalizations of the standard design include the use of microlenses with different focal lengths [48, 49] to increase the depth of field of the generalized plenoptic camera or the use microlenses with different apertures to obtain high-dynamic-range imaging [50].

8.6 High-Performance Computing with Plenoptic Cameras

Processing and rendering the plenoptic images require significant computational power and memory bandwidth. At the same time, interactive rendering performance is highly desirable so that users can explore the variety of images that can be rendered from a single plenoptic image. For example, it is desirable to analyze the refocused photographs as a whole to estimate depth or to select a group of them with focus on determined elements. Therefore, it is convenient to use

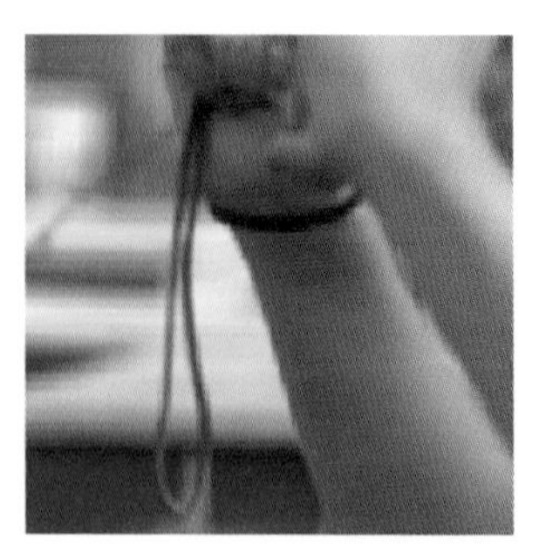

Figure 8.20 Refocusing from a generalized plenoptic camera.

fast algorithms and appropriate hardware platform such as FPGA or graphics processing unit (GPU). A GPU is a coprocessor that takes on graphical calculations and transformations so that the main CPU does not have to be burdened by them. The FPGAs are semiconductor devices that are based on a matrix of configurable logic blocks connected via programmable interconnects. This technology makes possible to insert the applications on a custom-built integrated circuits. Also, FPGA devices offer extremely high-performance signal processing through parallelism and highly flexible interconnection possibilities. In Ref. [44], a method is presented for simultaneously recovering superresolved depth and all-in-focus images from a standard plenoptic camera in near real time using GPUs. In Ref. [51], a GPU-based approach for light field processing and rendering for a generalized plenoptic camera is detailed. The rendering performance allows to process 39 Mpixel plenoptic data to 2 Mpixel images with frame rates in excess of 500 fps. Another application example for FPGAs is presented in Ref. [52]. The authors describe a fast, specialized hardware implementation of a superresolution algorithm for plenoptic cameras. Results show a time reduction factor of 50 in comparison with the solution on a conventional computer. Another algorithm designed for FPGA devices is shown in Ref. [53]. The authors describe a low-cost embedded hardware architecture for real-time refocusing based on a standard plenoptic camera. The video output is transmitted via High-Definition Multimedia Interface (HDMI) with a resolution of 720p at a frame rate of 60 fps conforming to the HD-ready standard.

8.7 Conclusions

During the last years, the field of computational photography continued developing new devices with unconventional structures that extend the capabilities of current commercial cameras. The design of these sensors is guided toward increasing the information that can be captured from the environment. The available visual information can be described through the plenoptic function and its simplified 4D light field representation. In this chapter, we have reviewed several devices to capture this visual information and shown how the plenoptic camera captures the light field by inserting a microlens array between the lens of the camera and the image sensor measuring the radiance and direction of all the light rays in a scene. We have explained how the plenoptic images can be used, for instance, to change the scene viewpoint, to refocus the image, to estimate the scene depth, or to obtain extended depth of field images. The drawback of the plenoptic camera in terms of spatial resolution loss has also been detailed, and some solutions to overcome these problems have also been presented. Finally, we have explored the computational part of the techniques presented in high-performance computing platforms such as GPUs and FPGAs.

The plenoptic camera has left the research laboratories, and now, several commercial versions can be found [4, 5, 13]. We expect that the extension of its use will increase the capabilities of current imaging sensors and drive imaging science to new directions.

References

1. Goldstein, E.B. (2009) *Encyclopedia of Perception*, SAGE Publications, Inc., Los Angeles, CA.
2. Duparré, J.W. and Wippermann, F.C. (2006) Micro-optical artificial compound eyes. *Bioinspir. Biomim.*, **1** (1), R1–16.
3. Ng, R. (2006) Digital light field photography. PhD dissertation, Stanford University.
4. Lytro. *https://www.lytro.com* (accessed 4 July 2014).
5. Raytrix. *http://www.raytrix.de/* (accessed 04 July 2014).
6. Neumann, J., Fermuller, C., and Aloimonos, Y. (2003) Eye design in the plenoptic space of light rays. Proceedings of the Ninth IEEE International Conference on Computer Vision, Vol. 2, pp. 1160–1167.
7. Adelson, E.H. and Bergen, J.R. (1991) The plenoptic function and the elements of early vision. Computational Models of Visual Processing, pp. 3–20.
8. Levoy, M. and Hanrahan, P. (1996) Light field rendering. in Proceedings of the 23rd Annual Conference on Computer Graphics and Interactive Techniques, New York, pp. 31–42.
9. Born, M. and Wolf, E. (1999) *Principles of Optics*, 7th edn, Cambridge University Press.
10. Liang, C.-K., Shih, Y.-C., and Chen, H.H. (2011) Light field analysis for modeling image formation. *IEEE Trans. Image Process.*, **20** (2), 446–460.
11. Bracewell, R.N. (1990) Numerical transforms. *Science*, **248** (4956), 697–704.
12. Wilburn, B., Joshi, N., Vaish, V., Talvala, E.-V., Antunez, E., Barth, A.A., Adams, A., Horowitz, M., and Levoy, M. (2005) High performance imaging using large camera arrays. ACM SIGGRAPH 2005 Papers, New York, pp. 765–776.
13. Liu, Y., Dai, Q., and Xu, W. (2006) A real time interactive dynamic light field transmission system. 2006 IEEE International Conference on Multimedia and Expo, pp. 2173–2176.
14. Pelican Imaging. *http://www.pelicanimaging.com/* (accessed 09 July 2014).
15. Gortler, S.J., Grzeszczuk, R., Szeliski, R., and Cohen, M.F. (1996) The lumigraph. Proceedings of the 23rd Annual Conference on Computer Graphics and Interactive Techniques, New York, pp. 43–54.
16. Veeraraghavan, A., Raskar, R., Agrawal, A., Mohan, A., and Tumblin, J. (2007) Dappled photography: mask enhanced cameras for heterodyned light fields and coded aperture refocusing. ACM SIGGRAPH 2007 Papers, New York.

17. Ives, F. (1903) Parallax stereogram and process of making same. US 725567.
18. Lippmann, G. (1908) Epreuves reversibles donnant la sensation du relief. *J. Phys.*, **7** (4), 821–825.
19. Adelson, E.H. and Wang, J.Y.A. (1992) Single lens stereo with a plenoptic camera. *IEEE Trans. Pattern Anal. Mach. Intell.*, **14** (2), 99–106.
20. Ng, R. (2005) Fourier slice photography. ACM SIGGRAPH 2005 Papers, New York, pp. 735–744.
21. Lumsdaine, A. and Georgiev, T. (2009) The focused plenoptic camera. 2009 IEEE International Conference on Computational Photography (ICCP), pp. 1–8.
22. Rodríguez-Ramos, J.M., Femenía Castellá, B., Pérez Nava, F., and Fumero, S. (2008) Wavefront and distance measurement using the CAFADIS camera. *Proc. SPIE*, Adaptive Optics Systems, **7015**, 70155Q–70155Q–10.
23. Nava, F.P., Luke, J.P., Marichal-Hernandez, J.G., Rosa, F., and Rodriguez-Ramos, J.M. (2008) A simulator for the CAFADIS real time 3DTV camera. 3DTV Conference: The True Vision – Capture, Transmission and Display of 3D Video, pp. 321–324.
24. Wanner, S., Meister, S., and Goldluecke, B. (2013) Datasets and benchmarks for densely sampled 4D light fields. Vision, Modelling and Visualization (VMV).
25. Chai, J.-X., Tong, X., Chan, S.-C., and Shum, H.-Y. (2000) Plenoptic sampling. Proceedings of the 27th Annual Conference on Computer Graphics and Interactive Techniques, New York, pp. 307–318.
26. Zhang, C. and Chen, T. (2003) Spectral analysis for sampling image-based rendering data. *IEEE Trans. Circuits Syst. Video Technol.*, **13** (11), 1038–1050.
27. Bolles, R.C., Baker, H.H., and Marimont, D.H. (1987) Epipolar plane image analysis: an approach to determining structure from motion. *Int. Comp. Vis.*, **1**, 1–7.
28. Arai, J., Kawai, H., and Okano, F. (2006) Microlens arrays for integral imaging system. *Appl. Opt.*, **45** (36), 9066–9078.
29. Yeom, J., Hong, K., Jeong, Y., Jang, C., and Lee, B. (2014) Solution for pseudoscopic problem in integral imaging using phase-conjugated reconstruction of lens-array holographic optical elements. *Opt. Express*, **22** (11), 13659–13670.
30. Okano, F., Arai, J., Hoshino, H., and Yuyama, I. (1999) Three-dimensional video system based on integral photography. *Opt. Eng.*, **38** (6), 1072–1077.
31. Naemura, T., Yoshida, T., and Harashima, H. (2001) 3-D computer graphics based on integral photography. *Opt. Express*, **8** (4), 255–262.
32. Georgeiv, T., Zheng, K.C., Curless, B., Salesin, D., Nayar, S., and Intwala, C. (2006) Spatio-angular resolution tradeoff in integral photography. Eurographics Symposium on Rendering, pp. 263–272.
33. Platt, B.C. and Shack, R. (2001) History and principles of Shack-Hartmann wavefront sensing. *J. Refract. Surg. (Thorofare, NJ 1995)*, **17** (5), S573–577.
34. Dansereau, D.G., Pizarro, O., and Williams, S.B. (2013) Decoding, calibration and rectification for lenselet-based plenoptic cameras. 2013 IEEE Conference on Computer Vision and Pattern Recognition (CVPR), pp. 1027–1034.
35. Research – Raytrix GmbH. *http://www.raytrix.de/index.php/Research.html* (accessed 28 November 2014).
36. Plenoptic Imaging | ACFR Marine. *http://www-personal.acfr.usyd.edu.au/ddan1654/LFSamplePack1-r2.zip* (accessed 28 November 2014).
37. Heidelberg Collaboratory for Image Processing. *http://hci.iwr.uni-heidelberg.de/HCI/Research/LightField/lf_benchmark.php* (accessed 28 November 2014).
38. Todor Georgiev – Photoshop – Adobe. *http://www.tgeorgiev.net/Gallery/* (accessed 28 November 2014).
39. Nava, F.P., Marichal-Hernández, J.G., and Rodríguez-Ramos, J.M. (2008) The discrete focal stack transform. Proceedings European Signal Processing Conference.
40. Tao, M.W., Hadap, S., Malik, J., and Ramamoorthi, R. (2013) Depth from combining defocus and correspondence

using light-field cameras. IEEE International Conference on Computer Vision, Los Alamitos, CA, pp. 673–680.
41. Schechner, Y.Y. and Kiryati, N. (2000) Depth from defocus vs. stereo: how different really are they? *Int. J. Comput. Vis.*, **39** (2), 141–162.
42. Wanner, S. and Goldluecke, B. (2014) Variational light field analysis for disparity estimation and super-resolution. *IEEE Trans. Pattern Anal. Mach. Intell.*, **36** (3), 606–619.
43. Drazic, V. (2010) Optimal depth resolution in plenoptic imaging. 2010 IEEE International Conference on Multimedia and Expo (ICME), pp. 1588–1593.
44. Perez Nava, F. and Luke, J.P. (2009) Simultaneous estimation of super-resolved depth and all-in-focus images from a plenoptic camera. 3DTV Conference: The True Vision – Capture, Transmission and Display of 3D Video, pp. 1–4.
45. Bishop, T.E. and Favaro, P. (2012) The light field camera: extended depth of field, aliasing, and superresolution. *IEEE Trans. Pattern Anal. Mach. Intell.*, **34** (5), 972–986.
46. Pérez, F., Pérez, A., Rodríguez, M., and Magdaleno, E. (2014) Super-resolved Fourier-slice refocusing in plenoptic cameras. *J. Math. Imaging Vis.*, **52** (2), June 2015, 200–217.
47. Lumsdaine, A., Georgiev, T.G., and Chunev, G. (2012) Spatial analysis of discrete plenoptic sampling. *Proc. SPIE*, Digital Photography VIII, **8299**, 829909–829909–10.
48. Perwass, C. and Wietzke, L. (2012) Single lens 3D-camera with extended depth-of-field. *Proc. SPIE*, Human Vision and Electronic Imaging XVII, **8291**, 829108–829108–15.
49. Georgiev, T. and Lumsdaine, A. (2012) The multifocus plenoptic camera. *Proc. SPIE*, **8299**, 829908–829908–11.
50. Georgiev, T.G., Lumsdaine, A., and Goma, S. (2009) High dynamic range image capture with plenoptic 2.0 camera. Frontiers in Optics 2009/Laser Science XXV/Fall 2009 OSA Optics & Photonics Technical Digest, p. SWA7P.
51. Lumsdaine, A., Chunev, G., and Georgiev, T. (2012) Plenoptic rendering with interactive performance using GPUs. *Proc. SPIE*, Image Processing: Algorithms and Systems X; and Parallel Processing for Imaging Applications II, **8295**, 829513–829513–15.
52. Pérez, J., Magdaleno, E., Pérez Nava, F., Rodríguez, M., Hernández, D., and Corrales, J. (2014) Super-resolution in plenoptic cameras using FPGAs. *Sensors*, **14** (5), 8669–8685.
53. Hahne, C. and Aggoun, A. (2014) Embedded FIR filter design for real-time refocusing using a standard plenoptic video camera. *Proc. SPIE*, Digital Photography X, **9023**, 902305.

Part III
Modelling

9
Probabilistic Inference and Bayesian Priors in Visual Perception

Grigorios Sotiropoulos and Peggy Seriès

9.1
Introduction

Throughout its life, an animal is faced with uncertainty about the external environment. There is often inherent ambiguity in the information entering the brain. In the case of vision, for example, the projection of a 3D physical stimulus onto the retina most often results in a loss of information about the true properties of the stimulus. Thus multiple physical stimuli can give rise to the same retinal image. For example, an object forming an elliptical pattern on the retina may indeed have an elliptical shape but it may also be a disc viewed with a slant (Figure 9.1). How does the visual system "choose" whether to see an ellipse or a slanted disc?

Even in the absence of external ambiguity, internal neural noise or physical limitations of sensory organs, such as limitations in the optics of the eye or in retinal resolution, may also result in information loss and prevent the brain from perceiving details in the sensory input that are necessary to determine the structure of the environment.

Confronted with these types of uncertainty, the brain must somehow make a guess about the external world; that is, it has to estimate the identity, location, and other properties of the objects that generate the sensory input. This estimation may rely on sensory cues but also on assumptions, or *expectations*, about the external world. Perception has thus been characterized as a form of "unconscious inference" – a hypothesis first proposed by Hermann von Helmholtz (1821–1894). According to this view, vision is an instance of inverse inference whereby the visual system estimates the true properties of the physical environment from their 2D projections on the retina, that is, inverts the process of the mapping of the visual inputs onto the retina, with the help of expectations about the properties of the stimulus.

Helmholtz's view of perception as unconscious inference has seen a resurgence in popularity in recent years in the form of the "Bayesian brain" hypothesis [2–4].

Biologically Inspired Computer Vision: Fundamentals and Applications, First Edition.
Edited by Gabriel Cristóbal, Laurent Perrinet, and Matthias Keil.

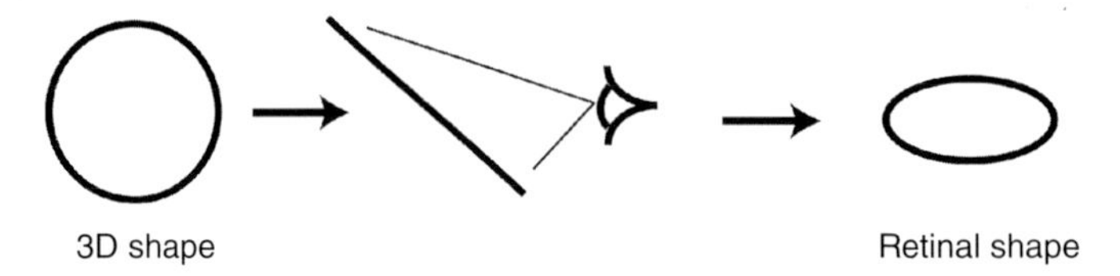

Figure 9.1 An elliptical pattern forming on the retina can be either an ellipse viewed upright or a circle viewed at an angle. Image adapted from Ref. [1] with permission.

In this chapter, we explain what it means to view visual perception as Bayesian inference. We review studies using this approach in human psychophysics. As we will see below, central to Bayesian inference is the notion of priors; we ask which priors are used in human visual perception and how they can be learned. Finally, we briefly address how Bayesian inference processes could be implemented in the brain, a question still open to debate.

9.2 Perception as Bayesian Inference

The "Bayesian brain hypothesis" proposes that the brain works by constantly forming hypotheses or "beliefs" about what is present in the world, and evaluates those hypotheses based on current evidence and prior knowledge. These hypotheses can be described mathematically as conditional probabilities: $P(\text{Hypothesis} \mid \text{Data})$ is the probability of a hypothesis given the data (i.e., the signals available to our senses). These probabilities can be computed using Bayes' rule, named after Thomas Bayes (1701–1761):

$$P(\text{Hypothesis} \mid \text{Data}) = \frac{P(\text{Data} \mid \text{Hypothesis})P(\text{Hypothesis})}{P(\text{Data})} \tag{9.1}$$

Using Bayes' rule to update beliefs is called *Bayesian inference*. For example, suppose you are trying to figure out whether the moving shadow following you is that of a lion. The data available is the visual information that you can gather by looking behind you. Bayesian inference states that the best way to form this probability $P(\text{Hypothesis} \mid \text{Data})$, called the *posterior probability*, is to multiply two other probabilities:

- $P(\text{Data} \mid \text{Hypothesis})$: your knowledge about the probability of the data given the hypothesis (how probable is it that the visual image looks the way it does now when you actually know there is a lion?), which is called the *likelihood*, multiplied by:
- $P(\text{Hypothesis})$ is the *prior probability*: our knowledge about the hypothesis before we could collect any information; here, for example, the probability that we may actually encounter a lion in our environment, independently of the visual inputs, a number which would be very different if we lived in Edinburgh rather than in Tanzania.

The denominator, P(Data), is only there to ensure the resulting probability is between 0 and 1 and can often be disregarded in the computations. The hypothesis can be about the presence of an object, as in the example above, or about the value of a given stimulus (e.g., "the speed of this lion is 40 km/h" – an estimation task) or anything more complex.

Bayesian inference as a model of how the brain works thus rests on critical assumptions that can be tested experimentally:

- The brain takes into account uncertainty and ambiguity by always keeping track of the probabilities of the different possible interpretations.
- The brain has built (presumably through development and experience) an internal model of the world in the form of prior beliefs and likelihoods that can be consulted to interpret new situations.
- The brain combines new evidence with prior beliefs in a principled way, through the application of Bayes' rule.

An observer who uses Bayesian inference is called an *ideal observer*.

9.2.1 Deciding on a Single Percept

The posterior probability distribution $P(H|D)$ obtained through Bayes' rule contains all the necessary information to make inferences about the hypothesis, H, by assigning a probability to each value of H. For example, the probability that we are being followed by a lion would be 10% and that there is no lion, 90%. But we only perceive one interpretation at a time (no lion), not a mixture of the two interpretations (90% lion and 10% no lion). How does the brain choose a single value of H, based on the posterior distribution? This is not clearly understood yet but Bayesian decision theory provides a framework for answering this question. If the goal of the animal is to have the fewest possible mismatches between perception and reality, the value of H that achieves this (call it H^*) should simply be the most probable value. This is called the *maximum a posteriori* (MAP) solution:

$$H^* = \arg\max_H P(H|D) \tag{9.2}$$

where D denotes the data. Another possibility is to use the *mean* of the posterior (which is generally different to the maximum for skewed or multimodal distributions). This solution minimizes the squared difference of the inferred and actual percept, $(H - H^*)^2$.

Taking either the maximum or the mean of the posterior is a deterministic solution: for a given posterior, this will lead to a solution H^* that is always the same. In perceptual experiments, however, there is very often trial-to-trial variability in subjects' responses. The origin of this variability is debated.

One way to account for it with a Bayesian model is to assume that perception and/or responses are corrupted by noise. Two types of noise can be included: (i) *perceptual* noise in the image itself or in the neural activity of the visual system and (ii) *decision* noise – an additional source of noise between perception and response – for example, neural noise in motor areas of the brain that introduces variability in reporting what was seen, even when the percept itself is noiseless.

Another way to model trial-to-trial variability is to assume a stochastic rule for choosing H^*. The most popular approach is *probability matching*, whereby H^* is simply a sample from the posterior. Thus across trials, the relative frequency of a particular percept is equal to its posterior probability. It can be shown that probability matching is not an optimal strategy under the standard loss criteria discussed above. However, the optimality of a decision rule based on the maximum or the mean of the posterior rests on the assumption that the posterior is correct and that the environment is static; when either of these is not true, probability matching can be more useful because it increases exploratory behavior and provides opportunity for learning. Probability matching and generally posterior sampling has also been proposed as a mechanism to explain multistable perception [5], whereby an ambiguous image results in two (or more) interpretations that spontaneously alternate in time. For example, looking at Figure 9.2, we might see a vase at one time but two opposing faces at another time. It has been proposed that such bistable perception results from sampling from a posterior distribution which would be bimodal, with two peaks corresponding to the two possible interpretations.

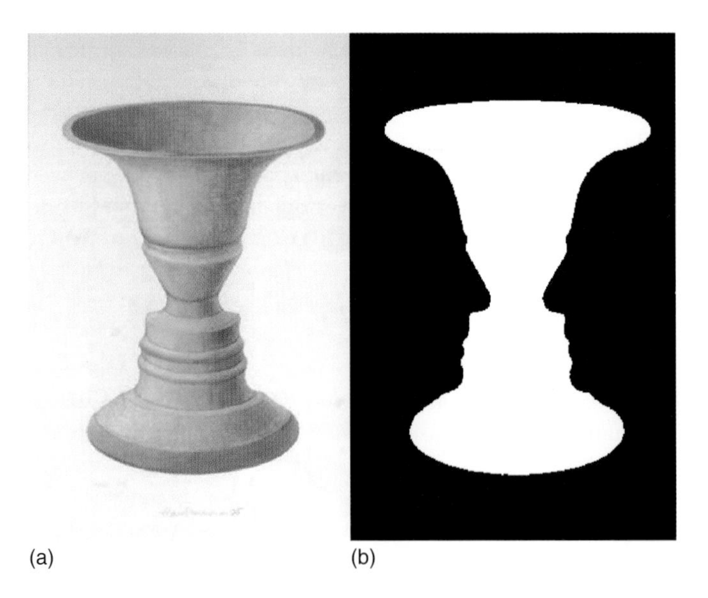

Figure 9.2 Rubin's vase. The black-and-white image (b) can be seen as a white vase or as two opposing black faces. The image (a) provide the context that primes the visual system to choose one of the interpretations for the ambiguous image: the vase, before it switches to the faces.

9.3
Perceptual Priors

As described above, the Bayesian brain hypothesis proposes that *a priori* knowledge is used in the perceptual inference and represented as a prior probability. Recently, a number of researchers have explored this idea: if the brain uses prior beliefs, what are those? And how do they influence perception?

Intuitively, it is when sensory data is limited or ambiguous that we rely on our prior knowledge. For example, if we wake up in the middle of the night and need to walk in total darkness, we automatically use our prior knowledge of the environment, or of similar environments, to guide our path. Mathematically, Bayes' rule similarly indicates that prior distributions should have maximum impact in situations of strong uncertainty. Thus, a good way to discover the brain's prior expectations is to study perception or cognition in situations where the current sensory inputs (the "evidence") is limited or ambiguous. Studying such situations reveals that our brain uses automatic expectations all the time.

9.3.1
Types of Prior Expectations

Visual illusions are a great example of this. Consider Figure 9.2, for example. The ambiguous "Rubin's vase" can be interpreted as either a vase, or two faces facing each other. However, in Figure 9.2, because of the spatial proximity of the ambiguous image on the right with the photo of the vase (a), you are more likely to perceive first a vase (b), which will then switch to a face. The same effect would be observed with temporal proximity, such as when Rubin's vase is presented shortly after the unambiguous image of a vase. These types of expectations, induced by cues that are local in space or time and have immediate and short-term effects, have been recently dubbed "contextual expectations" [6]. There are several different types of contextual expectations that affect perception. Some experiments manipulate expectations by explicitly giving information regarding the visual stimulus, for example, telling participants about the number of possible directions of motion that they are going to be exposed to. In yet other psychophysical experiments, expectations are formed implicitly and unconsciously, for example, by exposing participants to the association between the sense of rotation of a 3D cylinder and a particular signal, for example a sound [7]. When the sense of rotation is later made ambiguous by removing 3D information, the subjects' perception will be modulated by the presence of the sound.

Contextual expectations are not the only kind of expectations; another kind, conceptually more akin to Bayesian priors, are expectations based on general, or *prior*, knowledge about the world. These have been referred to as *structural expectations* [6, 8]. The expectation that shapes are commonly isotropic is one such example: when humans see an elliptical pattern such as that of Figure 9.1, they commonly assume that it is a circle viewed at a slant rather than an ellipse because circles, being isotropic, are considered more common than ellipses. Another well-known effect is that the human visual system is more sensitive to cardinal (horizontal and vertical) orientations, the so-called "oblique effect." This is thought to be due to an intrinsic expectation that cardinal orientations are more likely to occur than oblique orientations (see below). Another well-studied example of a structural expectation that has been formalized in Bayesian terms is that light comes from above us. The light-from-above prior is used by humans as well as other animals when inferring the properties of an object from its apparent shading. Figure 9.3(a), for example, is interpreted as one dimple in the middle of bumps. This is consistent with assuming that light comes from the top of the image. Turning the page upside down would lead to the opposite percept. In shape-from-shading judgments, as well as in figure–ground separation tasks, another expectation also influences perception—that objects tend to be convex rather than concave. For example, Figure 9.3(b) is typically interpreted as a set of black objects in white background (and not vice versa) because under this interpretation the objects are convex. A convexity prior seems to exist not just for objects within a scene (e.g., bumps on a surface) but also for the entire surface itself: subjects are better at local shape discrimination when the surface is globally convex rather than concave. A related, recently reported expectation is that depth (i.e., the distance between figure and ground) is greater when the figure is convex rather than concave. Other examples of expectations are that objects tend to be viewed from above; that objects are at a distance of 2–4 m from ourselves; that objects in nearby radial directions are at the same distance from ourselves; and that people's gaze is directed toward us (see [6] and citations therein). Finally, an expectation that has lead to numerous studies is the expectation that objects in the world tend to move slowly or to be still. In Bayesian models of motion perception, this is typically referred to as the *slow speed prior* (see below).

The distinction between contextual and structural expectations is not always clear-cut. For example, when you see a dark, pistol-shaped object in the bathroom after you have taken off your glasses and your vision is blurred, you will likely see that object as a hairdryer (Figure 9.3(c)). The exact same shape seen in a workshop will evoke the perception of a drill. The context, bathroom sink versus workbench, helps disambiguate the object – a contextual expectation. However, this disambiguation relies on prior knowledge that hair dryers are more common in bathrooms and drills are more common in workshops; these are structural expectations. In a Bayesian context, structural expectations are commonly described in terms of "priors."

9.3.2 Impact of Expectations

Expectations help us infer the state of the environment in the face of uncertainty. In the "bathroom sink versus workbench" example, expectations help disambiguate the dark object in the middle of each image in Figure 9.3. Apart from aiding with object identification or shape judgments, expectations can impact

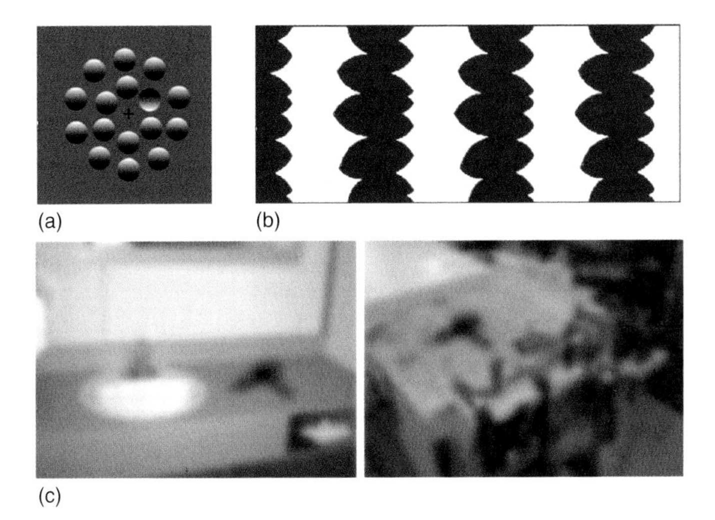

Figure 9.3 (a) Expectation that light comes from above. This image is interpreted as one dimple in the middle of bumps. This is consistent with assuming that light comes from the top of the image. Turning the page upside down would lead to the opposite percept. Image adapted from Ref. [9] with permission. (b) Convexity expectation for figure–ground separation. Black regions are most often seen as convex objects in a white background instead of white regions being seen as concave objects in a black background. Image adapted from Ref. [10] with permission. Interplay between contextual and structural expectations. The black object in (c) is typically perceived as a hair dryer because although it has a pistol-like shape (structural expectation), it appears to be in a bathroom (contextual expectation) and we know that hair dryers are typically found in bathrooms (structural expectation). The identical-looking black object in (d) is perceived as a drill because the context implies that the scene is a workshop. Image adapted from Ref. [11] with permission.

perception in several other ways. First, they can lead to an improvement in performances during detection tasks. For example, when either the speed or the direction of motion of a random-dot stimulus are expected, subjects are better and faster at detecting the presence of the stimulus in a two-interval forced choice task [12]. More interestingly perhaps, in some cases, expectations about a particular measurable property can also influence the perceived magnitude of that property, that is, the content of perception. A recent study, for example, showed that, when placed in a visual environment where some motion directions are more frequently presented than others, participants quickly and implicitly learn to expect those. These expectations affect their visual detection performance (they become better at detecting the expected directions at very low contrast) as well as their estimation performances (they show strong estimation biases, perceiving motion directions as being more similar to the expected directions). Moreover, in situations where nothing is shown, participants sometimes incorrectly report perceiving the expected motion directions – of the form of “hallucinations” of the prior distribution [13].

Another particularly interesting example of how expectations can affect the content of perception is the aforementioned expectation of slow speeds which we describe in more detail in the next section.

9.3.3
The Slow Speed Prior

Hans Wallach observed that a line moving behind a circular aperture, with its endpoints concealed such that its true direction cannot be recovered, always appears to move perpendicularly to its orientation [14]. This is known as the "aperture problem." Observing that the perpendicular direction corresponds to the interpretation of the retinal image sequence which has slowest speed, he hypothesized that the reason for this remarkably universal perception is an innate preference of the visual system for slow speeds. This idea has been formalized by Weiss *et al.* in 2002 [15]. These researchers showed that a Bayesian model incorporating a prior for slow speeds can explain not only the aperture problem but also a variety of perceptual phenomena, including a number of visual illusions and biases.

One effect that this model explains is the "Thompson effect"—the decrease in perceived speed of a moving stimulus when its contrast decreases [16]. In the model of Weiss *et al.*, the speed of a rigidly translating stimulus is determined by the integration of local motion signals under the assumptions of measurement noise and a prior that favors slow speeds. Given the noisy nature of eye optics and of neural activity, at low contrasts, the signal-to-noise ratio is lower than it is at high contrasts. This means that the speed measurements that the visual system performs are more noisy, which is reflected by a broader likelihood function.

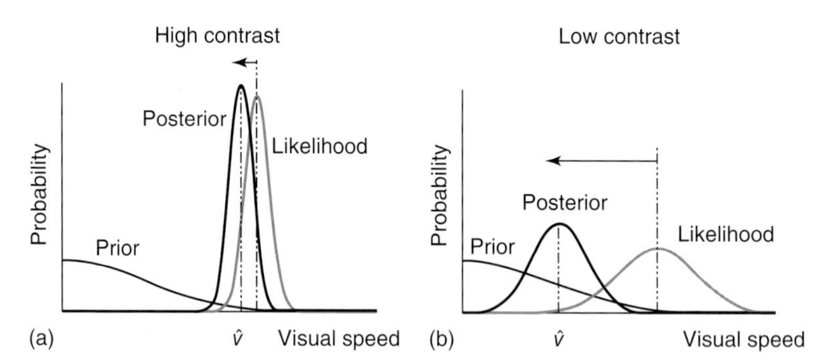

Figure 9.4 Influence of image contrast on the estimation of speed. (a) At high contrast, the measurements of the stimulus are precise, and thus lead to a sharp likelihood. Multiplication of the likelihood by the prior distribution leads to a posterior distribution that is similar to the likelihood, only slightly shifted toward the prior distribution. (b) At low contrast, on the other hand, the measurements are noisy and lead to a broad likelihood. Multiplication of the prior by the likelihood thus leads to a greater shift of the posterior distribution toward the prior distribution. This will result in an underestimation of speed at low contrast. Reproduced from Ref. [17] with permission.

According to Bayes' rule, the prior will thus have a greater influence on the posterior at low contrasts than at high contrasts, where the likelihood is sharper. It follows that perceived speed will be lower at low contrasts (see Figure 9.4). A real-world manifestation of the Thompson effect is the well-documented fact that drivers speed up in the fog [18]. It should be noted that contrast is not the only factor that can affect uncertainty: similar biases toward slow speeds can be observed when the duration of the stimulus is shortened [19].

The slow speed prior affects not only the perceived speed but also the perceived direction. The model of Weiss *et al.* accounts for the aperture problem as well as the directional biases observed with the motion of lines that are unoccluded but are presented at very low contrast. The rhombus illusion is a spectacular illustration of this. In this illusion (Figure 9.5), a thin, low-contrast rhombus that moves horizontally appears to move diagonally, whereas the same rhombus at a high

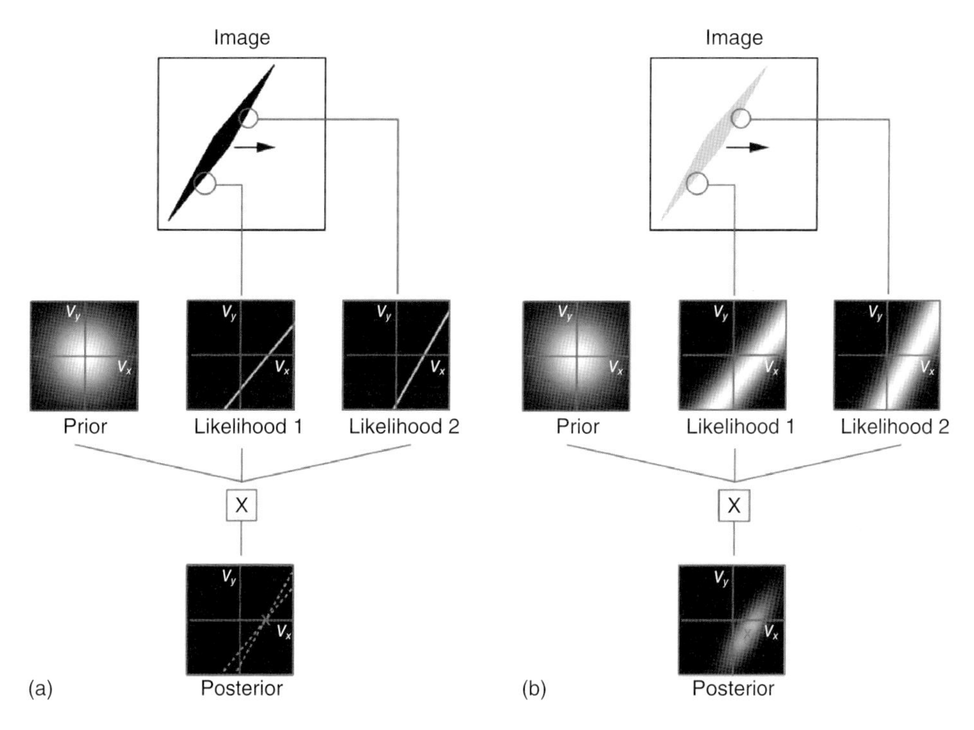

Figure 9.5 Influence of contrast on the perceived direction of a horizontally moving rhombus. (a). With a high-contrast rhombus the signal-to-noise ratio of the two local measurements (only two are shown for clarity) is high and thus the likelihood for each measurement in velocity space is sharp, tightly concentrated around a straight line, and dominates the prior, which is broader. The resulting posterior is thus mostly dictated by the combination of the likelihoods and favors the veridical direction (because the point where the two likelihood lines intersect has $v_y = 0$). (b) With a low-contrast rhombus, the constraint line of the likelihood is fuzzy, that is, the likelihood is broad and the prior exerts greater influence on the posterior, resulting in a posterior that favors an oblique direction. Image adapted from Ref. [15], with permission.

contrast appears to move in its true direction. This illusion is well accounted by the model. Weiss *et al.* make a convincing case that visual illusions might thus be due not to the limitations of a collection of imperfect hacks that the brain would use, as commonly thought, but would be instead "a result of a coherent computational strategy that is optimal under reasonable assumptions." See also Chapter 12 of this book for related probabilistic approaches to motion perception.

9.3.4
Expectations and Environmental Statistics

If our perceptual systems are to perform well, that is, if the interpretation of an ambiguous or noisy scene is to match the real world as closely as possible, our expectations need to accurately reflect the structure of the real world. In Bayesian terms, this means that our priors must closely approximate the statistics of the environment (see Chapter 4 of this book). Using the previous "bathroom sink versus workbench" example, we need to have a prior that favors the presence of drills in workshops. If that were not the case, we might interpret the object in the workshop as a hair dryer, which would lead to an incorrect inference more often than not because, statistically speaking, drills are more common in workshops than hair dryers are.

Our visual input—the images forming on the retina—although very varied, is only a small subset of the entire possible image space. Natural images are highly redundant, containing many statistical regularities that the visual system may exploit to make inferences about the world. If expectations are to facilitate vision, they should approximate environmental statistics and, indeed, there is considerable evidence that in many cases they do so. For example, as described above, it is known that, when assessing the orientation of visual lines, human participants show strong biases toward cardinal orientations (vertical and horizontal), indicating a tendency to expect cardinal orientations in the world. Moreover, this bias is known to depend on the uncertainty associated with the stimulus [20]. This expectation is consistent with natural scene statistics : FFT analysis, for example, shows stronger power at cardinal compared with oblique orientations in a variety of natural images [21, 22]. The link between natural images statistics and the use of Bayesian priors in behavioral tasks was demonstrated recently by Girshick *et al.* [23]. Girshick *et al.* studied the performances of participants comparing different orientations, and found that participants were strongly biased toward the cardinal axes when the stimuli were uncertain. They further measured the distribution of local orientations in a collection of photographs and found that it was strongly nonuniform, with a dominance of cardinal directions. Assuming that participants behaved as Bayesian observers, and using a methodology developed by Stocker and Simoncelli [17], they could extract the Bayesian priors that participants used in the discrimination task and found that the recovered priors matched the measured environmental distribution (Figure 9.6).

Similarly, a study measured the pairwise statistics of edge elements from contours found in natural images and found that the way humans group edge

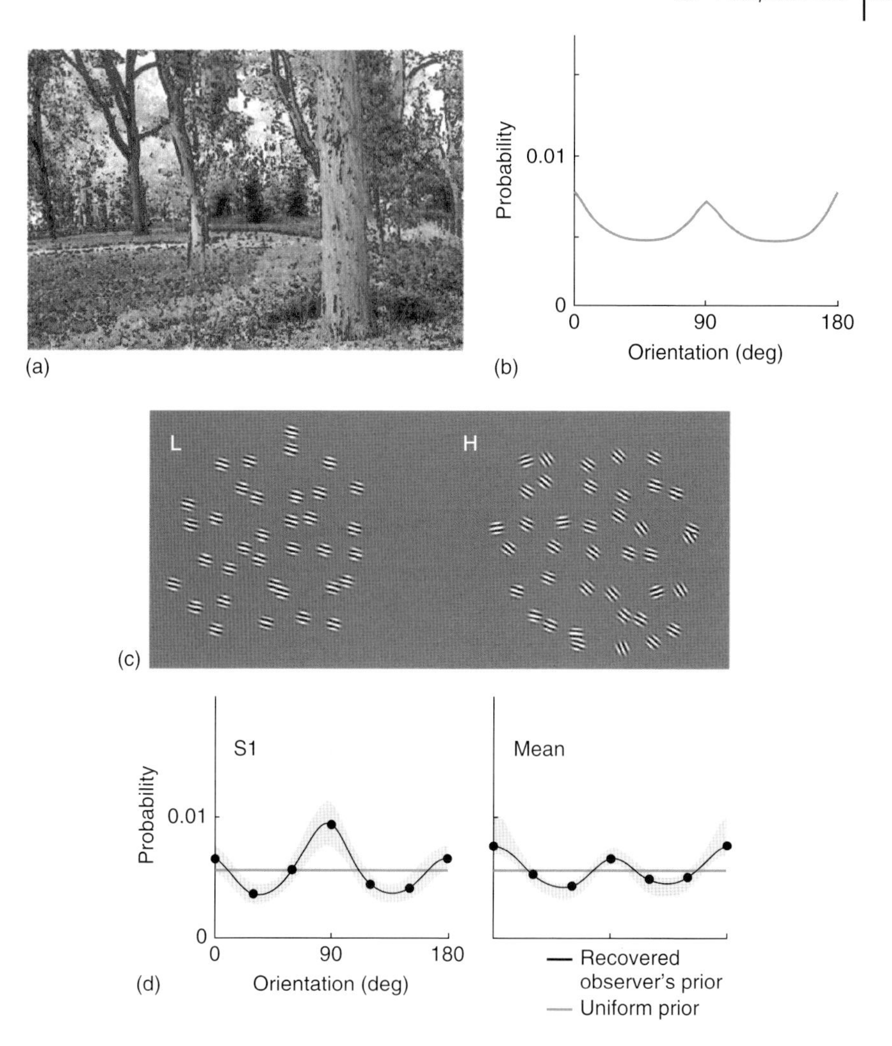

Figure 9.6 Expectations about visual orientations. (a) Example of a natural scene, with strongly oriented locations marked in red. (b) Distribution of orientations in natural scene photographs. (c) Participants viewed arrays of oriented Gabor functions and had to indicate whether the right stimulus was oriented counterclockwise or clockwise relative to the left stimulus. (d). The recovered priors, extracted from participants' performances, here shown for subject S1 and mean subject, are found to match the statistics of orientation in the natural scenes. Reproduced from Ref. [23] with permission.

elements of occluded contours matches the performance of an ideal observer Bayesian model, that is, a model with a prior reflecting the statistics of the natural image database [24]. This suggests that perceptual mechanisms of contour grouping are closely related to natural image statistics. The influence of the convexity expectation in figure–ground separation may also have a basis in natural scene statistics. As mentioned above, it was found that subjects expect greater distances between convex figures and background than between concave ones and background [25]. These expectations were in accord with the statistics of a collection of luminance and range images obtained from indoor and outdoor scenes.

As far as the slow speed prior is concerned, research related to the statistical structure of time-varying images is scarce. Inferring a distribution of object speeds from image sequences is complicated by the fact that retinal motion (the motion of patterns forming on the retina) does not correspond to object motion in a trivial way. One problem is that object speed can be inferred from retinal speed (the rate of movement of an object's projection on the retina) only if the distance of the object from the observer is known. Another problem is that retinal motion can be produced both by the independent motion of objects in the view field and by self-motion, head and eye movements. There are however some indications from studies looking at the properties of optical flow and the power spectrum of video clips that the distribution of retinal speeds can be described by a log-normal distribution or a power-law distribution favoring slow speeds [26, 27].

9.4 Outstanding Questions

While expectations have been studied for more than a century, the formalization of expectations as priors in Bayesian models of perception is relatively recent. Although by now there is a considerable body of theoretical work on Bayesian models that account for a wealth of perceptual data from human and nonhuman observers, there are a number of outstanding questions.

9.4.1 Are Long-Term Priors Plastic?

Are long-term priors hard-wired, or fixed after long-term exposure, or are they constantly updating through experience? This question was first addressed in the context of the light-from-above prior. In 1970, Hershberger showed that chickens reared in an environment illuminated from below did not differ from controls in their interpretation of shadows and depth [28]. They thus suggested that the prior that light comes from above is innate. This question was recently revisited in humans [29]. In their experiment, the authors first asked participants to make convex–concave judgments of bump–dimple stimuli at different orientations, and measured the light-from-above prior based on their responses. During a training phase, they then added new shape information via haptic (active touch) feedback, that disambiguated object shape but conflicted with the participants' initial interpretation, by corresponding to a light source shifted by 30° compared to the participants' baseline prior. When participants were finally tested again on visual only stimuli, their light direction prior had shifted significantly in the direction of the information provided during training. Adams *et al.* thus concluded that, unlike in chickens, the light-from-above prior could be updated in humans. We recently investigated this question in the context of the slow speed prior [30]. The aim of this study was to test whether expectations about the speed of visual stimuli could be implicitly changed solely through exposure to fast speeds and if so, whether this could result in a disappearance or reversal of the classically reported direction biases. Using a field of coherently moving lines presented at low contrast and short durations, this was found to be the case. After about 3×1 hour exposure to fast speeds on three consecutive days, directional biases disappeared and a few days later, they changed sign, becoming compatible with a speed prior centered on speeds faster than 0 deg/s.

9.4.2 How Specific are Priors?

Another crucial issue that needs to be clarified concerns the specificity of expectations. For example, is there only one speed prior, which is applied to all types of visual objects and stimuli? Or are there multiple priors specific to different objects and conditions? When new priors are learned in the context of a task, do they automatically transfer to different tasks? Adams *et al.* provided evidence that the visual system uses the same prior about light source position in quite different tasks, one involving shape and another requiring lightness judgments [29]. Similarly, Adams measured the light-from-above in different tasks: visual search, shape perception, and a novel reflectance-judgment task [9]. She found strong positive correlations between the light priors measured using all three tasks, suggesting a single mechanism used for "quick and dirty" visual search behavior, shape perception, and reflectance judgments. In the context of short-term statistical learning, and using a familiarization task with complex shapes, Turk-Browne and Scholl (2009) provided evidence for transfer of perceptual learning across space and time, suggesting that statistical learning leads to flexible representations [31]. However, the generality of these findings is still unclear and needs further exploration. A related question is to understand whether expectations learned in the laboratory can persist over time and for how long. Recent work suggests that contextual priors persist over time, but may remain context-dependent, with the experimental setup acting as a contextual cue [32].

9.4.3 Inference in Biological and Computer Vision

The Bayesian approach is not the first computational framework to describe vision, whether artificial or biological, in terms of inverse inference. In a series of landmark papers published in the 1980s, Tomaso Poggio and colleagues postulated that inverse inference in vision is an *ill-posed* problem [33]. The term "ill-posed" was first used in the field of partial differential equations several decades earlier to describe a problem that has either multiple solutions, no solution, or the solution does not depend continuously on the data. Poggio and colleagues suggested that to solve the problem of inverse inference, the class of admissible solutions must be restricted by introducing suitable *a priori* knowledge. The concept of *a priori* knowledge, or assumptions, had been used earlier by Tikhonov and Arsenin [34] as a general approach to solving ill-posed problems in the form of *regularization theory*. Let y be the data available and let z be the stimulus of interest, such that $f(z) = y$, where f is a known operator. The direct problem is to determine y from z; the inverse problem is to obtain z when y is given. In vision, most problems are ill-posed because z is typically not unique. Regularization works by restricting the space of solutions by adding the stabilizing functional, or constraint, $P(z)$. The solution is then found using

$$z^* = \arg\min_z(\| f(z) - y\|^2 + \lambda \| P(z)\|^2) \tag{9.3}$$

where z^* is the estimate of the stimulus property (or properties) z of interest; and λ is a constant that controls the relative weight given to the error in estimation versus the violation of the constraints when computing the solution. Thus z^* approximates the inverse mapping $f^{-1}(y)$ under the constraints P.

Since the work of Poggio and colleagues, regularization has been widely used in the computer vision community for a variety of problems, such as edge detection, optical flow (velocity field estimation), and shape from shading. These endeavors were met with success in many hard problems in vision—particularly problems that were hard to solve when the visual input were natural scenes rather than simplified artificial ones. This success lent popularity to the hypothesis, also central to the Bayesian approach, that the brain perceives the world by constraining the possible inverse inference solutions based on prior assumptions about the structure of the world. Bayesian inference and regularization have more in common than just this common concept, however. This becomes more apparent by taking the logarithm of the posterior in Bayes' rule equation (9.1):

$$\log P(z|y) \propto \log P(y|z) + \log P(z) - \log P(y)$$

The MAP solution of the log-posterior (which is the same as the MAP of the posterior, as logarithms are monotonic functions) thus becomes

$$z^* = \arg\min_z(-\log P(y|z) - \log P(z)) \tag{9.4}$$

where $P(y)$ can been ignored as it does not depend on z, and z^* is expressed in terms of minimizing the sum of negative logarithms (instead of maximizing the sum of positive logarithms). The similarities between Eqs. (9.3) and (9.4) are striking: in both equations, the solution is derived by minimizing the sum of two quantities: one that is based on a forward-inference model and another that is proportional to the deviation of the solution from the one expected by prior knowledge/constraints. Bayesian inference can thus be regarded as a stochastic version of regularization — the main difference being that Bayesian inference is a more general and powerful framework. This is due to two reasons. First, Bayesian inference, being a probabilistic *estimation theory*, provides the optimal way of dealing with noise in the data (y in our case), whereas regularization can be sensitive to noise. Second, in Bayesian inference, the posterior also represents the reliability of each possible solution (whereas regularization is akin to only representing the mode of a posterior). In other words, a Bayesian approach makes use of all available information in situations of uncertainty (such as noisy measurements) and evaluates the relative "goodness" of each possible solution, as opposed to merely yielding the best solution.

Probabilistic approaches have been used in a number of ways in computer vision. They typically consist of graphical models, whereby different levels of representation of visual features are linked to each other as nodes in a graph. *Particle filtering* and *Markov chain Monte Carlo* [35] are two examples of techniques whereby a probability distribution (rather than a single value) of a property is maintained and used in subsequent computations (e.g., as more data is received). An example use of such techniques in computer vision is the CONDENSATION algorithm, a particle filtering approach for the tracking of dynamic, changing objects. Interestingly, electrophysiological data from neurons in early visual areas (such as V1) has revealed long latencies of responses, an indication that there are multiple levels of information processing where low (early) levels interact with higher ones. Some authors have suggested that these interactions can be modeled with the algorithms of particle filtering and Bayesian belief propagation [36].

Prior assumptions in graphical models are most often modeled with *Markov random fields* (MRF), which capture dependencies between neighboring elements in an image (such as collinearity between adjacent line elements) and are flexible enough to allow the specification of various types of prior knowledge.

As a final note, although Bayesian inference is a more general framework than regularization, it is also more costly in computational resources. Furthermore, many problems in vision can be solved without the need for a full probabilistic approach such as Bayesian inference. Thus Bayesian inference is not always used in computer vision and may not always be used in biological vision either — or may be used in an approximate manner [37, 38], for example, through the representation of approximate, simpler versions of the true likelihood and prior distributions.

9.4.4 Conclusions

Bayesian models and probabilistic approaches have been increasingly popular in the machine vision literature [39]. At the same time, they appear to be very useful for describing human perception and behavior at the computational level. How these algorithms are implemented in the brain and relate to neural activity is still an open question and an active area of research.

While its popularity has been steadily increasing in recent years, the Bayesian approach has also received strong criticism. Whether the Bayesian approach can actually make predictions for neurobiology, for example, on which parts of the brain would be involved, or how neural activity could represent probabilities, is debated. It is yet unclear whether the Bayesian approach is only useful at the "computational level," to describe the computations performed by the brain overall, or whether it can be also useful at the "implementation level" to predict how those algorithms might be implemented in the neural tissue [40]. It has also been argued that the Bayesian framework is so general that it is difficult to falsify.

However, a number of models have been proposed that suggest how (approximate) Bayesian inference could be implemented in the neural substrate [36, 41]. Similarly, a number of suggestions have been made about how visual priors could be implemented. Priors could correspond to different aspects of brain organization and neural plasticity. Contextual priors could be found in the numerous feedback loops present in the cortex at all levels of the cortical hierarchy [36]. A natural way in which structural priors could be represented in the brain is in the selectivity of the neurons and the inhomogeneity of their preferred features: the neurons that are activated by the expected features of the environment would be present in larger numbers or be more strongly connected to higher processing stages than neurons representing non-expected features. For example, as discussed above, a Bayesian model with a prior on vertical and horizontal orientations (reflecting the fact that they are more frequent in the visual environment) can account for the observed perceptual biases toward cardinal orientations. These effects can also be simply accounted for in a model of the visual cortex where more neurons are sensitive to vertical and horizontal orientations than to other orientations. Another very interesting idea that has recently attracted much interest is that spontaneous activity in sensory cortex could also serve to implement prior distributions. It is very well known that the brain is characterized by ongoing activity, even in the absence of sensory stimulation. Spontaneous activity has been traditionally considered as being just "noise." However, the reason the brain is constantly active might be because it continuously generates predictions about the sensory inputs, based on its expectations. This would be computationally advantageous, driving the network closer to states that correspond to likely inputs, and thus shortening the reaction time of the system.

Finally, a promising line of work is interested in relating potential deficits in Bayesian inference with psychiatric disorders [42]. It is well known that the visual experience of patients suffering from mental disorders such as schizophrenia and autism are different from that of healthy controls. Recently, such differences have been tentatively explained in terms of differences in the inference process, in particular regarding the influence of the prior distributions compared to the likelihood. For example, priors that would be too strong could lead to hallucinations (such as in schizophrenia), while priors that would be too weak could lead to patients being less susceptible to visual illusions, and to the feeling of being overwhelmed by sensory information (such as in autism).

A more detailed assessment of Bayesian inference as a model of perception at the behavioral level, as well as a better understanding of its predictions at the neural level and their potential clinical applications are the focus of current experimental and theoretical research.

References

1. Knill, D.C. (2007) Learning Bayesian priors for depth perception. *J. Vis.*, **7**, 13.
2. Knill, D. and Richards, W. (1996) *Perception as Bayesian Inference*, Cambridge University Press.
3. Knill, D.C. and Pouget, A. (2004) The Bayesian brain: the role of uncertainty in neural coding and computation. *Trends Neurosci.*, **27**, 712–719.
4. Vilares, I. and Kording, K. (2011) Bayesian models: the structure of the world, uncertainty, behavior, and the brain. *Ann. N. Y. Acad. Sci.*, **1224**, 22–39.
5. Gershman, S.J., Vul, E., and Tenenbaum, J.B. (2012) Multistability and perceptual inference. *Neural Comput.*, **24**, 1–24.
6. Seriès, P. and Seitz, A.R. (2013) Learning what to expect (in visual perception). *Front. Human Neurosci.*, **7**, 668.
7. Haijiang, Q. *et al.* (2006) Demonstration of cue recruitment: change in visual appearance by means of Pavlovian conditioning. *Proc. Natl. Acad. Sci. U.S.A.*, **103**, 483.
8. Sotiropoulos, G. (2014) Acquisition and influence of expectations about visual speed. PhD thesis. Informatics, University of Edinburgh.
9. Adams, W.J. (2007) A common light-prior for visual search, shape, and reflectance judgments. *J. Vis.*, **7**, 11.
10. Peterson, M.A. and Salvagio, E. (2008) Inhibitory competition in figure-ground perception: context and convexity. *J. Vis.*, **8**, 4.
11. Bar, M. (2004) Visual objects in context. *Nat. Rev. Neurosci.*, **5**, 617–629.
12. Sekuler, R. and Ball, K. (1977) Mental set alters visibility of moving targets. *Science*, **198**, 60–62.
13. Chalk, M., Seitz, A.R., and Seriès, P. (2010) Rapidly learned stimulus expectations alter perception of motion. *J. Vis.*, **10**, 1–18.
14. Wuerger, S., Shapley, R., and Rubin, N. (1996) On the visually perceived direction of motion by Hans Wallach: 60 years later. *Perception*, **25**, 1317–1368.
15. Weiss, Y., Simoncelli, E.P., and Adelson, E.H. (2002) Motion illusions as optimal percepts. *Nat. Neurosci.*, **5**, 598–604.
16. Thompson, P. (1982) Perceived rate of movement depends on contrast. *Vision Res.*, **22**, 377–380.
17. Stocker, A.A. and Simoncelli, E.P. (2006) Noise characteristics and prior expectations in human visual speed perception. *Nat. Neurosci.*, **9**, 578–585, doi: 10.1038/nn1669.
18. Snowden, R.J., Stimpson, N., and Ruddle, R.A. (1998) Speed perception fogs up as visibility drops. *Nature*, **392**, 450.
19. Zhang, R., Kwon, O., and Tadin, D. (2013) Illusory movement of stationary stimuli in the visual periphery: evidence for a strong centrifugal prior in motion processing. *J. Neurosci.*, **33**, 4415–4423.
20. Tomassini, A., Morgan, M.J., and Solomon, J.A. (2010) Orientation uncertainty reduces perceived obliquity. *Vision Res.*, **50**, 541–547.
21. Keil, M.S. and Cristóbal, G. (2000) Separating the chaff from the wheat: possible origins of the oblique effect. *J. Opt. Soc. Am. A Opt. Image Sci. Vis.*, **17**, 697–710.
22. Nasr, S. and Tootell, R.B.H. (2012) A cardinal orientation bias in scene-selective visual cortex. *J. Neurosci.*, **32**, 14921–14926.
23. Girshick, A.R., Landy, M.S., and Simoncelli, E.P. (2011) Cardinal rules: visual orientation perception reflects knowledge of environmental statistics. *Nat. Neurosci.*, **14**, 926–932.
24. Geisler, W.S. and Perry, J.S. (2009) Contour statistics in natural images: grouping across occlusions. *Visual Neurosci.*, **26**, 109–121.
25. Burge, J., Fowlkes, C.C., and Banks, M.S. (2010) Natural-scene statistics predict

how the figure–ground cue of convexity affects human depth perception. *J. Neurosci.*, **30**, 7269–7280.
26. Dong, D.W. and Atick, J.J. (1995) Statistics of natural time-varying images. *Netw. Comput. Neural Syst.*, **6**, 345–358.
27. Calow, D. and Lappe, M. (2007) Local statistics of retinal optic flow for self-motion through natural sceneries. *Network*, **18**, 343–374.
28. Hershberger, W. (1970) Attached-shadow orientation perceived as depth by chickens reared in an environment illuminated from below. *J. Comp. Physiol. Psychol.*, **73**, 407.
29. Adams, W.J., Graf, E.W., and Ernst, M.O. (2004) Experience can change the 'light-from-above' prior. *Nat. Neurosci.*, **7**, 1057–1058.
30. Sotiropoulos, G., Seitz, A.R., and Seriès, P. (2011) Changing expectations about speed alters perceived motion direction. *Curr. Biol.*, **21**, R883–R884.
31. Turk-Browne, N.B. and Scholl, B.J. (2009) Flexible visual statistical learning: transfer across space and time. *J. Exp. Psychol. Hum. Percept Perform*, **35**, 195–202.
32. Kerrigan, I.S. and Adams, W.J. (2013) Learning different light prior distributions for different contexts. *Cognition*, **127**, 99–104.
33. Poggio, T., Torre, V., and Koch, C. (1985) Computational vision and regularization theory. *Nature*, **317**, 314–319.
34. Tikhonov, A.N. and Arsenin, V.Y. (1977) *Solutions of Ill-Posed Problems*, Winston.
35. Bishop, C.M. *et al.* (2006) *Pattern Recognition and Machine Learning*, vol. 1, Springer-Verlag, New York.
36. Lee, T.S. and Mumford, D. (2003) Hierarchical bayesian inference in the visual cortex. *J. Opt. Soc. Am. A Opt. Image Sci. Vis.*, **20**, 1434–1448.
37. Beck, J.M. *et al.* (2012) Not noisy, just wrong: the role of suboptimal inference in behavioral variability. *Neuron*, **74**, 30–39.
38. Ma, W.J. (2012) Organizing probabilistic models of perception. *Trends Cognit. Sci.*, **16**, 511–518, doi: 10.1016/j.tics.2012.08.010.
39. Prince, S. (2012) *Computer Vision: Models, Learning and Inference*, Cambridge University Press.
40. Colombo, M. and Seriès, P. (2012) Bayes in the brain: on Bayesian modelling in neuroscience. *Br. J. Philos. Sci.*, **63**, 697–723.
41. Fiser, J. *et al.* (2010) Statistically optimal perception and learning: from behavior to neural representations. *Trends Cognit. Sci.*, **14**, 119–130, doi: 10.1016/j.tics.2010.01.003.
42. Montague, P.R. *et al.* (2012) Computational psychiatry. *Trends Cognit. Sci.*, **16**, 72–80.

10 From Neuronal Models to Neuronal Dynamics and Image Processing

Matthias S. Keil

10.1 Introduction

Neurons make contacts with each other through synapses: the presynaptic neuron sends its output to the synapse via an axon and the postsynaptic neuron receives it via its dendritic tree ("dendrite"). Most neurons produce their output in the form of virtually binary events (action potentials or spikes). Some classes of neurons (e.g., in the retina) nevertheless do not generate spikes but show continuous responses. Dendrites were classically considered as passive cables whose only function is to transmit input signals to the soma (the cell body of the neuron). In this picture, a neuron integrates all input signals and generates a response if their sum exceeds a threshold. Therefore, the neuron is the site where computations take place, and information is stored across the network in synaptic weights ("connection strengths"). This "connectionist" point of view on neuronal functioning inspired neuronal networks learning algorithms such as error backpropagation [1], and the more recent deep-learning architectures [2]. Recent evidence, however, suggest that dendrites are excitable structures rather than passive cables, which can perform sophisticated computations [3–6]. This suggests that even single neurons can carry out far more complex computations than previously thought.

Naturally, a modeler has to choose a reasonable level of abstraction. A good model is not necessarily one which incorporates all possible details, because too many details can make it difficult to identify important mechanisms. The level of detail is related to the research question that we wish to answer, but also to our computational resources [7]. For example, if our interest was to understanding the neurophysiological details of a small circuit of neurons, then we probably would choose a Hodgkin–Huxley model for each neuron [8] and also include detailed models of dendrites, axons, and synaptic dynamic and learning. A disadvantage of the Hodgkin–Huxley model is its high computational complexity. It requires about 1200 floating point operations (FLOPS) for the simulation of 1 ms of time [9].

Biologically Inspired Computer Vision: Fundamentals and Applications, First Edition.
Edited by Gabriel Cristóbal, Laurent Perrinet, and Matthias Keil.

Therefore, if we want to simulate a large network with many neurons, then we may omit dendrites and axons, and use a simpler model such as the integrate-and-fire neuron model (which is presented in the following) or the Izhikevich model [10]. The Izhikevich model offers the same rich spiking dynamics as the full Hodgkin–Huxley model (e.g., bursting, tonic spiking), while having a computational complexity similar to the integrate-and-fire neuron model (about 13 and 5 FLOPS/1 ms, respectively). Further simplifications can be made if we aim at simulating psychophysical data, or at solving image processing tasks. For example, spiking mechanisms are often not necessary then, and we can compute a continuous response y (as opposed to binary spikes) from the neuron's state variable x (e.g., membrane potential) by thresholding (half-wave rectification, e.g., $y = max[x, 0]$) or via a sigmoidal function ("squashing function", e.g., $y = 1/[1 + \exp(-x)]$).

This chapter approaches neural modeling and computational neuroscience, respectively, in a tutorial-like fashion. This means that basic concepts are explained by way of simple examples, rather than providing an in-depth review on a specific topic. We proceed by first introducing a simple neuron model (the membrane potential equation), which is derived by considering a neuron as an electrical circuit. When the membrane potential equation is augmented with a spiking mechanism, it is known as the *leaky integrate-and-fire* model. "Leaky" means that the model forgets exponentially fast about past inputs. "Integrate" means that it sums its inputs, which can be excitatory (driving the neuron toward its response threshold) or inhibitory (drawing the state variable away from the response threshold). Finally, "fire" refers to the spiking mechanism.

The membrane potential equation is subsequently applied to three image processing tasks. Section 10.3 presents a reaction-diffusion-based model of the retina, which accounts for simple visual illusions and afterimages. Section 10.4 describes a model for segregating texture features that is inspired by computations in the primary visual cortex [11]. Finally, Section 10.5 introduces a model of a collision-sensitive neuron in the locust visual system [12]. The last section discusses a few examples where image processing (or computer vision) methods also could explain how the brain processes the corresponding information.

10.2
The Membrane Equation as a Neuron Model

In order to derive a simple yet powerful neuron model, imagine the neuron's cell membrane (comprising dendrites, soma, and axon) as a small sphere. The membrane is a bilayer of lipids, which is 30–50 Å thick (1 Å = 10^{-10} m = 0.1 nm). It isolates the extracellular space from a cell's interior and thus forms a barrier for different ion species, such as Na^+ (sodium), K^+ (potassium), and Cl^- (chloride). Now if the ionic concentrations in the extracellular fluid and the cytoplasm (the cell's interior) were all the same, then one would be probably dead. Neuronal signaling relies on the presence of ionic gradients. With all cells at rest (i.e., neither input nor output), about half of the brain's energy budget is consumed for moving Na^+

to the outside of the cell and K^+ inward (the $Na^+ - K^+$ pump). Some types of neurons also pump Cl^- outside (via the Cl^- transporter). The pumping mechanisms compensate for the ions that leak through the cell membrane in the respective reverse directions, driven by electrochemical gradients. At rest, the neuron therefore maintains a dynamic equilibrium.

We are now ready to model the neuron as an electrical circuit, where a capacitance (cell membrane) is connected in parallel with a (serially connected) resistance and a battery. The battery sets the cell's resting potential V_{rest}. In particular, all (neuron-, glia-, muscle-) cells have a negative resting potential, typically $V_{rest} = -65$ mV. The resting potential is the value of the membrane voltage V_m when all ionic concentrations are in their dynamic equilibrium. This is, of course, a simplification because we lump together the diffusion potentials (such as V^+_{Na} or V^+_K) of each ion species. The simplification comes at a cost, however, and our resulting neuron model will not be able to produce action potentials or spikes by itself (the binary events by which many neurons communicate with each other) without explicitly adding a spike-generating mechanism (more on that in Section 10.2.2). By Ohm's law, the current that leaks through the membrane is $I_{leak} = g_{leak}(V_m - V_{rest})$, where the *leakage conductance* g_{leak} is just the inverse of the membrane resistance. The charge Q which is kept apart by a cell's membrane with capacitance C is $Q(t) = CV_m(t)$. Whenever the neuron signals, the distribution of charges changes, and so does the membrane potential, and thus $dQ(t)/dt \equiv I_C = CdV_m(t)/dt$ will be nonzero. In other words, a current I_C will flow, carried by ions. Assuming some fixed value for I_C, the change in V_m will be slower for a higher capacitance C (better buffering capacity). *Kirchhoff's current law* is the equivalent of current conservation $I = I_C + I_{leak}$, or

$$C\frac{dV_m(t)}{dt} + g_{leak}(V_m - V_{rest}) = I \tag{10.1}$$

The right-hand side corresponds to current flows due to excitation and inhibition (this is discussed later in the following). Biologically, current flows occur across protein molecules that are embedded in the cell membrane. The various protein types implement specific functions such as ionic channels, enzymes, pumps, and receptors. These "gates" or "doors" through the cell membrane are highly specific, such that only particular information or substances (such as ions) can enter or exit the cell. Strictly speaking, each channel which is embedded in a neuron's cell membrane would correspond to an RC-circuit ($R = 1/g_{leak}$ is the resistance, C is the capacitance) such as Eq. (10.1). But fortunately, neurons are very small – this is what justifies the assumption that channels are uniformly distributed – and the potential does not vary across the cell membrane: The cell is said to be *isopotential*, and it can be adequately described by a *single* RC-compartment.

Let us assume that we have nothing better to do than waiting for a sufficiently long time such that the neuron reaches equilibrium. Then, by definition, V_m is constant, and thus $I_C = 0$. What remains from the last equation is just $(V_m - V_{rest})/R = I$, or $V_m = V_{rest} + RI$. In the absence of excitation and inhibition, we have $I = 0$ and the neuron will be at its resting potential. But how long do we have to wait until equilibrium is reached? To find this out, we just move all terms with V_m to one side of the equation, and the term which contains time to the other. This technique is known as *separation of variables,* and permits integration in order to convert the infinitesimal quantities dt and dV_m into "normal" variables (we formally rename the corresponding integration variables for time and voltage as T and V, respectively),

$$\int_{V_0}^{V_m} \frac{dV}{V - V_{\text{rest}} - RI} = -\frac{1}{RC}\int_0^t dT \tag{10.2}$$

where $V_0 \equiv V_m(t = 0)$. We further define $V_\infty \equiv RI$. Integration of the left-hand side gives $\log[(V_m - V_{\text{rest}} - V_\infty)/(V_0 - V_{\text{rest}} - V_\infty)]$. With the (membrane) time constant of the cell $\tau \equiv RC$, the above equation yields

$$V_m(t) = (V_0 - V_{\text{rest}} - V_\infty)e^{-t/\tau} + V_{\text{rest}} + V_\infty \tag{10.3}$$

It is easy to see that for $t \to \infty$ we get $V_m = V_{\text{rest}} + V_\infty$, where $V_\infty = 0$ in the absence of external currents I (this confirms our previous result). The time that we have to wait until this equilibrium is reached depends on τ: The higher the resistance R, and the bigger the capacitance C, the longer it will take (and vice versa).

The constant V_0 has to be selected according to the initial conditions of the problem. For example, if we assume that the neuron is at rest when we start the simulation, then $V_0 = V_{\text{rest}}$, and therefore

$$V_m(t) = V_\infty(1 - e^{-t/\tau}) + V_{\text{rest}} \tag{10.4}$$

10.2.1 Synaptic Inputs

Neurons are not loners but are massively connected to other neurons. The connection sites are called *synapses.* Synapses come in two flavors: electrical and chemical. Electrical synapses (also called *gap junctions*) can directly couple the membrane potential of neighboring neurons. In this way, distinct networks of specific neurons are formed, usually between neurons of the same type. Examples of electrically coupled neurons are retinal horizontal cells [13, 14], cortical low-threshold-spiking interneurons, and cortical fast-spiking interneurons [15, 16] (interneurons are inhibitory). Sometimes chemical and electrical synapses even combine to permit reliable and fast signal transmission, such as it is the case in the locust, where the lobula giant movement detector (LGMD) connects to the descending contralateral movement detector (DMCD) [17–20].

Chemical synapses are far more common than gap junctions. In one cubic millimeter of cortical gray matter, there are about one billion ($\approx 10^9$) chemical synapses (about 10^{15} in the whole human brain). Synapses are usually plastic. Whether they increase or decrease their connection strength to a postsynaptic neuron depends on causality. If a presynaptic neuron fires within some 5–40 ms before the postsynaptic neuron, the connection gets stronger (potentiation: "P") [21]. In contrast, if the presynaptic spike arrives after activation of the postsynaptic neuron, synaptic strength is decreased (depression: "D") [22]. This mechanism is known as *spike-time-dependent plasticity* (STDP), and can be identified with Hebbian learning [23, 24]. Synaptic potentiation is thought to be triggered by backpropagating calcium spikes in the dendrites of postsynaptic neuron [25] Synaptic plasticity can occur over several timescales, short term (ST) and long term (LT). Remember these acronyms if you see letter combinations such as "LTD," "LTP," "STP," or "STD".

An activation of fast, chemical synapses causes a rapid and transient voltage change in the postsynaptic neuron. These voltage changes are called *postsynaptic potentials* (PSPs). PSPs can be either inhibitory (IPSPs) or excitatory (EPSPs). Excitatory neurons depolarize their target neurons (V_m will get more positive as a consequence of the EPSP), whereas inhibitory neurons hyperpolarize their postsynaptic targets. How can we model synapses? PSPs are caused by a temporary increase in membrane conductance in series with a so-called *synaptic reversal battery E* (also synaptic reversal potential, or synaptic battery). The synaptic input defines the current on the right-hand side of Eq. (10.1),

$$I = \sum_{i=1}^{N} g_i(t) \cdot (E_i - V_m) \tag{10.5}$$

The last equation sums N synaptic inputs, each with conductance g_i and reversal potential E_i. Notice that whether an input g_i acts excitatory or inhibitory on the membrane potential V_m depends usually on whether the synaptic battery E_i is bigger or smaller than the resting potential V_{rest}. Just consider only one type of excitatory and inhibitory input. Then we can write

$$C\frac{dV_m(t)}{dt} = g_{\text{leak}}(V_{\text{rest}} - V_m) + g_{\text{exc}}(t) \cdot (V_{\text{exc}} - V_m) + g_{\text{inh}}(t) \cdot (V_{\text{inh}} - V_m) \tag{10.6}$$

(for all simulations, if not otherwise stated, we assume $C = 1$ and omit the physical units). How do we solve this equation? After converting the differential equation into a difference equation, the equation can be solved numerically with standard integration schemes, such as Euler's method, Runge–Kutta, Crank–Nicolson, or Adams–Bashforth (see, e.g., chapter 6 in Ref. [26] and chapter 17 in Ref. [27] for more details). Typically, model neurons are integrated with a step size of 1 ms or less, as this is the relevant timescale for neuronal signaling. If the simulation focuses more on perceptual dynamics (or biologically inspired image processing tasks), then one may choose a bigger integration time constant as well. The ideal integration method is stable, produces solutions with a high accuracy, and has a low computational complexity. In practice, of course, we have to make the one or the other trade-off.

Remember that $g_{\text{leak}} \equiv 1/R$ is called *leakage conductance*, which is just the inverse of the membrane resistance. For constant capacitance C, the leakage conductance determines the time constant $\tau \equiv C/g_{\text{leak}}$ of the neuron: bigger values of g_{leak} will make it "faster" (i.e., less memory on past inputs), while smaller values will cause a higher degree of lowpass filtering of the input. $V_{exc} > V_{rest}$ and $V_{inh} \leq V_{rest}$ are the excitatory and inhibitory synaptic batteries, respectively.

For $V_{exc} > V_{rest}$ the synaptic current (mainly Na^+ and K^+) is inward and negative by convention. The membrane thus gets depolarized. This is a signature of an EPSP. In the brain, the most common type of excitatory synapses release *glutamate* (a neurotransmitter).[1] The neurotransmitter diffuses across the synaptic cleft, and binds on glutamate-sensitive receptors in the postsynaptic cell membrane. As a consequence, ion channels will open, and Na^+ and K^+ (and also Ca^{2+} via voltage sensitive channels) will enter the cell.

Agonists are pharmacological substances that do not exist in the brain, but open these channels as well. For instance, the agonist NMDA[2] will open excitatory, voltage-sensitive NMDA-channels. AMPA[3] is another agonist that activates fast excitatory synapses. However, AMPA-synapses will remain silent in the presence of NMDA, and vice versa. Therefore one can imagine the ionic channels as locked doors. For their opening, the right key is necessary, which is either a specific neurotransmitter, or some "artificial" pharmacological agonist. The "locks" are the receptor sites to which a neurotransmitter or an agonist binds. The reversal potentials of the fast AMPA-synapse is about 80 – 100 mV above the resting potential. Usually, AMPA-channels co-occur with NMDA-channels, and this may enhance the computational power of a neuron [28].

For $V_{inh} < V_{rest}$ the membrane is hyperpolarized. For $V_{inh} \approx V_{rest}$, it gets less sensitive to depolarization, and accelerates the return to V_{rest} for any synaptic input. Presynaptic release of GABA[4] can activate three subtypes of receptors: $GABA_A$, $GABA_B$, and $GABA_C$ (as before, they are identified through the action of specific pharmalogicals)[5]

Why are the synaptic batteries also called reversal potentials ? For excitatory input, V_{exc} imposes an upper limit on V_m. This means that no matter how big g_{exc} will be, it can drive the neuron only up to V_{exc}. In order to understand this, consider $(V_{exc} - V_m)$, the so-called *driving potential*: if $V_m << V_{exc}$, then the driving potential is high, and the neuron depolarizes fast. The closer V_m gets to V_{exc}, the smaller the driving potential, until the excitatory current $g_{exc}(t) \cdot (V_{exc} - V_m)$ eventually approaches zero. (Analog considerations hold for the inhibitory input).

What is the value of V_m at equilibrium? Equilibrium means that $V_m(t)$ does not change with t, and then the left hand side of Eq. (10.6) is zero. Of course this implies that all excitatory and inhibitory inputs vary sufficiently slowly and we can consider them as being constant. Otherwise expressed, the neuron reaches

1) In the peripheral nervous system of vertebrates, excitatory synapses are activated instead by acetylcholine (ACh)
2) *N*-Methyl-D-aspartat
3) α-Amino-3-hydroxy-5-methyl-4-isoxazole propionic acid
4) From Ref. [29]: γ-aminobutyric acid
5) $GABA_A$, are ligand-gated ion channels permeable to Cl^-. Postsynaptic $GABA_B$ are heptahelical receptors coupled to inwardly rectifying K^+ channels. Finally, $GABA_C$ are ligand-gated Cl^- channels, which are primarily expressed in the retina.

the equilibrium before a significant change in $g_{exc}(t)$ or $g_{inh}(t)$ takes place. Then, solving Eq. (10.6) for V_m yields

$$V_m(t \to \infty) = \frac{g_{leak} V_{rest} + g_{exc} V_{exc} + g_{inh} V_{inh}}{g_{leak} + g_{exc} + g_{inh}} \tag{10.7}$$

The time until the equilibrium is reached depends not only on g_{leak}, but on all other active conductances. As a consequence, a neuron which receives continuous input from other neurons can react faster than a neuron which starts from V_{rest} [30]. (Ongoing cortical activity is the normal situation, where it is thought that excitation and inhibition are just balanced [31]).

A specially interesting case is defined by $V_{inh} = V_{rest}$, which is called *silent* or *shunting inhibition.* It is silent because it only becomes evident if the neuron is depolarized (and hyperpolarized if more than one type of inhibition is considered). Shunting inhibition decreases the time constant of the neuron, thus making it faster. In this way, the return to the resting potential is accelerated for excitatory and inhibitory input. Furthermore, *divisive inhibition* is a special form of shunting inhibition if $V_{rest} = 0$. With spiking neurons, however, pure divisive inhibition does not seem to exist. In that case, shunting inhibition is rather subtractive [32] and cannot act as a gain control mechanism. But in networks with balanced excitation and inhibition, the choice is ours: if we change the balance between excitation and inhibition, then the effect on a neuron's response will be additive and subtractive, respectively. If we leave the balance unchanged and increase or decrease excitation and inhibition in parallel, then a multiplicative or divisive effect on a neuron's response will occur [33].

10.2.2
Firing Spikes

Equation (10.6) represents the membrane potential of a neuron, but V_m could represent different quantities as well. For example, V_m could be interpreted directly as response probability if we set, for example, $V_{rest} = 0$, $V_{exc} = 1$, and $V_{inh} = 0$. Accordingly, in the latter case we have $0 \leq V_m \leq 1$. Another possibility is to set $V_{inh} = -1$. As neuronal responses are only positive, however, V_m has to be *half-wave rectified,* meaning that we take $[V_m]^+ \equiv \max(0, V_m)$ as the output of the neuron.[6] Naturally, half-wave rectification makes sense only if negative values of V_m can occur. Because of the absence of explicit spiking, the neuron's output represents a (mean) *firing rate,* usually interpreted as spikes per second.

When should one use Eq. (10.6) or (10.7)? This depends mainly on the purpose of the simulation. When the synaptic input consists of spikes, then one needs some mechanism to convert them into continuous quantities. The lowpass filtering characteristics of Eq. (10.6) will do that. For instance, spikes are necessary for implementing spike-time-dependent plasticity (STDP), which modifies synaptic

6) In this case, the *response threshold* $V_{thresh} = 0$. For arbitrary thresholds, the right hand side would become $\max(V_{thresh}, V_m)$.

strength dependent on pre- and postsynaptic activity. For some purposes (e.g., biologically inspired image processing), spikes are not strictly necessary, and one can use the (mean or instantaneous) firing rate (or activity, that is, the rectified membrane potential) of presynaptic neurons directly as input to postsynaptic neurons. In that case, either Eq. (10.6) or (10.7) could be used. The steady-state solution (Eq. (10.7)), however, has less computational complexity, as one does not need to integrate it numerically. Recall that when using the steady-state solution, one implicitly assumes that the synaptic input varies on a relatively slow timescale, such that the neuron can reach its equilibrium state at each instant.

It is straightforward to convert (10.6) into a *leaky integrate-and-fire neuron.* The term "leaky" refers to the leakage conductance g_{leak}, and the neuron would only be a perfect integrator for $g_{\text{leak}} = 0$ As soon as the membrane potential $V_m(t)$ crosses the neuron's response threshold V_{thresh}, we record a pulse with some amplitude in the neuron's response. Otherwise, the response is usually defined as being zero:

```
1    if V>Vthresh % Vthresh = response threshold
2        % response is a pulse with amplitude 'SpikeAmp'
3        response = SpikeAmp;
4        % reset membrane potential (afterhyperpolarization)
5        V = Vreset;
6    else
7        response = 0; % else, the neuron stays silent
8    end
```

In order to account for afterhyperpolarization (i.e., the refractory period), V_m is set to some value V_{reset} after each spike. Usually, $V_{\text{reset}} \leq V_{\text{rest}}$ is chosen. The refractory period of a neuron is the time that the ionic pumps need to reestablish the original ion charge distributions. Within the absolute refractory period, the neuron will not spike at all or spiking probability will be greatly reduced (relative refractory period). Other spiking mechanisms are conceivable as well. For example, we can define the model neuron's response as being identical to firing rate $[V_m]^+$, and add a spike to V_m whenever $V_m > V_{\text{thresh}}$. Then V_{thresh} would represent a *spiking threshold* :

```
1    if V>Vthresh % 'Vthresh' = spiking threshold
2        % add a spike with amplitude 'SpikeAmp' to current ...
             membrane potential 'V'
3        response = V + SpikeAmp;
4        % reset membrane potential (refractory period)
5        V = Vreset;
6    else
7        response = max(V,0); % otherwise, rate-code-like ...
              response (by half-wave rectification of 'V')
8    end
```

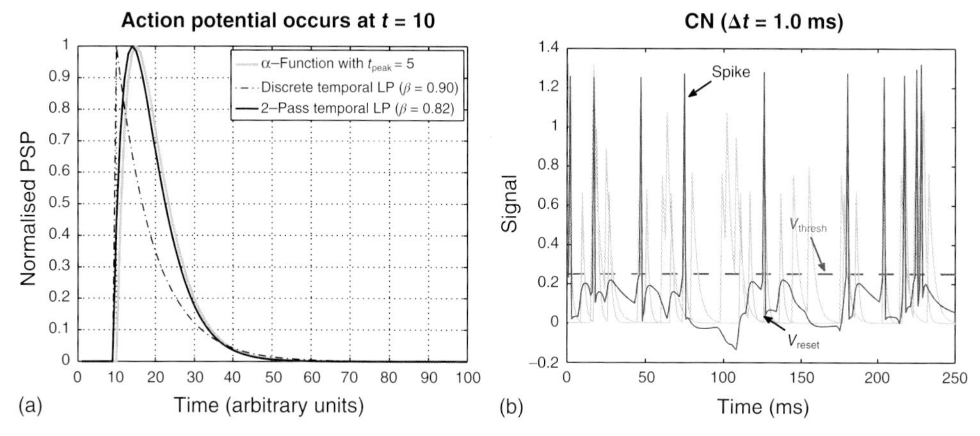

Figure 10.1 Spikes and postsynaptic potentials. (a) The figure shows three methods for converting a spike (value one at $t_0 = 10$, that is $\delta[t-10]$) into a postsynaptic potential (PSP). The α-function with $t_{peak} = 5$ (Eq. (10.8)) is represented by the gray curve. The result of lowpass filtering the spike once (via Eq. (10.10) with $\beta = 0.9$) is shown by the dashed line: The curve has a sudden rise and a gradual decay. Finally, applying two times the lowpass filter (each with $\beta = 0.82$) to the spike results in the black curve. Thus, a 2-pass lowpass filter can approximate the α-function reasonably well. (b) The figure shows PSPs and output of the model neuron Eq. (10.6) endowed with a spike mechanism. Excitatory ($g_{exc}(t)$, pale green curve) and inhibitory ($g_{inh}(t)$, pale red curve) PSPs cause corresponding fluctuations in the membrane potential V_m (black curve). As soon as the membrane potential crosses the threshold $V_{thresh} = 0.25$ (dashed horizontal line), a spike is added to V_m, after which the membrane potential is reset to $V_{reset} = 0$. The half-wave rectified membrane potential represents the neuron's output. The input to the model neuron were random spikes which were generated according to a Poisson process (with rates $\mu_{exc} = 100$ and $\mu_{inh} = 50$ spikes per second, respectively). The random spikes were converted into PSPs via simple lowpass filtering (Eq. (10.10)) with filter memories $\beta_{exc} = 0.5$ and $\beta_{inh} = 0.75$, respectively, and weight $w_{syn} = 75$. The integration method was Crank (not Jack)–Nicolson with step size $\Delta t = 1$ ms. The rest of the parameters of Eq. (10.6) were $V_{exc} = 3$, $V_{inh} = -1$, $V_{rest} = 0$, and $g_{leak} = 50$.

The response thus switches between a rate code ($V_m < V_{thresh}$) and a spike code ($V_m \geq V_{thresh}$) [34, 35]. A typical spike train produced by the latter mechanism is shown in Figure 10.1(b).

How are binary events such as spikes converted into PSPs? A PSP has a sharp rise and a smooth decay, and therefore is usually broader than the spike by which it was evoked. Assume that a spike arrives at time t_i at the postsynaptic neuron. Then the time course of the corresponding PSP (i.e., the excitatory or inhibitory input to the neuron) is adequately described by the so-called *α-function*:

$$\alpha(t, t_i) = \text{const} \cdot (t - t_i)e^{-(t-t_i)/t_{peak}}\Theta(t - t_i) \tag{10.8}$$

The constant is chosen such that $g_{syn}(t = t_{peak})$ matches the desired maximum of the PSP. The Heaviside function $\Theta(x)$ is zero for $x \leq 0$, and 1 otherwise. It makes sure that the PSP generated by the spike at t_i starts at time t_i. The total synaptic input $g_{syn}(t)$ into the neuron is $\sum_i \alpha(t, t_i)$, multiplied with a synaptic weight w_{syn}.

Instead of using the α-function, we can simplify matters (and accelerate our simulation) by assuming that each spike causes an instantaneous increase in $g_{exc}(t)$ or $g_{inh}(t)$, respectively, followed by an exponential decay. Doing so just amounts to adding one simple differential equation to our model neuron:

$$\tau \frac{dx}{dt} = -x + w_{syn} \cdot \delta[t - t_i] \tag{10.9}$$

The time constant τ determines the rate of the exponential decay (faster decay if smaller), w_{syn} is the synaptic weight, and $\delta[\cdot]$ is the Kronecker delta function, which is just one if its argument is zero: The ith spike increments x by w_{syn}. The last equation lowpass filters the spikes. An easy-to-compute discrete version can be obtained by converting the last equation into a finite difference equation, either by forward or backward differencing (details can be found in section S8 of Ref. [36]):

$$x_{n+1} = \beta x_n + w_{syn}(1 - \beta)\delta[t - t_i] \tag{10.10}$$

For forward differentiation, $\beta = 1 - \Delta t/\tau$, where Δt is the integration time constant that comes from approximating dx/dt by $(x_{n+1} - x_n)/\Delta t$. For backward differencing, $\beta = \tau/(\tau + \Delta t)$. The degree of lowpass filtering (filter memory on past inputs) is determined by β. For $\beta = 0$, the filter output x_{n+1} just reproduces the input spike pattern (the filter is said to have no memory on past inputs). For $\beta = 1$, the filter ignores any input spikes, and stays forever at the value with which it was initialized ("infinite memory"). For any value between zero and one, filtering takes place, where filtering gets stronger with increasing β.

When one or more spikes are filtered by Eq. (10.10), then we see a sudden increase in x at time t_i, followed by a gradual decay (Figure 10.1(a), dashed curve). This sudden increase stands in contrast to the gradual increase of the PSP as predicted by Eq. (10.8) (Figure 10.1(a), gray curve). A better approximation to the shape of a PSP results from applying lowpass filtering twice, as shown by the black curve in Figure 10.1(a). This is tantamount to cascade two times equation (10.10) for each synaptic input Eq. (10.6).

10.3 Application 1: A Dynamical Retinal Model

The retina is a powerful computational device. It transforms light intensities of different wavelengths – as captured by cones (and rods for low-light vision) – into an *efficient* representation which is sent to the brain by the axons of the retinal ganglion cells [37]. The term "efficient" refers to *redundancy reduction* ("decorrelation") in the stimulus on the one hand, and coding efficiency at the level of ganglion cells, on the other. Decorrelation means that predictable intensity levels in time and space are suppressed in the responses of ganglion cells [38, 39]. For example, a digital photograph of a clear blue sky has a lot of spatial redundancy, because if we select a blue pixel, it is highly probable that its neighbors are blue pixels as well [40]. *Coding efficiency* is linked to metabolic energy consumption. Energy consumption increases faster than information transmission capacity, and organisms therefore seem to have evolved to a trade-off between increasing their evolutionary fitness and saving energy [41]. Retinal ganglion cells show efficient coding in the sense that noisy or energetically expensive coding symbols are less "used" [37, 42]. Often, the spatial aspects of visual information processing by the retina are grossly approximated by employing the *difference-of-Gaussian* ("DoG") model (one Gaussian is slightly broader than the other) [43]. The Gaussians are typically two-dimensional, isotropic, and centered at identical spatial coordinates. The resulting DoG model is a convolution kernel with positive values in the center surrounded by negative values. In mathematical terms, the DoG-kernel is a filter that takes the second derivative of an image. In signal processing terms, it is a bandpass filter (or a highpass filter if a small 3×3 kernel is used). In this way, the *center–surround antagonism* of (foveal) retinal ganglion cells can be modeled [44]: ON-center ganglion cells respond when the center is more illuminated than the surround (i.e., positive values after convolving the DoG kernel with an image). OFF-cells respond when the surround receives more light intensity than the center (i.e., negative values after convolution). The DoG model thus assumes symmetric ON- and OFF-responses; this is again a simplification: differences between biological ON- and OFF ganglion cells include receptive field size, response kinetics, nonlinearities, and light–dark adaptation [45–47]. Naturally, convolving an image with a DoG filter can neither account for adaptation nor for dynamical aspects of retinal information processing. On the other hand, however, many retinal models which target the explanation of physiological or psychophysical data are not suitable for image processing tasks. So, biologically inspired image processing means that the model should solve an image processing task (e.g., boundary extraction, dynamic range reduction), while at the same time it should produce predictions or have features which are by and large consistent with psychophysics and biology (e.g., brightness illusions, kernels mimicking receptive fields). In this spirit, we now introduce a simple dynamical model for retinal processing, which reproduces some interesting brightness illusions and could even account for afterimages. An afterimage is an illusory percept where one continues to see a stimulus which is physically not present any more (e.g., a spot after looking into a bright light source). Unfortunately, the author was unable to find a version of the model which could be strictly based on Eq. (10.6). Instead of that, here is an even simpler version that is based on the temporal lowpass filter (Eq. (10.10)). Let $I_t(x, y)$ be a gray-level image with luminance values between zero (dark) and one (white). The number of iterations is denoted by t (discrete time). Then

$$u_{t+1} = \beta_1 u_t + (1 - \beta_1)(I_t - v_t) \tag{10.11}$$

$$v_{t+1} = \beta_2 v_t + (1 - \beta_2)([u_t]^+ + I_t) + D \cdot \vec{\nabla}^2 v_t \tag{10.12}$$

where $[u_t(x, y)]^+ =$ ON-cell responses, $[-u_t(x, y)]^+ =$ OFF-cell responses, $D =$ const is the diffusion coefficient, $\vec{\nabla}^2 v_t(x, y) \equiv \mathrm{div}(\mathrm{grad}(v_t))$ is the diffusion operator (Laplacian), which was discretized as a 3×3 convolution kernel with -1 in the center, and 0.25 in north, east, south, and west pixels. Corner pixels were

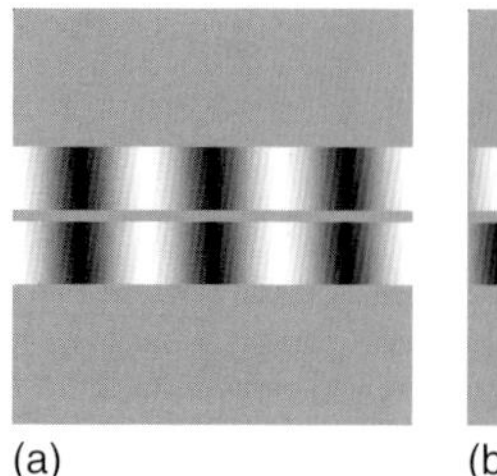
(a)

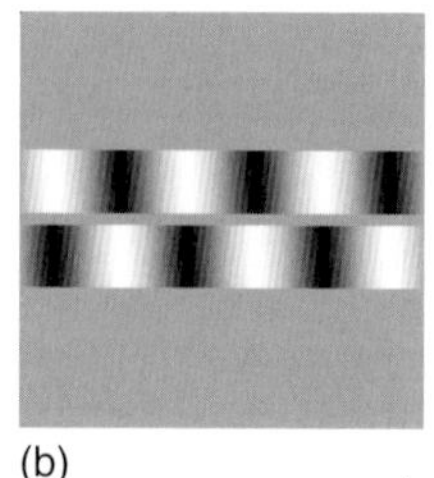
(b)

(c)

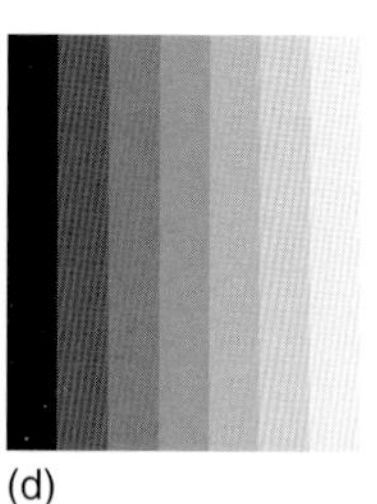
(d)

Figure 10.2 Test images All images have 256 rows and 256 columns. (a) The upper and the lower grating ("inducers") are separated by a small stripe, which is called the *test stripe*. Although the test stripe has the same luminance throughout, humans perceive a wavelike pattern with opposite brightness than the inducers, that is, where the inducers are white, the test stripe appears darker and vice versa. (b) When the inducer gratings have an opposite phase (i.e., white stands *vis-á-vis* black), then the illusory luminance variation across the test stripe is weak or absent. (c) A real-world image or photograph ("camera"). (d) A luminance staircase, which is used to illustrate afterimages in Figure 10.4(c) and (d).

zero. Thus, whereas the receptive field center is just one pixel, the surround is dynamically constructed by diffusion. Diffusion length (and thus surround size) depends on the filter memory constant β_2 and the diffusion coefficient D (bigger values will produce a larger surround area).

Figure 10.2 shows four test images which were used for testing the dynamic retina model. Figure 10.2(a) shows a visual illusion ("grating induction"), where observers perceive an illusory modulation of luminance between the two gratings, although luminance is actually constant. Figure 10.3(a) shows that the dynamic retina predicts a wavelike activity pattern within the grating via $u_t(x, y)$. Because ON-responses represent brightness (perceived luminance) and OFF-responses represent darkness (perceived inverse luminance) , the dynamic retina correctly predicts grating induction. Does it also account for the absence of grating induction in Figure 10.2(b)? The corresponding simulation is shown in Figure 10.3(b), where the amplitude of the wave-like pattern is strongly reduced, and the frequency has doubled. Thus, the absence of grating induction is adequately predicted.

In its equilibrium state, the dynamic retina performs contrast enhancement or boundary detection, respectively. This is illustrated with the ON- and OFF-responses (Figure 10.4(a) and (b), respectively) to Figure 10.2(c). In comparison to an ordinary DoG-filter, however, the responses of the dynamic retina are asymmetric, with somewhat higher OFF-responses to a luminance step (not shown). A nice feature of the dynamic retina is the prediction of after images. This is illustrated by computing first the responses to a luminance staircase (Figure 10.2(d)), and then replacing the staircase image by the image of the cameraman (Figure 10.2(c)). Figure 10.4(c) and (d) shows corresponding responses immediately after the images were swapped. Although the camera man image is now assigned to I_t, a slightly blurred afterimage of the staircase still appears in

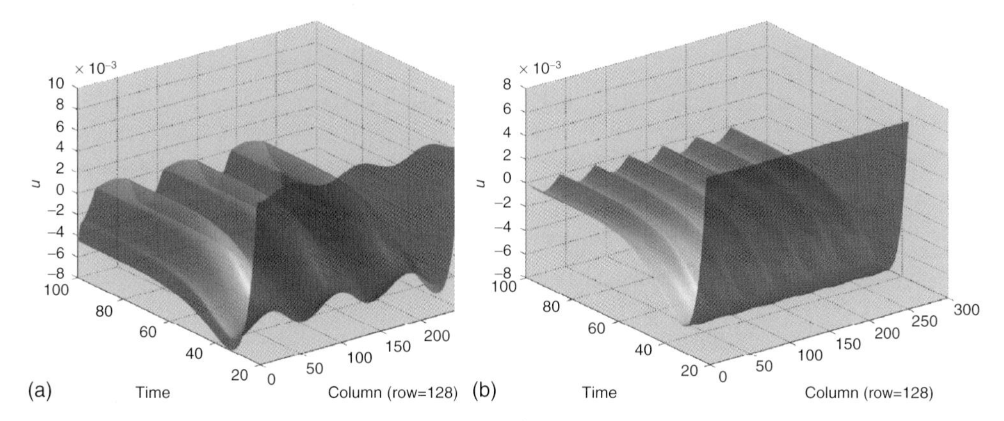

Figure 10.3 Simulation of grating induction. Does the dynamic retina (Eqs. (10.11) and (10.12)) predict the illusory luminance variation across the test stripe (the small stripe which separates the two gratings in Figure 10.2(a) and (b))? (a) Here, the image of Figure 10.2(a) was assigned to I_t. The plot shows the temporal evolution of the horizontal line centered at the test stripe, that is, it shows all columns $1 \leq x \leq 256$ of $u_t(x, y_0)$ for the fixed row number $y_0 = 128$ at different instances in time t. Time increases toward the background. If values $u_t(x, y_0) > 0$ (ON-responses) are interpreted as brightness, and values $u_t(x, y_0) < 0$ (OFF-responses) as darkness, then the wave pattern adequately predicts the grating induction effect. (b) If the image of Figure 10.2(b) is assigned to I_t (where human observers usually do not perceive grating induction), then the wavelike pattern will have twice the frequency of the inducer gratings, and moreover a strongly reduced amplitude. Thus, the dynamic retina correctly predicts a greatly reduced brightness (and darkness) modulation across the test stripe.

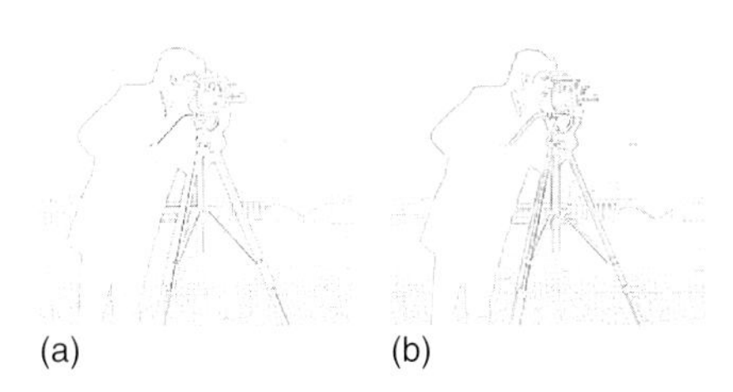

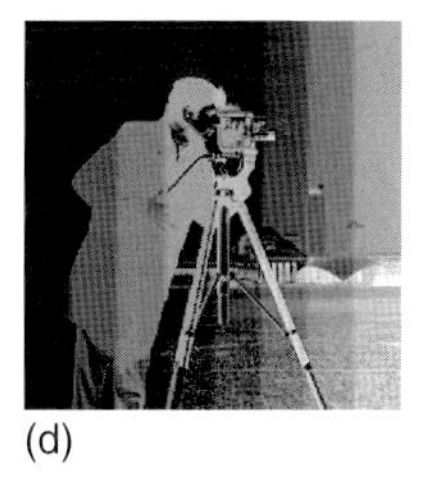

(a) (b) (c) (d)

Figure 10.4 Snapshots of the dynamic retina. (a) This is the ON-response (Eq. (10.11) $[u_t(x, y)]^+$) after $t = 200$ iterations of the dynamic retina (Eqs. 10.11 and 10.12), where the image of Figure 10.2(c) was assigned to $I_t(x, y)$. Darker gray levels indicate higher values of $[u_t(x, y)]^+$). (b) Here the corresponding OFF-responses ($[-u_t(x, y)]^+$) are shown, where darker gray levels indicate higher OFF-responses. (c) Until $t = 200$, the luminance staircase (Figure 10.2(d)) was assigned to $I_t(x, y)$. Then the image was replaced by the image of the camera man. This simulates a retinal saccade. As a consequence, a ghost image of the luminance staircase is visible in both ON- and OFF-responses (approximately until $t = 250$). From $t = 260$ on, the ON- and OFF-responses are indistinguishable from (a) and (b). Here, *brighter* gray levels indicate higher values of $[u_t(x, y)]^+$. (d) Corresponding OFF-responses to (c). Again, *brighter* gray levels indicate higher values of $[-u_t(x, y)]^+$. All simulations were performed with filter memory constants $\beta_1 = 0.9$, $\beta_2 = 0.85$, and diffusion coefficient $D = 0.25$.

u_t. The persistence of the afterimage depends on the luminance values: Higher intensities in the first image, and lower intensities in the second image will promote a prolonged effect.

10.4
Application 2: Texture Segregation

Networks based on Eq. (10.6) can be used for biologically plausible image processing tasks. Nevertheless, the specific task that one wishes to achieve may imply modifications of Eq. (10.6). As an example we outline a corresponding network for segregating texture from grayscale image ("texture system"). We omit many mathematical details at this point (cf. Chapter 4 in Ref. [11]), because they probably would make reading too cumbersome.

The texture system forms part of a theory for explaining early visual information processing [11]. The essential proposal is that simple cells in V1 (primary visual cortex) segregate the visual input into texture, surfaces, and (slowly varying) luminance gradients. This idea emerges quite naturally from considering how the symmetry and scale of simple cells relate to features in the visual world. Simple cells with small and odd-symmetric receptive fields (RFs) respond preferably to contours that are caused by changes in material properties of objects such as reflectance. Object surfaces are delimited by odd-symmetric contours. Likewise, even-symmetrical simple cells respond particularly well to lines and points, which we call *texture in this context*. Texture features are often superimposed on object surfaces. Texture features may be variable and be rather irrelevant in the recognition of a certain object (e.g., if the object is covered by grains of sand), but also may correspond to an identifying feature (e.g., tree bark). Finally, simple cells at coarse resolutions (i.e., those with big RFs of both symmetries) are supposed to detect shallow luminance gradients. Luminance gradients are a pictorial depth cue for resolving the three-dimensional layout of a visual scene. However, they should be ignored for determining the material properties of object surfaces.

The first computational step in each of the three systems consists in detecting the respective features. Following feature detection, representations of surfaces [48], gradients [49–51], and texture [11], respectively, are eventually build by each corresponding system. Our normal visual perception would then be the result of superimposing all three representations (brightness perception). Having three separate representations (instead of merely a single one) has the advantage that higher-level cortical information processing circuits could selectively suppress or reactivate texture and/or gradient representations in addition to surface representations. This flexibility allows for the different requirements for deriving the material properties of objects. For instance, surface representations are directly linked to the perception of reflectance ("lightness") and object recognition, respectively. In contrast, the computation of surface curvature and the interpretation of the three-dimensional scene structure relies on gradients (e.g., shape from shading) and/or texture representations (texture compression with distance).

How do we identify texture features? We start with processing a grayscale image with a retinal model which is based on a modification of Eq. (10.6):

$$\frac{dV_{ij}(t)}{dt} = g_{\text{leak}}(V_{\text{rest}} - V_{ij}) + \zeta(\mathscr{E}_{ij} - \mathscr{I}_{ij}) + g_{ij,si} \cdot (E_{si} - V_{ij}) \tag{10.13}$$

$\mathscr{E}_{ij}$ is just the input image itself – so the center kernel (also known as the *receptive field*) is just one pixel. $\mathscr{I}_{ij}$ is the result of convolving the input image with a 3×3 surround kernel that has $1/4$ at north, east, west, and south positions. Elsewhere it is zero. Thus, $\mathscr{E}_{ij} - \mathscr{I}_{ij}$ approximates the (negative) second spatial derivative of the image. Accordingly, we can define two types of ganglion cell responses: ON-cells are defined by $\zeta = 1$ and respond preferably to (spatial) increments in luminance. OFF-cells have $\zeta = -1$ and prefer decrements in luminance. Figure 10.5(b) shows the output of Eq. (10.13), which is the half-wave rectified membrane potential $[V_{ij}]^+ \equiv \max(V_{ij}, 0)$. Biological ganglion cell responses saturate with increasing contrast. This is modeled here by $g_{ij,si} \cdot (E_{si} - V_{ij})$, where $g_{ij,si}$ corresponds to self-inhibition – the more center and surround are activated, the stronger. Mathematically, $g_{ij,si} = \xi \cdot [\zeta(\mathscr{E}_{ij} - \mathscr{I}_{ij})]^+$, with a constant $\xi > 0$ that determines how fast the responses saturate. Why did we not simply set $g_{\text{exc}} = \mathscr{E}_{ij}$ (and $V_{\text{exc}} = 1$), and $g_{\text{inh}} = \mathscr{I}_{ij}$ (and $V_{\text{inh}} = -1$) in Eq. (10.6) (vice versa for an OFF-cell)? This is because in the latter case the ON- and OFF-response amplitudes would be different for a luminance step (cf. Eq. (10.7)). Luminance steps are odd-symmetric features and are not texture. Thus, we want to suppress them, and suppression is easier if ON- and OFF-response amplitudes (to odd-symmetric features) are equal.

In response to even-symmetric features (i.e., texture features) we can distinguish two (quasi one-dimensional) response patterns. A black line on a white background produces an ON–OFF–ON (i.e., LDL) response: a central OFF-response, and two flanking ON-responses with much smaller amplitudes. Analogously, a bright line on a dark background will trigger an OFF–ON–OFF or

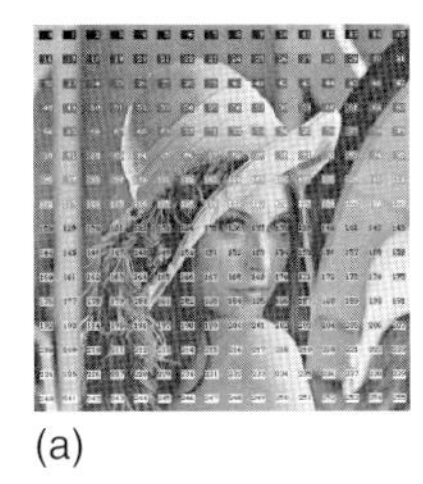

(a)

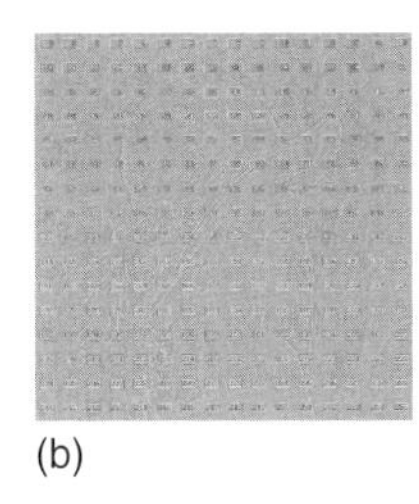

(b)

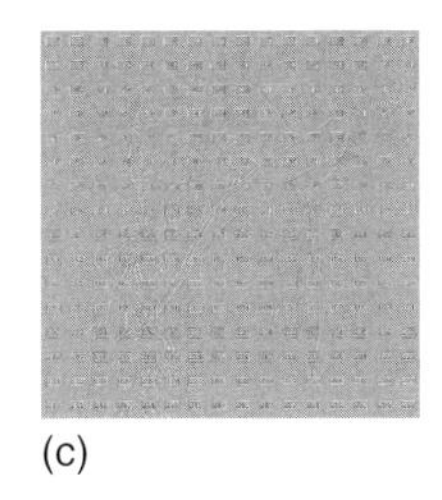

(c)

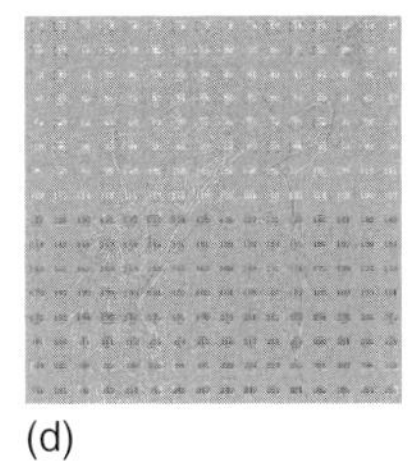

(d)

Figure 10.5 Texture segregation. Illustration of processing a grayscale image with the texture system. (a) Input image "Lena" with 512×512 pixels and superimposed numbers. (b) The output of the retina (Eq. (10.13)). ON-activity is white, while OFF is black. (c) The analysis of the retinal image proceeds along four orientation channels. The image shows an intermediate result after summing across the four orientations (texture brightness in white, texture darkness in black). After this, a local WTA-competition suppresses residual features that are not desired, leading to the texture representation. (d) This is the texture representation of the input image and represents the final output of the texture system. As before, texture brightness is white and texture darkness is black.

DLD response pattern. ON- and OFF-responses to lines (and edges) vary essentially in one dimension. Accordingly, we analyze them along four orientations. Orientation-selective responses (without blurring) are established by convolving the OFF-channel with an oriented Gaussian kernel, and subtract it from the (not blurred) ON-channel. This defines texture brightness. The channel for texture darkness is defined by subtracting blurred ON-responses from OFF-responses. Subsequently, even-symmetric response patterns are further enhanced with respect to surface features. In order to boost a DLD pattern, the left-OFF is multiplied with its central-ON and its right-OFF (analogous for LDL patterns). Note that surface response pattern (LD or DL, respectively) ideally has only one flanking response. Dendritic trees are a plausible neurophysiological candidate for implementing such a "logic" AND gate (simultaneous left *and* central *and* right response) [6].

In the subsequent stage the orientated texture responses are summed across orientations, leaving nonoriented responses (Figure 10.5(c)). By means of a winner-takes-all (WTA) competition between adjacent (nonoriented) texture brightness ("L") and texture darkness ("D"), it is now possible to suppress the residual surface features on the one hand, and the flanking responses from the texture features, on the other. For example, an LDL response pattern will generate a competition between L and D (on the left side) and D and L (right side). Since for a texture feature the central response is bigger, it will survive the competition with the flanking responses. The flanking responses, however, will not. A surface response (say DL) will not survive either, because D- and L-responses have equal amplitudes. The local spatial WTA-competition is established with a nonlinear diffusion paradigm [52]. The final output of the texture system (i.e., a texture representation) is computed according to Eq. (10.6), where texture brightness acts excitatory, and texture darkness inhibitory. An illustration of a texture representation is shown in Figure 10.5(d).

10.5
Application 3: Detection of Collision Threats

Many animals show avoidance reactions in response to rapidly approaching objects or other animals [53, 54]. Visual collision detection also has attracted attention from engineering because of its prospective applications, for example, in robotics or in driver assistant systems. It is widely accepted that visual collision detection in biology is mainly based on two *angular variables*: (i) The *angular size* $\Theta(t)$ of an approaching object, and (ii) its *angular velocity* or rate of expansion $\dot{\Theta}(t)$ (the dot denotes derivative in time). If an object approaches an observer with constant velocity, then both angular variables show a nearly exponential increase with time. Biological collision avoidance does not stop here, but computes mathematical functions (here referred to as *optical variables*) of $\Theta(t)$ and $\dot{\Theta}(t)$. Accordingly, three principal classes of collision-sensitive neurons have been identified [55]. These classes of neurons can be found in animals as different as insects or birds. Therefore, evolution came up with similar computational principles that are shared across many species [36].

A particularly well-studied neuron is the LGMD neuron of the locust visual system, because the neuron is relatively big and easy to access. Responses of the LGMD to object approaches can be described by the so-called *eta-function*: $\eta(t) \propto \dot{\Theta} \cdot \exp(-\alpha\Theta)$ [56]. There is some evidence that the LGMD biophysically implements $\log \eta$ [57] (but see Ref. [58, 59]): logarithmic encoding converts the product into a sum, with $\log \dot{\Theta}$ representing excitation and $-\alpha\Theta$ inhibition. One distinctive property of the eta-function is a response maximum before collision would occur. The time of the response maximum is determined by the constant α and always occurs at the fixed angular size $2 \cdot \arctan(1/\alpha)$.

So much for the theory – but how can the angular variables be computed from a sequence of image frames? The model which is presented below (first proposed in Ref. [12]) does not compute them explicitly, although its output resembles the eta-function. However, the eta-function is not explicitly computed either. Without going too far into an ongoing debate on the biophysical details of the computations which are carried out by the LGMD [53, 58, 60], the model rests on lateral inhibition in order to suppress self-motion and background movement [61].

The first stage of the model computes the difference between two consecutive image frames $I_t(x, y)$ and $I_{t-1}(x, y)$ (assuming grayscale videos):

$$\frac{dp}{dt} = -g_{\text{leak}} \cdot p + I_t \cdot (1 - p) - I_{t-1} \cdot (1 + p) \tag{10.14}$$

where $p \equiv p_t(x, y)$ – we omit spatial indices (x, y) and time t for convenience. The last equation and all subsequent ones derives directly from equation (10.6). Further processing in the model proceeds along two parallel pathways. The ON-pathway is defined by the positive values of p, that is $p^\circ = \max(p, 0) \equiv [p]^+$. The OFF-pathway is defined by $p^\bullet = [-p]^+$. In the absence of background movement, $\sum_{x,y}(p^\circ + p^\bullet)$ is related to angular velocity: if an approaching object is yet far away, then the sum will increase very slowly. In the last phase of an approach (shortly before collision), however, the sum increases steeply.

The second stage has two diffusion layers $s \in \{s^\circ, s^\bullet\}$ (one for ON, another one for OFF) that implement lateral inhibition:

$$\frac{ds}{dt} = g_{\text{leak}}(V_{\text{rest}} - s) + g_{\text{exc}}(1 - s) + D \cdot \vec{\nabla}^2 s \tag{10.15}$$

where D is the diffusion coefficient (cf. Eq. (10.12)). The diffusion layers act inhibitory (see below) and serve to attenuate background movement and translatory motion as caused by self-movement. If an approaching object is sufficiently far away, then the spatial variations (as a function of time) in p are small. Similarly, translatory motion at low speed will also generate small spatial displacements. Activity propagation proceeds at constant speed and thus acts as a predictor for small movement patterns. But diffusion cannot keep up with those spatial variations as they are generated in the late phase of an approach. The output of the diffusion layer is $s^\circ = [s^\circ]^+$ and $s^\bullet$, respectively (the tilde denotes half-wave rectified variables).

The inputs into the diffusion layers $g^{\circ}_{\text{exc}} = 250 \cdot v^{\circ}$ and $g^{\bullet}_{\text{exc}} = 250 \cdot v^{\bullet}$, respectively, are brought about by feeding back activity from the third stage of the model:

$$\frac{dv}{dt} = g_{\text{leak}}(V_{\text{rest}} - v) + g_{\text{exc}}(1 - v) - g_{\text{inh}}(0.25 + v) \tag{10.16}$$

with excitatory input $g^{\circ}_{exc} = 250p^{\circ} \cdot \exp\,(-500s^{\circ})$ and inhibitory input $g^{\circ}_{\text{inh}} = 500s^{\circ}$ ($g^{\bullet}_{\text{exc}}$ and $g^{\bullet}_{\text{inh}}$ analogously). Hence, v receives two types of inhibition from s. First, s directly inhibits v via the inhibitory input g_{inh}. Second, s gates the excitatory input g_{exc}: activity from the first stage is attenuated at those positions where $s > 0$. This means that v can only decrease where $\tilde{s} > 0$ and feedback from v to s assures also that the activity in s will not grow further then. In this way, the situation that the diffusion layers continuously accumulate activity and eventually "drown" (i.e., $s_t(x, y) > 0$ at all positions (x, y)) is largely avoided. Drowning otherwise would occur in the presence of strong background motion, making the model essentially blind to object approaches. The use of two diffusion layers contributes to a further reduction of drowning.

The fifth and final stage of the model represents the LGMD neuron and spatially sums the output from the previous stage:

$$\frac{dl}{dt} = g_{\text{leak}}(V_{\text{rest}} - l) + g_{\text{exc}}(1 - l) \tag{10.17}$$

where $l \in \{l^{\circ}, l^{\bullet}\}$ and $g^{\circ}_{\text{exc}} = \gamma \cdot \sum_{x,y} \tilde{v}^{\circ}(x, y)$ and analogously for $g^{\bullet}_{exc}$. γ is a synaptic weight. Notice that, whereas p, s and v are two-dimensional variables in space, l is a scalar. The final model output corresponds to the two half-wave rectified LGMD activities l°_t and $l^{\bullet}_t$, respectively. Figure 10.6 shows representative frames from two video sequences that were used as input to the model. Figure 10.7 shows the corresponding output as computed by the last equation. Simulated LGMD responses are nice and clean in the absence of background movement (Figure 10.7(a)). The presence of background movement, on the other hand, produces spurious LGMD activation before collision occurs (Figure 10.7(b)).

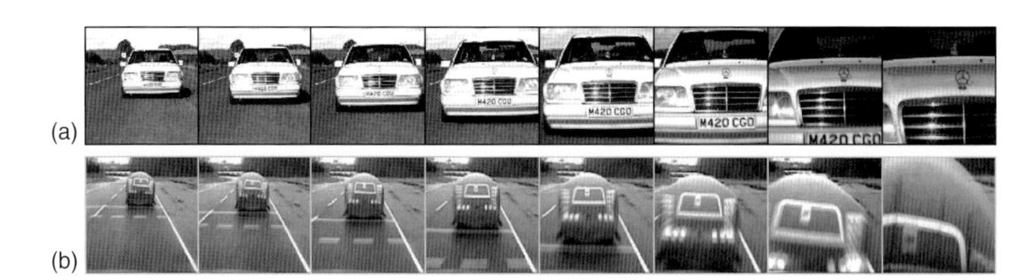

Figure 10.6 Video sequences showing object approaches. Two video sequences which served as input $I_t(x, y)$ to Eq. (10.14). (a) Representative frames of a video where a car drives to a still observer. Except for some camera shake, there is no background motion present in this video. The car does actually not collide with the observer. (b) The video frames show a car (representing the observer) driving against a static obstacle. This sequence implies background motion. Here the observer actually collides with the balloon car, which flies through the air after the impact.

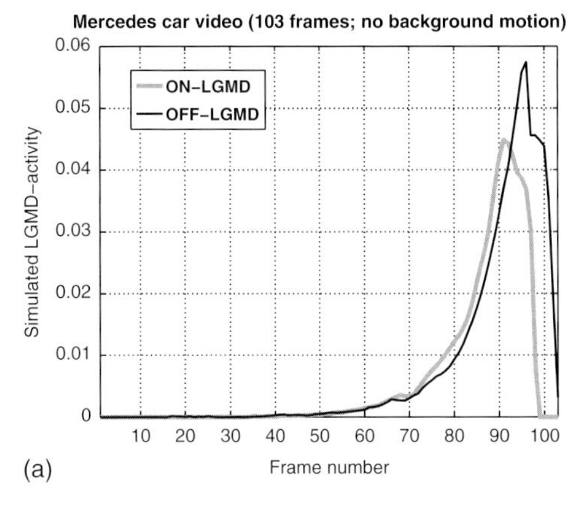

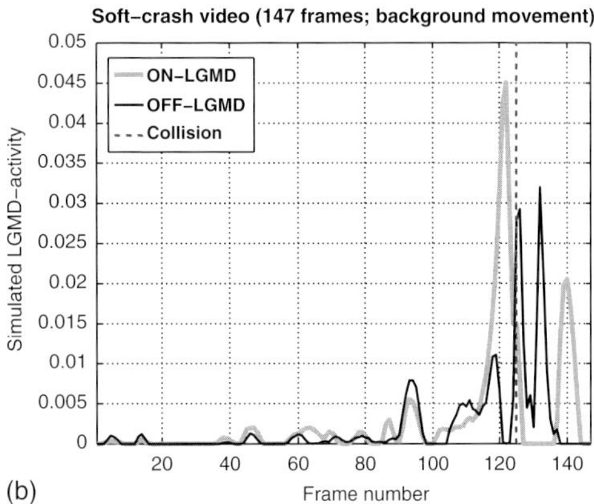

Figure 10.7 Simulated LDMD responses. Both figures show the rectified LGMD activities I_t° (gray curves; label "ON-LGMD") and $I_t^{\bullet}$ (black curves; label "OFF-LGMD") as computed by Eq. (10.17). LGMD activities are one-dimensional signals that vary with time t (the abscissa shows the frame number instead of time). (a) Responses to the video shown in Figure 10.6 (a). The observer does not move and no background motion is generated. Both LDMD responses peak before collision would occur. (b) Responses to the video shown in Figure 10.6 (b). Here the observer moves, and the resulting background motion causes spurious LGMD activity with small amplitude before collision. The collision time is indicated by the dashed vertical line. ON-LGMD activity peaks a few frames before collision. OFF-LGMD activity after collision is generated by the balloon car swirling through the air while it is moving away from the observer.

10.6 Conclusions

Neurodynamical models (which are suitable for processing real-world images) and PDE (partial differential equation)-based image processing algorithms typically differ in the way they are designed, and in their respective predictions with regard to neuroscience. PDE-based image processing algorithms derive often from an optimization principle (such as minimizing an energy functional). As a result, a set of differential equations is usually obtained, which evolves over time [62]. Many of these algorithms are designed *ad hoc* for specific image processing tasks (e.g., optical flow computation [63], segmentation [64], or denoising [65]), but they are usually not designed according to neuronal circuits. Similarly, they usually do not predict psychophysical results. Some remarkable exceptions, however, do exist. For example, the color enhancement algorithm described in reference [66] is based on a perceptually motivated energy functional, which includes a contrast term (enforcing local contrast enhancement) and a dispersion term (implementing the gray-world assumption and enforcing fidelity to the data). In a similar vein, the Retinex algorithm – for estimating perceived reflectance (lightness) – could also be casted into a variational framework [67]. An algorithm for tone mapping of high dynamic range images was originally motivated by compressing high contrasts while preserving low contrasts [68]. Although the authors did not explicitly acknowledge any inspiration from neuroscience, it is nevertheless striking how the algorithm resembles filling-in architectures [48]. Filling-in has been proposed as a mechanism for computing smooth representations of object surfaces in the visual system. Smooth representations imply that surfaces are “tagged” with perceptual values such as color, movement direction, depth, or lightness [69]. Filling-in is often modeled by (lateral) activity propagation within compartments, which are defined through contrast boundaries [70]. A recently proposed filling-in-based computer vision algorithm identifies regions with a coherent movement direction from a (initially noisy) optic flow field [71]. The algorithm proposes also a solution to the so-called aperture problem, and is based on corresponding computations of the brain’s visual system [72]. A further method that resembles the filling-in process is image inpainting (e.g., Ref. [73]). Image inpainting completes missing regions by propagating the structure of the surround into that region. Although image inpainting has not been related to neurophysiological principles, similar (slow) filling-in effects seem to exist also in the brain (e.g., texture filling-in Ref. [74]).

Acknowledgments

MSK acknowledges support from a *Ramon & Cajal* grant from the Spanish government, the *“Retención de talento”* program from the University of Barcelona, and the national grants *DPI2010-21513* and *PSI2013-41568-P*.

References

1. Rumelhart, D.E., Hinton, G.E., and Williams, R.J. (1986) Learning representations by back-propagating errors. *Nature*, **323**, 533–536.
2. Hinton, G.E. and Salakhutdinov, R.R. (2006) Reducing the dimensionality of data with neural networks. *Science*, **313**, 504–507.
3. Mel, B. (1994) Information processing in dendritic trees. *Neural Comput.*, **6**, 1031–1085.
4. Häusser, M., Spruston, N., and Stuart, G. (2000) Diversity and dynamics of dendritic signalling. *Science*, **290**, 739–744.
5. Segev, I. and London, M. (2000) Untangling dendrites with quantitative models. *Science*, **290**, 744–750.
6. London, M. and Häusser, M. (2005) Dendritic computation. *Annu. Rev. Neurosci.*, **28**, 503–532.
7. Brette, R. *et al.* (2007) Simulation of networks of spiking neurons: a review of tools and strategies. *J. Comput. Neurosci.*, **23**, 349–398.
8. Hodkin, A. and Huxley, A. (1952) A quantitative description of membrane current and its application to conduction and excitation in nerve. *J. Physiol.*, **117**, 500–544.
9. Izhikevich, E. (2004) Which model to use for cortical spiking neurons? *IEEE Trans. Neural Netw.*, **15**, 1063–1070.
10. Izhikevich, E. (2003) Simple model of spiking neurons. *IEEE Trans. Neural Netw.*, **14**, 1569–1572.
11. Keil, M. (2003) Neural architectures for unifying brightness perception and image processing. PhD thesis. Universität Ulm, Faculty for Computer Science, Ulm, Germany, *http://vts.uni-ulm.de/doc.asp?id=3042* (accessed 5 May 2015).

12. Keil, M., Roca-Morena, E., and Rodríguez-Vázquez, A. (2004) A neural model of the locust visual system for detection of object approaches with real-world scenes. *Proceedings of the 4th IASTED International Conference*, vol. 5119, Marbella, Spain, pp. 340–345.
13. Kolb, H. (1977) The organization of the outer plexiform layer in the retina of the cat: electron microscopic observations. *J. Neurocytol.*, **6**, 131–153.
14. Nelson, R. *et al.* (1985) Spectral mechanisms in cat horizontal cells, in *Neurocircuitry of the Retina: A Cajal Memorial* (ed. A. Gallego), Elsevier, New York, pp. 109–121.
15. Galarreta, M. and Hestrin, S. (1999) A network of fast-spiking cells in the neocortex connected by electrical synapses. *Nature*, **402**, 72–75.
16. Gibson, J., Beierlein, M., and Connors, B. (1999) Two networks of electrically coupled inhibitory neurons in neocortex. *Nature*, **402**, 75–79.
17. O'Shea, M. and Rowell, C. (1975) A spike-transmitting electrical synapse between visual interneurons in the locust movement detector system. *J. Comp. Physiol.*, **97**, 875–885.
18. Rind, F. (1984) A chemical synapse between two motion detecting neurons in the locust brain. *J. Exp. Biol.*, **110**, 143–167.
19. Killmann, F. and Schürmann, F. (1985) Both electrical and chemical transmission between the lobula giant movement detector and the descending contralateral movement detector neurons of locusts are supported by electron microscopy. *J. Neurocytol.*, **14**, 637–652.
20. Killmann, F., Gras, H., and Schürmann, F. (1999) Types, numbers and distribution of synapses on the dendritic tree of an identified visual interneuron in the brain of the locust. *Cell Tissue Res.*, **296**, 645–665.
21. Markram, H., Lubke, M., Frotscher, J., and Sakmann, B. (1997) Regulation of synaptic efficacy by coincidence of postsynaptic APs and EPSPs. *Science*, **275**, 213–215.
22. Wang, H.-X. *et al.* (2005) Coactivation and timing-dependent integration of synaptic potentiation and depression. *Nat. Neurosci.*, **8**, 187–193.
23. Hebb, D. (1949) *The Organization of Behavior*, John Wiley & Sons, Inc., New York.
24. Song, S., Miller, K., and Abbott, L. (2000) Competitive Hebbian learning through spike-timing-dependent synaptic plasticity. *Nat. Neurosci.*, **3**, 919–926.
25. Magee, J. and Johnston, D.A. (1997) Synaptically controlled, associative signal for Hebbian plasticity in hippocampal neurons. *Science*, **275**, 209–213.
26. Strang, G. (2007) *Computational Science and Engineering*, Wellesley-Cambride Press, Wellesley, MA.
27. Press, W. *et al.* (2007) *Numerical Recipes - The Art of Scientific Computing*, 3rd edn, Cambridge University Press, New York.
28. Mel, B. (1993) Synaptic integration in an excitable dendritic tree. *J. Neurophysiol.*, **70**, 1086–1101.
29. Jonas, P. and Buzsaki, G. (2007) Neural inhibition. *Scholarpedia*, **2**, 3286.
30. van Vreeswijk, C. and Sompolinsky, H. (1996) Chaos in neuronal networks with balanced excitatory and inhibitory activity. *Science*, **274**, 1724–1726.
31. Haider, B. *et al.* (2006) Neocortical network activity in vivo is generated through a dynamic balance of excitation and inhibition. *J. Neurosci.*, **26**, 4535–4545.
32. Holt, G. and Koch, C. (1997) Shunting inhibition does not have a divisive effect on firing rates. *Neural Comput.*, **9**, 1001–1013.
33. Abbott, L. and Chance, F. (2005) Drivers and modulators from push-pull balanced synaptic input. *Curr. Biol.*, 147–155, doi: 10.1016/S0079-6123(05)49011-1.
34. Huxter, J., Burgess, N., and O'Keefe, J. (2003) Independent rate and temporal coding in hippocampal pyramidal cells. *Nature*, **425**, 828–831.
35. Alle, H. and Geiger, J. (2006) Combined analog and action potential coding in hippocampal mossy fibers. *Science*, **311**, 1290–1293.
36. Keil, M. and López-Moliner, J. (2012) Unifying time to contact estimation and collision avoidance across species.

PLoS Comput. Biol., **8**, e1002625, doi: 10.1371/journal.pcbi.1002625.
37. Pitkov, X. and Meister, M. (2012) Decorrelation and efficient coding by retinal ganglion cells. *Nat. Neurosci.*, **15**, 497–643.
38. Hosoya, T., Baccus, S., and Meister, M. (2005) Dynamic predictive coding by the retina. *Nature*, **436**, 71–77.
39. Doi, E. *et al.* (2012) Efficient coding of spatial information in the primate retina. *J. Neurosci.*, **32**, 16256–16264.
40. Attneave, F. (1954) Some informational aspects of visual perception. *Psychol. Rev.*, **61**, 183–193.
41. Niven, J., Anderson, J., and Laughlin, S. (2007) Fly photoreceptors demonstrate energy-information trade-offs in neural coding. *PLoS Biol.*, **5**, e116.
42. Balasubramanian, V. and Berry, M. (2002) A test of metabolically efficient coding in the retina. *Netw. Comput. Neural Syst.*, **13**, 531–552.
43. Rodieck, R.W. (1965) Quantitative analysis of cat retinal ganglion cell response to visual stimuli. *Vision Res.*, **5**, 583–601.
44. Kuffler, S. (1953) Discharge patterns and functional organization of mammalian retina. *J. Neurophysiol.*, **16**, 37–68.
45. Chichilnisky, E. and Kalmar, R. (2002) Functional asymmetries in ON and OFF ganglion cells of primate retina. *J. Neurosci.*, **22**, 2737–2747.
46. Kaplan, E. and Benardete, E. (2001) The dynamics of primate retinal ganglion cell. *Prog. Brain Res.*, **134**, 17–34.
47. Pandarinath, J., an Victor, C., and Nirenberg, S. (2010) Symmetry breakdown in the ON and OFF pathways of the retina at night: functional implications. *J. Neurosci.*, **30**, 10006–10014.
48. Keil, M. *et al.* (2005) Recovering real-world images from single-scale boundaries with a novel filling-in architecture. *Neural Netw.*, **18**, 1319–1331.
49. Keil, M., Cristóbal, G., and Neumann, H. (2006) Gradient representation and perception in the early visual system - a novel account to Mach band formation. *Vision Res.*, **46**, 2659–2674.
50. Keil, M. (2006) Smooth gradient representations as a unifying account of Chevreul's illusion, Mach bands, and a variant of the Ehrenstein disk. *Neural Comput.*, **18**, 871–903.
51. Keil, M. (2007) Gradient representations and the perception of luminosity. *Vision Res.*, **47**, 3360–3372.
52. Keil, M. (2008) Local to global normalization dynamic by nonlinear local interactions. *Physica D*, **237**, 732–744.
53. Rind, F. and Simmons, P. (1999) Seeing what is coming: building collision-sensitive neurons. *Trends Neurosci.*, **22**, 215–220.
54. Rind, F. and Santer, D. (2004) Collision avoidance and a looming sensitive neuron: size matters but biggest is not necessarily best. *Proc. R. Soc. London, Ser. B*, **271**, S27–S29, doi: 10.1098/rsbl.2003.0096.
55. Sun, H. and Frost, B. (1998) Computation of different optical variables of looming objects in pigeon nucleus rotundus neurons. *Nat. Neurosci.*, **1**, 296–303.
56. Hatsopoulos, N., Gabbiani, F., and Laurent, G. (1995) Elementary computation of object approach by a wide-field visual neuron. *Science*, **270**, 1000–1003.
57. Gabbiani, F. *et al.* (2002) Multiplicative computation in a visual neuron sensitive to looming. *Nature*, **420**, 320–324.
58. Keil, M. (2011) Emergence of multiplication in a biophysical model of a wide-field visual neuron for computing object approaches: dynamics, peaks, & fits, *Advances in Neural Information Processing Systems*, vol. **24** (eds J. Shawe-Taylor, R. Zemel, P. Bartlett, F. Pereira, and K. Weinberger), pp. 469–477, *http://arxiv.org/abs/1110.0433*.
59. Keil, M. (2012) The role of neural noise in perceiving looming objects and eluding predators. *Perception*, **S41**, 162.
60. Gabbiani, F. *et al.* (1999) The many ways of building collision-sensitive neurons. *Trends Neurosci.*, **22**, 437–438.
61. Rind, F. and Bramwell, D. (1996) Neural network based on the input organization of an identified neuron signaling implending collision. *J. Neurophysiol.*, **75**, 967–985.
62. Chan, T., Shen, J., and Vese, L. (2003) Variational PDE models in image processing. *Not. Am. Math. Soc.*, **50** (1), 14–26.

63. Aubert, G., Deriche, R., and Kornprobst, P. (1999) Computing optical flow via variational techniques. *SIAM J. Appl. Math.*, **60**, 156–182.
64. Vitti, A. (2012) The Mumford–Shah variational model for image segmentation: an overview of the theory, implementation and use. *ISPRS J. Photogramm. Remote Sens.*, **69**, 50–64.
65. Rudin, L., Osher, S., and Fatemi, E. (1992) Nonlinear total variation based noise removal algorithm. *Physica D*, **60**, 259–268.
66. Palma-Amestoy, R. *et al.* (2009) A perceptually inspired variational framework for color enhancement. *IEEE Trans. Pattern Anal. Mach. Intell.*, **31**, 458–474.
67. Kimmel, R. *et al.* (2003) A variational framework for retinex. *Int. J. Comput. Vision*, **52**, 7–23.
68. Fattal, R., Lischinski, D., and Werman, M. (2002) Gradient domain high dynamic range compression, in *SIGGRAPH '02: ACM SIGGRAPH 2002 Conference Abstracts and Applications*, ACM, New York.
69. Komatsu, H. (2006) The neural mechanisms of perceptual filling-in. *Nat. Rev. Neurosci.*, **7**, 220–231.
70. Grossberg, S. and Mingolla, E. (1987) Neural dynamics of surface perception: boundary webs, illuminants, and shape-from-shading. *Computer Vision Graph. Image Process.*, **37**, 116–165.
71. Bayerl, P. and Neumann, H. (2007) A fast biologically inspired algorithm for recurrent motion estimation. *IEEE Trans. Pattern Anal. Mach. Intell.*, **3**, 904–910.
72. Bayerl, P. and Neumann, H. (2004) Disambiguating visual motion through contextual feedback modulation. *Neural Comput.*, **16**, 2041–2066.
73. Bugeau, A. *et al.* (2010) A comprehensive framework for image inpainting. *IEEE Trans. Image Process.*, **19**, 2634–2645.
74. Motoyoshi, I. (1999) Texture filling-in and texture segregation revealed by transient masking. *Vision Res.*, **39**, 1285–1291.

11
Computational Models of Visual Attention and Applications

Olivier Le Meur and Matei Mancas

11.1
Introduction

Our visual environment contains much more information than we are able to perceive at once. To deal with this large amount of data, human beings have developed biological mechanisms to optimize the visual processing. Visual attention is probably the most important one of these. It allows us to concentrate our biological resources over the most important parts of the visual field.

Visual attention may be differentiated into covert and overt visual attention. Covert attention is defined as paying attention without moving the eyes and could be referred to the act of mentally focusing on a particular area. Overtattention, which involves eye movements, is used both to direct the gaze toward interesting spatial locations and to explore complex visual scenes [1]. As these overt shifts of attention are mainly associated with the execution of saccadic eye movements, this kind of attention is often compared to a window to the mind. Saccade targeting is influenced by top-down factors (the task at hand, behavioral goals, motivational state) and bottom-up factors (both the local and global spatial properties of the visual scene). The bottom-up mechanism, also called *stimulus-driven selection*, occurs when a target item effortlessly attracts the gaze.

In this chapter, we present models of bottom-up visual attention and their applications. In the first section, we present some models of visual attention. A taxonomy proposed by Borji and Itti [2] is discussed. We will describe more accurately cognitive models, that is, models replicating the behavior of the human visual system (HVS). In the second part, we describe how saliency information can be used. We will see that saliency maps can be used not only in classical image and video applications such as compression but also to envision new applications. In the last section, we will draw some conclusions.

Biologically Inspired Computer Vision: Fundamentals and Applications, First Edition.
Edited by Gabriel Cristóbal, Laurent Perrinet, and Matthias Keil.

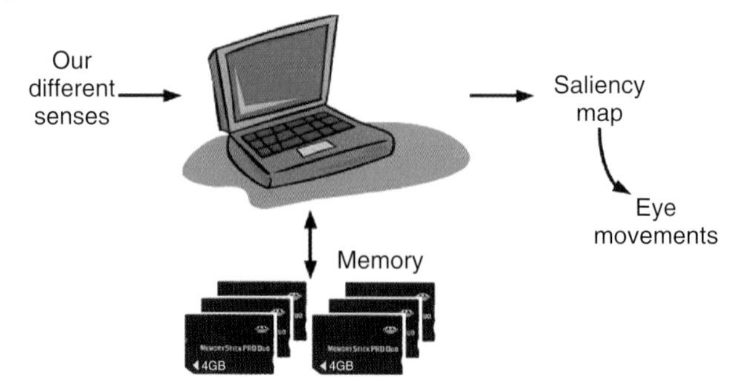

Figure 11.1 The brain as a computer, an unrealistic but convenient hypothesis.

11.2 Models of Visual Attention

Computational saliency models are designed to predict where people look within a visual scene. Most of them are based on the assumption that there exists a unique saliency map in the brain. This saliency map, also called a *master map*, aims at indicating where the most visually important areas are located. This is a comfortable view for computer scientists as the brain is compared to a computer as illustrated by Figure 11.1. The inputs would come from our different senses, whereas our knowledge would be stored in the memory. The output would be the saliency map, which is used to guide the deployment of attention over the visual space.

From this assumption which is more than questionable, a number of saliency models have been proposed. In Section 11.2.1, we present a taxonomy and briefly describe the most influential computational models of visual attention.

11.2.1 Taxonomy

Since 1998, the year in which the most influential computational and biologically plausible model of bottom-up visual attention was published by Itti *et al.* [3], there has been a growing interest in the subject. Indeed, several models, more or less biological and based on different mathematical tools, have been investigated. We proposed in 2009 first saliency taxonomy of models taxonomy [4], which has been significantly improved and extended by Borji and Itti [2].

This taxonomy is composed of eight categories as illustrated by Figure 11.2 (extracted from Ref. [2]). A comprehensive description of these categories is given in Ref. [2]. Here we just give the main features of the four most important categories:

- **Cognitive models.** Models belonging to this category rely on two seminal works: the feature integration theory (FIT) [5] and a biological plausible architecture [6]. The former relies on the fact that some visual features (commonly

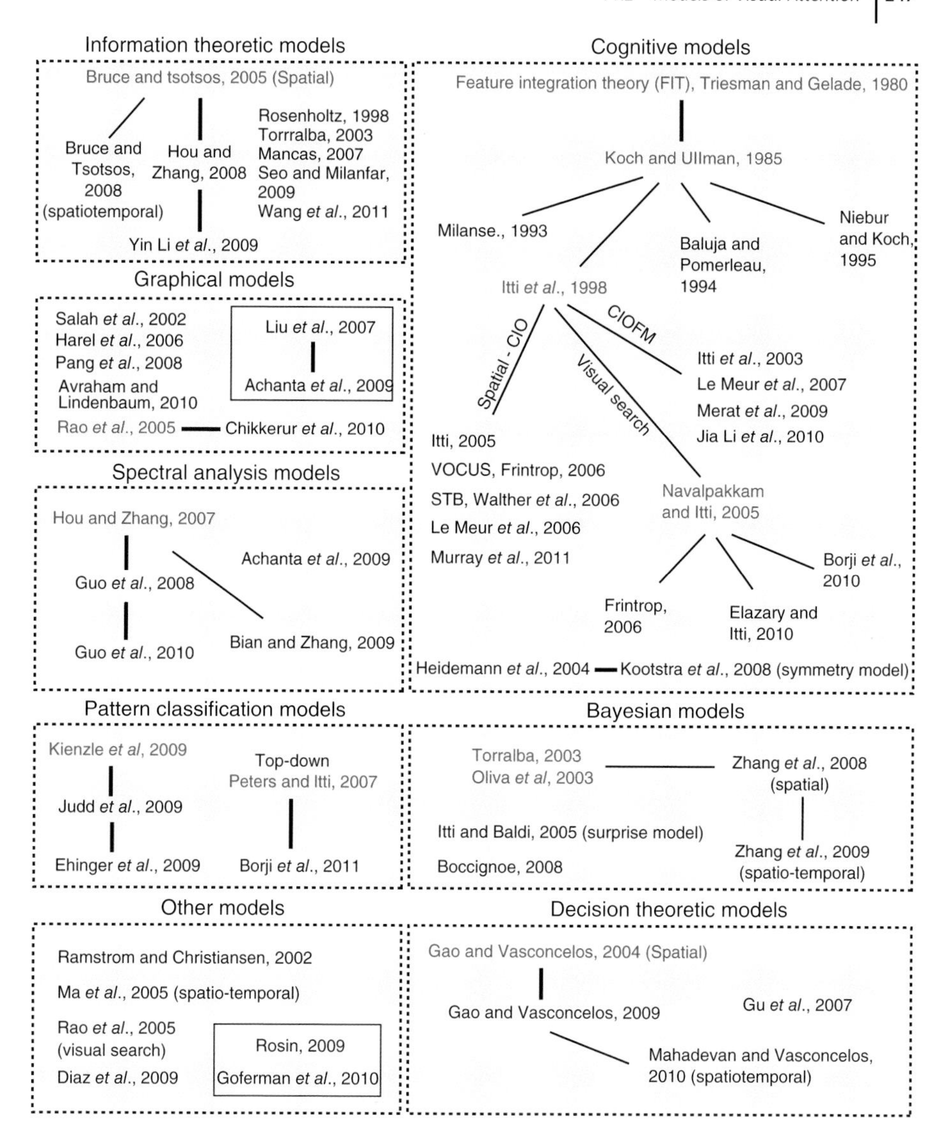

Figure 11.2 Taxonomy of a computational model of visual attention. Courtesy of Borji and Itti [2].

called *early visual features* [7]) are extracted automatically, unconsciously, effortlessly, and very early in the perceptual process. These features such as color, orientation, and shape, to name a few, are automatically separated in parallel throughout the entire visual field. From the FIT, the first biological conceptual architecture has been proposed by Koch and Ullman [6]. This allows the computation of a saliency map based on the assumption that there exists in the brain a single topographic saliency map. Models of this category follow a three-step approach:

- From an input picture, a set of visual features which are known to influence our overt bottom-up visual attention are extracted in a massively parallel manner. These features may be color, orientation, direction of movement, disparity, and so on. Each feature is represented in a topographical map called the *feature map*.
- A filtering operation is then applied on these maps in order to filter out most of the visually irrelevant information; a particular location of the visual field is considered irrelevant when this location does not differ from its spatial neighborhood.
- Finally, these maps are mixed together to form a saliency map.

In this category, we find the model of Le Meur *et al.* [8, 9], Marat *et al.* [10], and so on. We will elaborate further on the Itti and Le Meur models in Section 11.3.

- **Information theoretic models.** These models are grounded in a probabilistic approach. The assumption is that a rare event is more salient than a non-rare event. The mathematical tool that can simply simulate this behavior is self-information. Self-information is a measure of the amount of information carried out by an event. For a discrete random variable X, defined by $\mathcal{A} = x_1, \ldots, x_N\}$ and a probability density function, the amount of information of the event $X = x_i$ is given by $I(X = x_i) = -\log_2(p(X = x_i))$ bit/symbol. The first model based on this approach was proposed by Oliva *et al.* [11]. Bottom-up saliency is given by

$$S = \frac{1}{p(F|G)} \tag{11.1}$$

where, F denotes a vector of local visual features observed at a given location while G represents the same visual features but computed over the whole image. When the probability to observe F given G is low, the saliency S tends to infinity. This approach has been reused and adapted by a number of authors. The main modification is related to the support used to compute the probability density function:

  - Oliva *et al.* [11] determine the probability density function over the whole picture. More recently, Mancas [12] and Riche *et al.* [13] have computed the saliency map by using a multiscale spatial rarity concept.
  - In References [14] and [15], the saliency depends on the local neighborhood from which the probability density function is estimated. The self-information [14] or the mutual information [15] between the probability density functions of the current location and its neighborhood are used to deduce the saliency value.
  - A probability density function is learned on a number of natural image patches. Features extracted at a given location are then compared to this prior knowledge in order to infer the saliency value [16].

- **Bayesian models.** The Bayesian framework is an elegant method to combine current sensory information and prior knowledge concerning the environment (see also Chapter 9). The former is simply the bottom-up saliency which is directly computed from the low-level visual information, whereas the latter is related to the visual inference, also called *prior knowledge*. This refers to the statistic of visual features in a natural scene, its layout, the scene's category, or its spectral signature. This prior knowledge, which is daily shaped by our visual environment, is one of the most important factors influencing our perception. It acts like visual priming that facilitates perception of the scene and steers our gaze to specific parts.There exist a number of models using prior information, the most well known being the theory of surprise [17], Zhang's model [16], and the model suggested by Torralba *et al.* [18].

- **Spectral analysis models.** This kind of model was proposed in 2007 by Hou and Zhang [19]. The saliency is derived from the freqsency domain on the basis of the following assumption: *the statistical singularities in the spectrum may be responsible for anomalous regions in the image, where proto-objects are popped up*. From this assumption, they defined the spectral residual of an image which is the difference on a log amplitude scale between the amplitude spectrum of the image and its lowpass filtered version. This residual is considered as being the innovation of the image in the frequency domain. The saliency map in the spatial domain is obtained by applying the inverse Fourier transform. The whole process for an image I is given below:

$$\begin{aligned}
\text{Compute the amplitude spectrum:} \quad & \mathscr{A}(\mathbf{f}) = \mathscr{R}(\mathscr{F}[I(\mathbf{x})]) \\
\text{Compute the phase spectrum:} \quad & \mathscr{P}(\mathbf{f}) = \phi(\mathscr{F}[I(\mathbf{x})]) \\
\text{Logarithmic amplitude scaling:} \quad & \mathcal{L}(\mathbf{f}) = \log(\mathscr{A}(\mathbf{f})) \\
\text{Spectral residual:} \quad & \mathscr{E}(\mathbf{f}) = \mathcal{L}(\mathbf{f}) - h(\mathbf{f}) * \mathcal{L}(\mathbf{f}) \\
\text{Saliency computation:} \quad & \mathscr{S}(\mathbf{x}) = g(\mathbf{x}) * \mathscr{F}^{-1}[\exp(\mathscr{E}(\mathbf{f}) + \mathscr{P}(\mathbf{f}))]^2
\end{aligned}$$

  where, $\mathbf{f}$ is the radial frequency. $\mathscr{F}$ and $\mathscr{F}^{-1}$ represent the direct and inverse Fourier transform, respectively. $\mathscr{A}$ and $\mathscr{P}$ are the amplitude and phase spectrum obtained through $\mathscr{R}$ and ϕ, respectively. h and g are two lowpass filters. This first approach has been further extended or modified by taking into account the phase spectrum instead of the amplitude one [20], quaternion representation, and a multiresolution approach [21].

Figure 11.3 illustrates predicted saliency maps computed by different models. The brighter the pixel value is, the higher the saliency.

There exist several methods to evaluate the degree of similarity between the prediction computed by a model and the ground truth. We can classify the metrics into three categories. As a comprehensive review of these metrics is beyond the scope of this chapter, we briefly described these categories below:

- Scanpath-based metrics perform the comparison between two scanpaths. We remind that a scanpath is a series of fixations and saccades.
- Saliency maps-based metrics involve two saliency maps. The most commonly used method is to compare a human saliency map, computed from the eye-tracking data, with a predicted saliency map.

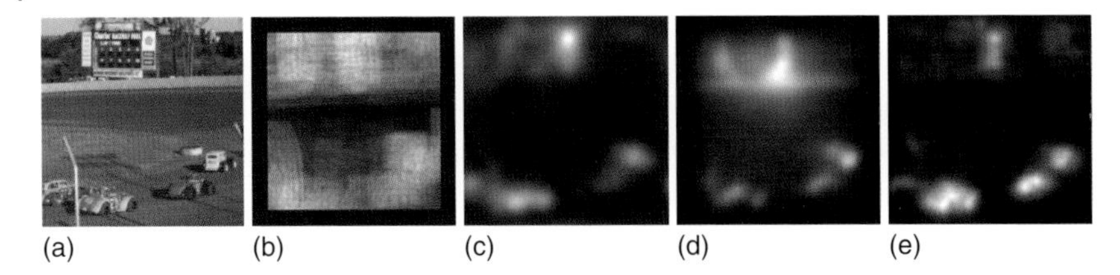

Figure 11.3 (a) Original pictures. (b)–(f) Predicted saliency maps. AIM: attention based on information maximization [14]; AWS: adaptive whitening saliency model [22]; GBVS: graph-based visual saliency [23]; RARE2012: model based on the rarity concept [13].

- Hybrid metrics involve the visual fixations and a predicted saliency map. This kind of metric aims at evaluating the saliency located at the spatial locations of visual fixations.

All these methods are described in Ref. [24].

11.3 A Closer Look at Cognitive Models

As briefly presented in the previous section, cognitive saliency models are inspired by the behavior of visual cells and more generally by the properties of the HVS. The modeling strives to reproduce biological mechanisms as faithfully as possible. In the following subsections, we describe two cognitive models: the first is the well-known model proposed by Itti *et al.* [3]. The second is an extension of Itti's model proposed by Le Meur *et al.* [8]. We will conclude this section by emphasizing the strengths and limitations of current models.

11.3.1 Itti *et al.*'s Model [3]

Figure 11.4 shows the architecture of Itti's model. This model could be decomposed into three main steps: topographic feature extraction, within-map saliency competition, and computation of the saliency map, which represents local conspicuity over the entire scene.

The input is first decomposed into three independent channels, namely, colors, intensity, and orientation. A pyramidal representation of these channels is created using Gaussian pyramids with a depth of nine scales.

To determine the local salience of each feature, a center–surround filter is used. The center–surround organization simulates the receptive fields of visual cells. These two regions provide opposite responses. This is implemented as the difference between a fine and a coarse scale for a given feature. This filter is insensitive to uniform illumination and strongly responds on contrast. In total, 42 feature maps (6 for intensity, 12 for color, and 24 for orientation) are obtained.

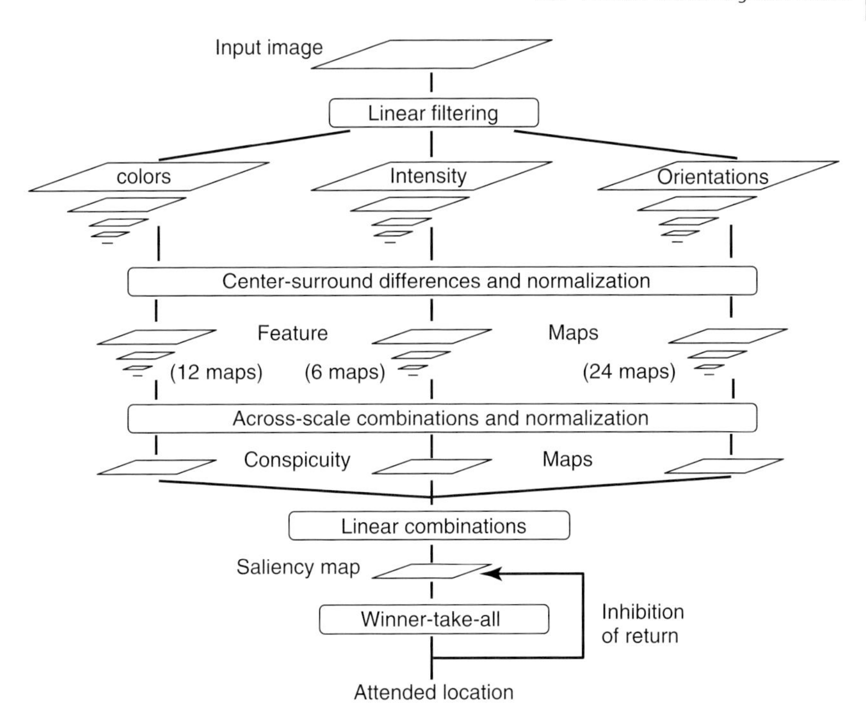

Figure 11.4 Architecture of Itti *et al.*'s model. The input picture is decomposed into three independent channels (colors, intensity, and orientation) representing early visual features. A Gaussian pyramid is computed on each channel. Center–surround differences and across-scale combinations are applied on the pyramid's scales to infer the saliency map.

The final saliency map is then computed by combining the feature maps. Several feature combination strategies were proposed: naive summation, learned linear combination, and contents-based, global, nonlinear amplification and iterative localized interactions. These strategies are defined in Ref. [25].

Figure 11.5 illustrates feature maps and the saliency map computed by Itti's model on a given image. Figure 11.5 (f) and (g) represent the visual scanpaths inferred from the saliency map, following a winner-take-all approach. The first one is composed of the first two fixations, whereas the second is composed of five fixations.

11.3.2 Le Meur *et al.*'s Model [8]

11.3.2.1 Motivations

In 2006, Le Meur *et al.* [8] proposed an extension of L. Itti's model. The motivations were twofold:

The first one was simply to improve and deal with the issues of Itti's model [3]. Its most important drawback relates to the combination and normalization of

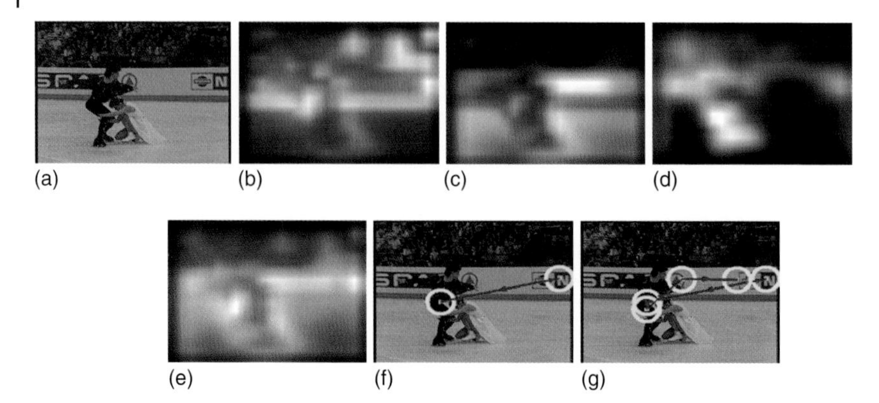

Figure 11.5 Example of feature maps and saliency map computed by Itti's model. (a) Input image; (b)–(d) represent the color, intensity, and orientation feature maps, respectively; (e) saliency map; (f) and (g) represent the visual scanpath with two and five fixations, respectively.

the feature maps which come from different visual modalities. In other words, the question is how to combine color, luminance, and orientation information to get a saliency value. A simple and efficient method is to normalize all feature maps in the same dynamic range (e.g., between 0 and 255) and to sum them into the saliency map. Although efficient, this approach does not take into account the relative importance and the intrinsic features of one dimension compared to another.

The second motivation was the willingness to incorporate into the model important properties of the HVS. For instance, we do not perceive all information present in the visual field with the same accuracy. Contrast sensitivity functions (CSFs) and visual masking are then at the heart of Le Meur's model.

11.3.2.2 Global Architecture

Figure 11.6 illustrates the global architecture of Le Meur *et al.*'s model. The input picture is first transformed into an opponent-color space from which three components $\{A, Cr_1, Cr_2\}$, representing the achromatic, the reddish-greenish, and the bluish-yellowish signals respectively are obtained. Figure 11.7 gives an example of these three components.

CSFs and visual masking are then applied in the frequency domain on the three components of the color space. The former normalizes the dynamic range of $\{A, Cr_1, Cr_2\}$ in terms of visibility threshold. In this model, the CSF proposed by Daly [26] is used to normalize the Fourier spectrum of the achromatic component. This CSF model is a function of many parameters, including radial spatial frequency, orientation, luminance levels, image size, image eccentricity, and viewing distance. This model behaves as an anisotropic bandpass filter, with greater sensitivity to horizontal and vertical spatial frequencies than to diagonal frequencies. Figure 11.8(a) shows the transfer function of the anisotropic 2D CSF used to normalize the achromatic component. Regarding the color components,

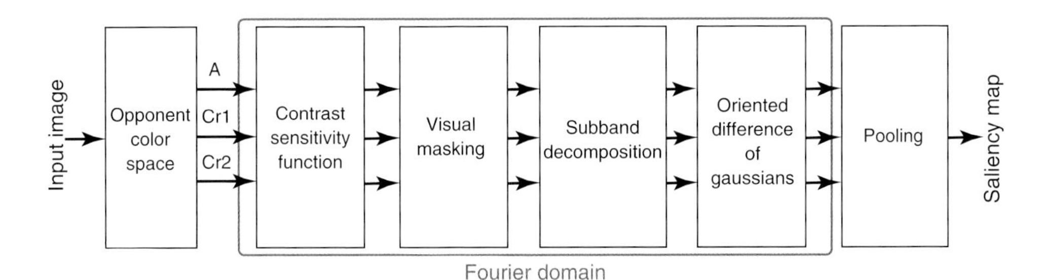

Figure 11.6 Architecture of Le Meur *et al.*'s model. The input picture is decomposed into one achromatic (A) and two chromatic components (Cr_1 and Cr_2). The Fourier transform is then used to encode all of the spatial frequencies present in these three components. Several filters are eventually applied on the magnitude spectrum to get the saliency map.

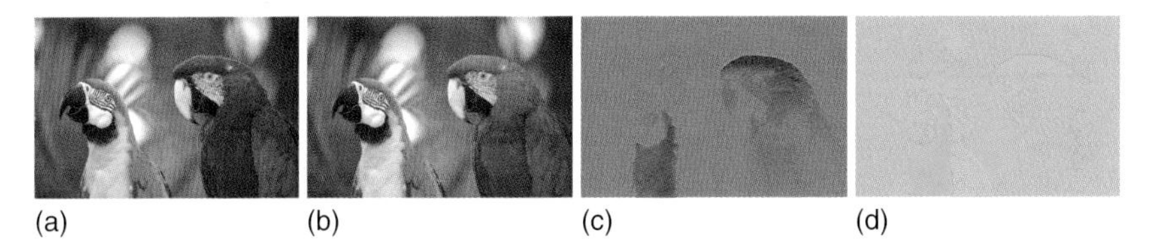

Figure 11.7 Projection of the input color image into an opponent-color space. (a) Input image; (b) achromatic component; (c) Cr_1 channel (reddish-greenish); (d) Cr_2 channel (bluish-yellowish).

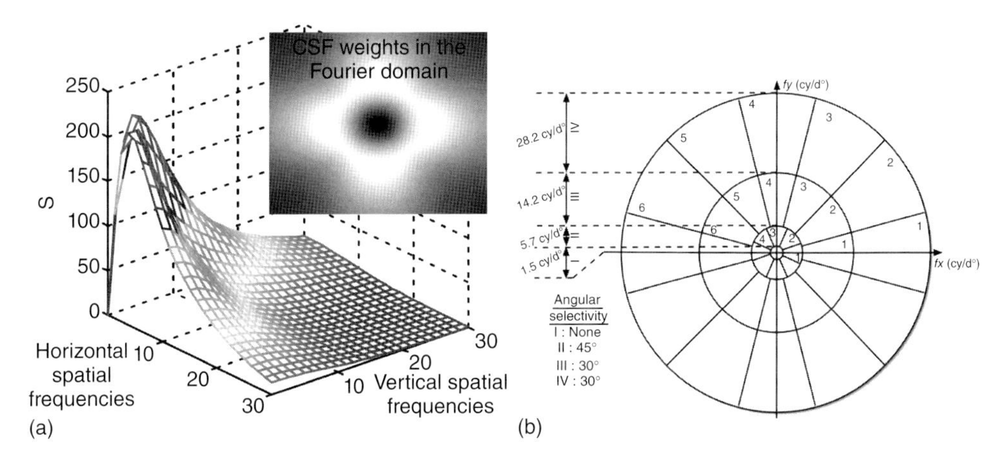

Figure 11.8 Normalization and perceptual subband decomposition in the Fourier domain: (a) anisotropic CSF proposed by Ref. [26]. This CSF is applied on the achromatic component. The inset represents the CSF weights in the Fourier domain (the center of the image represents the lowest radial frequencies, whereas the four corners indicate the highest radial frequencies). (b) The amplitude spectrum of the achromatic component is decomposed into 17 subbands [27].

the anisotropic CSFs defined in Ref. [27] are used. They are defined as follows:

$$S_{Cr_1} = \frac{33}{1+\left(\frac{w}{5.52}\right)^{1}.72} \times (1 - 0.27\sin(2\theta)) \tag{11.2}$$

$$S_{Cr_2} = \frac{5}{1+\left(\frac{w}{4.12}\right)^{1}.64} \times (1 - 0.24\sin(2\theta)) \tag{11.3}$$

where, w is the radial pulsation (expressed in degrees of visual angle) and θ the orientation (expressed in degrees).

Once all visual features are expressed in terms of the visibility threshold, visual masking is applied in order to take into account the influence of the spatial context. This aims to increase or to decrease the visibility threshold. For instance, the visibility threshold of a given area tends to increase when its local neighborhood is spatially complex. All details can be found in Refs [8, 27].

The 2D spatial frequency domain is then decomposed into a number of subbands which may be regarded as the *neural image* corresponding to a population of visual cells tuned to both a range of spatial frequencies and orientations. These decompositions, which have been defined thanks to psychophysics experiments, leads to 17 subbands for the achromatic component and 5 subbands for chromatic components. Figure 11.8 (b) gives the radial frequencies and the angular selectivity of the 17 subbands obtained from the achromatic component decomposition. The decomposition is organized into four crowns, namely I, II, III, and IV:

- Crown I represents the lowest frequencies of the achromatic component;
- Crown II is decomposed into four subbands with an angular selectivity equal to 45°;
- Crowns III and IV are decomposed into six subbands with an angular selectivity equal to 30°.

Figure 11.9 presents the achromatic subbands of crowns I and II for the input picture illustrated in Figure 11.7 (b).

An oriented center–surround filter is then used to filter out redundant information and this behaves as within-map competition to infer the local conspicuity. This filter is implemented in the model as a difference of Gaussians, also called a *Mexican hat*. The difference of Gaussians is indeed a classical method for simulating the behavior of visual cells.

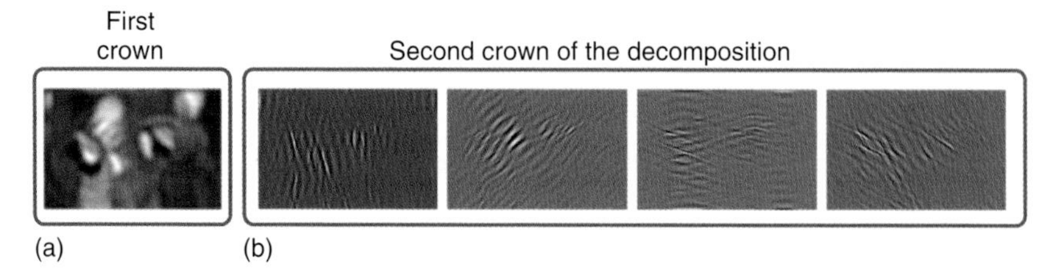

Figure 11.9 Achromatic subbands of the achromatic component of image illustrated on Figure 11.7. Only the subbands corresponding to the first crown (a) and the second crown (b) are shown.

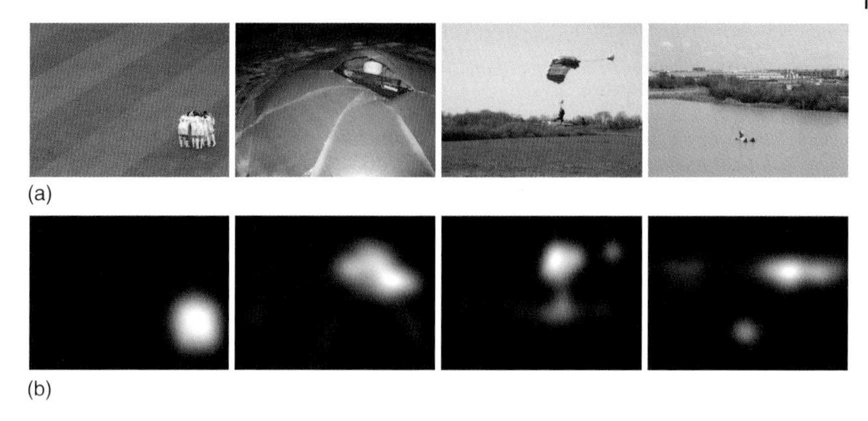

Figure 11.10 Examples of saliency maps predicted by the proposed model. (a) Original images. (b) Predicted saliency maps.

The filtered subbands are then combined into a unique saliency map. There exist a number of pooling strategies described in Ref. [28]. The simplest one consists in normalizing and summing all subbands into the final saliency map. This method is called *NS,* standing for *normalization and sum.*

Figure 11.10 illustrates saliency maps computed by the proposed method (the fusion called NS is used here). Saliency models perform well for this kind of images for which there is a salient object on a simple background. The model performance significantly decreases in the presence of high-level information [29] such as faces (*whether they are human or animal, real or cartoon, frontal or sideways, masked or not, in-focus or blurred*), text (whether it is font, size, or quality), and horizon line, which strongly attracts our attention [30]. The model performance is also low when the dispersion between observers is high.

11.3.3 Limitations

Although computational models of visual attention have substantial power to predict where we look within a scene, improvement is still required. As mentioned previously, computational models of visual attention perform quite well for simple images in which one region stands out from the background. However, when the scene contains high-level information, such as face, text, or horizon line, it becomes much more difficult to accurately predict the salient areas. Some models already embed specific detectors such as face [31], text [31], or horizon line [31, 32]. These detectors allow to improve undeniably the performance of models to predict salient areas. As well discussed in Ref. [29], knowledge of and the ability to define the high-level visual features required to improve models is one of the major challenges faced by researchers studying visual attention.

As mentioned in Section 11.2, there exist a number of models which are more or less biologically plausible. They all output a 2D static saliency map. Although this representation is a convenient way to indicate where we look within a scene, some important aspects of our visual system are clearly overlooked. When viewing a scene, our eyes alternate between fixations and saccades, jumping from one specific location to another. This visual exploration within a visual scene is a highly dynamic process in which time plays an important role. However, most computational implementations of human visual attention could be boiled down to a simple nondynamic map of interest. The next generation of visual attention models should be able to consider, at least, the temporal dimension in order to account for the complexity of our visual system.

11.4 Applications

The applications of saliency maps are numerous. In this section, a non-exhaustive list of saliency-based applications is first given. After a brief description of these applications, we will emphasize two of them: prediction of the picture's memorability and quality estimation.

11.4.1 Saliency-Based Applications: A Brief Review

We present in the following a list of saliency-based applications.

- **Computer graphics.** In this field, saliency can serve several goals including rendering and performing artistic effects. Concerning the former, the idea is to render salient areas with higher accuracy than non-salient parts [33, 34]. The latter consists in modifying an image by removing details while preserving salient areas. DeCarlo and Santella [35] used saliency maps for this purpose.
- **Compression.** Predicting where people look within a scene can be used to locally adapt the visual compression. The simplest way is to allocate more bit budget to salient parts compared to non-salient areas [36]. This saliency-based bit budget allocation could potentially improve on the overall level of observer satisfaction. This reallocation is more efficient when the target bit budget is rather low. At medium to high bit rate, the saliency-based allocation is more questionable.
- **Extraction of object of interest.** This application consists in extracting in an automatic manner the most interesting object in an image or video sequence. From an input image, an object with well-defined boundaries is detected based on its saliency. A renewed interest in this subject can be observed since 2010 [37, 38]. A number of datasets serving as ground truth have been recently released and can be used to benchmark methods.
- **Images optimization for communication, advertisement and marketing.** These are websites, ads, or other documents need to make the important information visible enough to be seen in a very short time. Attention models can

help to find the best configuration of a website for example. Some approaches use saliency to optimize the location of an item in a gallery and others to optimize the location (and time in case of videos) where/when to introduce ads in a document.

Besides these applications, we would like to mention robotics, recognition and retrieval, video summarization, and medical and security applications, for the sake of completeness.

11.4.2 Predicting Memorability of Pictures

The study of image memorability in computer science is a recent topic [39–41]. From the first attempts, it appears that it is possible to predict the degree of an image's memorability quite well. In this section, we present the concept of memorability of pictures, the relationship between memorability and eye movement, and finally the computational models predicting the extent to which a picture is memorable.

11.4.2.1 Memorability Definition

Humans have an amazing visual memory. Only a few seconds is enough to memorize an image [42]. However, not all images are equally memorable. Some are very easy to memorize and to recall, whereas the memorization task appears to be much more difficult for other pictures. Isola *et al.* [39] were the first to build a large dataset of pictures associated with their own memorability score. The score varies between 0 and 1: 0 indicates that the picture is not memorable at all while 1 indicates the highest score of memorability. Memorability has been quantified by performing a visual memory game. Six hundred and sixty five participants were involved in a test to score the memorability of 2222 images. This dataset is freely available on the author's website.

From this large amount of data, Isola *et al.* [39] investigated the contributions of different factors and envisioned the first computational model for predicting memorability scores.

11.4.2.2 Memorability and Eye Movement

Mancas and Le Meur [41] performed an eye-tracking experiment in order to investigate whether the memorability of a picture has an influence on visual deployment. For this, 135 pictures were extracted from the dataset proposed by Isola *et al.* [39]. They are organized into three classes of statistically different memorability, each composed of 45 pictures. The first class consists of the most memorable pictures ($C1$, score 0.82 ± 0.05), the second of typical memorability ($C2$, score 0.68 ± 0.04), and the third of the least memorable images ($C3$, score 0.51 ± 0.08).

As visual attention might be a step toward memory, image memorability should influence the intrinsic parameters of eye movements such as the duration of visual

Figure 11.11 (a) Original pictures; (b) fixation map (a green circle represents the first fixation of observers); (c) saliency map; and (d) heat map. From top to bottom, the memorability score is 0.346, 0.346, 0.897, and 0.903, respectively (from a low to high memorability).

fixations, the congruency between observers, and the saccade lengths. From the collected eye-tracking data, the visual behavior of participants is analyzed according to the picture's memorability. Figure 11.11 illustrates this point. Four pictures are depicted; the first two pictures have a low memorability score whereas this score is high for the last two pictures. The first picture has a memorability score of 0.346, whereas the last one has a memorability score of 0.903. The average fixation durations for these two pictures are 391 and 278 ms, respectively. The average lengths of saccades are 2.39 and 2.99 degree of visual angle, respectively.

From the proposed experiment in Ref. [41], several conclusions have been drawn. First, the fixation durations increase with the degree of memorability of the pictures. This trend is especially noticeable just after the stimuli onset. Fixations are the longest when observers watch memorable pictures. A statistically significant difference is found between fixation durations when the top 20 most memorable and the bottom 20 least memorable pictures are considered. This difference is confirmed for different viewing times.

The congruency between observers watching the same stimulus is the second indicator that has been analyzed. It indicates the degree of similarity between observers' fixations. A high congruency would mean that observers look at the same regions of the stimuli. Otherwise, the congruency is low. Generally, the consistency between visual fixations of different participants is high just after the stimulus onset but progressively decreases over time [43]. To quantify inter-observer congruency, two metrics can be used: receiver operating characteristic (ROC) [24] or a bounding box approach [44]. The former is a parametric approach contrary to the latter. The main drawback of the bounding box approach is its high sensitivity to outliers. A value of 1 indicates a perfect similarity between observers, whereas the value 0 corresponds to the minimal congruency. Results of Ref. [41] indicate that the congruency is highest on the class $C1$ (especially after the stimuli onset (first two fixations)). The difference between congruency of class $C1$ and $C2$ is not statistically significant. However, there is a significant difference between congruency of pictures belonging to $C1$ and $C3$. This indicates that pictures of classes $C1$ and $C2$ are composed of more salient areas which would attract more attention from the observers.

These results show that memorability and attention are linked. It would then be reasonable to use attention-based visual features to predict the memorability of pictures.

11.4.2.3 **Computational Models**

As mentioned earlier, Isola *et al.* [39] were the first to propose a computation model for predicting the memorability score of an image. The authors used a mixture of several low-level features which were automatically extracted. A support vector regression classifier was used to infer the relationship between those features and memorability scores. The best result was achieved by mixing together GIST [45], SIFT [46], histogram of oriented gradient (HOG) [47], structure similarity index (SSIM) [48], and pixel histograms (PH).

Mancas and Le Meur [41] improved Isola's framework by considering saliency-based features, namely, saliency coverage and visibility of structure. Saliency coverage, which describes the spatial computational saliency density distribution, could be approximated by the mean of normalized saliency maps (computed by the RARE model [13]). A low coverage would indicate that there is at least one salient region in the image. A high coverage may indicate that there is nothing in the scene visually important as most of the pixels are attended. The second feature related to the visibility of structure is obtained by applying a lowpass filter several times on images with kernels of increasing sizes as in Gaussian pyramids (see Ref. [41] for more details). By using saliency-based features, the performance in terms of linear correlation increases by 2% while reducing the number of features required to perform the learning (86% less features).

The same year, Celikkale *et al.* [49] extended the work by Isola *et al.* [39] by proposing an attention-driven spatial pooling strategy. Instead of considering all the features (SIFT, HOG, etc.) with an equal contribution, their idea is to emphasize features of salient areas. This saliency-based pooling strategy improves memorability prediction. Two levels of saliency were used: a bottom-up saliency and an object-level saliency. A linear correlation coefficient of 0.47 was obtained.

11.4.3 Quality Metric

11.4.3.1 Introduction

The most relevant quality metrics (IQM (image quality metric) or VQM (video quality metric)) use HVS properties to predict accurately the quality score that an observer would have given. Hierarchical perceptual decomposition, CSFs, visual masking, and so on are the common components of a perceptual metric. These operations simulate different levels of human perception and are now well mastered. In this section, we present quality metrics using visual attention.

Assessing the quality of an image or video sequence is a complex process, involving visual perception as well as visual attention. It is actually incorrect to think that all areas of a picture or video sequence are accurately inspected during a quality assessment task. People preferentially and unconsciously focus on regions of interest. Our sensitivity to distortions might be significantly increased on these regions compared to non-salient ones. Even though we are aware of this, very few IQM or VQM approaches use a saliency map to give more importance to distortions occurring on the salient parts.

Before describing saliency-based quality metrics, we need to understand more accurately the visual strategy deployed by observers while assessing the quality of an image or video sequences.

11.4.3.2 Eye Movement During a Quality Task

The use of the saliency map in a quality metric raises two main issues.

The first issue deals with the way we compute the saliency map. A bottom-up saliency model is classically used for this purpose. This kind of model, as those presented in the previous section, makes the assumption that observers watch the scene without performing any task. We then have a paradoxical situation. Indeed we are seeking to know where observers look within the scene while they perform a quality task and not when they freely view the scene. So the question is whether a bottom-up saliency map could be used to weight distortions or not. To make this point clear, Le Meur *et al.* [50] investigated the influence of quality assessment task on the visual deployment. Two eye-tracking experiments were carried out: one in free-viewing task and the second during a quality task. A first analysis performed on the fixation durations does not reveal a significant difference between the two conditions. A second test consisted in comparing the human saliency maps. The degree of similarity between these maps was evaluated by using a ROC analysis and by computing the area under the ROC curve. The results indicate that the degree of similarity between the two maps is very high. These two results suggest eye movements are not significantly influenced by the quality task instruction.

The second issue is related to the presence of strong visual coding impairments which could disturb the deployment of visual attention in a free-viewing task. In other words, should we compute the saliency map from the original unimpaired image or from the impaired image. Le Meur *et al.* [51] investigated this point by performing eye-tracking experiments on video sequences with and without video-coding artifacts. Observers were asked to watch the video clips without specific instruction. They found the visual deployment is almost the same in both cases. This conclusion is interesting, knowing that the distortions of the video clips were estimated as being as visually annoying by a panel of observers.

To conclude, these two experiments indicate that the use of a bottom-up visual attention makes sense in a context of quality assessment.

11.4.3.3 Saliency-Based Quality Metrics

Quality metrics are composed of several stages. The last stage is called *pooling*. This stage aims at computing the final quality score from a 2D distortion (or error) map. For most saliency-based metrics [52–56], the use of the saliency map consists in modifying the pooling strategy. The degree of saliency of a given pixel is used as a weight, giving more or less importance to the error occurring on this pixel location.

The difference between these methods concerns the way the weights are defined. As presented in Ninassi *et al.* [53], different methods to compute the weights can be used:

$$
\begin{aligned}
w_0(x, y, t) &= 1 \\
w_1(x, y, t) &= SM_n(x, y, t) \\
w_2(x, y, t) &= 1 + SM_n(x, y, t) \\
w_3(x, y, t) &= SM(x, y, t) \\
w_4(x, y, t) &= 1 + SM(x, y, t) \\
w_5(x, y, t) &= SM_b(x, y, t) \\
w_6(x, y, t) &= 1 + SM_b(x, y, t)
\end{aligned}
\tag{11.4}
$$

where $SM(x, y, t)$ is the unnormalized human saliency map, $SM_n(x, y, t)$ is the human saliency map normalized in the range $[0, 1]$, and $SM_b(x, y, t)$ is a binarized human saliency map. The weighting function w_0 is the baseline quality metrics in which the pooling is not modified. The functions w_1, w_3, and w_5 give more importance to the salient areas than the others. The offset value of 1 in the weighting functions w_2, w_4, and w_6 allows to take into account distortions also appearing on the non-salient areas.

The use of a saliency map in the pooling stage provides contrasting results. In Reference [53], the use of a saliency map does not improve the performance of the quality metric. On the other hand, Akamine and Farias [56] showed that the performance of very simple metrics (Peak Signal to Noise-ratio (PSNR) and MSE) has been improved by the use of saliency information. However, for the SSIM metric [57], the saliency does not allow to improve the metric performance. In addition, they showed that the performance improvement depends on the saliency model used to generate the saliency map and on the distortion type (white noise, JPEG distortions).

Many authors working in this field consider that visual attention is important for assessing the quality of an image. However, there are still a number of open issues as demonstrated by Ninassi *et al.* [53, 56]. New strategies to incorporate visual attention into quality metrics as well as a better understanding of the interactions between saliency and distortion need to be addressed.

11.5 Conclusion

During the last two decades, significant progress has been made in the area of visual attention. Although the picture is much clearer, there are still a number of hurdles to overcome. For instance, the eye-tracking datasets used for evaluating the performance of computational models are more or less corrupted by biases. Among them, the central bias, which is the tendency of observers to look near the screen center, is probably the most important [24, 58]. The central bias, which is extremely difficult to cancel or to remove, is a fundamental flaw which can significantly undermine conclusions of some studies and the model's performance.

Regarding the applications, we are still in the early stages of use of saliency maps in computer vision applications. This is a promising avenue for improving existing image and video applications, and for the creation of new applications. Indeed, several factors are nowadays turning saliency computation from labs to industry:

- The models' accuracy has drastically increased in two decades, both concerning bottom-up saliency and top-down information and learning. The results of recent models are far better than the first results obtained in 1998.
- Models working both on videos and images are more and more numerous and provide more and more realistic results.
- The combined enhancement of computing hardware and algorithms optimization have led to real-time or almost real-time good-quality saliency computation.

While some industries have already begun to use attention maps (marketing), others (TV, multimedia) now make use of such algorithms. Video surveillance and video summarization will also enter the game of using saliency maps shortly. This move from labs to industry will further encourage research on the topic toward understanding human attention, memory, and human motivation.

References

1. Findlay, J.M. (1997) Saccade target selection during visual search. *Vision Res.*, **37**, 617–631.
2. Borji, A. and Itti, L. (2013) State-of-the-art in visual attention modeling. *IEEE Trans. Pattern Anal. Mach. Intell.*, **35** (1), 185–207.
3. Itti, L., Koch, C., and Niebur, E. (1998) A model for saliency-based visual attention for rapid scene analysis. *IEEE Trans. Pattern Anal. Mach. Intell.*, **20**, 1254–1259.
4. Le Meur, O. and Le Callet, P. (2009) What we see is most likely to be what matters: visual attention and applications. ICIP, pp. 3085–3088.
5. Treisman, A. and Gelade, G. (1980) A feature-integration theory of

attention. *Cognit. Psychol.*, **12** (1), 97–136.
6. Koch, C. and Ullman, S. (1985) Shifts in selective visual attention: towards the underlying neural circuitry. *Hum. Neurobiol.*, **4**, 219–227.
7. Wolfe, J.M. and Horowitz, T.S. (2004) What attributes guide the deployment of visual attention and how do they do it? *Nat. Rev. Neurosci.*, **5** (6), 495–501.
8. Le Meur, O., Le Callet, P., Barba, D., and Thoreau, D. (2006) A coherent computational approach to model the bottom-up visual attention. *IEEE Trans. Pattern Anal. Mach. Intell.*, **28** (5), 802–817.
9. Le Meur, O., Le Callet, P., and Barba, D. (2007) Predicting visual fixations on video based on low-level visual features. *Vision Res.*, **47**, 2483–2498.
10. Marat, S., Ho-Phuoc, T., Granjon, L., Guyader, N., Pellerin, D., and Guérin-Dugué, A. (2009) Modeling spatio-temporal saliency to predict gaze direction for short videos. *Int. J. Comput. Vis.*, **82**, 231–243.
11. Oliva, A., Torralba, A., Castelhano, M., and Henderson, J. (2003) Top-down control of visual attention in object detection. IEEE ICIP.
12. Mancas, M. (2007) *Computational Attention Towards Attentive Computers*, Presses University de Louvain.
13. Riche, N., Mancas, M., Duvinage, M., Mibulumukini, M., Gosselin, B., and Dutoit, T. (2013) Rare2012: a multi-scale rarity-based saliency detection with its comparative statistical analysis. *Signal Process. Image Commun.*, **28** (6), 642–658, doi: *http://dx.doi.org/10.1016/j.image.2013.03.009.*
14. Bruce, N. and Tsotsos, J. (2009) Saliency, attention and visual search: an information theoretic approach. *J. Vis.*, **9**, 1–24.
15. Gao, D. and Vasconcelos, N. (2009) Bottom-up saliency is a discriminant process. ICCV.
16. Zhang, L., Tong, M.H., Marks, T.K., Shan, H., and Cottrell, G.W. (2008) SUN: a Bayesian framework for salience using natural statistics. *J. Vis.*, **8** (7), 1–20.
17. Itti, L. and Baldi, P. (2005) Bayesian surprise attracts human attention. Neural Information Processing Systems.
18. Torralba, A., Oliva, A., Castelhano, M., and Henderson, J. (2006) Contextual guidance of eye movements and attention in real-world scenes: the role of global features in object search. *Psychol. Rev.*, **113** (4), 766–786.
19. Hou, X. and Zhang, L. (2007) Saliency detection: a spectral residual approach. IEEE Conf. on CVPR, 1–8.
20. Guo, C., Ma, Q., and Zhang, L. (2008) Spatio-temporal saliency detection using phase spectrum of quaternion Fourier transform. Computer Vision and Pattern Recognition, 1–8.
21. Guo, C. and Zhang, L. (2010) A novel multiresolution spatiotemporal saliency detection model and its application in image and video compression. *IEEE Trans. Image Process.*, **19** (1), 185–198.
22. Garcia-Diaz, A., Fdez-Vidal, X.R., Pardo, X.M., and Dosil, R. (2012) Saliency from hierarchical adaptation through decorrelation and variance normalization. *Image Vision Comput.*, **30** (1), 51–64, doi: *http://dx.doi.org/10.1016/j.imavis.2011.11.007.*
23. Harel, J., Koch, C., and Perona, P. (2006) Graph-based visual saliency. Proceedings of Neural Information Processing Systems (NIPS).
24. Le Meur, O. and Baccino, T. (2013) Methods for comparing scanpaths and saliency maps: strengths and weaknesses. *Behav. Res. Methods*, **1**, 1–16.
25. Itti, L. (2000) Models of bottom-up and top-down visual attention. PhD thesis. California Institute of Technology.
26. Daly, S. (1993) *Digital Images and Human Vision*, MIT Press, Cambridge, MA, pp. 179–206.
27. Le Callet, P. (2001) Critères objectifs avec référence de qualité visuelle des images couleurs. PhD thesis. Université de Nantes.
28. Chamaret, C., Le Meur, O., and Chevet, J. (2010) Spatio-temporal combination of saliency maps and eye-tracking assessment of different strategies. ICIP, pp. 1077–1080.
29. Judd, T., Durand, F., and Torralba, A. (2012) A Benchmark of Computational Models of Saliency to Predict Human Fixations. Technical Report (CSAIL-TR-2012-001), MIT.

30. Foulsham, T., Kingstone, A., and Underwood, G. (2008) Turning the world around: patterns in saccade direction vary with picture orientation. *Vision Res.*, **48**, 1777–1790.
31. Judd, T., Ehinger, K., Durand, F., and Torralba, A. (2009) Learning to predict where people look. ICCV.
32. Le Meur, O. (2011) Predicting saliency using two contextual priors: the dominant depth and the horizon line. IEEE Int. Conf. on Multimedia and Expo ICME, 1–6.
33. Kim, Y., Varshney, A., Jacobs, D.W., and Guimbretière, F. (2010) Mesh saliency and human eye fixations. *ACM Trans. Appl. Percept.*, **7** (2), 1–13.
34. Song, R., Liu, Y., Martin, R.R., and Rosin, P.L. (2014) Mesh saliency via spectral processing. *ACM Trans. Graph.*, **33** (1), 6:1–6:17, doi: 10.1145/2530691.
35. DeCarlo, D. and Santella, A. (2002) Stylization and abstraction of photographs. *ACM Trans. Graph.*, **21** (3), 769–776, doi: 10.1145/566654.566650.
36. Li, Z., Qin, S., and Itti, L. (2011) Visual attention guided bit allocation in video compression. *Image Vision Comput.*, **29** (1), 1–14.
37. Liu, Z., Zou, W., and Le Meur, O. (2014) Saliency tree: a novel saliency detection framework. *IEEE Trans. Image Process.*, **23** (5), 1937–1952.
38. Liu, Z., Zhang, X., Luo, S., and Le Meur, O. (2014) Superpixel-based spatiotemporal saliency detection. *IEEE Trans. Circ. Syst. Video Technol.*, doi: 10.1109/TCSVT.2014.23086422014.2308642.
39. Isola, P., Xiao, J., Torralba, A., and Oliva, A. (2011) What makes an image memorable?. IEEE Conference on Computer Vision and Pattern Recognition (CVPR), pp. 145–152.
40. Khosla, A., Xiao, J., Torralba, A., and Oliva, A. (2012) Memorability of image regions. Advances in Neural Information Processing Systems (NIPS), Lake Tahoe, CA, USA.
41. Mancas, M. and Le Meur, O. (2013) Memorability of natural scene: the role of attention. ICIP.
42. Standing, L. (1973) Learning 10,000 pictures. *Q. J. Exp. Psychol.*, **25**, 207–222.
43. Tatler, B., Baddeley, R.J., and Gilchrist, I. (2005) Visual correlates of fixation selection: effects of scale and time. *Vision Res.*, **45**, 643–659.
44. Carmi, R. and Itti, L. (2006) Visual causes versus correlates of attentional selection in dynamic scenes. *Vision Res.*, **46** (26), 4333–4345.
45. Oliva, A. and Torralba, A. (2001) Modeling the shape of the scene: a holistic representation of the spatial envelope. *Int. J. Comput. Vision*, **42** (3), 145–175.
46. Lazebnik, S., Schmid, C., and Ponce, J. (2006) Beyond bags of features: spatial pyramid matching for recognizing natural scene categories. 2006 IEEE Computer Society Conference on Computer Vision and Pattern Recognition, vol. **2**, IEEE, pp. 2169–2178.
47. Dalal, N. and Triggs, B. (2005) Histograms of oriented gradients for human detection. IEEE Computer Society Conference on Computer Vision and Pattern Recognition, 2005. CVPR 2005, vol. 1, IEEE, pp. 886–893.
48. Shechtman, E. and Irani, M. (2007) Matching local self-similarities across images and videos. IEEE Conference on Computer Vision and Pattern Recognition, 2007. CVPR'07, IEEE, pp. 1–8.
49. Celikkale, B., Erdem, A., and Erdem, E. (2013) Visual attention-driven spatial pooling for image memorability. 2013 IEEE Computer Society Conference on Computer Vision and Pattern Recognition Workshops (CVPRW), IEEE, pp. 1–8.
50. Le Meur, O., Ninassi, A., Le Callet, P., and Barba, D. (2010) Overt visual attention for free-viewing and quality assessment tasks: impact of the regions of interest on a video quality metric. *Signal Process. Image Commun.*, **25** (7), 547–558.
51. Le Meur, O., Ninassi, A., Le Callet, P., and Barba, D. (2010) Do video coding impairments disturb the visual attention deployment? *Signal Process. Image Commun.*, **25** (8), 597–609.
52. Ninassi, A., Le Meur, O., Le Callet, P., and Barba, D. (2007) Does where you gaze on an image affect your perception of quality? Applying visual attention to image quality metric, ICIP. 169–172.

53. Ninassi, A., Le Meur, O., Le Callet, P., and Barba, D. (2009) Considering temporal variations of spatial visual distortions in video quality assessment. *IEEE J. Sel. Top. Signal Process.*, **3** (2), 253–265.
54. Liu, H. and Heynderickx, I. (2011) Visual attention in objective image quality assessment: based on eye-tracking data. *IEEE Trans. Circ. Syst. Video Technol.*, **21** (7), 971–982.
55. Guo, A., Zhao, D., Liu, S., Fan, X., and Gao, W. (2011) Visual attention based image quality assessment. IEEE International Conference on Image Processing, pp. 3297–3300.
56. Akamine, W.Y.L. and Farias, M.C.Q. (2014) Incorporating visual attention models into image quality metrics. Proc. SPIE-IS&T Electronic Imaging, 9016,90160O-1-9.
57. Wang, Z., Bovik, A.C., Sheikh, H.R., and Simoncelli, E. (2004) Image quality assessment: from error visibility to structural similarity. *IEEE Trans. Image Process.*, **13** (4), 600–612.
58. Tatler, B.W. (2007) The central fixation bias in scene viewing: selecting an optimal viewing position independently of motor biases and image feature distributions. *J. Vis.*, **7** (14), 4.1–417.

12
Visual Motion Processing and Human Tracking Behavior

Anna Montagnini, Laurent U. Perrinet, and Guillaume S. Masson

12.1
Introduction

Vision is our primary source of information about the 3D layout of our environment or the occurrence of events around us. In nature, visual motion is abundant and is generated by a large set of sources, such as the movement of another animal, be it predator or prey, or our own movements. Primates possess high-performance systems for motion processing. Such a system is closely linked to ocular tracking behavior: a combination of smooth pursuit and saccadic eye movements is engaged to stabilize the retinal image of a selected moving object within the high-acuity region of the retina, the fovea. During smooth phases of tracking, eye velocity nearly perfectly matches the target velocity, minimizing retinal slip. Fast saccadic eye movements interrupt these slow phases to correct position errors between the retinal location of the target image and the fovea. Thus, smooth pursuit and saccadic eye movements are largely controlled by velocity and position information, respectively [1]. Smooth pursuit eye movements (which are the focus of this chapter) are traditionally considered to be the product of a *reflexive* sensorimotor circuitry. Two important observations support this assumption: first, it is not possible to generate smooth pursuit at will across a stationary scene, and, second, it is not possible to fully suppress pursuit in a scene consisting solely of moving targets [2, 3].

In nonhuman and human primates, visual motion information is extracted through a cascade of cortical areas spanning the dorsal stream of the cortical visual pathways (for reviews, see Ref. [4]). Direction- and speed-selective cells are found in abundance in two key areas lying at the junction between the occipital and parietal lobes. The middle temporal (MT) and medio-superior temporal (MST) areas can decode local motion information of a single object at multiple scales, isolate it from its visual background, and reconstruct its trajectory [5, 6]. Interestingly, neuronal activities in MT and MST areas have been strongly related to the initiation and maintenance of tracking eye movements [7]. Dozens of experimental studies have shown that local direction and speed information for pursuit are encoded by MT populations [8] and transmitted to MST populations

Biologically Inspired Computer Vision: Fundamentals and Applications, First Edition.
Edited by Gabriel Cristóbal, Laurent Perrinet, and Matthias Keil.

where non-retinal information are integrated to form an internal model of object motion [9]. Such a model is then forwarded to both frontal areas involved in pursuit control, such as frontal eye fields (FEF) and supplementary eye fields (SEF), as well as to the brainstem and cerebellar oculomotor system (see Ref. [10] for a review).

What can we learn about visual motion processing by investigating smooth pursuit responses to moving objects? Since the pioneering work of Hassenstein and Reichardt [11], behavioral studies of tracking behavioral responses have been highly influential upon theoretical approaches to motion detection mechanisms (see Chapter 17, this book), as illustrated by the fly vision literature (see Refs [12, 13] for recent reviews) and its application in bioinspired vision hardwares (e.g., Refs [14, 15]). Because of their strong *stimulus-driven* nature, tracking eye movements have been used similarly in primates to probe the properties of fundamental aspects of motion processing, from detection to pattern motion integration (see a series of recent reviews in Refs [2, 16, 17]) and the coupling between the active pursuit of motion and motion perception [18]. One particular interest of tracking responses is that they are time-continuous, smooth, measurable movements that reflect the temporal dynamics of sensory processing in the changes of eye velocity [16, 19]. Here, we would like to focus on a particular aspect of motion processing in the context of sensorimotor transformation: uncertainty. Uncertainty arises from both random processes, such as the noise reflecting unpredictable fluctuations on the velocity signal, as well as from nonstochastic processes such as ambiguity when reconstructing global motion from local information. We will show that both sensory noise and ambiguities impact the sensorimotor transformation as seen from the variability of eye movements and their course toward a steady, optimal solution.

Understanding the effects of various sources of noise and ambiguities can change our views on the two faces of the sensorimotor transformation. First, we can better understand how visual motion information is encoded in neural populations and the relationships between these population activities and behaviors [17, 20]. Second, it may change our view about how the brain controls eye movements and opens the door to new theoretical approaches based on inference rather than linear control systems. Albeit still within the framework of stimulus-driven motor control, such new point of view can help us elucidate how higher-level cognitive processes, such as prediction, can dynamically interact with sensory processing to produce a versatile, adaptive behavior by which we can catch a flying ball despite its complex and even sometimes partially occluded trajectory.

This chapter is divided into four parts. First, we examine how noise and ambiguity both affect the pursuit initiation. In particular, we focus on the temporal dynamics of uncertainty processing and the estimate of the optimal solution for motion tracking. Second, we summarize how nonsensory, predictive signals can help maintain a good performance when sensory evidences become highly unreliable and, on the contrary, when the future sensory inputs become highly predictable. Herein, we illustrate these aspects with behavioral results gathered in both human and nonhuman primates. Third, we show that these different results on visually guided and predictive smooth pursuit dynamics can be reconciled within a single Bayesian framework. Last, we propose a biologically plausible architecture implementing a hierarchical inference network for a closed-loop, visuomotor control of tracking eye movements.

12.2 Pursuit Initiation: Facing Uncertainties

The ultimate goal of pursuit eye movements is to reduce the retinal slip of image motion down to nearly zero such that fine details of the moving pattern can be analyzed by spatial vision mechanisms. Overall, the pursuit system acts as a negative feedback loop where the eye velocity matches target velocity (such that the *pursuit gain* is close to 1) to cancel the image motion on the retina. However, because of the delays due to both sensory and motor processing, the initial rising phase of eye velocity, known as *pursuit initiation*, is *open-loop*. This means that during this short period of time (less than about 100 ms), no information about eye movements is available to the system and the eye velocity depends only on the properties of the target motion presented to the subject. This short temporal window is ideal to probe how visual motion information is processed and transformed into an eye movement command [17, 21]. It becomes possible to map the different phases of visual motion processing to the changes in initial eye velocity and therefore to dissect out the contribution of spatial and temporal mechanisms of direction and speed decoding [19]. However, this picture becomes more complicated as soon as one considers more naturalistic and noisy conditions for motion tracking, whereby, for instance, the motion of a complex-shaped extended object has to be estimated, or when several objects move in different directions. We will not focus, here, on the last problem, which involves the complex and time-demanding computational task of object segmentation [16, 22]. In the next subsections, we focus instead on the nature of the noise affecting pursuit initiation and the uncertainty related to limitations in processing spatially localized motion information.

12.2.1 Where Is the Noise? Motion-Tracking Precision and Accuracy

Human motion tracking is variable across repeated trials and the possibility to use tracking behavior (at least during the initiation phase) to characterize visual motion processing across time and across all kinds of physical properties of the moving stimuli relies strongly on the assumption that oculomotor noise does not override the details of visual motion processing. In order to characterize the variability of macaques' pursuit responses to visual motion signals, Ref. [23] analyzed the monkeys' pursuit eye movements during the initiation phase – here, the first 125 ms after pursuit onset. On the basis of a principal component analysis of pursuit covariance matrix, they concluded that pursuit variability was mostly due to sensory fluctuations in estimating target motion parameters such as onset time, direction, and speed, accounting for around 92% of the pursuit variability. In a follow-up study, they estimated the time course of the pursuit system's sensitivity to small changes in target direction, speed, and onset time [24]. This analysis was based on pursuit variability during the first 300 ms after target motion onset. Discrimination thresholds (inverse of sensitivity) decreased rapidly during open-loop pursuit and, in the case of motion direction, followed a similar time course to the one obtained from the analysis of neuronal activity in the MT area [25].

Noise inherent to the kinematic parameters of the moving target is not the only source of uncertainty for visual motion processing. A very well-known and puzzling finding in visual motion psychophysics is that the speed of low-contrast moving stimuli is most often underestimated as compared to high-contrast stimuli moving with exactly the same motion properties (see Chapter 9). In parallel with the perceptual misjudgment, previous studies have shown that tracking quality is systematically degraded (with longer onset latencies, lower acceleration at initiation, and lower pursuit gain) with low-contrast stimuli [26]. This reduction of motion estimation and tracking accuracy when decreasing the luminance-contrast of the moving stimulus has been interpreted as evidence in favor of the fact that, when visual information is corrupted, the motion-tracking system relies more on internal models of motion, or motion *priors* [27].

12.2.2
Where Is the Target Really Going?

The external world is largely responsible for variability. Another major source of uncertainty when considering sensory processing is ambiguity: a single retinal image can correspond to many different physical arrangements of the objects in the environment. In the motion domain, a well-known example of such input ambiguity is called "the aperture problem." When seen through a small aperture, the motion of an elongated edge (i.e., a one-dimensional -1D- change in luminance, see Figure 12.1(a), middle panel) is highly ambiguous. The same local motion signal can be generated by an infinite number of physical translations of the edge. Hans Wallach (see Ref. [28] for an English translation of the original publication in German) was the first psychologist to recognize this problem and to propose a solution for it. A spatial integration of motion information provided by edges with different orientations can be used to recover the true velocity of the pattern. Moreover, two-dimensional (2D) features such as corners or line-endings (whereby luminance variations occur along two dimensions, see Figure 12.1(a), right panel, for an example) can also be extracted through the same small aperture as their motion is no longer ambiguous. Again, 1D and 2D motion signals can be integrated to reconstruct the two-dimensional velocity vector of the moving pattern. After several decades of intensive research at both physiological and behavioral levels, it remains largely unclear what computational rules are used by the brain to solve the 2D motion integration problem (see Ref. [29] for a collection of review articles).

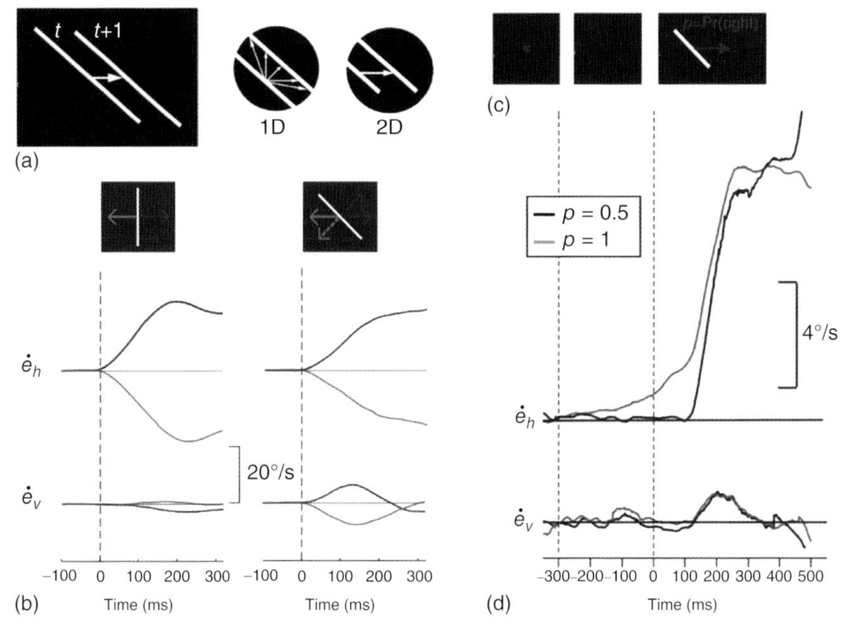

Figure 12.1 Smooth pursuit's account for the dynamic solution of motion ambiguity and motion prediction. (a) A tilted bar translating horizontally in time (left panel) carries both ambiguous 1D motion cues (middle panel), and nonambiguous 2D motion cues (rightmost panel). (b) Example of average horizontal ($\dot{e}_h$) and vertical ($\dot{e}_v$) smooth pursuit eye velocity while tracking a vertical (left) or a tilted bar (right) translating horizontally, either to the right (red curves) or to the left (green curves). Velocity curves are aligned on smooth pursuit onset. (c) Schematic description of a trial in the experiment on anticipatory smooth pursuit: after a fixation display, a fixed duration blank precedes the motion onset of a tilted line moving rightward (with probability p) or leftward (with probability $1-p$). (d) Example of average horizontal ($\dot{e}_h$) and vertical ($\dot{e}_v$) smooth pursuit eye velocity in the anticipation experiment for two predictability conditions, $p = 0.5$ (unpredictable, black curves) and $p = 1$ (completely predictable, gray curves).

Indeed, several computational rules for motion integration have been proposed over the last 40 years (see Ref. [5] for a review). In a limited number of cases, a simple vector averaging of the velocity vectors corresponding to the different 1D edge motions can be sufficient. A more generic solution, the intersection-of-constraints (IOC) is a geometrical solution that can always recover the exact global velocity vector from at least two moving edges with different orientations [30, 31]. However, the fact that perceived direction does not always correspond to the IOC solution (for instance, for very short stimulus duration [32] or when a single 1D motion signal is present [33, 34]) has supported the role of local 2D features in motion integration.

Several feedforward computational models have been proposed to implement these different rules [35, 36]. All these feedforward models have the same architecture. Motion integration is seen as a two-stage computation. The first stage, corresponding to cortical area V1 in primates, extracts local motion information through a set of oriented spatiotemporal filters. This corresponds to the fact that most V1 neurons respond to the direction orthogonal to the orientation of an edge drifting across their receptive field [37]. The local motion analyzers feed a second, integrative stage where pattern motion direction is computed. This integrative stage is thought to correspond to the extra-striate cortical MT area in primates. MT neurons have large receptive fields, they are strongly direction selective and a large fraction of them can unambiguously signal the pattern motion direction, regardless of the orientation of their 1D components [37, 38]. Different nonlinear combinations of local 1D motion signals can be used to extract either local 2D motion cues or global 2D motion velocity vectors. Another solution proposed by Perrinet and Masson [39] is to consider that local motion analyzers are modulated by motion coherency [40]. This theoretical model shows the emergence of similar 2D motion detectors. These two-stage frameworks can be integrated into more complex models where local motion information is diffused across some retinotopic maps.

12.2.3 Human Smooth Pursuit as Dynamic Readout of the Neural Solution to the Aperture Problem

Behavioral measures do not allow capture of the detailed temporal dynamics of the neuronal activity underlying motion estimate. However, smooth pursuit recordings do still carry the signature of the dynamic transition between the initial motion estimate dominated by the vector average of local 1D cues and the later estimate of global object motion. In other terms, human tracking data provides a continuous (delayed and lowpass filtered) dynamic readout of the neuronal solution to the aperture problem. Experiments in our and other groups [41–43] have consistently demonstrated, in both humans and monkeys, that tracking is transiently biased at initiation toward the direction orthogonal to the moving edge (or the vector average if multiple moving edges are present), when such direction does not coincide with the global motion direction. After some time (typically 200–300 ms), such bias is extinguished and the tracking direction converges to the object's global motion. In the example illustrated in Figure 12.1, a tilted bar translates horizontally, thereby carrying locally ambiguous edge-related information (middle panel of part a). A transient nonzero vertical smooth pursuit velocity (lower right panel of Figure 12.1(b)) reflects the initial aperture-induced bias, which is different from the case where local and global motion are coherent (as for the pursuit of a horizontally moving vertical bar, see Figure 12.1(b), leftmost panels).

The size of the transient directional tracking bias and the time needed for converging to the global motion solution depend on several properties of the visual moving stimulus [42] including stimulus luminance contrast [44, 45]. In Section 12.4.1, we will see that this tracking dynamics is consistent with a simple Bayesian recurrent model (or equivalently a Kalman filter [44]), which takes into account the uncertainty associated with the visual moving stimulus and combines it with prior knowledge about visual motion (see also Chapter 9).

12.3 Predicting Future and On-Going Target Motion

Smooth pursuit eye movements can also rely on prediction of target movement to accurately follow the target despite a possible major disruption of the sensory evidence. Prediction allows also to compensate for processing delays, an unavoidable problem of sensory-to-motor transformations.

12.3.1 Anticipatory Smooth Tracking

The exploitation of statistical regularities in the sensory world and/or of cognitive information ahead of a sensory event is a common trait of adaptive and efficient cognitive systems that can, on the basis of such predictive information, anticipate choices and actions. It is a well-established fact that humans cannot generate smooth eye movements at will: for instance, it is impossible to smoothly track an imaginary target with the eyes, except in the special condition in which an unseen target is self-moved through the smooth displacement of one's own finger [46]. In addition, smooth pursuit eye movements do necessarily lag unpredictable visual target motion by a (short) time delay. In spite of this, it was already known many years ago that, when tracking regular periodic motion, pursuit sensorimotor delay can be nulled and a perfect synchronicity between target and eye motion is possible (see Refs [3, 47] for detailed reviews). Furthermore, when the direction of motion of a moving target is known in advance (for instance, because motion properties are the same across many repeated experimental trials), anticipatory smooth eye movements are observed in advance of the target motion onset [48], as illustrated in the upper panel of Figure 12.1(d). Interestingly, relatively complicated motion patterns such as piecewise linear trajectories [49] or accelerated motion [50] can be also anticipated. Finally, probabilistic knowledge about target motion direction or speed [51], and even subjectively experienced regularities extracted from the previous few trials [52] can modulate anticipatory smooth pursuit in a systematic way. Recently, several researchers have tested the role of higher-level cognitive cues for anticipatory smooth pursuit, leading to a rather diverse set of results. Although verbal or pictorial abstract cues indicating the direction of the upcoming motion seem to have a rather weak (although non-inexistent) influence on anticipatory smooth pursuit, other cues are more easily and immediately interpreted and used for motion anticipation [3]. For instance, a barrier blocking one of two branches in a line-drawing illustrating an inverted-y-shaped tube, where the visual target was about to move [53], leads to robust anticipatory smooth tracking in the direction of the other, unblocked branch.

12.3.2
If You Don't See It, You Can Still Predict (and Track) It

While walking on a busy street downtown, we may track a moving car with our gaze and, even when it is hidden behind a truck driving in front of it, we can still closely follow its motion and have our gaze next to the car's position at its reappearance. In the lab, researchers have shown [54] that during the transient disappearance of a moving target, human subjects are capable of continuing to track the hidden motion with their gaze, although with a lower gain (see Figure 12.2 (a) and (b)). During blanking, indeed, after an initial drop, eye velocity can be steadily maintained, typically at about 70% of pre-blanking target velocity, although higher eye speed can be achieved with training [55]. In addition, when the blank duration is fixed, an anticipatory reacceleration of the gaze rotation is observed ahead of target reappearance [56]. Extra-retinal, predictive information

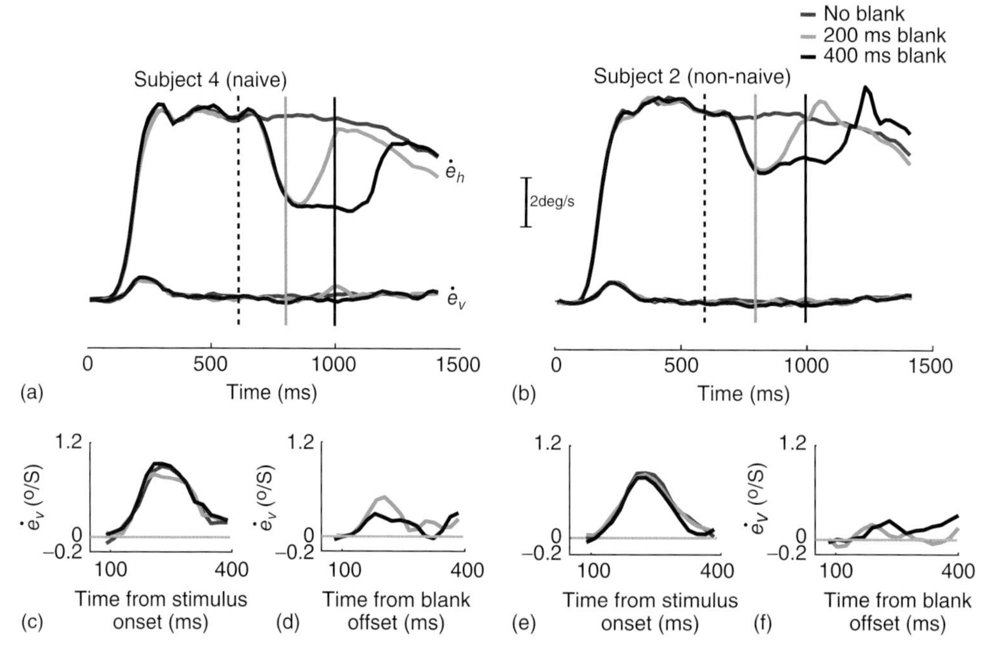

Figure 12.2 Examples of human smooth pursuit traces (one different participant on each column, a naive one on the left and a non-naive one on the right side) during horizontal motion of a tilted bar which is transiently blanked during steady-state pursuit. (a) and (b): Average horizontal ($\dot{e}_h$) and vertical ($\dot{e}_v$) eye velocity. Different blanking conditions are depicted by different colors, as from the figure legend. The vertical dashed line indicates the blank onset; vertical full colored lines indicate the end of the blanking epoch for each blanking duration represented. (c) and (e) Zoom on the aperture-induced bias of vertical eye velocity at target motion onset, for all blanking conditions. (d) and (f) Zoom on the aperture-induced bias of vertical eye velocity at target reappearance after blank (time is shifted so that 0 corresponds to blank offset), for all blanking conditions.

is clearly called into play to drive ocular tracking in the absence of a visual target. The true nature of the drive for such predictive eye velocity is still debated (see Ref. [2] for a review). Previous studies have proposed that it could either be a copy of the oculomotor command (an efference copy) serving as a positive feedback [55, 57] or a sample of visual motion being held in working memory [56]. In all cases, a rather implausible "switch-like" mechanism was assumed, in order to account for the change of regime between the visual- and prediction-driven tracking.

While the phenomenology of human smooth pursuit during the transient absence of a visual target is well investigated (see, e.g., Refs [56, 58, 59]), less is known about its functional characterization, and about how the extra-retinal signals implicated in motion tracking without visual input interact with retinal signals across time. In particular, as motion estimation for tracking is affected by sensory noise and computational limitations (see Sections 12.2.1 and 12.2.2), do we rely on extra-retinal predictive information in a way that depends on sensory uncertainty? In the past two decades, the literature on optimal cue combinations in multisensory integration has provided evidence for a weighted sum of different sources of information (such as visual and auditory [60], or visual and haptic cues [61], whereby each source is weighted according to its reliability (defined as inverse variance). Recently, [62] have shown that a predictive term on the smoothness of the trajectory is sufficient to account for the motion extrapolation part of the motion; however, it lacked a mechanism to weight retinal and extra retinal information. In another recent study, we have tested the hypothesis that visual and predictive information for motion tracking are weighted according to their reliability and dynamically integrated to provide the observed oculomotor command [63].

In order to do so, we have analyzed human smooth pursuit during the tracking of a horizontally-moving tilted bar that could be transiently hidden at different moments (blanking paradigm), either early, during pursuit initiation, or late, during steady-state tracking. By comparing the early and late blanking conditions we found two interesting results: first, the perturbation of pursuit velocity caused by the disappearance of the target was more dramatic for the late than the early blanking, both in terms of relative velocity drop and presence of an anticipatory acceleration before target reappearance. Second, a small, but significant, aperture-induced tracking bias (as described in Section 12.2.3) was observed at target reappearance after late but not early blanking. Interestingly, these two measures (the size of the tracking velocity reduction after blank onset and the size of the aperture bias after target disappearance) turned out to be significantly correlated across subjects for the late blanking conditions.

We interpreted the ensemble of these results as evidence in favor of dynamic optimal integration of visual and predictive information: at pursuit initiation, sensory variability is strong and predictive cues related to target- or gaze-motion dominate, leading to a relative reduction of both the effects of target blanking and of the aperture-bias. On the contrary, later on, sensory information becomes more reliable and the sudden disappearance of a visible moving target leads to a more dramatic disruption of motion tracking; coherently with this, the (re)estimation of motion is more strongly affected by the inherent ambiguity of the stimulus (a tilted bar). Finally, the observed correlation of these two quantities across different human observers indicates that the same computational mechanism (i.e., optimal dynamic integration of visual and predictive information) is scaled at the individual level in such a way that some people rely more strongly than others on predictive cues rather than on intrinsically noisy sensory evidence. Incidentally, in our sample of human volunteers, the expert subjects seemed to rely more on predictive information than the naive ones.

In Section 12.4.2, we illustrate a model which is based on hierarchical Bayesian inference and is capable to qualitatively capture the human behavior in our blanking paradigm. A second important question is whether and how predictive information is affected by uncertainty as well. We start to address this question in the next section.

12.4 Dynamic Integration of Retinal and Extra-Retinal Motion Information: Computational Models

12.4.1 A Bayesian Approach for Open-Loop Motion Tracking

Visual image noise and motion ambiguity, the two sources of uncertainty for motion estimate described in Sections 12.2.1 and 12.2.2 can be well integrated within a Bayesian [27, 44, 64, 65] or, equivalently, a Kalman-filtering framework [66, 67], whereby estimated motion is the solution of a dynamical statistical inference problem [68]. In these models, the information from different visual cues (such as local 1D and 2D motions) can be represented as probability distributions by their likelihood functions. Bayesian models also allow the inclusion of prior constraints related to experience, expectancy bias, and all possible sources of extrasensory information. On the ground of a statistical predominance of static or slowly moving objects in nature, the most common assumption used in models of motion perception is a preference for slow speeds, typically referred to as a *low-speed prior* (represented in Figure 12.3 (a)). The effects of priors are especially salient when signal uncertainty is high (see Chapter 9).

The sensory likelihood functions can be derived for simple objects with the help of a few reasonable assumptions. For instance, the motion cue associated with a nonambiguous 2D feature would be approximated by a Gaussian likelihood centered on the true stimulus velocity and with a variance proportional to visual noise (e.g., inversely related to its visibility, see Figure 12.3)(c). On the other hand, edge-related ambiguous information would be represented by an elongated velocity distribution parallel to the orientation of the moving edge, with an infinite variance along the edge direction reflecting the aperture ambiguity (Figure 12.3(b)). Weiss

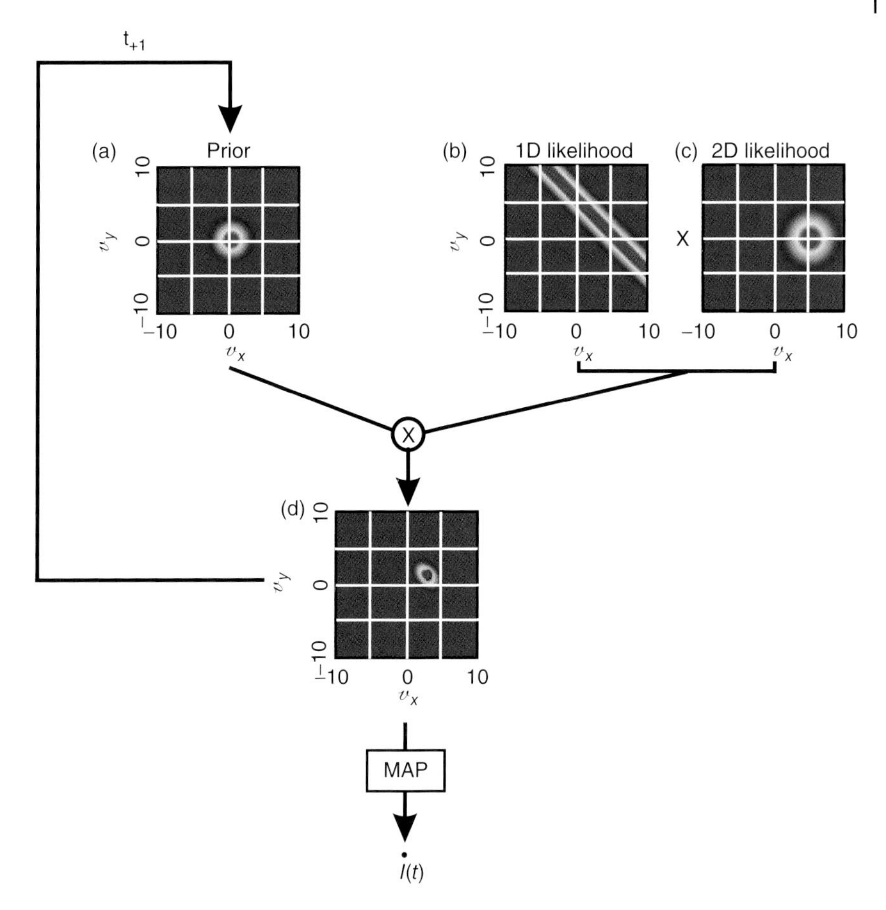

Figure 12.3 A Bayesian recurrent module for the aperture problem and its dynamic solution. (a) the prior and the two independent 1D (b) and 2D (c) likelihood functions (for a tilted line moving rightward at 5°/s) in the velocity space are multiplied to obtain the posterior velocity distribution (d). The inferred image motion is estimated as the velocity corresponding to the posterior maximum (MAP). Probability density functions are color-coded, such that dark red corresponds to the highest probability and dark blue to the lowest one.

and colleagues [27] have shown that, by combining a low-speed prior with an elongated velocity likelihood distribution parallel to the orientation of the moving line, it is possible to predict the aperture-induced bias (as illustrated in Figure 12.3(d)). By introducing the independent contribution of the nonambiguous 2D likelihood, as well as a recurrent optimal update of the prior (with a feedback from the posterior [44], see Figure 12.3), and cascading this recurrent network with a realistic model of smooth pursuit generation [45], we have managed to reproduce the complete dynamics of the solution of the aperture problem for motion tracking, as observed in human smooth pursuit traces.

12.4.2
Bayesian (or Kalman-Filtering) Approach for Smooth Pursuit: Hierarchical Models

Beyond the inferential processing of visual uncertainties, which mostly affect smooth pursuit initiation, we have seen in Section 12.3.2 that predictive cues can efficiently drive motion tracking when the visual information is deteriorated or missing. This flexible control of motion tracking has traditionally been modeled [55–57, 59] in terms of two independent modules, one processing visual motion and the other maintaining an internal memory of target motion. The weak point of these classical models is that they did not provide a biologically plausible mechanism for the interaction or alternation between the two types of control: a somewhat *ad hoc* switch was usually assumed for this purpose.

Inference is very likely to occur at different spatial, temporal, and neurofunctional scales. Sources of uncertainty can indeed affect different steps of the sensorimotor process in the brain. Recent work in our group [63] and other groups [67] has attempted to model several aspects of human motion tracking within a single framework, that of Bayesian inference, by postulating the existence of multiple inferential modules organized in a functional hierarchy and interacting according to the rules of optimal cue combination [68, 69]. Here we outline an example of this modeling approach applied to the processing of visual ambiguous motion information under different conditions of target visibility (in the blanking experiment).

In order to explain the data summarized in Section 12.3.2 for the transient blanking of a translating tilted bar, we designed a two-module hierarchical Bayesian recurrent model, illustrated in Figure 12.4. The first module, the retinal recurrent network (panel Figure 12.4a), implements the dynamic inferential process which is responsible for solving the ambiguity at pursuit initiation (see Section 12.2.3) and it only differs from the model scheme in Figure 12.3 by the introduction of processing delays estimated from the literature in monkey electrophysiology [70]. The second module (Figure 12.4b), the extra-retinal recurrent network, implements a dynamic short-term memory buffer for motion tracking. Crucially, the respective outputs of the retinal and extra-retinal recurrent modules are optimally combined in the Bayesian sense, so that their mean is weighted with their reliability (inverse variance) before the combination. By cascading the two Bayesian modules with a standard model [71] for the transformation of the target velocity estimate into eye velocity (Figure 12.4 c) and adding some feedback connections (also standard in the models of smooth pursuit to mimic the closed-loop phase), the model is capable, with few free parameters, to simulate motion tracking curves that resemble qualitatively the ones observed for human subjects both during visual pursuit and during blanking. Note that one single crucial free parameter, representing a scaling factor for the rapidity with which the sensory variance increases during target blank, is responsible for the main effects described in Section 12.3.2 on the tracking behavior during the blank and immediately after the end of it.

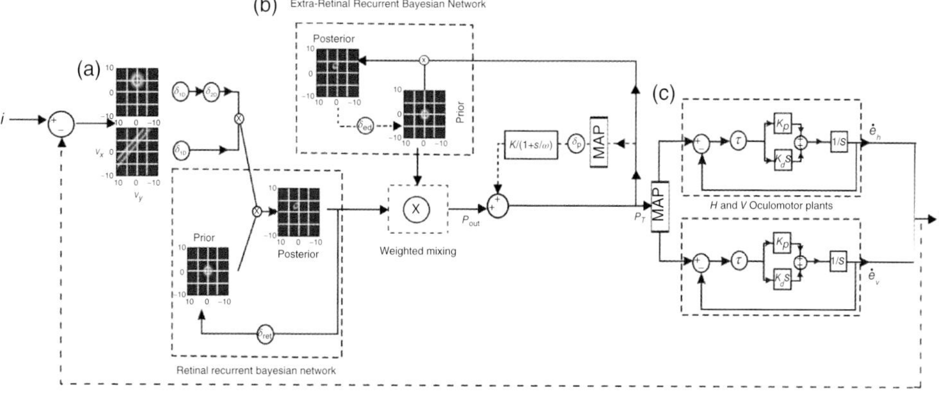

Figure 12.4 Two-stages hierarchical Bayesian model for human smooth pursuit in the blanking experiment. The retinal recurrent loop (a) is the same as in Figure 12.3, with the additional inclusion of physiological delays. The posterior from the retinal recurrent loop and prior from the extra-retinal Bayesian network (b) are combined to form the postsensory output (P_{out}). The maximum a posteriori of the probability (P_T) of target velocity in space serves as an input to both the positive feedback system as well as the oculomotor plants (c). The output of the oculomotor plant is subtracted from the target velocity to form the image's retinal velocity (physical feedback loop shown as broken line). During the transient blank when there is no target on the retina, the physical feedback loop is not functional so that the retinal recurrent block does not decode any motion. The output of the positive feedback system (shown by the broken line) is added to the postsensory output (P_{out}) only when the physical feedback loop is functional. The probability distribution of target velocity in space (P_T) is provided as an input to the extra-retinal recurrent Bayesian network where it is combined with a prior to obtain a posterior which is used to update the prior.

Orban de Xivry *et al.* [67] have proposed an integrated Kalman filter model based on two filters, the first one extracting a motion estimate from noisy visual motion input, similar to a slightly simplified version of the previously described Bayesian retinal recurrent module. The second filter (referred to as a *predictive pathway*) provides a prediction for the upcoming target velocity on the basis of long-term experience (i.e., from previous trials). Importantly, the implementation of a long-term memory for a dynamic representation of target motion (always associated with its uncertainty) allows to reproduce the observed phenomenon of anticipatory tracking when target motion properties are repeated across trials (see Section 12.3.1. However, in Section 12.6 we mention some results that challenge the current integrated models of hierarchical inference for motion tracking.

12.4.3
A Bayesian Approach for Smooth Pursuit: Dealing with Delays

Recently, we considered optimal motor control and the particular problems caused by the inevitable delay between the emission of motor commands and their sensory consequences [72]. This is a generic problem that we illustrate within the context of oculomotor control where it is particularly insightful (see, for instance, Ref. [73] for a review). Although focusing on oculomotor control, the more general contribution of this work is to treat motor control as a pure inference problem. This allows us to use standard (Bayesian filtering) schemes to resolve the problem of sensorimotor delays – by absorbing them into a generative (or forward) model. A generative model is a set of parameterized equations which describe our knowledge about the dynamics of the environment. Furthermore, this principled and generic solution has some degree of biological plausibility because the resulting active (Bayesian) filtering is formally identical to predictive coding, which has become an established metaphor for neuronal message passing in the brain (see Ref. [74], for instance). It uses oculomotor control as a vehicle to illustrate the basic idea using a series of generative models of eye movements – that address increasingly complicated aspects of oculomotor control. In short, we offer a general solution to the problem of sensorimotor delays in motor control – using established models of message passing in the brain – and demonstrate the implications of this solution in the particular setting of oculomotor control.

Specifically, we considered delays in the visuo-oculomotor loop and their implications for active inference. Active inference uses a generalization of Kalman filtering to provide Bayes' optimal estimates of hidden states and action (such that our model is a particular hidden Markov model) in generalized coordinates of motion. Representing hidden states in generalized coordinates provides a simple way of compensating for both sensory and oculomotor delays. The efficacy of this scheme is illustrated using numerical simulations of pursuit initiation responses, with and without compensation. We then considered an extension of the generative model to simulate smooth pursuit eye movements – in which the visuo-oculomotor system believes both the target and its center of gaze are attracted to a (hidden) point moving in the visual field, similar to what was proposed above in Section 12.4.3. Finally, the generative model is equipped with a hierarchical structure, so that it can recognize and remember unseen (occluded) trajectories and emit anticipatory responses (see Section 12.4.2).

We show in Figure 12.5 the results of this model for a two-layered hierarchical generative model. The hidden causes are informed by the dynamics of hidden states at the second level: these hidden states model underlying periodic dynamics using a simple periodic attractor that produces sinusoidal fluctuations of arbitrary amplitude or phase and a frequency that is determined by a second-level hidden cause with a prior expectation of a frequency of η (in Hz). It is somewhat similar to a control system model that attempts to achieve zero-latency target tracking by fitting the trajectory to a (known) periodic signal [75]. Our formulation ensures a Bayes' optimal estimate of periodic motion in terms of a posterior belief about its frequency. In these simulations, we used a fixed Gaussian prior centered on the correct frequency with a period of 512 *ms*. This prior reproduces a typical experimental setting in which the oscillatory nature of the trajectory is known, but its amplitude and phase (onset) are unknown. Indeed, it has been shown that

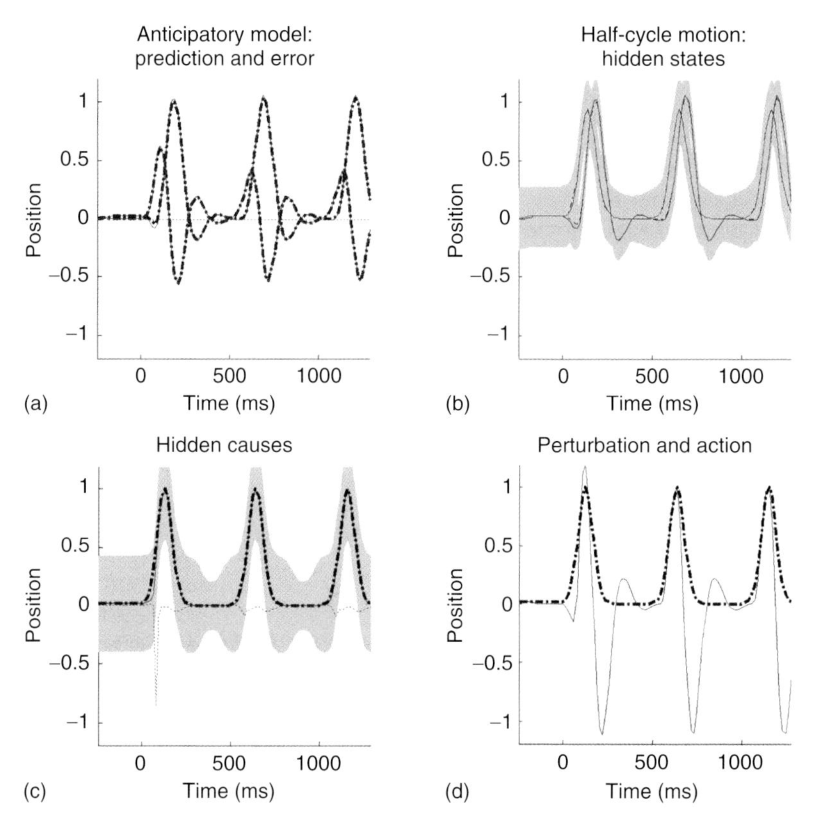

Figure 12.5 This figure reports the simulation of smooth pursuit when the target motion is hemi-sinusoidal, as would happen for a pendulum that would be stopped at each half cycle left of the vertical (broken black lines in panel (d). We report the horizontal excursions of oculomotor angle in retinal space (a), (b) and the angular position of the target in an intrinsic frame of reference (visual space), (c), (d). Panel (d) shows the true value of the displacement in visual space (broken black lines) and the action (blue line) which is responsible for oculomotor displacements. Panel (a) shows in retinal space the predicted sensory input (colored lines) and sensory prediction errors (dotted red lines) along with the true values (broken black lines). The latter is effectively the distance of the target from the center of gaze and reports the spatial lag of the target that is being followed (solid red line). One can see clearly the initial displacement of the target that is suppressed after a few hundred milliseconds. The sensory predictions are based upon the conditional expectations of hidden oculomotor (blue line) and target (red line) angular displacements shown in panel (b). The gray regions correspond to 90% Bayesian confidence intervals and the broken lines show the true values of these hidden states. The generative model used here has been equipped with a second hierarchical level that contains hidden states, modeling latent periodic behavior of the (hidden) causes of target motion (states not shown here). The hidden cause of these displacements is shown with its conditional expectation in panel (c). The true cause and action are shown in panel (d). The action (blue line) is responsible for oculomotor displacements and is driven by the proprioceptive prediction errors.

anticipatory responses are cofounded when randomizing the intercycle interval [54]. In principle, we could have considered many other forms of generative model, such as models with prior beliefs about continuous acceleration [76]. With this addition, the improvement in pursuit accuracy apparent at the onset of the second cycle is consistent with what was observed empirically [77].

This is because the model has an internal representation of latent causes of target motion that can be called upon even when these causes are not expressed explicitly in the target trajectory. These simulations speak to a straightforward and neurobiologically plausible solution to the generic problem of integrating information from different sources with different temporal delays and the particular difficulties encountered when a system – such as the oculomotor system – tries to control its environment with delayed signals. Neurobiologically, the application of delay operators just means changing synaptic connection strengths to take different mixtures of generalized sensations and their prediction errors.

12.5 Reacting, Inferring, Predicting: A Neural Workspace

We have proposed herein a hierarchical inference network that can both estimate the direction and speed of a moving object despite the inherent ambiguities present in the images and predict the target trajectory from accumulated retinal and extra-retinal evidence. What could be the biologically plausible implementation of such a hierarchy? What are its advantages for a living organism?

The fact that the pursuit system can be separated into two distinct blocks has been proposed by many others (see for recent reviews [17, 47, 78, 79]. Such structure is rooted on the need to mix retinal and extra-retinal information to ensure stability of pursuit, as originally proposed by Ref. [80]. Elaborations of this concept have formed the basis of a number of models based on a negative feedback control system [81, 82]. However, the fact that a simple efference copy feedback loop cannot account for anticipatory responses during target blanking as well as for the role of expectation about future target trajectory [47] or reinforcement learning during blanking [55] has called into question the validity of this simplistic approach (see Ref. [79] for a recent review). This has led to more complex models where an internal model of target motion is reconstructed from an early sampling and storage of target velocity and an efference copy of the eye's velocity signal. Several memory components have been proposed to serve different aspects of cognitive control of smooth pursuit and anticipatory responses [79, 83]. The hierarchical Bayesian model presented above (see Section 12.4.2) follows the same structure with two main differences. First, the sensory processing itself is seen as a dynamical network, whereas most of the models cited in this section have oversimplified target velocity representation. Second, we collapse all the different internal blocks and loops into a single inference loop representing the perceived target motion.

We have proposed earlier that these two inference loops might be implemented by two large-scale brain networks [78]. The dynamical visual inference loop is based on the properties of primate visual areas V1 and MT where local and global motion signals have been clearly identified, respectively (see Ref. [29] for a series of recent review articles). MT neurons solve the aperture problem with a slow temporal dynamics. When presented with a set of elongated, tilted bars, their initial preferred direction matches the motion direction orthogonal to the bar orientation. From there, that preferred direction gradually rotates toward the true, 2D translation of the bar so that after about 120 ms, the MT population signals the correct pattern motion direction, independently of the bar orientation [70]. Several models have proposed that such dynamics is due to recurrent interactions between the V1 and MT cortical areas (e.g., [84–86]) and we have shown that the dynamical Bayesian model presented above give a good description of such neuronal dynamics and its perceptual and behavioral counterparts [44] (see also Chapter 10). Such recurrent network can be upscaled to include other cortical visual areas involved in shape processing (e.g., areas V2, V3, V4) to further improve form–motion integration and select one target among several distractors or the visual background [84]. The V1–MT loop exhibits however two fundamental properties with respect to the tracking of the object's motion. First, neuronal responses stop immediately when the retinal input disappears as during object blanking. Second, the loop remains largely immune to higher, cognitive inputs. For instance, the slow-speed prior used in our Bayesian model can hardly be changed by training in human observers [48].

In nonhuman primates, the medial superior temporal (MST) cortical area is essential for pursuit. It receives MT inputs about target direction and speed of visual pattern and represents the target's velocity vector. In the context of smooth pursuit control, neuronal activities in the MST area show several interesting features. First, during pursuit, if the target is blanked, the neuronal activity is maintained throughout the course of the blanking. This is clearly different from the MT area where neurons stop firing in the occurrence of even a brief disappearance of the target [9]. Second, both monkeys and humans can track imaginary large line-drawing targets where the central foveal part is absent [87]. MST neurons, but not MT cells, can signal the motion direction of these parafoveal targets, despite the fact visual edges fall outside their receptive fields. Thus, in the MST area, neurons are found whose activities are not different during pursuit of real (i.e., complete) or imaginary (i.e., parafoveal) targets [29, 88]. Lastly, many MST neurons can encode target motion veridically during eye movements in contrast to MT cells [89]. The above evidence strongly suggests that MST neurons integrate both retinal and extra-retinal information to reconstruct the perceived motion of the target. However, despite the fact that MT and MST areas are strongly, and recurrently connected, the strong difference between MT and MST neuronal responses during pursuit seems to indicate that extra-retinal signals are not back propagated to early visual processing areas. Interestingly, several recent models have articulated their two-stage computational approach with this architecture [90, 91] in order to model the dynamics of primate smooth-pursuit.

From MST, output signals are sent in two directions. One signal reaches the brainstem oculomotor structures through the pontine nuclei, the cerebellar floccular region, and the vestibular nuclei [92]. This cortico-subcortical pathway conveys the visual drive needed for pursuit initiation and maintenance. The second signal reaches the frontal cortex that includes the caudal parts of the FEF and the SEF (see Refs [10, 79] for recent reviews). FEF neurons share many properties of MST cells. In particular, they integrate both retinal and extra-retinal information during pursuit so that responses remain sustained during blanking or when simulated with imaginary targets [78, 88]. Moreover, both FEF and MST cells show a buildup of activity during anticipatory pursuit [93, 94]. Thus, FEF and MST appear to be strongly coupled to build an internal representation of target motion that can be used during steady-state tracking as well as during early phases of pursuit. Moreover, FEF area issues pursuit commands that are sent to the brainstem nucleus reticularis tegmenti pontis (NRTP) and the cerebellar vermis lobules before reaching the pursuit oculomotor structures. Several authors have proposed that such parieto-frontal loops might implement the prediction component of the pursuit responses (see Ref. [78] for review). Other have proposed to restrict its role to the computation of the perceived motion signal that drive the pursuit response [79], while higher signals related to prediction might be computed in more anterior areas such as SEF and prefrontal cortex (PFC).

Prediction is influenced by many cognitive inputs (cues, working memory, target selection) [47]. Accordingly, prediction-related neuronal responses during pursuit have been reported in the SEF area [95] and the caudal part of FEF [94]. Moreover, SEF activity facilitates anticipatory pursuit responses to highly predictable targets [96]. The group of Fukushima have identified several sub-populations of neurons in both areas that can encode directional visual motion memory, independently of movement preparation signals (see Ref. [79] for a complete review). However, FEF neurons more often mix predictive and motor preparation signals, while SEF cells more specifically encode a visual motion memory signal. This is consistent with the fact that many neurons in the PFC have been linked to temporal storage of sensory signals [97]. Thus, a working memory of target motion might be formed in the SEF area by integrating multiple inputs from parietal (MST) and prefrontal (FEF, PFC) cortical areas. Fukushima *et al.* [79] proposed that a recurrent network made of these areas (MST, FEF, SEF, PFC) might signal future target motion using prediction, timing, and expectation, as well as experience gained over trials.

All these studies define a neuronal workspace for our hierarchical inference model, as illustrated in Figure 12.6. Two main loops seem to be at work. A first loop predicts the optimal target motion direction and speed from sensory evidence (image motion computation, in red). It uses sensory priors such as the "smooth and slow motion prior" used for both perception and pursuit initiation that are largely immune to higher influence. By doing so, the sensory loop can preserve its ability to quickly react to a new sensory event and avoid the inertia of prediction systems. This loop would correspond to the reactive pathway of the pursuit

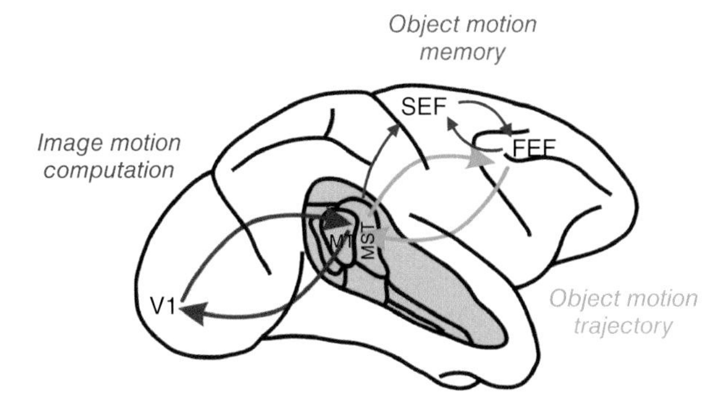

Figure 12.6 A lateral view of the macaque cortex. The neural network corresponding to our hierarchical Bayesian model of smooth pursuit is made of three main corticocortical loops. The first loop between primary visual cortex (V1) and the mediotemporal (MT) area computes image motion and infers the optimal low-level solution for object motion direction and speed. Its main output is the medio-superior temporal (MST) area that acts as a gear between the sensory loop and the object motion computation loop. Retinal and extra-retinal signals are integrated in both MST and FEF areas. Such dynamical integration computes the perceived trajectory of the moving object and implements an online prediction that can be used on the course of a tracking eye movement to compensate for target perturbation such as transient blanking. FEF and MST area signals are sent to the supplementary eye field (SEF) and the interconnected prefrontal cortical areas. This third loop can elaborate a motion memory of the target trajectory and is interfaced with higher cognitive processes such as cue instruction or reinforcement learning. It also implements off-line predictions that can be used across trials, in particular to drive anticipatory responses to highly predictable targets.

model proposed by Barnes [47] and Fukushima *et al.* [79]. On the ground of some behavioral evidence [48], we believe that target motion prediction cannot easily overcome the aperture problem (see also the open questions discussed in Section 12.6), providing a strong indication that this sensory loop is largely not penetrable to cognitive influences. The second loop involves the MST and FEF areas to compute and store online target motion by taking into account both sensory and efference copy signals (object motion computation, green). A key aspect is that MST must act as a gear to prevent predictive or memory-related signals to backpropagate downstream to the sensory loop. We propose to distinguish between online prediction involving these two areas during an ongoing event such as target blanking and off-line prediction. The latter is based on a memory of target motion that spans across trials and might be used to both trigger anticipatory responses or drive responses based on cues. It might most certainly involve a dense, recurrent prefrontal network articulated around SEF and that offers a critical interface with the cognitive processes interfering with pursuit control (object memory loop, in blue).

This architecture presents many advantages. First, it preserves the brain to quickly react to a new event as a brutal change in target motion direction. Second, it ensures maintaining pursuit in a wide variety of conditions with good stability, and by constructing an internal model of target motion, it allows a tight coordination between pursuit and saccades [98]. Third, it provides an interface with higher cognitive aspects of sensorimotor transformation. Several questions remain however unanswered. Because of the strong changes seen in prediction with different behavioral contexts, several models, including the one presented here, postulate the existence of hard switches that can turn on or off the contribution of a particular model component. We need a better theoretical approach about decision making between these different loops. The Bayesian approach proposed here, similar to the Kalman filter models, opens the door to better understanding these transitions. It proposes that each signal (sensory evidence, motion memory, prediction) can be weighted from its reliability. Such a unifying theoretical approach can then be used to design new behavioral and physiological experiments.

12.6 Conclusion

Recent experimental evidence points to the need to revise our view of the primates' smooth pursuit system. Rather than a reflexive velocity-matching, negative-feedback loop, the human motion tracking system seems to be grounded on a complex set of dynamic functions that subserve a quick and accurate adaptive behavior even in visually challenging situations. By analyzing tracking eye movements produced with a simple, unique, and highly visible moving target, many of these notions could not have clearly emerged and it is now clear that testing more naturalistic visual motion contexts and carefully taking into account the sources of uncertainty at different scales is a crucial step toward understanding biological motion tracking. The approach highlighted here opens the door to several questions.

First, we have focused herein on luminance-based motion processing mechanisms. Such inputs can be well extracted by a bank of filters sensitive to motion energy at multiple scales. The human visual motion system is however more versatile and psychophysical studies have demonstrated that motion perception is based on cues that can be defined in many different ways. These empirical studies have led to the three-system theory of human visual motion perception by Lu and Sperling [99]. Besides the first-order system that responds to moving luminance patterns, a second-order system responds to moving modulations of feature types (i.e., stimuli where the luminance is the same everywhere but an area of higher contrast or of flicker moves). A third-order system slowly computes the motion of marked locations in a "salience map" where locations of important visual features in the visual space (i.e., a figure) are highlighted respective to their "background" (see Chapter 11). The contribution of the first-order motion to the initiation of tracking responses have been largely investigated (see Refs [17, 19] for reviews). More recent studies have pointed out that feature tracking mechanisms (i.e., a second-order system) are critical for finely adjusting this initial eye acceleration to object speed when reaching steady-state tracking velocity [100]. Whether and how the third-order motion system is involved in the attentional modulation of tracking is currently under investigation by many research groups. Altogether, these psychophysical and theoretical studies point toward the need for more complex front-end layers of visuomotor models so that artificial systems would become more versatile and adapt to complex, ambiguous environments.

Second, although the need for hierarchical, multiscale inferential models is now apparent, current models will have to meet the challenge of explaining a rich and complex set of behavioral data. Just to detail an example, somewhat unexpectedly, we have found that predicting 2D motion trajectories does not help solving the aperture problem [48]. Indeed, when human subjects are exposed to repeated motion conditions for a horizontally translating tilted bar across several experimental trials and they develop anticipatory smooth pursuit in the expected direction, yet their pursuit traces reflect the aperture-induced bias at initiation, as illustrated in Figure 12.1(d). It is important to notice that the robust optimal cue combination of the output of sensory and predictive motion processing modules postulated in the previous section [63, 67] are not capable, at this stage, to explain this phenomenon. Other crucial issues deserve to be better understood and modeled in this new framework, for instance, the dynamic interaction between global motion estimate and object segmentation, the precise functional and behavioral relationship between smooth tracking and discrete jump-like saccadic movements, or the role of high-level cognitive cues in modulating motion tracking [3].

12.6.1 Interest for Computer Vision

Machine-based motion tracking, similar to human motion tracking, needs to be robust to visual perturbations. As we have outlined above, including some form of predictive information might be extremely helpful to stabilize object tracking, in much the same way as what happens during transient target blanking for human smooth pursuit. Importantly, human tracking seems to be equipped with a more complete and advanced software for motion tracking, namely, one that allows us (i) to exploit the learned regularities of motion trajectories to anticipate and compensate for sensorimotor delays and consequent mismatch between gaze and target position in the event of a complex trajectory; (ii) to keep tracking an object temporarily occluded by other objects; and (iii) to start tracking in the dark when the future motion of an object is predictable, or even partly predictable. Some of these predictive cues might depend upon a high-level cognitive representation of kinematic rules or arbitrary stimulus–response associations. This type of apparatus is probably much too advanced to be implemented into a common automatic motion tracking device, but still it may be a source of inspiration for future advanced developments for machine vision and tracking.

Acknowledgments

The authors were supported by EC IP project FP7-269921, "BrainScaleS" and the French ANR-BSHS2-2013-006, 'Speed'. They are grateful to Amarender Bogadhi, Laurent Madelain, Frederic Chavane and all the colleagues in the InVibe team for many enriching discussions. L.U.P wishes to thank Karl Friston and Rick Adams and the The Wellcome Trust Centre for Neuroimaging, University College London, for their essential contribution in the closed-loop delayed model. The chapter was written when G.S.M was an invited fellow from the CONICYT program, Chili. Correspondence and requests for materials should be addressed to A.M. (email:Anna.Montagnini@univ-amu.fr).

References

1. Rashbass, C. (1961) The relationship between saccadic and smooth tracking eye movements. *J. Physiol.*, **159**, 326–338.
2. Kowler, E. (2011) Eye movements: the past 25years. *Vision Res.*, **51** (13), 1457–1483, doi: 10.1016/j.visres.2010.12.014
3. Kowler, E., Aitkin, C.D., Ross, N.M., Santos, E.M., and Zhao, M. (2014) Davida teller award lecture 2013: the importance of prediction and anticipation in the control of smooth pursuit eye movements. *J. Vis.*, **14** (5), 10, doi: 10.1167/14.5.10
4. Maunsell, J.H. and Newsome, W.T. (1987) Visual processing in monkey extrastriate cortex. *Annu. Rev. Neurosci.*, **10**, 363–401, doi: 10.1146/annurev.ne.10.030187.002051
5. Bradley, D.C. and Goyal, M.S. (2008) Velocity computation in the primate visual system. *Nat. Rev. Neurosci.*, **9** (9), 686–695, doi: 10.1038/nrn2472
6. Born, R.T. and Bradley, D.C. (2005) Structure and function of visual area MT. *Annu. Rev. Neurosci.*, **28**, 157–189, doi: 10.1146/annurev.neuro.26.041002.131052
7. Newsome, W.T. and Wurtz, R.H. (1988) Probing visual cortical function with discrete chemical lesions. *Trends Neurosci.*, **11** (9), 394–400.
8. Lisberger, S.G. and Movshon, J.A. (1999) Visual motion analysis for pursuit eye movements in area MT of macaque monkeys. *J. Neurosci.*, **19** (6), 2224–2246.
9. Newsome, W.T., Wurtz, R.H., and Komatsu, H. (1988) Relation of cortical areas MT and MST to pursuit eye movements. II. Differentiation of retinal from extraretinal inputs. *J. Neurophysiol.*, **60** (2), 604–620.
10. Krauzlis, R.J. (2004) Recasting the smooth pursuit eye movement system. *J. Neurophysiol.*, **91** (2), 591–603, doi: 10.1152/jn.00801.2003
11. Hassenstein, B. and Reichardt, W. (1956) Systemtheoretische analyze der zeit-, reihenfolgen- und vorzeichenauswer- tung bei der bewegungsperzeption des russelkafers chlorophanus. *Z. Naturforsch.*, **11b**, 513–524.
12. Borst, A. (2014) Fly visual course control: behavior, algorithms and circuits. *Nat. Rev. Neurosci.*, **15**, 590–599.
13. Borst, A., Haag, J., and Reiff, D. (2010) Fly motion vision. *Annu. Rev. Neurosci.*, **33**, 49–70.
14. Harrison, R. and Koch, C. (2000) A robust analog VLSI Reichardt motion sensor. *Analog Integr. Circ. Signal Process.*, **24**, 213–229.
15. Kohler, T., Röchter, F., Lindemann, J., and Möller, R. (2009) Bio-inspired motion detection in a FPGA-based smart camera. *Bioinspiration Biomimetics*, **4**, 015 008.
16. Masson, G.S. (2004) From 1D to 2D via 3D: dynamics of surface motion

segmentation for ocular tracking in primates. *J. Physiol. Paris*, **98** (1-3), 35–52, doi: 10.1016/j.jphysparis.2004.03.017
17. Lisberger, S.G. (2010) Visual guidance of smooth-pursuit eye movements: sensation, action, and what happens in between. *Neuron*, **66** (4), 477–491, doi: 10.1016/j.neuron.2010.03.027
18. Spering, M. and Montagnini, A. (2011) Do we track what we see? Common versus independent processing for motion perception and smooth pursuit eye movements: a review. *Vision Res.*, **51** (8), 836–852, doi: 10.1016/j.visres.2010.10.017
19. Masson, G.S. and Perrinet, L.U. (2012) The behavioral receptive field underlying motion integration for primate tracking eye movements. *Neurosci. Biobehav. Rev.*, **36** (1), 1–25, doi: 10.1016/j.neubiorev.2011.03.009
20. Osborne, L.C. (2011) Computation and physiology of sensory-motor processing in eye movements. *Curr. Opin. Neurobiol.*, **21** (4), 623–628, doi: 10.1016/j.conb.2011.05.023
21. Lisberger, S.G., Morris, E.J., and Tychsen, L. (1987) Visual motion processing and sensory-motor integration for smooth pursuit eye movements. *Annu. Rev. Neurosci.*, **10**, doi: 10.1146/annurev.ne.10.030187.000525
22. Schütz, A.C., Braun, D.I., Movshon, J.A., and Gegenfurtner, K.R. (2010) Does the noise matter? Effects of different kinematogram types on smooth pursuit eye movements and perception. *J. Vis.*, **10** (13), 26, doi: 10.1167/10.13.26
23. Osborne, L.C., Lisberger, S.G., and Bialek, W. (2005) A sensory source for motor variation. *Nature*, **437** (7057), 412–416, doi: 10.1038/nature03961
24. Osborne, L.C., Hohl, S.S., Bialek, W., and Lisberger, S.G. (2007) Time course of precision in smooth-pursuit eye movements of monkeys. *J. Neurosci.*, **27** (11), 2987–2998, doi: 10.1523/JNEUROSCI.5072-06.2007
25. Osborne, L.C., Bialek, W., and Lisberger, S.G. (2004) Time course of information about motion direction in visual area MT of macaque monkeys. *J. Neurosci.*, **24** (13), 3210–3222, doi: 10.1523/JNEUROSCI.5305-03.2004
26. Spering, M., Kerzel, D., Braun, D.I., Hawken, M., and Gegenfurtner, K. (2005) Effects of contrast on smooth pursuit eye movements. *J. Vis.*, **20** (5), 455–465, *http://w.journalofvision.org/content/5/5/6.full.pdf.*
27. Weiss, Y., Simoncelli, E.P., and Adelson, E.H. (2002) Motion illusions as optimal percepts. *Nat. Neurosci.*, **5** (6), 598–604, doi: 10.1038/nn858
28. Wuerger, S., Shapley, R., and Rubin, N. (1996) "On the visually perceived direction of motion" by Hans Wallach: 60 years later. *Perception*, **25** (11), 1317–1367, doi: 10.1068/p251317
29. Masson, G.S. and Ilg, U.J. (eds) (2010) *Dynamics of Visual Motion Processing: Neuronal, Behavioral and Computational Approaches*, Springer-Verlag.
30. Fennema, C. and Thompson, W. (1979) Velocity determination in scenes containing several moving images. *Comput. Graph. Image Process.*, **9**, 301–315.
31. Adelson, E.H. and Movshon, J.A. (1982) Phenomenal coherence of moving visual patterns. *Nature*, **300** (5892), 523–525.
32. Yo, C. and Wilson, H.R. (1992) Perceived direction of moving two-dimensional patterns depends on duration, contrast and eccentricity. *Vision Res.*, **32** (1), 135–147.
33. Lorenceau, J., Shiffrar, M., Wells, N., and Castet, E. (1993) Different motion sensitive units are involved in recovering the direction of moving lines. *Vision Res.*, **33** (9), 1207–1217.
34. Gorea, A. and Lorenceau, J. (1991) Directional performances with moving plaids: component-related and plaid-related processing modes coexist. *Spat. Vis.*, **5** (4), 231–252.
35. Wilson, H.R., Ferrera, V.P., and Yo, C. (1992) A psychophysically motivated model for two-dimensional motion perception. *Visual Neurosci.*, **9** (1), 79–97.
36. Löffler, G. and Orbach, H.S. (1999) Computing feature motion without feature detectors: a model for terminator motion without end-stopped cells. *Vision Res.*, **39** (4), 859–871.
37. Albright, T.D. (1984) Direction and orientation selectivity of neurons in visual

area MT of the macaque. *J. Neurophysiol.*, **52** (6), 1106–1130.
38. Movshon, J.A., Adelson, E.H., Gizzi, M.S., and Newsome, W.T. (1985) The analysis of moving visual patterns, in *Pattern Recognition Mechanisms*, vol. 54 (eds C. Chagas, R. Gattass, and C. Gross), Vatican Press, Rome, pp. 117–151.
39. Perrinet, L.U. and Masson, G.S. (2012) Motion-Based prediction is sufficient to solve the aperture problem. *Neural Comput.*, **24** (10), 2726–2750, doi: 10.1162/NECO_a_00332
40. Burgi, P.Y., Yuille, A.L., and Grzywacz, N.M. (2000) Probabilistic motion estimation based on temporal coherence. *Neural Comput.*, **12** (8), 1839–1867, *http://portal.acm.org/citation.cfm?id=1121336.*
41. Masson, G.S. and Stone, L.S. (2002) From following edges to pursuing objects. *J. Neurophysiol.*, **88** (5), 2869–2873, doi: 10.1152/jn.00987.2001
42. Wallace, J.M., Stone, L.S., and Masson, G.S. (2005) Object motion computation for the initiation of smooth pursuit eye movements in humans. *J. Neurophysiol.*, **93** (4), 2279–2293, doi: 10.1152/jn.01042.2004
43. Born, R.T., Pack, C.C., Ponce, C.R., and Yi, S. (2006) Temporal evolution of 2-dimensional direction signals used to guide eye movements. *J. Neurophysiol.*, **95** (1), 284–300, doi: 10.1152/jn.01329.2005
44. Montagnini, A., Mamassian, P., Perrinet, L.U., Castet, E., and Masson, G.S. (2007) Bayesian modeling of dynamic motion integration. *J. Physiol. Paris*, **101** (1-3), 64–77, doi: 10.1016/j.jphysparis.2007.10.013
45. Bogadhi, A.R., Montagnini, A., Mamassian, P., Perrinet, L.U., and Masson, G.S. (2011) Pursuing motion illusions: a realistic oculomotor framework for Bayesian inference. *Vision Res.*, **51** (8), 867–880, doi: 10.1016/j.visres.2010.10.021
46. Gauthier, G.M. and Hofferer, J.M. (1976) Eye tracking of self-moved targets in the absence of vision. *Exp. Brain Res.*, **26** (2), 121–139.
47. Barnes, G.R. (2008) Cognitive processes involved in smooth pursuit eye movements. *Brain Cogn.*, **68** (3), 309–326, doi: 10.1016/j.bandc.2008.08.020
48. Montagnini, A., Spering, M., and Masson, G.S. (2006) Predicting 2D target velocity cannot help 2D motion integration for smooth pursuit initiation. *J. Neurophysiol.*, **96** (6), 3545–3550, doi: 10.1152/jn.00563.2006
49. Barnes, G.R. and Schmid, A.M. (2002) Sequence learning in human ocular smooth pursuit. *Exp. Brain Res.*, **144** (3), 322–335, doi: 10.1007/s00221-002-1050-8
50. Bennett, S.J., Orban de Xivry, J.J., Barnes, G.R., and Lefèvre, P. (2007) Target acceleration can be extracted and represented within the predictive drive to ocular pursuit. *J. Neurophysiol.*, **98** (3), 1405–1414, doi: 10.1152/jn.00132.2007
51. Montagnini, A., Souto, D., and Masson, G.S. (2010) Anticipatory eye-movements under uncertainty: a window onto the internal representation of a visuomotor prior. *J. Vis.*, **10** (7), 554.
52. Kowler, E., Martins, A.J., and Pavel, M. (1984) The effect of expectations on slow oculomotor control–IV. anticipatory smooth eye movements depend on prior target motions. *Vision Res.*, **24** (3), 197–210.
53. Kowler, E. (1989) Cognitive expectations, not habits, control anticipatory smooth oculomotor pursuit. *Vision Res.*, **29** (9), 1049–1057.
54. Becker, W. and Fuchs, A.F. (1985) Prediction in the oculomotor system: smooth pursuit during transient disappearance of a visual target. *Exp. Brain Res.*, **57** (3), 562–575, doi: 10.1007/BF00237843
55. Madelain, L. and Krauzlis, R.J. (2003) Effects of learning on smooth pursuit during transient disappearance of a visual target. *J. Neurophysiol.*, **90** (2), 972–982, doi: 10.1152/jn.00869.2002
56. Bennett, S.J. and Barnes, G.R. (2003) Human ocular pursuit during the transient disappearance of a visual target. *J. Neurophysiol.*, **90** (4), 2504–2520, doi: 10.1152/jn.01145.2002

57. Churchland, M.M., Chou, I.H.H., and Lisberger, S.G. (2003) Evidence for object permanence in the smooth-pursuit eye movements of monkeys. *J. Neurophysiol.*, **90** (4), 2205–2218, doi: 10.1152/jn.01056.2002
58. Orban de Xivry, J.J., Bennett, S.J., Lefèvre, P., and Barnes, G.R. (2006) Evidence for synergy between saccades and smooth pursuit during transient target disappearance. *J. Neurophysiol.*, **95** (1), 418–427, doi: 10.1152/jn.00596.2005
59. Orban de Xivry, J.J., Missal, M., and Lefèvre, P. (2008) A dynamic representation of target motion drives predictive smooth pursuit during target blanking. *J. Vis.*, **8** (15), 6.1–13, doi: 10.1167/8.15.6
60. Alais, D. and Burr, D. (2004) The ventriloquist effect results from near-optimal bimodal integration. *Curr. Biol.*, **14** (3), 257–262, doi: 10.1016/j.cub.2004.01.029
61. Ernst, M.O. and Banks, M.S. (2002) Humans integrate visual and haptic information in a statistically optimal fashion. *Nature*, **415** (6870), 429–433, doi: 10.1038/415429a
62. Khoei, M., Masson, G., and Perrinet, L.U. (2013) Motion-based prediction explains the role of tracking in motion extrapolation. *J. Physiol. Paris*, **107** (5), 409–420, doi: 10.1016/j.jphysparis.2013.08.001
63. Bogadhi, A., Montagnini, A., and Masson, G. (2013) Dynamic interaction between retinal and extraretinal signals in motion integration for smooth pursuit. *J. Vis.*, **13** (13), 5, doi: 10.1167/13.13.5
64. Stocker, A.A. and Simoncelli, E.P. (2006) Noise characteristics and prior expectations in human visual speed perception. *Nat. Neurosci.*, **9** (4), 578–585, doi: 10.1038/nn1669
65. Perrinet, L.U. and Masson, G.S. (2007) Modeling spatial integration in the ocular following response using a probabilistic framework. *J. Physiol. Paris*, **101** (1-3), 46–55, doi: 10.1016/j.jphysparis.2007.10.011
66. Dimova, K. and Denham, M. (2009) A neurally plausible model of the dynamics of motion integration in smooth eye pursuit based on recursive Bayesian estimation. *Biol. Cybern.*, **100** (3), 185–201, doi: 10.1007/s00422-009-0291-z
67. Orban de Xivry, J.J., Coppe, S., Blohm, G., and Lefèvre, P. (2013) Kalman filtering naturally accounts for visually guided and predictive smooth pursuit dynamics. *J. Neurosci.*, **33** (44), 17 301–17 313, doi: 10.1523/JNEUROSCI.2321-13.2013
68. Kalman, R.E. (1960) A new approach to linear filtering and prediction problems. *Trans. ASME–J. Basic Eng.*, **82** (Series D), 35–45.
69. Fetsch, C.R., Pouget, A., DeAngelis, G.C., and Angelaki, D.E. (2012) Neural correlates of reliability-based cue weighting during multisensory integration. *Nat. Neurosci.*, **15** (1), 146–154, doi: 10.1038/nn.2983
70. Pack, C.C. and Born, R.T. (2001) Temporal dynamics of a neural solution to the aperture problem in visual area MT of macaque brain. *Nature*, **409**, 1040–1042.
71. Goldreich, D., Krauzlis, R.J., and Lisberger, S.G. (1992) Effect of changing feedback delay on spontaneous oscillations in smooth pursuit eye movements of monkeys. *J. Neurophysiol.*, **67** (3), 625–638, *http://view.ncbi.nlm.nih.gov/pubmed/1578248.*
72. Perrinet, L.U., Adams, R.A., and Friston, K.J. (2014) Active inference, eye movements and oculomotor delays. *Biol. Cybern.*, **108** (6), 777–801, doi: 10.1007/s00422-014-0620-8
73. Nijhawan, R. (2008) Visual prediction: psychophysics and neurophysiology of compensation for time delays. *Behav. Brain Sci.*, **31** (02), 179–198, doi: 10.1017/s0140525x08003804
74. Bastos, A.M., Usrey, W.M., Adams, R.A., Mangun, G.R., Fries, P., and Friston, K.J. (2012) Canonical microcircuits for predictive coding. *Neuron*, **76** (4), 695–711, doi: 10.1016/j.neuron.2012.10.038
75. Bahill, A.T. and McDonald, J.D. (1983) Model emulates human smooth pursuit system producing zero-latency target tracking. *Biol. Cybern.*, **48** (3), 213–222,

http://view.ncbi.nlm.nih.gov/pubmed/6639984.

76. Bennett, S.J., Orban de Xivry, J.J., Lefèvre, P., and Barnes, G.R. (2010) Oculomotor prediction of accelerative target motion during occlusion: long-term and short-term effects. *Exp. Brain Res.*, **204** (4), 493–504, doi: 10.1007/s00221-010-2313-4
77. Barnes, G.R., Barnes, D.M., and Chakraborti, S.R. (2000) Ocular pursuit responses to repeated, single-cycle sinusoids reveal behavior compatible with predictive pursuit. *J. Neurophysiol.*, **84** (5), 2340–2355, *http://jn.physiology.org/content/84/5/2340.abstract.*
78. Masson, G.S., Montagnini, A., and Ilg, U.J. (2010) When the brain meets the eye: tracking object motion, *Biological Motion Processing*, vol. 8 (eds G.S. Masson and U.J. Ilg), Springer-Verlag, pp. 161–188.
79. Fukushima, K., Fukushima, J., Warabi, T., and Barnes, G.R. (2013) Cognitive processes involved in smooth pursuit eye movements: behavioral evidence, neural substrate and clinical correlation. *Front. Syst. Neurosci.*, **7** (4), 1–28.
80. Yasui, S. and Young, L. (1975) Perceived visual motion as effective stimulus to pursuit eye movement system. *Science*, **190**, 906–908.
81. Robinson, D.A., Gordon, J.L., and Gordon, S.E. (1986) A model of the smooth pursuit eye movement system. *Biol. Cybern.*, **55** (1), 43–57, doi: 10.1007/bf00363977
82. Krauzlis, R.J. and Lisberger, S.G. (1994) A model of visually-guided smooth pursuit eye movements based on behavioral observations. *J. Comput. Neurosci.*, **1**, 265–283.
83. Barnes, G.R. and Collins, C. (2011) The influence of cues and stimulus history on the non-linear frequency characteristics of the pursuit response to randomized target motion. *Exp. Brain Res.*, **212**, 225–240.
84. Tlapale, E., Masson, G.S., and Kornprobst, P. (2010) Modelling the dynamics of motion integration with a new luminance-gated diffusion mechanism. *Vision Res.*, **50** (17), 1676–1692, doi: 10.1016/j.visres.2010.05.022
85. Bayerl, P. and Neunmann, H. (2004) Disambiguating visual motion through contextual feedback modulation. *Neural Comput.*, **16**, 2041–2066.
86. Berzhanskaya, J., Grossberg, S., and Mingolla, E. (2007) Laminar cortical dynamics of visual form and motion interactions during coherent object motion perception. *Spat. Vis.*, **20**, 337–395.
87. Ilg, U.J. and Thier, P. (1999) Eye movements of rhesus monkeys directed towards imaginary targets. *Vision Res.*, **39**, 2143–2150.
88. Ilg, U.J. and Thier, P.P. (2003) Visual tracking neurons in primate area MST are activated by Smooth-Pursuit eye movements of an "imaginary" target. *J. Neurophysiol.*, **90** (3), 1489–1502, doi: 10.1152/jn.00272.2003
89. Chukoskie, L. and Movshon, J.A. (2008) Modulation of visual signals in macaque MT and MST neurons during pursuit eye movements. *J. Neurophysiol.*, **102**, 3225–3233.
90. Pack, C., Grossberg, S., and Mingolla, E. (2001) A neural model of smooth pursuit control and motion. *J. Cogn. Neurosci.*, **13** (1), 102–120.
91. Grossberg, S., Srihasam, K., and Bullock, D. (2012) Neural dynamics of saccadic and smooth pursuit coordination during visual tracking of unpredictably moving targets. *Neural Netw.*, **27**, 1–20.
92. Leigh, R. and Zee, D. (2006) *The neurology of eye movements*, 4th edn, Oxford University Press, New York.
93. Ilg, U.J. (2003) Visual tracking neurons in area MST are activated during anticipatory pursuit eye movements. *Neuroreport*, **14**, 2219–2223.
94. Fukushima, K., Yamatobe, T., Shinmei, Y., and Fukushima, J. (2002) Predictive responses of periarcuate pursuit neurons to visual target motion. *Exp. Brain Res.*, **145**, 104–120.
95. Heinen, S. (1995) Single neuron activity in the dorsolateral frontal cortex during smooth pursuit eye movements. *Exp. Brain Res.*, **104**, 357–361.

96. Missal, M. and Heinen, S. (2004) Supplementary eye fields stimulation facilitates anticipatory pursuit. *J. Neurophysiol.*, **92** (2), 1257–1262.
97. Goldman-Rakic, P. (1995) Cellular basis of working memory. *Neuron*, **14**, 477–485.
98. Orban de Xivry, J. and Lefèvre, P. (2004) Saccades and pursuit: two outcomes of a single sensorimotor process. *J. Physiol.*, **584**, 11–23.
99. Lu, Z.L. and Sperling, G. (2001) Three-systems theory of human visual motion perception: review and update. *J. Opt. Soc. Am. A*, **18**, 2331–2370.
100. Wilmer, J. and Nakayama, K. (2007) Two distinct visual motion mechanisms for smooth pursuit: evidence from individual differences. *Neuron*, **54**, 987–1000.

13
Cortical Networks of Visual Recognition

Christian Thériault, Nicolas Thome, and Matthieu Cord

13.1
Introduction

Human visual recognition is a far from trivial feat of nature. The light patterns projected on the human retina are always changing, and an object will never create exactly the same pattern twice. Objects move and transform constantly, appearing under unlimited number of aspects. Yet, the human visual system can recognize objects in milliseconds. It is thus natural for computer vision models to draw inspiration from the human visual cortex.

Most of what is understood about the visual cortex, and particularly about how it is able to achieve object recognition, comes from neurophysiological and psychophysical studies. A global picture emerges from over six decades of studies – the visual cortex is mainly organized as a parallel and massively distributed network of self-repeating local operations. Neurophysiological data and models of cortical circuitry have shed light on the processes by which feedforward (bottom-up) activation can generate early neural response of visual recognition [1, 2]. Contextual modulation and attentional mechanisms, through lateral and cortical feedback (top-down) connections, are clearly essential to the full visual recognition process [3–8]. Nevertheless, basic feedforward models [9–11] without feedback connections already display interesting levels of recognition, and provide a simple design around which the full functioning of the visual cortex can be studied. The circuitry of the visual cortex has also been studied in the language of differential geometry, which provides a natural connection between local neural operations, global activation, and perceptual phenomena [12–16].

This chapter introduces basic concepts on visual recognition by the cortex and some of its models. The global organization of the visual cortex is presented in Section 13.2. Local operations are followed in Section 13.4 by a special emphasis on operations in the *primary visual cortex*. Object recognition models are presented in Section 13.5, with a detailed description of a general model in Section 13.6. Section 13.7 focuses on a mathematical abstraction which corresponds to the structure of the primary visual cortex and which provides a model of contour emergence. Section 13.8 presents psychophysical and biological

Biologically Inspired Computer Vision: Fundamentals and Applications, First Edition.
Edited by Gabriel Cristóbal, Laurent Perrinet, and Matthias Keil.

bases supporting such a model. The importance of feedback connections is discussed in Section 13.9, and the chapter concludes with the role of transformations (i.e., motion) in learning-invariant representations of objects.

13.2
Global Organization of the Visual Cortex

The brain is a dynamical system in which specialized areas receive and send connections to other multiple areas – brain functions emerge from the interaction of subpopulations of specialized neurons. As illustrated in Figure 13.1, the visual cortex makes no exception, and contains interacting subpopulations of neurons tuned to process visual information (shape, color, motion, etc.).

The laminar structure of the cortex enables researchers to distinguish between *lateral connections* inside each area, *feedback connections* between areas, and *feedforward connections* streaming up from retinal inputs [17]. Up to now, the two most studied cortical visual areas have been the V1 area, the first area receiving extracortical inputs from the lateral geniculate nucleus (LGN), and the V5/MT (medial temporal) area.

A general consensus is that neurons in V1 behave as spatiotemporal filters, selective to local changes in color, spatial orientations, spatial frequencies, motion direction, and speed [18, 19]. The V5/MT area shares connections with V1 and also contains populations of neurons sensitive to motion speed and direction (see Chapter 12). Its contribution to motion perception beyond what is already observed in V1 is still not fully determined [20] and may involve the processing of motion over a broader range of spatiotemporal patterns compared to V1 [18, 21, 22].

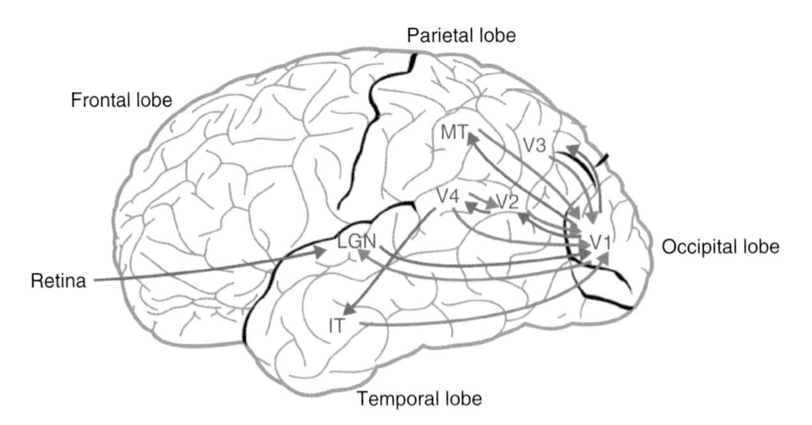

Figure 13.1 Basic organization of the visual cortex. Arrows illustrate the connections between various specialized subpopulations of neurons. Green and blue arrows indicate feedback and feedforward connections, respectively. The lateral geniculate nucleus (LGN), an extracortical area, is represented in gray.

Beginning at V1, there are interacting streams of visual information – the *ventral pathway*, which begins at V1 and goes into the temporal lobe, and the *dorsal pathway*, which also begins at V1 but goes into the parietal lobe [23]. The latter is traditionally associated with spatial information (i.e., "where") while the former is traditionally associated with object recognition (i.e., "what"). However, mounting evidence indicates significant connections between the two pathways and the dynamics of their interaction is now recognized [17, 24–27].

Feedforward activation in the ventral pathway is correlated with the rapid (i.e., ∼100–200 ms) ability of humans to recognize visual objects [1, 28], but recurrent feedback activity between these areas should not be ruled out, even during rapid recognition [29]. The role of later stages beyond V1, (i.e., V2, V3, and V4) remains less understood, although neurons in the IT (inferior temporal) area have been shown to respond to more complex and global patterns, independently of spatial position and pose (i.e., full objects, faces, etc.) [1, 23].

13.3
Local Operations: Receptive Fields

Neurons in the visual cortex follow a *retinotopic* organization – local regions of the visual field, called *receptive fields*, are mapped to corresponding neurons [30]. In some areas, such as V1, the retinotopic mapping is continuous [31] – adjacent neurons have adjacent overlapping receptive fields. At different stages in the visual pathway, neurons have receptive fields which vary in size and complexity [23]. As presented in Section 13.4, neurons in V1 respond to local changes (i.e., the derivative of light intensity) over orientations and spatiotemporal frequencies. These small, local, receptive fields are integrated into larger receptive fields by neurons in the later stages of the V1→IT pathway (Figure 13.2). This gives rise to representations of more complex patterns along the pathway, such as in the V4 [32, 33]. In the later stages, such as the IT area, neurons have receptive sizes which

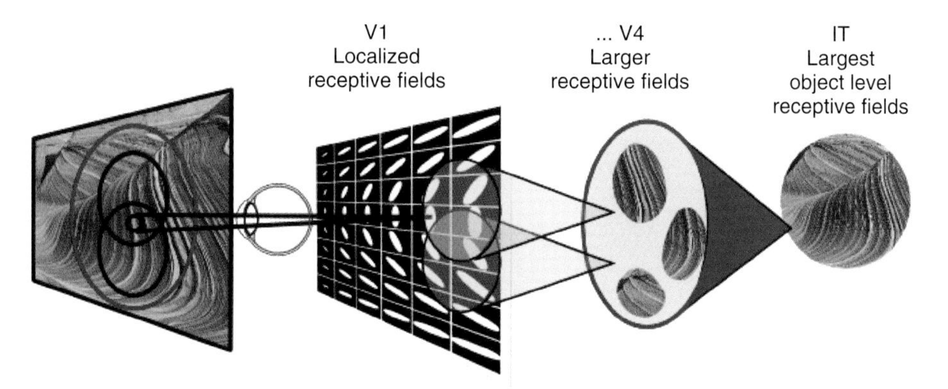

Figure 13.2 Receptive field organization. Neurons along the visual pathway V1→IT have receptive fields which vary in size and complexity.

cover the entire visual field and respond to the identity of objects rather than their position [1, 34].

13.4 Local Operations in V1

Together with the work of Ref. [35], neurophysiological studies [36–38] have established a common finding about the spatial domain profile of receptive fields in the V1 area. This visual area is organized into a tilling of *hypercolumns*, in which the columns are composed of cells sensitive to spatial orientations with ocular dominance. Brain imaging techniques [39] have since revealed that orientation columns are spatially organized into a pinwheel crystal as illustrated in Figure 13.3.

Cells inside orientation columns are referred to as *simple cells*. In Chapter 14, the conditions under which the profile of simple cells naturally emerge from unsupervised learning are presented. The profile of simple cells can be modeled by Gaussian-modulated filters – Gaussian derivatives [41, 42] or Gabor filters [43–45]. Gaussian derivatives describe the V1 area in terms of differential geometry [41], whereas Gabor filters present the V1 area as a spatiotemporal frequency analyzer [37, 46]. Both filters are mathematically equivalent and differ mostly in terminology. As illustrated in Figure 13.4, the first- and second-order Gaussian derivatives give good approximations of Gabor filters with odd and

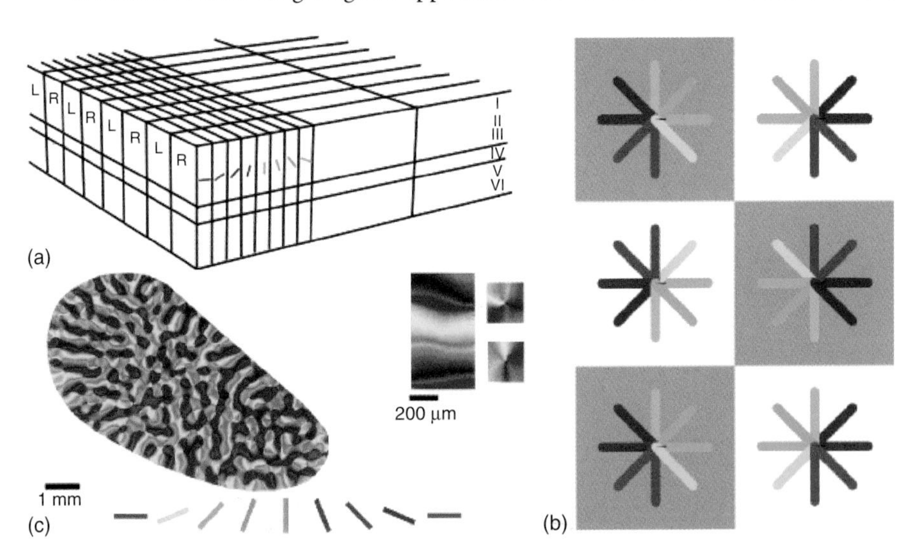

Figure 13.3 Primary visual cortex organization. (a) Hypercolumn structure showing the orientation and ocular dominance axes (image reproduced from Ref. [40]). (b) Idealized crystal pinwheel organization of hypercolumns in visual area V1 (image taken from Ref. [14]). (c) Brain imaging of visual area V1 of the tree shrew showing the pinwheel organization of orientation columns (image adapted from Ref. [39]).

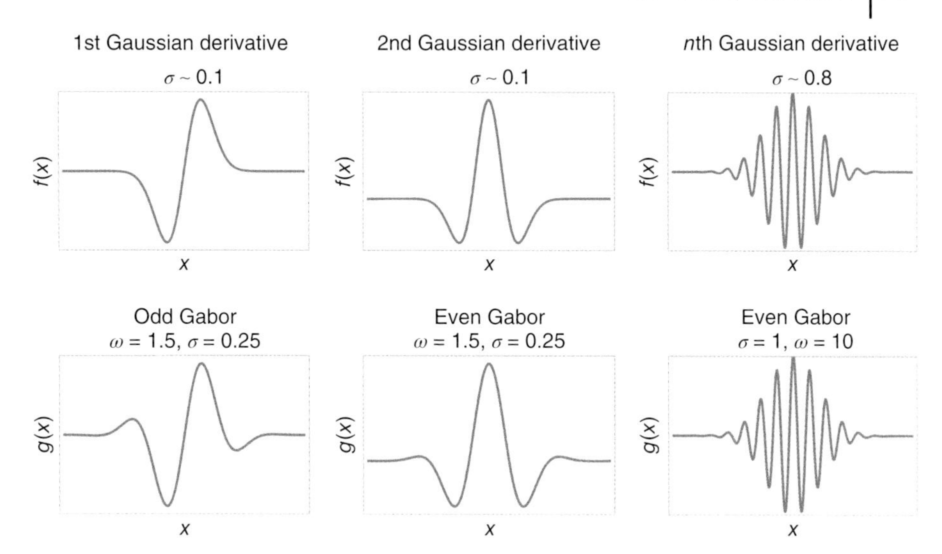

Figure 13.4 One-dimensional simple cell profiles. The curves shown are the graphs of local operators modeling the sensitivity profile of neurons in the primary visual cortex. Gabor filter and Gaussian derivative are local approximations of each other and are asymptotically equivalent for high-order of differentiation.

even phase, respectively. In fact, Gaussian derivatives are asymptotically equal (for high-order derivatives) to Gabor filters [41]. The one-dimensional profile for the Gaussian derivatives and the (odd-phase) Gabor filter are respectively given by

$$f(x) = \frac{\partial^n}{\partial x^n} G(x) \quad \text{and} \quad g(x) = G(x) \sin(2\pi\omega x) \tag{13.1}$$

where $G(x) = \frac{1}{\sqrt{2\pi\sigma^2}} e^{-\frac{x^2}{2\sigma^2}}$ is the Gaussian envelope of scale σ, and where ω gives the spatial frequency of the Gabor filter. In Reference [47], derivatives are further normalized with respect to σ to obtain scale invariance – filters at different scales will give the same maximum output. The even-phase Gabor filters is simply defined by a cosine function instead of a sine function. In Reference [42], cortical receptive field with order of differentiation as high as $n = 10$ are reported, with the vast majority being $n \leq 4$.

Two-dimensional filters for the first-order Gaussian derivative and the odd-phase Gabor are respectively given by

$$f(x,y) = \frac{\partial}{\partial y} G(x,y) \quad \text{and} \quad g(x,y) = G(x,y) \sin(2\pi\omega y) \tag{13.2}$$

where the Gaussian envelope is $G(x,y) = \frac{1}{\sqrt{2\pi\sigma^2}} e^{-\frac{x^2+y^2}{2\sigma^2}}$. For $u = x\cos\theta + y\sin\theta$ and $v = -x\sin\theta + y\cos\theta$, the filters $f_{\sigma,\theta}(u,v)$ and $g_{\sigma,\theta}(u,v)$ correspond to a rotation of the axes through an angle θ at scale σ, and model the orientation selectivity of simple cells in the hypercolumns. Specifically, when applied to an

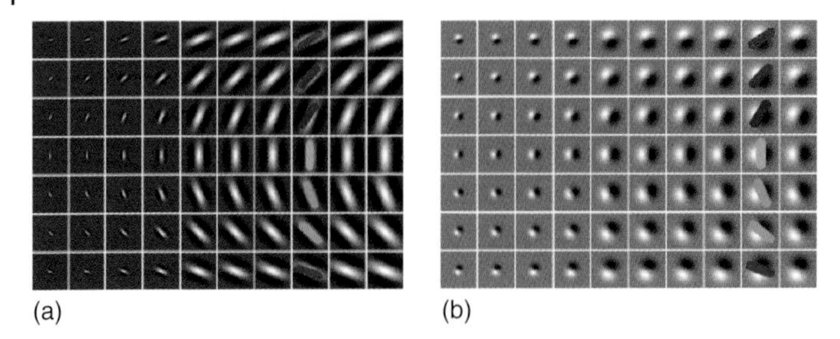

Figure 13.5 Two-dimensional simple cell profiles. The figure illustrates two-dimensional Gabor filters (a) and first-order Gaussian derivatives (b) at various scales and orientations. Colors can be put in relation to Figure 13.3.

image I, the Gaussian derivative $f_{\sigma,\theta}(u, v)$ gives the directional derivative, in the direction $v = (-\sin\theta, \cos\theta)$, of the image smoothed by the Gaussian G at scale σ

$$f_{\sigma,\theta} * I = \left[-\sin\theta\frac{\partial}{\partial x} + \cos\theta\frac{\partial}{\partial y}\right] G * I \tag{13.3}$$

where $*$ denotes the convolution product. The odd-phase Gabor $g_{\sigma,\theta}$ gives the same directional derivative up to a multiplicative constant. Figure 13.5 illustrates examples of first-order Gaussian derivatives and even-phase Gabor filters at various scales and orientations. When representing cell activations in V1, the outputs of these filters can then be propagated synchronously or asynchronously [48] through a multilayer architecture simulating the basic feedforward principles of the visual cortex, as in Section 13.6.

Neurophysiological studies also identified *complex cells* in the V1→V4 pathway. These cells also respond to specific frequencies and orientations, but their spatial sensitivity profiles are not localized as with simple cells, as illustrated in Figure 13.6. Complex cells display invariance (tolerance) to the exact position of the visual patterns at the scale of the receptive field. By allowing local shifts in the exact position of patterns, they may play a role in our ability to recognize objects invariantly with respect to transformations. Complex cells can be modeled by a MAX or soft-MAX operations applied to incoming simple cells [49]. Section 13.6

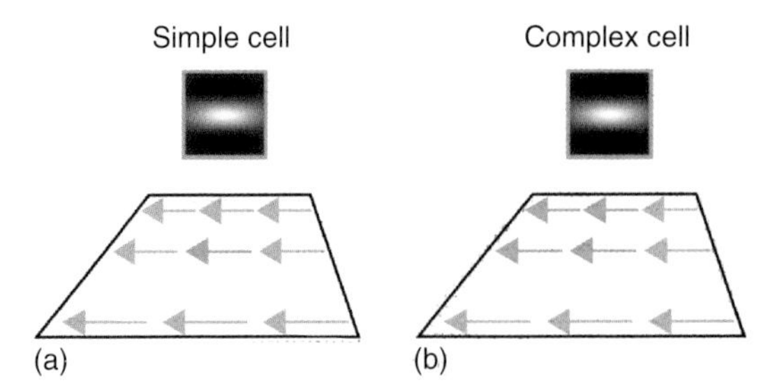

Figure 13.6 Complex cells. (a) The simple cell responds selectively to a local spatial derivative (blue arrow). (b) The complex cell responds selectively to the same spatial pattern, but displays invariance to its exact position. Complex cells are believed to gain their invariance by integrating (summing) over simple cells of the same selectivity.

presents a model of such simple and complex cells network using the MAX operation in a multilayer architecture.

13.5
Multilayer Models

On the basis of the above considerations, the vast majority of biologically inspired models are multilayer networks where the layers represent the various stages of processing corresponding to physiological data obtained about the mammalian visual pathways.

For most multilayer networks, the first layers are modeled by the operations of *simple cells* and *complex cells* in the primary visual cortex V1, as presented in the preceding section. However, the bulk of neurophysiological data clearly indicate important areas for visual recognition beyond area V1. These stages of processing are less understood. Multilayer networks usually model these stages with a series of layers which generate increasing complexity and invariance of representations [9, 11] (Figure 13.7). These representations are most often learned from experience by defining learning rules which modify the connections between layers.

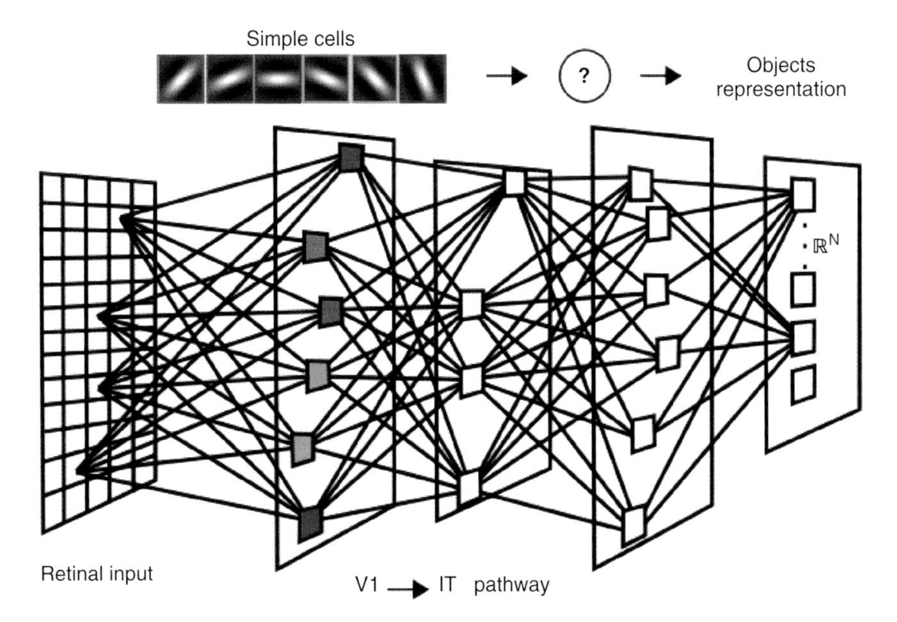

Figure 13.7 Multilayer architecture. The cortical visual pathway is usually modeled by multilayer networks. Synaptic connections for the early stages, such as V1 (first layer), are usually sensitive to local changes in spatial orientations and frequencies. Connections in the latter stages is an active research topic.

Supervised learning can be used with the objective to minimize error in object representations while retaining some degree of invariance to basic visual transforms [50, 51]. Networks using supervised learning are most often based on the error gradient *backpropagation* through the layers [52].

Unsupervised learning uses statistical regularities of the visual world to learn useful representations. One basic principle is to use temporal correlations between views of a transforming object. The hypothesis, in this case, is that we do not change the world by moving in it. Stated differently, neural representations should change on a slower timescale than the retinal input [53]. Such networks [23, 54, 55] are often based on the so called *trace rule* which minimizes variations of neural response to objects undergoing transformations. Other authors [48] have shown that the temporal aspect of neural spikes across layers can be used to define unsupervised learning of relevant object features. Unsupervised learning has also been used in *generative models* to find the appropriate representation space for each layer as the initial condition for supervised learning [56, 57].

13.6
A Basic Introductory Model

A simple and efficient model of the basic feedforward mechanisms of the visual pathway V1→IT is given by the HMAX network and its variations [9–11]. In its most basic form, this network does not model full connectivity of the visual cortex, for instance, long-range horizontal connections described in the next section – the basic HMAX model is in essence a purely feedforward network. Nevertheless, it is based on hypercolumn organization and provides a starting point to model the visual cortical pathway.

The layers of the HMAX model are decomposed into parallel sublayers called *feature maps*. The activation of one feature map corresponds to the mapping of one filter (one feature) at all positions of the visual field. On the first layer, the features are defined by the outputs of simple cells given by Eq. (13.1). Each simple cell calculates a directional derivative and corresponds geometrically to an orientation and an amplitude (i.e., a vector). The feature maps on the first layer of the HMAX are therefore vector fields, corresponding to cross sections of the hypercolumns in V1 – the selection of one directional derivative per position is, by definition, a vector field expressing the local action of a transformation.

As illustrated in Figure 13.8, the first layer in HMAX enables the expression of various *group actions*, particularly those corresponding to basic transformations imposed on the retinal image, for which visual recognition is known to be invariant. Pooling over the parameters of a transformation group (i.e., pooling over translations) generates representations which are invariant to such transformation [58]. In the basic HMAX, mapping and pooling over translation fields of multiple orientations gives the model some degree of invariance to local translations, and consequently to shape deformations.

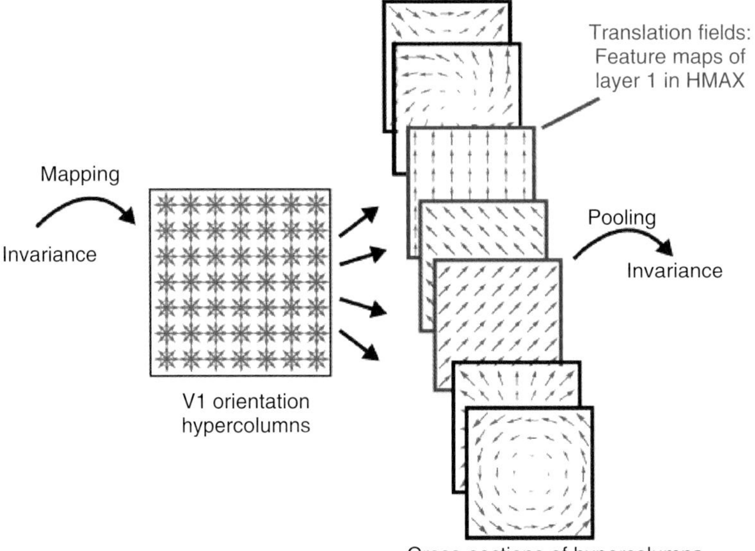

Figure 13.8 Sublayers in HMAX. The figure illustrates cross sections of the orientation hypercolumns in V1. Each cross section selects one simple cell per position, thereby defining a vector field. The first layer of HMAX is composed of sublayers defining translation fields (that is, the same orientation at all positions). By pooling over translation fields at various orientations, the HMAX model displays tolerance to local translations, which results in degrees of invariance to shape distortions. For illustration, the figure shows the image of the word *Invariance*, with its component translated. By pooling the local maximum values of translation fields mapped onto the image, the HMAX model tolerates local translations, and will produce a representation which is invariant to those local transformations.

The overall HMAX model follows a series of convolution/pooling steps as in Refs [9, 59] and illustrated in Figure 13.9. Each convolution step yields a set of feature maps and each pooling step provides tolerance (invariance) to variations in these feature maps. Each step is detailed below.

Layer 1 (simple cells). Each feature map $\mathbf{L1}_{\sigma,\theta}$ is activated by the convolution of the input image with a set of simple cell filters $g_{\sigma,\theta}$ with orientations θ and scales σ as defined in Eq. (13.2). Given an image I, Layer 1 at orientation θ and scale σ is given by the absolute value of the convolution product

$$\mathbf{L1}_{\sigma,\theta} = |g_{\sigma,\theta} * I| \tag{13.4}$$

The layer can further be self-organized [60] through a process called *lateral inhibition* which appears throughout the cortex. Lateral inhibition is the process by which a neuron suppresses the activation of its neighbors through inhibitory connections. This competition between neurons creates a form of refinement or sharpening of the signal by filtering out components of smaller amplitude relative to their surroundings. It is also considered as a biological mechanism for neural sparsity and corresponds to a *subtractive normalization* [61]. Sparse neural firing

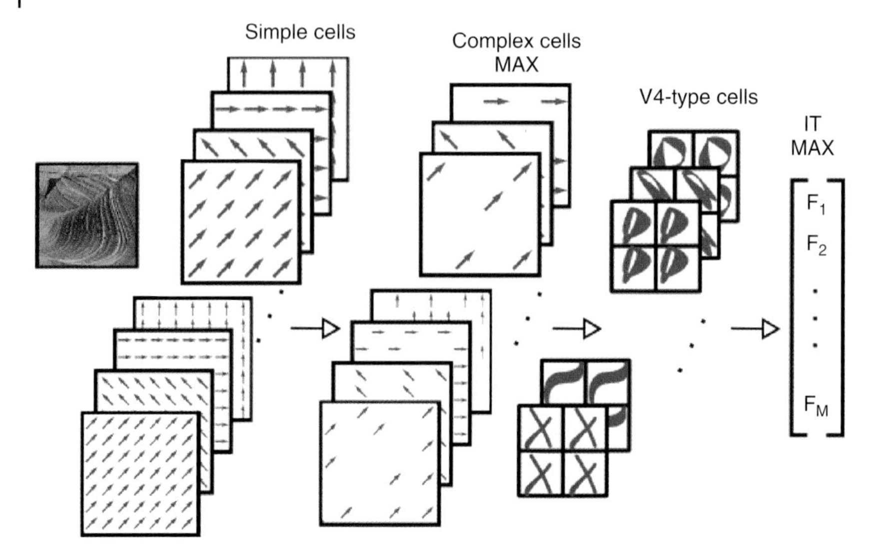

Figure 13.9 General HMAX network. The network alternates layers of feature mapping (convolution) and layers of feature pooling (MAX function). The convolution layers generate specific feature information, whereas the pooling layers results in degrees of invariance by relaxing the configuration of these features.

has been shown to improve the original HMAX architecture on classification simulations [62]. The effect of inhibitory connections between neighboring neurons can be implemented by taking the convolution product of maps in layer 1 with the filter defined in Eq. 13.5 – an inhibitory surround with an excitatory center. A form of *divisive normalization* [63, 64] can also be used (see layer 3, below). More refinement can be obtained by applying inhibition (or suppressing) the weaker orientations at each position [10, 11]. This can be accomplished by a one-dimensional version of the lateral inhibition filter (Figure 13.10) in Eq. 13.5.

$$I(x,y) = \begin{cases} \delta e^{-\frac{x^2+y^2}{\sigma^2}} & : x \neq 0, y \neq 0 \\ 1 & : x = 0, y = 0 \end{cases} \tag{13.5}$$

where σ defines the width of inhibition and δ defines the contrast.

Layer 2 (complex cells). Each feature map $\mathbf{L2}_{\sigma,\theta}$ models the operations of complex cells in the visual cortex, illustrated in Figure 13.6. The output of complex cell selects the maximum value on a local neighborhood of simple cells. Maximum pooling over local neighborhoods results in invariance to local translations and thereby to global deformations [50] – it provides *elasticity* (illustrated in Figure 13.8) to the configuration of features of layer 1. Specifically, the second layer partitions each $\mathbf{L1}_{\sigma,\theta}$ map into small neighborhoods $\mathbf{u}_{i,j}$ and selects the maximum value inside each $\mathbf{u}_{i,j}$ such that

$$\mathbf{L2}_{\sigma,\theta}(i,j) = \max_{\mathbf{u}_{i,j} \in \mathbf{L1}_{\sigma,\theta}} \mathbf{u}_{i,j} \tag{13.6}$$

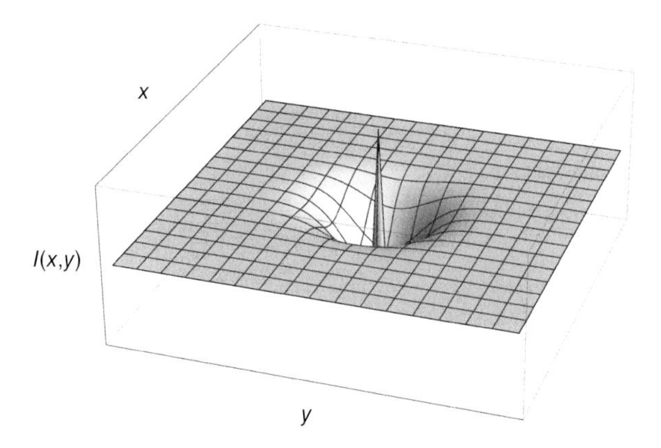

Figure 13.10 Lateral inhibition filter. A surround inhibition and an excitatory center.

Some degree of local scale invariance is also achieved by keeping only the maximum output over two adjacent scales at each position (i, j).

Layer 3 (V4-type cells). Layer $\mathbf{L}3$ at scale σ is obtained by applying filters α^m to layer 2.

$$\mathbf{L}3^m_\sigma = \alpha^m * \mathbf{L}2_\sigma \tag{13.7}$$

Each α^m filter represents a V4-type cell – a configuration of multiple orientations representing more elaborated visual patterns than simple cells. In Reference [11], the α^m also cover multiple scales, which give each filter the possibility of responding selectively to more complex patterns.

To implement a form of divisive normalization [63], the filters, and their input, can be normalized to unit length. This normalization will ensure that only the geometrical aspect of the inputs are considered, and not the contrast or the luminance level. The filters α^m are learned from training images by sampling subsets of layer $\mathbf{L}2$. This simple learning procedure consists in sampling subparts of layer $\mathbf{L}2$ and storing them as the connections weights of the filters α^m. The training procedure, as done in Ref. [11], is illustrated in Figure 13.11. In the figure, a local subset of spatial size $n \times n$ is selected from layer $\mathbf{L}2$. The sample covers all orientations θ and a range of scales $\Delta\sigma$. To make the filter more selective, one possibility is to keep only the strongest coefficient in scales and orientations at each position, setting all other coefficients to zero. This sampling is repeated for a total of M filters. In Reference [58], it is shown that for a system invariant to various geometrical transformations, such as the HMAX, this type of learning requires fewer samples in order to display classification properties.

Layer 4 (IT-like cells). To gain global invariance, and to represent patterns of neural activation in higher visual areas such as IT, the final representation is activated by pooling the maximum output of $\mathbf{L}3^m_\sigma$ across positions and scales. Pooling over the entire visual space, spatially and over all scales, guarantees invariance to position and size. However, important spatial and scale information for recognition might be lost by such global pooling.

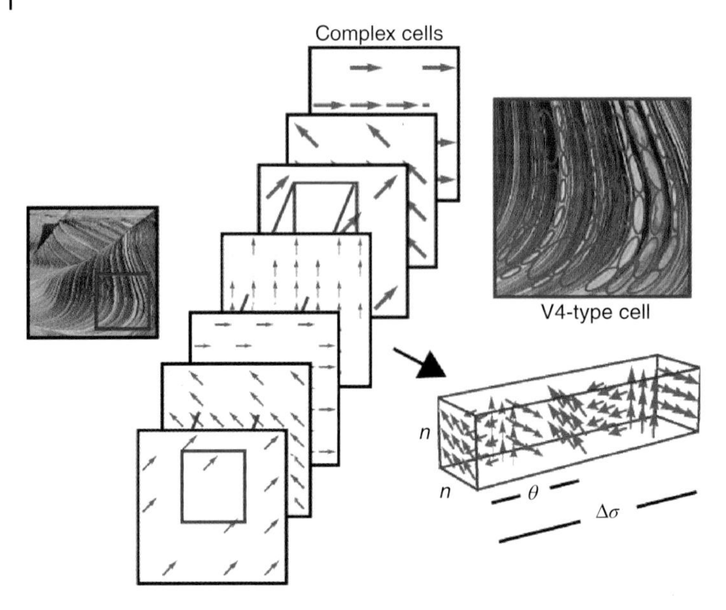

Figure 13.11 V4-type cells. On layer 3, the cells represent configurations of complex cells, of various orientations and scales, sampled during training. The ellipses represent the receptive fields of simple cells on the first layer.

In Reference [11], to maintain some spatial and scale information in the final representation, a set of concentric pooling regions is established around the spatial position and the scale at which each filter was sampled during training. As shown in Figure 13.12, each pooling region is defined by a radius R_i centered on the training position (x_m, y_m) and covers scales in the range $\sigma^m \pm 1$.

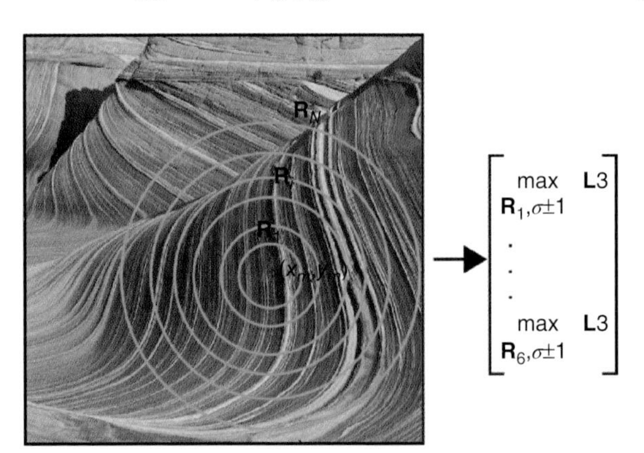

Figure 13.12 Multiresolution pooling. The training spatial position and scale of each filter is stored during training. For each new image, concentric pooling regions are centered on this coordinate. The maximum value is pooled from each pooling radius and across scales $\sigma^m \pm 1$. This ensures that some spatial and scale information are kept in the final representation.

This pooling procedure is applied for all M filters and the results are concatenated into a final **L4** vector representation (Eq. 13.8). Each element of the **L4** vector represents the maximum activation level of each filter inside each search region. One hypothesis discussed in Ref. [1] is that local selectivity, combined with invariant pooling, may be a general principle by which the cortex performs visual classification. A combination of selectivity and invariance (i.e., relaxation of spatial configurations by maximum pooling), gained locally on each layer of the network, progressively maps the inputs to a representation space (i.e., a vector space), where the visual classes are more easily separable – where the class manifolds are *untangled*. As such, the final vector representation of Eq. 13.8 generates a space inside which classification of complex visual inputs can be performed relatively well with a vector classifier [11].

$$\mathbf{L4} = \begin{bmatrix} \max\limits_{\mathbf{R}_1,\, \sigma^1 \pm 1} \mathbf{L3}^1 \\ \vdots \\ \max\limits_{\mathbf{R}_1,\, \sigma^M \pm 1} \mathbf{L3}^M \\ \max\limits_{\mathbf{R}_6,\, \sigma^1 \pm 1} \mathbf{L3}^1 \\ \vdots \\ \max\limits_{\mathbf{R}_6,\, \sigma^M \pm 1} \mathbf{L3}^M \end{bmatrix} \tag{13.8}$$

The operations of layer 1 of the basic HMAX network presented above is a highly simplified, and not entirely faithful, model of the hypercolumns' organization in the primary visual cortex. For one, the actual hypercolumns' structure is not explicitly represented, it is only implied by its decomposition into translation vector fields at multiple orientations. Also, there are no long-range horizontal connections between the hypercolumns corresponding to known neurophysiological data. The next section presents an idealized mathematical model of V1 which represents the hypercolumns explicitly. This model gives the hypercolumns of V1 a one-to-one correspondence with a mathematical structure and gives a formal expression of its horizontal connections.

13.7 Idealized Mathematical Model of V1: Fiber Bundle

There exists a mathematical model of visual area V1 which gives a theoretical formulation of the way the local operations of simple cells defined by Eq. 13.2 can merge into global percepts, and more precisely, into visual contours. As seen in the previous section, the HMAX is founded on a simplified model of the hypercolumns of V1. However, neurophysiological studies have clearly identified the important role, in the perception of shapes, of horizontal long-range connections between the hypercolumns [65]. When taken individually, the hypercolumns define the local orientations in the visual field. But how can these

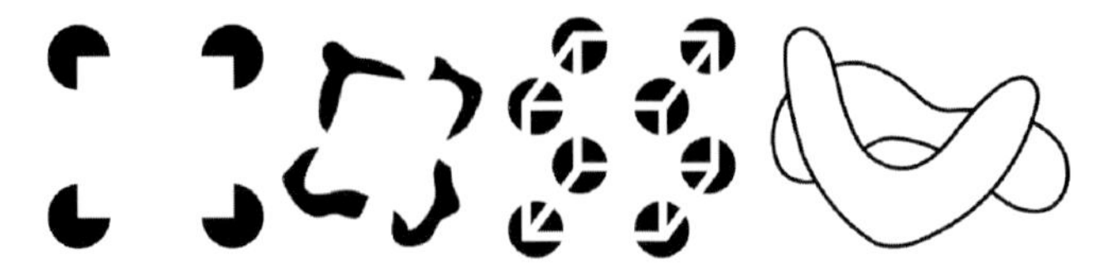

Figure 13.13 Contour completion. Understanding the principles by which the brain is able to spontaneously generate or complete visual contours sheds light on the importance of top-down and lateral processes in shape recognition (image reproduced from Ref. [66]).

local orientations dynamically interact such that a global percept emerges? The language of differential geometry provides a natural answer to this question. Indeed, there is a mathematical structure which, at a certain level, gives an abstraction of the physical structure of the primary visual cortex. It also provides, in the spirit of Gestalt psychology, a top-down definition of the way global shapes are generated by the visual system (Figure 13.13).

To understand the abstract model of V1 described below, one can first note the similarity between the retinotopic organization of overlapping receptive fields on the retinal plane and the mathematical structure of a *differentiable manifold* – a smooth structure which locally resembles the Euclidean plane $\mathbb{R}^2$. One can also note that the repetition of hypercolumns in V1 provide a copy of the space of orientations over each position of the retinal plane. As introduced in the seminal work of Hoffman [12, 67, 68] and illustrated in Figure 13.14, this structure is the physical realization of a *fiber bundle.*

A fiber bundle $E = M \times F$ is a manifold M on which is attached, at every position, the entire copy of another manifold called the fiber F. The analogy with the visual cortex is direct – M is the retinal plane and the fibers F are the hypercolumns of simple cells. If simple cells are directional derivatives (Eq. 13.3), then all orientations $[0, 2\pi]$ are distinguishable, and V1 is abstracted as $\mathbb{R}^2 \times \mathbb{S}^1$ [15, 16, 40, 69] where $\mathbb{R}^2$ corresponds to the retinal plane, and where the unit circle $\mathbb{S}^1$ corresponds to the group of rotations in the plane. This fiber bundle, also called the *unit tangent bundle,* is isomorphic to the *Euclidean motion group SE*(2), also called the *Roto-translation group.* If the simple cells are second-order derivatives (i.e., even-phase Gabors), then only the angles $[0, \pi]$ are distinguishable, and V1 is represented by $\mathbb{R} \times \mathbb{P}^1$, where $\mathbb{P}^1$ is the space of all lines through the origin [70].

As illustrated in Figure 13.14, when a visual contour, expressed as a regular parameterized curve $\gamma(t) = [x(t), y(t)]$, makes *contact* with the simple cells, it is lifted in the hypercolumns to a curve $\gamma^*(t) = [x(t), y(t), \theta(t)]$. In the new space $\mathbb{R}^2 \times \mathbb{S}^1$, the orientations θ are explicitly represented. Because of this, the problem of contour representation and contour completion in V1 is different than if it is expressed directly on the retinal image. Indeed, the type of curves in $\mathbb{R}^2 \times \mathbb{S}^1$ that can represent visual contours is very restricted. Every curve on the retinal plane corresponds to a lifted curve in $\mathbb{R}^2 \times \mathbb{S}^1$, but the converse is not true – only special curves in $\mathbb{R}^2 \times \mathbb{S}^1$ can be curves in $\mathbb{R}^2$. For instance, the reader can verify that a straight line in $\mathbb{R}^2 \times \mathbb{S}^1$ is not a curve in $\mathbb{R}^2$. The *Frobenius integrability*

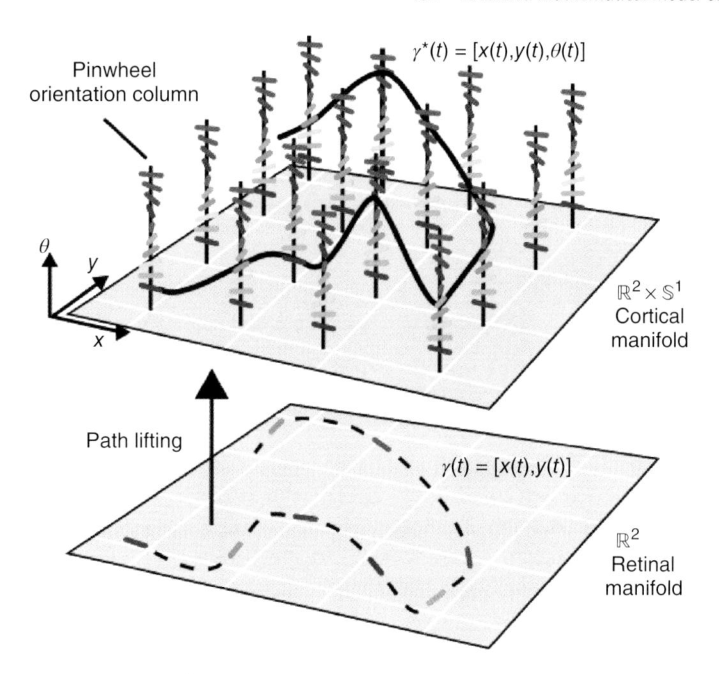

Figure 13.14 A fiber bundle abstraction of V1. Contour elements (dotted line) on the retinal plan ($\mathbb{R}^2$) are lifted in the cortical fiber bundle ($\mathbb{R}^2 \times \mathbb{S}^1$) by making contact with simple cells (figure inspired from Refs [16, 40]).

theorem says that for a curve in $\mathbb{R}^2 \times \mathbb{S}^1$ to be a curve in $\mathbb{R}^2$ we must have

$$\tan(\theta) = \frac{\dot{y(t)}}{\dot{x(t)}} = \mathrm{dy}/\mathrm{dx} \tag{13.9}$$

It follows that it is admissible for a curve in V1 to represent a visual contour only if condition 13.9 is satisfied. Admissible curves in $\mathbb{R}^2 \times \mathbb{S}^1$, illustrated in Figure 13.15, can be defined as the family of integral curves of $X_1 + kX_2$, where

$$X_1 = (\cos\theta, \sin\theta, 0), \quad X_2 = (0, 0, 1) \tag{13.10}$$

are vector fields generating a space in $\mathbb{R}^2 \times \mathbb{S}^1$ which is orthogonal to the directional derivative expressed by simple cells (Eq. 13.2), and where k gives the curvature of the curve projected on the retinal plane [15, 40].

In References [15, 40], individual points at each hypercolumn are connected into global curves through the fan of integral curves, which together give a model of the lateral connectivity in V1, described in Section 13.8. This pattern of connectivity over the columns of V1 is illustrated in Figure 13.16.

In References [16, 69], following earlier works [71–73], contour completion of the type shown in Figure 13.13 is defined by selecting, among the admissible curves of Eq. 13.9, the shortest path (i.e., geodesic) in $\mathbb{R}^2 \times \mathbb{S}^1$ between two visible boundary points. This gives a top-down formalism for perceptual phenomena such as contour completion and saliency (pop-out). As suggested by Hoffman

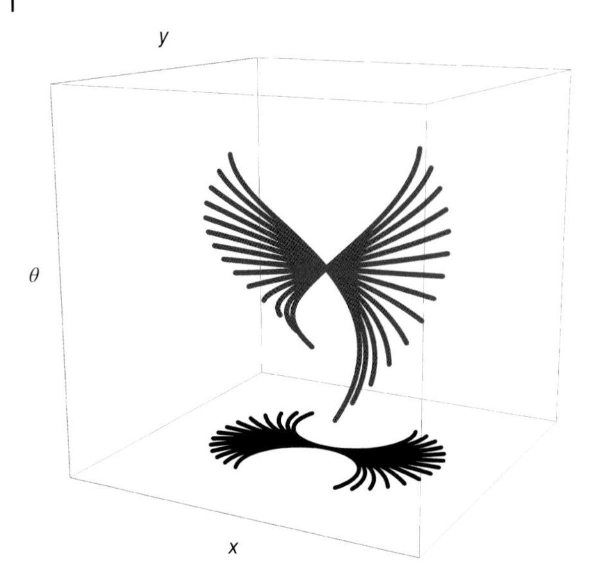

Figure 13.15 Family of integral curves in V1. The figure shows admissible V1 curves projected (in blue) on the retinal plane over one hypercolumn in $\mathbb{R}^2 \times \mathbb{S}^1$ (image adapted from Ref. [40]).

Figure 13.16 Horizontal connections. The fan of integral curves of the vector fields defined by Eq. (13.10) gives a connection between individual hypercolumns. It connects local tangent vectors at each hypercolumn into global curves (image adapted from Ref. [40]).

[12, 68], contour completion and saliency is more the exception than the rule in the visual world. When the premises of the Frobenius integrability theorem are not satisfied (when orientations of simple cells do not line up), texture is perceived instead of contours. In other words, contour formation is not possible in V1 when the above integrability condition is not satisfied – in this frequent situation, it is

not a contour that is perceived, but a texture. Using these principles, Hoffman defines visual contours of arbitrary complexity as the integral curves of an algebra of vector fields (i.e., Lie algebra) satisfying the above integrability conditions.

Of particular interest is the fact that the integral curves defined above are shown to correspond with long-range horizontal connections on the visual cortex. These connections are known as the *association field* [65] of perceptual contours and are described in the next section.

13.8
Horizontal Connections and the Association Field

Neurophysiological studies have shown that the outputs of simple cells (i.e., the shape of the filters in Figure 13.5) are modulated by contextual spatial surrounding – the global visual field modulates the local responses through long-range horizontal feedback in the primary visual cortex [74, 75]. In particular, in Ref. [65], psychophysical results suggest that long-range connections exist between simple cells oriented along particular paths, as illustrated in Figure 13.17. These connections may define what is referred to as an *association field* – grouping (associating) visual elements according to alignment constraints on position and orientation. By selecting configurations of simple cells, the V4-type cells modeled in Section 13.6 share some relations with the association field. However, these configurations are not explicitly defined by principles of grouping and alignment. As shown in Refs [14, 15, 40], the association field can be put in direct correspondence with the integral curves (see Figure 13.15) of the cortical fields expressed by simple cells and may well be the biological substrate for perceptual contours and phenomena (i.e., Gestalt) such as illusory contours, surface fill-in, and contour saliency.

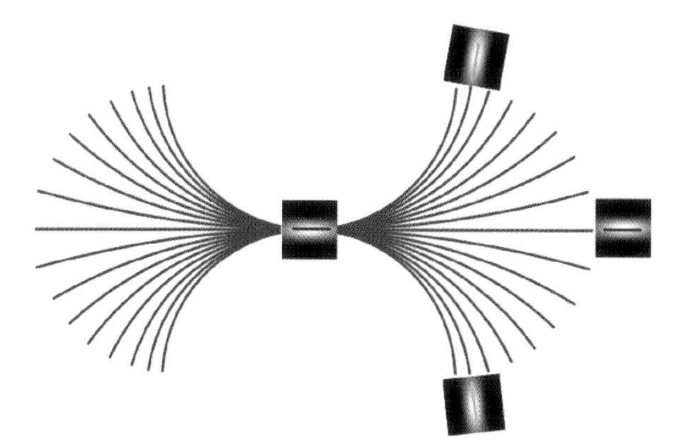

Figure 13.17 Association field. Long-range horizontal connections are the basis for the association field in Ref. [65]. The center of the figure represents a simple cell in V1. The curve displayed represents the visual path defined by the horizontal connections to other simple cells. The possible contours are shown to correspond to the family of integral curves as defined in Section 13.7, where the center cell gives the initial condition.

13.9
Feedback and Attentional Mechanisms

Studies also show that attentional mechanisms, regulated through cortical feedback, modulate neural processing in regions as early as V1. The spatial pattern of modulation in Ref. [76] reveals an attentional process which is consistent with an object-based attention window but is inconsistent with a simple circumscribed attentional processes. This study suggests that neural processing in visual area V1 is not only driven by inputs from the retina but is often influenced by attentional process mediated by top-down connections. Although rapid visual recognition has sometimes been modeled by purely feedforward networks (Section 13.6), studies and models [3] suggest that attentional mechanisms, mediated by top-down feedback, are essential for a full understanding of visual recognition mechanisms in the cortex. The question remains an open discussion, but it can be argued that there is indeed enough time for recurrent activation (feedback) to occur during rapid visual recognition ($\sim$150 ms) [29]. For a description of bottom-up attentional visual models see Chapter 11.

13.10
Temporal Considerations, Transformations and Invariance

Object recognition may appear as a static process which involves the identification of visual inputs at one given instant. However, recognizing a visual object is everything but a static process and is fundamentally involved with transformations. The retinal image is never quite stable and visual perception vanishes in less than 80 ms in the absence of motion [77]. Ecologically speaking, invariance to transformations generated by motion is essential for visual recognition to occur, and transformations of the retinal image are present at every instant from birth. Self-produced movements, such as head and eyes movements, are correlated with transformations of the retinal image and must have profound effect on the neural coding of the visual field [78].

The involvement of motion in the perception of shapes is displayed clearly by phenomena such as *structure from motion* [79] and the segmentation of shapes from motion [20]. The connectivity of the visual cortex also suggests a role for motion in object recognition. The ventral pathway, traditionally associated with object recognition, and the dorsal pathway, often associated with motion perception and spatial localization, are known to be significantly interacting [26, 27].

In neural network modeling of visual recognition, motion and the temporal dimension of the neural response to an object undergoing transformations has been extensively modeled [23, 53, 54, 80, 81]. One unsupervised learning principle, common to all of these models, is to remove temporal variations in the neural response in the presence of transforming objects, as done by Mitchison [80] with a *differential synapse*. Another unsupervised learning rule based on the same principle is known as the *trace rule* [55]. This is a *Hebbian* learning rule which keeps

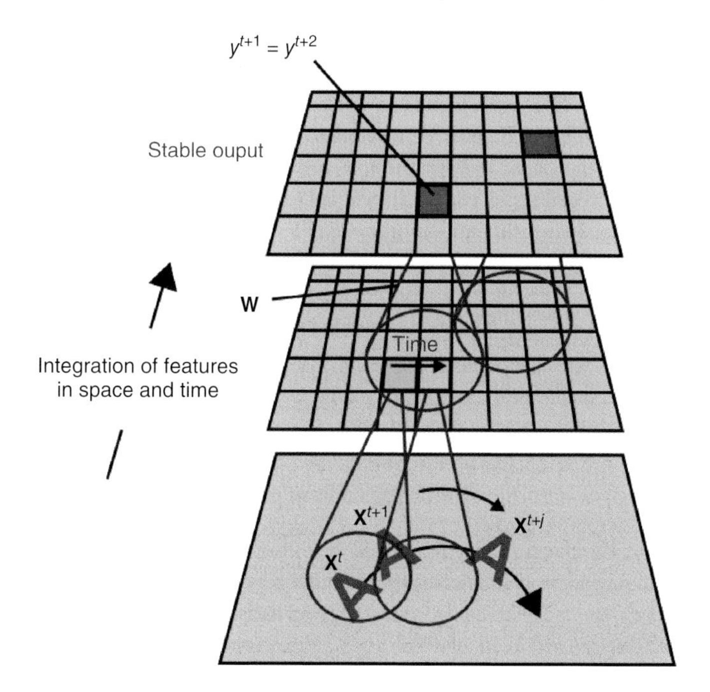

Figure 13.18 Temporal association. By keeping a temporal trace of recent activation, the trace learning rule enables neurons to correlate their activation with a transformation sequence passing through their receptive fields. This generates a neural response which is stable or invariant to transformation of objects inside their receptive fields.

a temporal trace of the neuron's output in response to a moving object passing inside its receptive field (see Figure 13.18). The inclusion of an activation trace in the learning rule drives neurons to respond invariantly to temporally close inputs (i.e., positions of a moving object). Inside an architecture with overlapping receptive fields, this allows a smooth mapping of transformations. This learning rule has also been used for the last layer of the HMAX model to gain translation invariance [81]. Interestingly, the trace learning rule can be mathematically related [82] to temporal difference (TD) learning [83], which minimizes TDs in response to successive inputs. For instance, one version of the trace learning rule modifies the connection weight vector w of output neuron $y^t = f(wx^t)$ at time t in response to input x^t (Figure 13.18) such that

$$\Delta w = \alpha(\beta y^{t-1} - y^t)x^t \tag{13.11}$$

where y^{t-1} is an exponential trace of the neuron's past activation.

A similar principle is also the basis of *slow feature analysis* [53] in which neurons learn to output a signal at a lower frequency than the incoming retinal signal, producing a stable response to transforming objects. The slow feature analysis has been integrated in the HMAX architecture in the context of recognition of dynamic scenes [84, 85]. The authors applied the slow feature analysis to the output of simple cells responding to image sequences in videos. This generates more stable representations of motion in comparison with the raw simple cell outputs. All these unsupervised learning rules, from the point of view of visual recognition, implicitly suggest that, for obvious ecological reasons, the identity of objects should not change because of relative motion between the observer and the objects, and that the brain uses spatiotemporal correlations in the signal to maintain a coherent representation.

13.11 Conclusion

The average number of synapses per neuron in the brain is close to 10,000, with a total estimate of 86 billion neurons in the human brain [86]. In particular, the primary visual cortex averages 140 millions neurons [87]. This roughly amounts to 1.4 trillion synaptic connections in and out of V1 alone. Interestingly, there is a large connection ratio between V1 and the retina. In particular, with an estimate of one million retinal ganglion cells [88], there is a connectivity ratio well over one hundred V1 neurons for each retinal ganglion cell.

With such massive connectivity, it is clear that the representation, the sampling, and the filtering (i.e., averaging) capacity of the visual cortex is far beyond neural network models which may be simulated on today's basic computers. The large number of V1 neurons allocated on each retinal fiber suggests a significant amount of *image processing*. For instance, the simple calculation of derivatives by the local operators defined in Eq. 13.2 is plagued with discretization noise when applied directly on pixelized images, unless very large images are used with a large Gaussian envelope defining the filters. This makes theoretical principles, such as differentiation, difficult to express in practice. It is tempting to suggest that visual cortex evolved with such an astronomical number of connections to deal with noise and irregularities of the visual world and to create a nearly smooth signal from the discretized retinal receptors. A massive number of horizontal connections between V1 cells, as presented in Sections 13.7 and 13.8, seem to be part of nature's solution to clean, segment, and amplify coherent visual structures in an otherwise cluttered environment.

Models such as the HMAX illustrate the basic functioning of simple and complex cells in V1. Without extensive horizontal and feedback connections, this model does not represent the full capacity of the visual cortical pathway. Nevertheless, it seems quite certain by now that one axiom of processing in V1 is based on local operations such as the ones described in Sections 13.4 and 13.6. A code for the model presented in Section 13.6 is available at *http://www-poleia.lip6.fr/cord/Projects/Projects.html* *http://www-poleia.lip6.fr/cord/Projects/Projects.html*. The next step in such a model could be to integrate feedback and horizontal connections, as defined in a more theoretical context in Section 13.7. It remains a challenge to implement such formal theoretical models on full-scale natural scenes. Research would most likely progress in the right direction by generating effort in bridging the gap between what is observed in the brain and models of visual recognition.

References

1. DiCarlo, J.J., Zoccolan, D., and Rust, N.C. (2012) How does the brain solve visual object recognition? *Neuron*, **73**, 415–434.
2. Serre, T., Kouh, M., Cadieu, C., Knoblich, U., Kreiman, G., and Poggio, T. (2005) A Theory of Object Recognition: Computations and Circuits in the Feedforward Path of the Ventral Stream in Primate Visual Cortex. AI Memo 2005-036/CBCL Memo 259. Massachusetts Institute of Technology.
3. Lee, T.S. and Mumford, D. (2003) Hierarchical Bayesian inference in the visual cortex. *J. Opt. Soc. Am. A Opt. Image Sci. Vis.*, **20**, 1434–1448.
4. Borji, A. and Itti, L. (2013) State-of-the-art in visual attention modeling. *IEEE Trans. Pattern Anal. Mach. Intell.*, **35** (1), 185–207.
5. Itti, L. and Koch, C. (2001) Computational modelling of visual attention. *Nat. Rev. Neurosci.*, **2** (3), 194–203.
6. Siagian, C. and Itti, L. (2007) Rapid biologically-inspired scene classification using features shared with visual attention. *IEEE Trans. Pattern Anal. Mach. Intell.*, **29** (2), 300–312.
7. Zhang, Y., Meyers, E.M., Bichot, N.P., Serre, T., Poggio, T.A., and Desimone, R. (2011) Object decoding with attention in inferior temporal cortex. *Proc. Natl. Acad. Sci. U.S.A.*, **108** (21), 8850–8855.
8. Chikkerur, S., Serre, T., Tan, C., and Poggio, T. (2010) What and where: a Bayesian inference theory of attention. *Vision Res.*, **50**, 2233–2247.
9. Serre, T., Wolf, L., Bileschi, S., Riesenhuber, M., and Poggio, T. (2007) Robust object recognition with cortex-like mechanisms. *IEEE Trans. Pattern Anal. Mach. Intell.*, **29**, 411–426.
10. Mutch, J. and Lowe, D.G. (2008) Object class recognition and localization using sparse features with limited receptive fields. *Int. J. Comput. Vision*, **80**, 45–57.
11. Theriault, C., Thome, N., and Cord, M. (2013) Extended coding and pooling in the HMAX model. *IEEE Trans. Image Process.*, **22** (2), 764–777.
12. Hoffman, W. (1989) The visual cortex is a contact bundle. *Appl. Math. Comput.*, **32**, 137–167.
13. Koenderink, J. (1988) Operational significance of receptive field assemblies. *Biol. Cybern.*, **58**, 163–171.
14. Petitot, J. (2003) The neurogeometry of pinwheels as a sub-Riemannian contact structure. *J. Physiol. Paris*, **97**, 265–309.
15. Citti, G. and Sarti, A. (2006) A cortical based model of perceptual completion in the roto-translation space. *J. Math. Imaging Vision*, **24** (3), 307–326.
16. Ben-Yosef, G. and Ben-Shahar, O. (2012) A tangent bundle theory for visual curve completion. *IEEE Trans. Pattern Anal. Mach. Intell.*, **34** (7), 1263–1280.
17. Felleman, D.J. and Essen, D.C.V. (1991) Distributed hierarchical processing in the primate cerebral cortex. *Cereb. Cortex*, **1** (1), 1–47.
18. Priebe, N.J., Lisberger, S.G., and Movshon, J.A. (2006) Tuning for spatiotemporal frequency and speed in directionally selective neurons of macaque striate cortex. *J. Neurosci.*, **26** (11), 2941–2950.
19. Lennie, P. and Movshon, J. (2005) Coding of color and form in the geniculostriate visual pathway (invited review). *J. Opt. Soc. Am. A Opt. Image Sci. Vis.*, **22**, 2013–2033.
20. Born, R.T. and Bradley, D.C. (2005) Structure and function of visual area MT. *Annu. Rev. Neurosci.*, **28** (1), 157–189.
21. Christopher, C.P., Richard, T.B., and Margaret, S.L. (2003) Two-dimensional substructure of stereo and motion interactions in macaque visual cortex. *Neuron*, **37**, 525–535.
22. Pack, C.C. and Born, R.T. (2001) Temporal dynamics of a neural solution to the aperture problem in visual area MT of macaque brain. *Nature*, **409** (6823), 1040–1042.

23. Rolls, E. and Deco, G. (2002) *Computational Neuroscience of Vision*, 1st edn, Oxford University Press.
24. Cloutman, L.L. (2013) Interaction between dorsal and ventral processing streams: where, when and how? *Brain Lang.*, **127** (2), 251–263.
25. Zanon, M., Busan, P., Monti, F., Pizzolato, G., and Battaglini, P. (2010) Cortical connections between dorsal and ventral visual streams in humans: evidence by TMS/EEG co-registration. *Brain Topogr.*, **22** (4), 307–317.
26. Schenk, T. and McIntosh, R. (2010) Do we have independent visual streams for perception and action? *Cogn. Neurosci.*, **1** (1), 52–62.
27. McIntosh, R. and Schenk, T. (2009) Two visual streams for perception and action: current trends. *Neuropsychologia*, **47** (6), 1391–1396.
28. Thorpe, S., Fize, D., and Marlot, C. (1996) Speed of processing in the human visual system. *Nature*, **381**, 520–522.
29. Johnson, J.S. and Olshausen, B.A. (2003) Timecourse of neural signatures of object recognition. *J. Vis.*, **3** (7), 499–512.
30. Hubel, D. (1962) The visual cortex of the brain. *Sci. Am.*, **209** (5), 54–62.
31. Adams, D. and Horton, J. (2003) A precise retinotopic map of primate striate cortex generated from the representation of angioscotomas. *J. Neurosci.*, **23** (9), 3371–3389.
32. Roe, A.W., Chelazzi, L., Connor, C.E., Conway, B.R., Fujita, I., Gallant, J.L., Lu, H., and Vanduffel, W. (2012) Toward a unified theory of visual area V4. *Neuron*, **14** (1), 12–29.
33. David, S.V., Hayden, B.Y., and Gallant, J.L. (2006) Spectral receptive field properties explain shape selectivity in area V4. *J. Neurophysiol.*, **96** (6), 3492–3505.
34. Rolls, E.T., Aggelopoulos, N.C., and Zheng, F. (2003) The receptive fields of inferior temporal cortex neurons in natural scenes. *J. Neurosci.*, **23**, 339–348.
35. Hubel, D. and Wiesel, T. (1959) Receptive fields of single neurones in the cat's striate cortex. *J. Physiol.*, **148**, 574–591.
36. De Valois, R.L. and De Valois, K.K. (1988) *Spatial Vision*, *Oxford Psychology Series*, Oxford University Press, (1990 [printing]), New York, Oxford.
37. Maffei, L., Hertz, B.G., and Fiorentini, A. (1973) The visual cortex as a spatial frequency analyser. *Vision Res.*, **13**, 1255–1267.
38. Campbell, F.W., Cooper, G.F., and Cugell, E.C. (1969) The spatial selectivity of the visual cells of the cat. *J. Physiol. (London)*, **203**, 223–235.
39. Bosking, W., Zhang, Y., Schofield, B., and Fitzpatrick, D. (1997) Orientation selectivity and the arrangement of horizontal connections in tree shrew striate cortex. *J. Neurosci.*, **17**, 2112–2127.
40. Sanguinetti, G. (2011) Invariant models of vision between phenomenology, image statistics and neurosciences. PhD thesis. Universidad de la República, Facultad de Ingeniería, Uruguay.
41. Koenderink, J. and van Doorn, A. (1987) Representation of local geometry in the visual system. *Biol. Cybern.*, **55**, 367–375.
42. Young, R.A. (1987) The Gaussian derivative model for spatial vision: I. Retinal mechanisms. *J. Physiol.*, **2**, 273–293.
43. Gabor, D. (1946) Theory of communication. *J. Inst. Electr. Eng.*, **93**, 429–457.
44. Daugman, J.G. (1980) Two-dimensional spectral analysis of cortical receptive field profiles. *Vision Res.*, **20** (10), 847–856.
45. Daugman, J.G. (1985) Uncertainty relation for resolution in space, spatial frequency, and orientation optimized by two-dimensional visual cortical filters. *J. Opt. Soc. Am. A*, **2** (7), 1160–1169.
46. Simoncelli, E. and Heeger, D. (1998) A model of neural responses in visual area MT. *Vision Res.*, **38**, 743–761.
47. Lindeberg, T. (1998) Feature detection with automatic scale selection. *Int. J. Comput. Vision*, **30**, 77–116.
48. Masquelier, T. and Thorpe, S.J. (2007) Unsupervised learning of visual features through spike timing dependent plasticity. *PLoS Comput. Biol.*, **3** (2), e31.
49. Lampl, I., Ferster, D., Poggio, T., Riesenhuber, M., Ferster, D., and Poggio, T. (2004) Intracellular measurements of spatial integration and the MAX operation in complex cells of the cat

primary visual cortex. *J. Neurophysiol.*, **92**, 2704–2713.

50. Fukushima, K. and Miyake, S. (1982) Neocognitron: a new algorithm for pattern recognition tolerant of deformations and shifts in position. *Pattern Recognit.*, **15** (6), 455–469.
51. Fukushima, K. (2003) Neocognitron for handwritten digit recognition. *Neurocomputing*, **51**, 161–180.
52. Lecun, Y., Bottou, L., Bengio, Y., and Haffner, P. (1998) Gradient-based learning applied to document recognition. *Proceedings of the IEEE*, pp. 2278–2324.
53. Wiskott, L. and Sejnowski, T. (2002) Slow feature analysis: unsupervised learning of invariances. *Neural Comput.*, **14**, 715–770.
54. Rolls, E.T. and Milward, T.T. (2000) A model of invariant object recognition in the visual system: learning rules, activation functions, lateral inhibition, and information-based performance measures. *Neural Comput.*, **12**, 2547–2572.
55. Foldiak, P. (1991) Learning invariance from transformation sequences. *Neural Comput.*, **3** (2), 194–200.
56. Vincent, P., Larochelle, H., Lajoie, I., Bengio, Y., and Manzagol, P.A. (2010) Stacked denoising autoencoders: learning useful representations in a deep network with a local denoising criterion. *J. Mach. Learn. Res.*, **11**, 3371–3408.
57. Goh, H., Thome, N., Cord, M., and Lim, J.H. (2012) Unsupervised and supervised visual codes with restricted Boltzmann machines. *European Conference on Computer Vision (ECCV 2012).*
58. Fabio, A., Joel, Z.L., Lorenzo, R., Jim, M., Andrea, T., and Poggio, T. (2014) Unsupervised Learning of Invariant Representations with Low Sample Complexity: The Magic of Sensory Cortex or a New Framework for Machine Learning? Technical Memo CBMM Memo No. 001, Center for Brains, Mind and Machines, MIT, Cambridge, MA.
59. Riesenhuber, M. and Poggio, T. (1999) Hierarchical models of object recognition in cortex. *Nat. Neurosci.*, **2**, 1019–1025.
60. Kohonen, T. (2001) *Self-Organizing Maps, Springer Series in Information Sciences*, vol. 30, 3rd edn, Springer-Verlag, Berlin.
61. Jarrett, K., Kavukcuoglu, K., Ranzato, M., and LeCun, Y. (2009) What is the best multi-stage architecture for object recognition? ICCV, IEEE, pp. 2146–2153.
62. Hu, X., Zhang, J., Li, J., and Zhang, B. (2014) Sparsity-regularized HMAX for visual recognition. *PLoS ONE*, **9** (1), e81 813.
63. Pinto, N., Cox, D.D., and DiCarlo, J.J. (2008) Why is real-world visual object recognition hard? *PLoS Comput. Biol.*, **4** (1), e27.
64. Lyu, S. and Simoncelli, E.P. (2008) Nonlinear image representation using divisive normalization. CVPR, IEEE Computer Society.
65. Field, D.J., Hayes, A., and Hess, R.F. (1993) Contour integration by the human visual system: evidence for a local association field. *Vision Res.*, **33**, 173–193.
66. Thériault, C. (2006) Mémoire visuelle et géométrie différentielle : forces invariantes des champs de vecteurs élémentaires. PhD thesis. Université du Québec à Montréal.
67. Hoffman, W. (1966) The Lie algebra of visual perception. *Math. Psychol.*, **3**, 65–98.
68. Hoffman, W. (1970) Higher visual perception as prolongation of the basic Lie transformation group. *Math. Biosci.*, **6**, 437–471.
69. Ben-Yosef, G. and Ben-Shahar, O. (2012) Tangent bundle curve completion with locally connected parallel networks. *Neural Comput.*, **24** (12), 3277–3316.
70. Petitot, J. and Tondut, Y. (1999) Vers une neurogéométrie. fibrations corticales, structures de contact et contours subjectifs modaux. *Math. Sci. Hum.*, **145**, 5–101.
71. Ullman, S. (1976) Filling-in the gaps: the shape of subjective contours and a model for their generation. *Biol. Cybern.*, **25**, 1–6.
72. Horn, B.K.P. (1983) The curve of least energy. *ACM Trans. Math. Softw.*, **9** (4), 441–460.

73. Mumford, D. (1994) Elastica and computer vision, in *Algebraic Geometry and its Applications* (ed. C.L. Bajaj), Springer-Verlag, New York.
74. Gilbert, C.D., Das, A., and Westheimer, G. (1996) Spatial integration and cortical dynamics. *Proc. Natl. Acad. Sci. U.S.A.*, **93**, 615–622.
75. Bosking, W., Zhang, Y., Schofield, B., and Fitzpatrick, D. (1997) Orientation selectivity and the arrangement of horizontal connections in tree shrew striate cortex. *J. Neurosci.*, **17** (6), 2112–2127.
76. Somers, D.C., Dale, A.M., and Tootell, R.B.H. (1999) Functional MRI reveals spatially specific attentional modulation in human primary visual cortex. *Proc. Natl. Acad. Sci. U.S.A.*, **96**, 1663–1668.
77. Coppola, D. and Purves, D. (1996) The extraordinarily rapid disappearance of entopic images. *Proc. Natl. Acad. Sci. U.S.A.*, **93**, 8001–8004.
78. Dodwell, P. (1983) The Lie transformation group of visual perception. *Percept. Psychophys.*, **34**, 1–16.
79. Grunewald, A., Bradley, D.C., and Andersen, R.A. (2002) Neural correlates of structure-from-motion perception in macaque V1 and MT. *J. Neurosci.*, **22**, 6195–6207.
80. Mitchison, G. (1991) Removing time variation with the anti-Hebbian differential synapse. *Neural Comput.*, **3** (3), 312–320.
81. Isik, L., Leibo, J., and Poggio, T. (2012) Learning and disrupting invariance in visual recognition with a temporal association rule. *Front. Comput. Neurosci.*, **6** (7), 37.
82. Rolls, E.T. and Stringer, S.M. (2001) Invariant object recognition in the visual system with error correction and temporal difference learning. *Network*, **12**, 111–130.
83. Sutton, R.S. (1988) Learning to predict by the methods of temporal differences, in *Machine Learning*, Kluwer Academic Publishers, pp. 9–44.
84. Theriault, C., Thome, N., Cord, M., and Perez, P. (2014) Perceptual principles for video classification with slow feature analysis. *J. Sel. Top. Signal Process.*, **8** (3), 428–437.
85. Theriault, C., Thome, N., and Cord, M. (2013) Dynamic scene classification: learning motion descriptors with slow features analysis. IEEE CVPR.
86. Azevedo, F.A.C., Carvalho, L.R.B., Grinberg, L.T., Farfel, J.M., Ferretti, R.E.L., Leite, R.E.P., Filho, W.J., Lent, R., and Herculano-Houzel, S. (2009) Equal numbers of neuronal and nonneuronal cells make the human brain an isometrically scaled-up primate brain. *J. Comp. Neurol.*, **513** (5), 532–541.
87. Leuba, G. and Kraftsik, R. (1994) Changes in volume, surface estimate, three-dimensional shape and total number of neurons of the human primary visual cortex from midgestation until old age. *Anat. Embryol.*, **190** (4), 351–366.
88. Curcio, C.A. and Allen, K.A. (1990) Topography of ganglion cells in human retina. *J. Comp. Neurol.*, **300** (1), 5–25.

14
Sparse Models for Computer Vision

Laurent U. Perrinet

14.1
Motivation

14.1.1
Efficiency and Sparseness in Biological Representations of Natural Images

The central nervous system is a dynamical, adaptive organ which constantly evolves to provide optimal decisions[1] for interacting with the environment. Early visual pathways provide with a powerful system for probing and modeling these mechanisms. For instance, the primary visual cortex of primates (V1) is absolutely central for most visual tasks. There, it is observed that some neurons from the input layer of V1 present selectivity for localized, edge-like features—as represented by their "receptive fields" [2]. Crucially, there is experimental evidence for *sparse* firing in the neocortex [3, 4] and in particular in V1. A representation is sparse when each input signal is associated with a relatively small subset of simultaneously activated neurons within a whole population. For instance, orientation selectivity of simple cells is sharper than the selectivity that would be predicted by linear filtering. Such a procedure produces a rough "sketch" of the image on the surface of V1 that is believed to serve as a "blackboard" for higher-level cortical areas [5]. However, it is still largely unknown how neural computations act in V1 to represent the image. More specifically, what is the role of sparseness—as a generic neural signature—in the global function of neural computations?

A popular view is that such a population of neurons operates such that relevant sensory information from the retinothalamic pathway is transformed (or "coded") efficiently. Such efficient representation will allow decisions to be taken optimally in higher-level areas. In this framework, optimality is defined in terms of information theory [6–8]. For instance, the representation produced by the neural activity

1) Decisions are defined in their broader sense of elementary choices operated in the system at associative or motor levels [1].

Biologically Inspired Computer Vision: Fundamentals and Applications, First Edition.
Edited by Gabriel Cristóbal, Laurent Perrinet, and Matthias Keil.

in V1 is sparse: it is believed that this reduces redundancies and allows to better segregate edges in the image [9, 10]. This optimization is operated given biological constraints, such as the limited bandwidth of information transfer to higher processing stages or the limited amount of metabolic resources (energy or wiring length). More generally, it allows to increase the storage capacity of associative memories before memory patterns start to interfere with each other [11]. Moreover, it is now widely accepted that this redundancy reduction is achieved in a neural population through lateral interactions. Indeed, a link between anatomical data and a functional connectivity between neighboring representations of edges has been found by Bosking *et al.* [12], although their conclusions were more recently refined to show that this process may be more complex [13]. By linking neighboring neurons representing similar features, one allows thus a more efficient representation in V1. As computer vision systems are subject to similar constraints, applying such a paradigm therefore seems a promising approach toward more biomimetic algorithms.

It is believed that such a property reflects the efficient match of the representation with the statistics of natural scenes, that is, with behaviorally relevant sensory inputs. Indeed, sparse representations are prominently observed for cortical responses to natural stimuli [14–17]. As the function of neural systems mostly emerges from unsupervised learning, it follows that these are adapted to the inputs which are behaviorally the most common and important. More generally, by being adapted to natural scenes, this shows that sparseness is a neural signature of an underlying optimization process. In fact, one goal of neural computation in low-level sensory areas such as V1 is to provide relevant predictions [18, 19]. This is crucial for living beings as they are often confronted with noise (internal to the brain or external, such as in low light conditions), ambiguities (such as inferring a three dimensional slant from a bidimensional retinal image). In addition, the system has to compensate for inevitable delays, such as the delay from light stimulation to activation in V1 which is estimated to be of 50 ms in humans. For instance, a tennis ball moving at 20 m/s at 1 m in the frontal plane elicits an input activation in V1 corresponding to around 45° of visual angle behind its physical position [20]. Thus, to be able to translate such knowledge to the computer vision community, it is crucial to better understand *why* the neural processes that produce sparse coding are efficient.

14.1.2 Sparseness Induces Neural Organization

A breakthrough in the modeling of the representation in V1 was the discovery that sparseness is sufficient to induce the emergence of receptive fields similar to V1 simple cells [21]. This reflects the fact that, at the learning timescale, coding is optimized relative to the statistics of natural scenes such that independent components of the input are represented [22, 23]. The emergence of edge-like simple cell receptive fields in the input layer of area V1 of primates may thus be considered as a coupled coding and learning optimization problem: at the coding timescale, the sparseness of the representation is optimized for any given input while at the learning timescale, synaptic weights are tuned to achieve on average an optimal representation efficiency over natural scenes. This theory has allowed to connect the different fields by providing a link among information theory models, neuromimetic models, and physiological observations.

In practice, most sparse unsupervised learning models aim at optimizing a cost defined on prior assumptions on the sparseness of the representation. These sparse learning algorithms have been applied both for images [21, 24–29] and sounds [30, 31]. Sparse coding may also be relevant to the amount of energy the brain needs to use to sustain its function. The total neural activity generated in a brain area is inversely related to the sparseness of the code, therefore the total energy consumption decreases with increasing sparseness. As a matter of fact, the probability distribution functions of neural activity observed experimentally can be approximated by so-called exponential distributions, which have the property of maximizing information transmission for a given mean level of activity [32]. To solve such constraints, some models thus directly compute a sparseness cost based on the representation's distribution. For instance, the kurtosis corresponds to the fourth statistical moment (the first three moments being in order the mean, variance, and skewness) and measures how the statistics deviates from a Gaussian: a positive kurtosis measures whether this distribution has a "heavier tail" than a Gaussian for a similar variance—and thus corresponds to a sparser distribution. On the basis of such observations, other similar statistical measures of sparseness have been derived in the neuroscience literature [15].

A more general approach is to derive a representation cost. For instance, learning is accomplished in the SparseNet algorithmic framework [22] on image patches taken from natural images as a sequence of coding and learning steps. First, sparse coding is achieved using a gradient descent over a convex cost. We will see later in this chapter how this cost is derived from a prior on the probability distribution function of the coefficients and how it favors the sparseness of the representation. At this step, the coding is performed using the current state of the "dictionary" of receptive fields. Then, knowing this sparse solution, learning is defined as slowly changing the dictionary using Hebbian learning [33]. As we will see later, the parameterization of the prior has a major impact on the results of the sparse coding and thus on the emergence of edge-like receptive fields and requires proper tuning. Yet, this class of models provides a simple solution to the problem of sparse representation in V1.

However, these models are quite abstract and assume that neural computations may estimate some rather complex measures such as gradients—a problem that may also be faced by neuromorphic systems. Efficient, realistic implementations have been proposed which show that imposing sparseness may indeed guide neural organization in neural network models, see for instance, Refs [34, 35]. In addition, it has also been shown that in a neuromorphic model, an efficient coding hypothesis links sparsity and selectivity of neural responses [36]. More generally, such neural signatures are reminiscent of the shaping of neural activity to account for contextual influences. For instance, it is observed that—depending on the context outside the receptive field of a neuron in area V1—the tuning curve may demonstrate a modulation of its orientation selectivity. This was accounted, for instance, as a way to optimize the coding efficiency of a population of neighboring neurons [37]. As such, sparseness is a relevant neural signature for a large class of neural computations implementing efficient coding.

14.1.3 Outline: Sparse Models for Computer Vision

As a consequence, sparse models provide a fruitful approach for computer vision. It should be noted that other popular approaches for taking advantage of sparse representations exist. The most popular is compressed sensing [38], for which it has been proved, assuming sparseness in the input, that it is possible to reconstruct the input from a sparse choice of linear coefficients computed from randomly drawn basis functions. Note, in addition, that some studies also focus on temporal sparseness. Indeed, by computing for a given neuron the relative numbers of active events relative to a given time window, one computes the so-called lifetime sparseness (see, for instance, Ref. [4]). We will see below that this measure may be related to population sparseness. For a review of sparse modeling approaches, we refer to Ref. [39]. Herein, we focus on the particular subset of such models on the basis of their biological relevance.

Indeed, we will rather focus on biomimetic sparse models as tools to shape future computer vision algorithms [40, 41]. In particular, we will not review models which mimic neural activity, but rather on algorithms which mimic their efficiency, bearing in mind the constraints that are linked to neural systems (no central clock, internal noise, parallel processing, metabolic cost, wiring length). For this purpose, we will complement some previous studies [26, 29, 42, 43] (for a review, see Ref. [44]) by putting these results in light of most recent theoretical and physiological findings.

This chapter is organized as follows. First, in Section 14.2 we outline how we may implement the unsupervised learning algorithm at a local scale for image patches. Then we extend in Section 14.3 such an approach to full-scale natural images by defining the *SparseLets* framework. Such formalism is then extended in Section 14.4 to include context modulation, for instance, from higher-order areas. These different algorithms (from the local scale of image patches to more global scales) are each be accompanied by a supporting implementation (with the source code) for which we show example usage and results. In particular, we highlight novel results and then draw some conclusions on the perspective of sparse models for computer vision. More specifically, we propose that bioinspired approaches may be applied to computer vision using predictive coding schemes, sparse models being one simple and efficient instance of such schemes.

14.2
What Is Sparseness? Application to Image Patches

14.2.1
Definitions of Sparseness

In low-level sensory areas, the goal of neural computations is to generate efficient intermediate *representations* as we have seen that this allows more efficient decision making. Classically, a representation is defined as the inversion of an internal generative model of the sensory world, that is, inferring the sources that generated the input signal. Formally, as in Ref. [22], we define a generative linear model (GLM) for describing natural, static, grayscale image patches $\mathbf{I}$ (represented by column vectors of dimension L pixels), by setting a "dictionary" of M images (also called "atoms" or "filters") as the $L \times M$ matrix $\Phi = \{\Phi_i\}_{1\leq i\leq M}$. Knowing the associated "sources" as a vector of coefficients $\mathbf{a} = \{a_i\}_{1\leq i\leq M}$, the image is defined using matrix notation as a sum of weighted atoms:

$$\mathbf{I} = \Phi\mathbf{a} + \mathbf{n} \tag{14.1}$$

where $\mathbf{n}$ is a Gaussian additive noise image. This noise, as in Ref. [22], is scaled to a variance of σ_n^2 to achieve decorrelation by applying principal component analysis to the raw input images, without loss of generality as this preprocessing is invertible. Generally, the dictionary Φ may be much larger than the dimension of the input space (i.e., $M \gg L$) and it is then said to be *overcomplete*. However, given an overcomplete dictionary, the inversion of the GLM leads to a combinatorial search and typically, there may exist many coding solutions $\mathbf{a}$ to Eq. (14.1) for one given input $\mathbf{I}$. The goal of efficient coding is to find, given the dictionary Φ and for any observed signal $\mathbf{I}$, the "best" representation vector, that is, as close as possible to the sources that generated the signal. Assuming that for simplicity, each individual coefficient is represented in the neural activity of a single neuron, this would justify the fact that this activity is sparse. It is therefore necessary to define an efficiency criterion in order to choose between these different solutions.

Using the GLM, we will infer the "best" coding vector as the most probable. In particular, from the physics of the synthesis of natural images, we know *a priori* that image representations are sparse: they are most likely generated by a small number of features relatively to the dimension M of the representation space. Similarly to [30], this can be formalized in the probabilistic framework defined by the GLM (see Eq. (14.1)), by assuming that knowing the prior distribution of the coefficients a_i for natural images, the representation cost of $\mathbf{a}$ for one given natural image is

$$\begin{aligned}\mathscr{C}(\mathbf{a}|\mathbf{I},\Phi) &\stackrel{\text{def}}{=} -\log P(\mathbf{a}|\mathbf{I},\Phi) = \log P(\mathbf{I}) - \log P(\mathbf{I}|\mathbf{a},\Phi) - \log P(\mathbf{a}|\Phi) \\ &= \log P(\mathbf{I}) + \frac{1}{2\sigma_n^2} \| \mathbf{I} - \Phi\mathbf{a}\|^2 - \sum_i \log P(a_i|\Phi)\end{aligned} \tag{14.2}$$

where $P(\mathbf{I})$ is the partition function which is independent of the coding (and that we thus ignore in the following) and $\| \cdot \|$ is the L_2-norm in image space. This efficiency cost is measured in bits if the logarithm is of base 2, as we will assume without loss of generality hereafter. For any representation $\mathbf{a}$, the cost value corresponds to the description length [45]: on the right-hand side of Eq. (14.2), the second term corresponds to the information from the image which is not coded by the representation (reconstruction cost) and thus to the information that can be at best encoded using entropic coding pixel by pixel (i.e., the negative log-likelihood $-\log P(\mathbf{I}|\mathbf{a}, \Phi)$ in Bayesian terminology, see Chapter 9 for Bayesian models applied to computer vision). The third term $S(\mathbf{a}|\Phi) = -\sum_i \log P(a_i|\Phi)$ is the representation or sparseness cost: it quantifies representation efficiency as the coding length of each coefficient of $\mathbf{a}$ which would be achieved by entropic coding knowing the prior and assuming that they are independent. The rightmost penalty term (see Eq. (14.2)) gives thus a definition of sparseness $S(\mathbf{a}|\Phi)$ as the sum of the log prior of coefficients.

In practice, the sparseness of coefficients for natural images is often defined by an *ad hoc* parameterization of the shape of the prior. For instance, the parameterization in Ref. [22] yields the coding cost:

$$\mathscr{C}_1(\mathbf{a}|\mathbf{I}, \Phi) = \frac{1}{2\sigma_n^2} \| \mathbf{I} - \Phi\mathbf{a}\|^2 + \beta \sum_i \log \left(1 + \frac{a_i^2}{\sigma^2}\right) \tag{14.3}$$

where β corresponds to the steepness of the prior and σ to its scaling (see Figure 14.2 from Ref. [46]). This choice is often favored because it results in a convex cost for which known numerical optimization methods such as conjugate gradient may be used. In particular, these terms may be put in parallel to regularization terms that are used in computer vision. For instance, an L2-norm penalty term corresponds to Tikhonov regularization [47] or an L1-norm term corresponds to the Lasso method. See Chapter 4 for a review of possible parameterization of this norm, for instance by using nested L_p norms. Classical implementation of sparse coding relies therefore on a parametric measure of sparseness.

Let us now derive another measure of sparseness. Indeed, a nonparametric form of sparseness cost may be defined by considering that neurons representing the vector $\mathbf{a}$ are either active or inactive. In fact, the spiking nature of neural information demonstrates that the transition from an inactive to an active state is far more significant at the coding timescale than smooth changes of the firing rate. This is, for instance, perfectly illustrated by the binary nature of the neural code in the auditory cortex of rats [16]. Binary codes also emerge as optimal neural codes for rapid signal transmission [48, 49]. This is also relevant for neuromorphic systems which transmit discrete events (such as a network packet). With a binary event-based code, the cost is only incremented when a new neuron gets active, regardless of the analog value. Stating that an active neuron carries a bounded amount of information of λ bits, an upper bound for the representation cost of neural activity on the receiver end is proportional to the count of active neurons, that is, to the ℓ_0 pseudo-norm $\| \mathbf{a}\|_0$:

$$\mathscr{C}_0(\mathbf{a}|\mathbf{I}, \Phi) = \frac{1}{2\sigma_n^2} \| \mathbf{I} - \Phi\mathbf{a}\|^2 + \lambda \| \mathbf{a}\|_0 \tag{14.4}$$

This cost is similar to information criteria such as the Akaike information criteria [50] or distortion rate [[51], p. 571]. This simple nonparametric cost has the advantage of being dynamic: the number of active cells for one given signal grows in time with the number of spikes reaching the target population. But Eq. (14.4) defines a harder cost to optimize (in comparison to Eq. (14.3) for instance) because the hard ℓ_0 pseudo-norm sparseness leads to a nonconvex optimization problem which is *NP-complete* with respect to the dimension M of the dictionary [[51], p. 418].

14.2.2
Learning to Be Sparse: The SparseNet Algorithm

We have seen above that we may define different models for measuring sparseness depending on our prior assumption on the distribution of coefficients. Note first that, assuming that the statistics are stationary (more generally ergodic), then these measures of sparseness across a population should necessarily imply a lifetime sparseness for any neuron. Such a property is essential to extend results from electrophysiology. Indeed, it is easier to record a restricted number of cells than a full population (see, for instance, Ref. [4]). However, the main property in terms of efficiency is that the representation should be sparse at any given time, that is, in our setting, at the presentation of each novel image.

Now that we have defined sparseness, how could we use it to induce neural organization? Indeed, given a sparse coding strategy that optimizes any representation efficiency cost as defined above, we may derive an unsupervised learning model by optimizing the dictionary Φ over natural scenes. On the one hand, the flexibility in the definition of the sparseness cost leads to a wide variety of proposed *sparse coding* solutions (for a review, see Ref. [52]) such as numerical optimization [22], non-negative matrix factorization [53, 54], or matching pursuit [26, 27, 29, 31]. They are all derived from correlation-based inhibition as this is necessary to remove redundancies from the linear representation. This is consistent with the observation that lateral interactions are necessary for the formation of elongated receptive fields [8, 55].

On the other hand, these methods share the same GLM model (see Eq. (14.1)) and once the sparse coding algorithm is chosen, the learning scheme is similar. As a consequence, after every coding sweep, we increased the efficiency of the dictionary Φ with respect to Eq. (14.2). This is achieved using the online gradient descent approach given the current sparse solution, $\forall i$:

$$\Phi_i \leftarrow \Phi_i + \eta \cdot a_i \cdot (\mathbf{I} - \Phi\mathbf{a}) \tag{14.5}$$

where η is the learning rate. Similar to Eq. (17) in Ref. [22] or to Eq. (2) in Ref. [31], the relation is a linear "Hebbian" rule [33] because it enhances the weight of neurons proportionally to the correlation between pre- and postsynaptic neurons. Note that there is no learning for nonactivated coefficients. The novelty of this formulation compared to other linear Hebbian learning rule such as Ref. [56] is to take advantage of the sparse representation, hence the name sparse Hebbian learning (SHL).

The class of SHL algorithms are unstable without homeostasis, that is, without a process that maintains the system in a certain equilibrium. In fact, starting with a random dictionary, the first filters to learn are more likely to correspond to salient features [26] and are therefore more likely to be selected again in subsequent learning steps. In SPARSENET, the homeostatic gain control is implemented by adaptively tuning the norm of the filters. This method equalizes the variance of coefficients across neurons using a geometric stochastic learning rule. The underlying heuristic is that this introduces a bias in the choice of the active coefficients. In fact, if a neuron is not selected often, the geometric homeostasis will decrease the norm of the corresponding filter, and therefore—from Eq. (14.1) and the conjugate gradient optimization—this will increase the value of the associated scalar. Finally, as the prior functions defined in Eq. (14.3) are identical for all neurons, this will increase the relative probability that the neuron is selected with a higher relative value. The parameters of this homeostatic rule have a great importance for the convergence of the global algorithm. In Reference [29], we have derived a more general homeostasis mechanism derived from the optimization of the representation efficiency through histogram equalization which we will describe later (see Section 14.4.1).

14.2.3
Results: Efficiency of Different Learning Strategies

The different SHL algorithms simply differ in the coding step. This implies that they only differ by first how sparseness is defined at a functional level and second how the inverse problem corresponding to the coding step is solved at the algorithmic level. Most of the schemes cited above use a less strict, parametric definition of sparseness (such as the convex L_1-norm), but for which a mathematical formulation of the optimization problem exists. Few studies such as Refs [57, 58] use the stricter ℓ_0 pseudo-norm as the coding problem gets more difficult. A thorough comparison of these different strategies was recently presented in Ref. [59]. See also Ref. [60] for properties of the coding solutions to the ℓ_0 pseudo-norm. Similarly, in Ref. [29], we preferred to retrieve an approximate solution to the coding problem to have a better match with the measure of efficiency Eq. (14.4).

Such an algorithmic framework is implemented in the SHL-scripts package.[2] These scripts allow the retrieval of the database of natural images and the replication of the results of Ref. [29] reported in this section. With a correct tuning of parameters, we observed that different coding schemes show qualitatively a

2) These scripts are available at *https://github.com/meduz/SHL_scripts*.

similar emergence of edge-like filters. The specific coding algorithm used to obtain this sparseness appears to be of secondary importance as long as it is adapted to the data and yields sufficiently efficient sparse representation vectors. However, resulting dictionaries vary qualitatively among these schemes and it was unclear which algorithm is the most efficient and what was the individual role of the different mechanisms that constitute SHL schemes. At the learning level, we have shown that the homeostasis mechanism had a great influence on the qualitative distribution of learned filters [29].

The results are shown in Figure 14.1. This figure represents the qualitative results of the formation of edge-like filters (receptive fields). More importantly, it shows the quantitative results as the average decrease of the squared error as a function of the sparseness. This gives a direct access to the cost as computed in Eq. (14.4). These results are comparable with the sparsenet algorithm. Moreover, this solution, by giving direct access to the atoms (filters) that are chosen, provides us with a more direct tool to manipulate sparse components. One

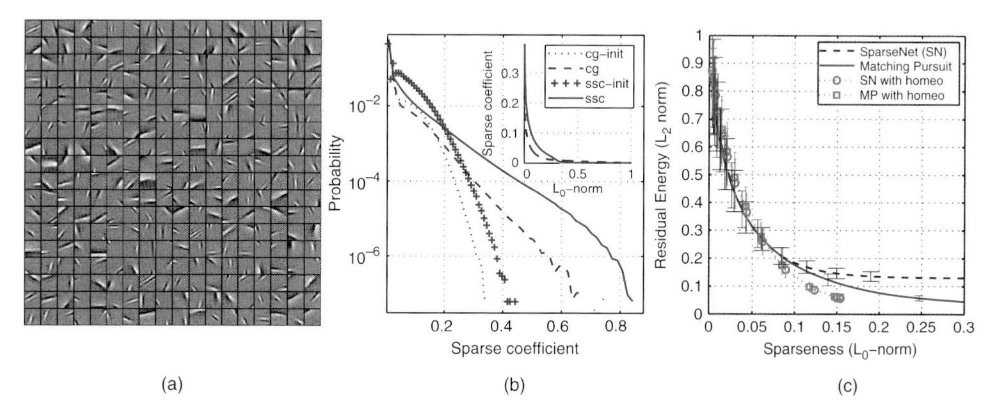

Figure 14.1 Learning a sparse code using sparse Hebbian learning. (a) We show the results at convergence (20,000 learning steps) of a sparse model with unsupervised learning algorithm which progressively optimizes the relative number of active (nonzero) coefficients (ℓ_0 pseudo-norm) [29]. Filters of the same size as the image patches are presented in a matrix (separated by a black border). Note that their position in the matrix is arbitrary as in ICA. These results show that sparseness induces the emergence of edge-like receptive fields similar to those observed in the primary visual area of primates. (b) We show the probability distribution function of sparse coefficients obtained by our method compared to Ref. [21] with first random dictionaries (respectively, "ssc-init" and "cg-init") and second with the dictionaries obtained after convergence of respective learning schemes (respectively, "ssc" and "cg"). At convergence, sparse coefficients are more sparsely distributed than initially, with more kurtotic probability distribution functions for "ssc" in both cases, as can be seen in the "longer tails" of the distribution. (c) We evaluate the coding efficiency of both methods with or without cooperative homeostasis by plotting the average residual error (L_2 norm) as a function of the ℓ_0 pseudo-norm. This provides a measure of the coding efficiency for each dictionary over the set of image patches (error bars represent one standard deviation). The best results are those providing a lower error for a given sparsity (better compression) or a lower sparseness for the same error.

further advantage consists in the fact that this unsupervised learning model is nonparametric (compare with Eq. (14.3)) and thus does not need to be parametrically tuned. Results show the role of homeostasis on the unsupervised algorithm. In particular, using the comparison of coding and decoding efficiency with and without this specific homeostasis, we have proven that cooperative homeostasis optimized overall representation efficiency (see also Section 14.4.1).

It is at this point important to note that in this algorithm, we achieve an exponential convergence of the squared error [[51], p. 422], but also that this curve can be directly derived from the coefficients' values. Indeed, for N coefficients (i.e., $\| \mathbf{a} \|_0 = N$), we have the squared error equal to

$$E_N \stackrel{\text{def}}{=} \| \mathbf{I} - \Phi\mathbf{a} \|^2 / \| \mathbf{I} \|^2 = 1 - \sum_{1 \leq k \leq N} a_k^2 / \| \mathbf{I} \|^2 \tag{14.6}$$

As a consequence, the sparser the distributions of coefficients, the quicker is the decrease of the residual energy. In the following section, we describe different variations of this algorithm. To compare their respective efficiency, we plot the decrease of the coefficients along with the decrease of the residual's energy. Using such tools, we explore whether such a property extends to full-scale images and not only to image patches, an important condition for using sparse models in computer vision algorithms.

14.3 SparseLets: A Multiscale, Sparse, Biologically Inspired Representation of Natural Images

14.3.1 Motivation: Architecture of the Primary Visual Cortex

Our goal here is to build practical algorithms of sparse coding for computer vision. We have seen above that it is possible to build an adaptive model of sparse coding that we applied to 12×12 image patches. Invariably, this has shown that the independent components of image patches are edge-like filters, such as are found in simple cells of V1. This model has shown that for randomly chosen image patches, these may be described by a sparse vector of coefficients. Extending this result to full-field natural images, we can expect that this sparseness would increase by a degree of order. In fact, except in a densely cluttered image such as a close-up of a texture, natural images tend to have wide areas which are void (such as the sky, walls, or uniformly filled areas). However, applying directly the sparsenet algorithm to full-field images is impossible in practice as its computer simulation would require too much memory to store the overcomplete set of filters. However, it is still possible to define *a priori* these filters and herein, we focus on a full-field sparse coding method whose filters are inspired by the architecture of the primary visual cortex.

The first step of our method involves defining the dictionary of templates (or filters) for detecting edges. We use a log-Gabor representation, which is well

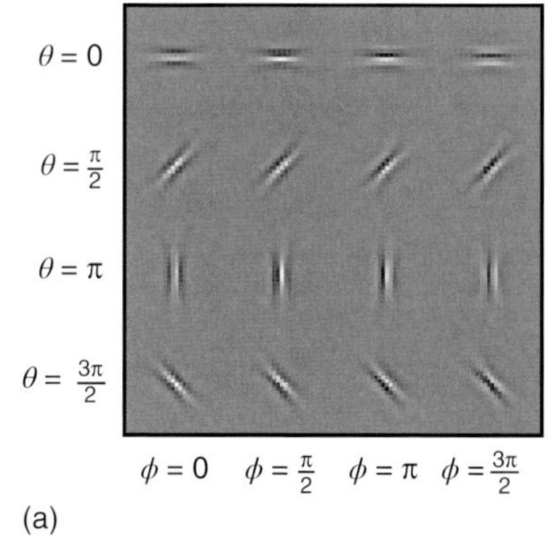

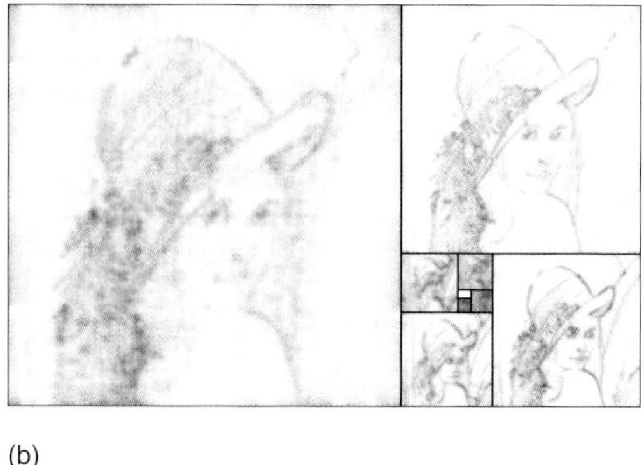

Figure 14.2 The log-Gabor pyramid (a) A set of log-Gabor filters showing in different rows different orientations and in different columns, different phases. Here we have only shown one scale. Note the similarity with Gabor filters. (b) Using this set of filters, one can define a linear representation that is rotation-, scaling- and translation-invariant. Here we show a tiling of the different scales according to a golden pyramid [43]. The hue gives the orientation while the value gives the absolute value (white denotes a low coefficient). Note the redundancy of the linear representation, for instance, at different scales.

suited to represent a wide range of natural images [42]. This representation gives a generic model of edges parameterized by their shape, orientation, and scale. We set the range of these parameters to match with what has been reported for simple-cell responses in macaque primary visual cortex (V1). Indeed, log-Gabor filters are similar to standard Gabors and both are well fitted to V1 simple cells [61]. Log-Gabors are known to produce a sparse set of linear coefficients [62]. Similarly to Gabors, these filters are defined by Gaussians in Fourier space, but their specificity is that log-Gabors have Gaussians envelopes in log-polar frequency space. This is consistent with physiological measurements which indicate that V1 cell responses are symmetric on the log frequency scale. They have multiple advantages over Gaussians: in particular, they have no DC component, and more generally, their envelopes more broadly cover the frequency space [63]. In this chapter, we set the bandwidth of the Fourier representation of the filters to 0.4 and $\pi/8$, respectively in the log-frequency and polar coordinates to get a family of relatively elongated (and thus selective) filters (see Ref. [63] and Figure 14.2(a) for examples of such edges). Before to the analysis of each image, we used the spectral whitening filter described by Olshausen and Field [22] to provide a good balance of the energy of output coefficients [26, 42]. Such a representation is implemented in the `LogGabor` package.[3]

This transform is linear and can be performed by a simple convolution repeated for every edge type. Following [63], convolutions were performed in the Fourier (frequency) domain for computational efficiency. The Fourier transform allows for

3) Python scripts are available at *https://github.com/meduz/LogGabor* and documented at *https://pythonhosted.org/LogGabor*.

a convenient definition of the edge filter characteristics, and convolution in the spatial domain is equivalent to a simple multiplication in the frequency domain. By multiplying the envelope of the filter and the Fourier transform of the image, one may obtain a filtered spectral image that may be converted to a filtered spatial image using the inverse Fourier transform. We exploited the fact that by omitting the symmetrical lobe of the envelope of the filter in the frequency domain, we obtain quadrature filters. Indeed, the output of this procedure gives a complex number whose real part corresponds to the response to the symmetrical part of the edge, while the imaginary part corresponds to the asymmetrical part of the edge (see Ref. [63] for more details). More generally, the modulus of this complex number gives the energy response to the edge—as can be compared to the response of complex cells in area V1, while its argument gives the exact phase of the filter (from symmetric to nonsymmetric). This property further expands the richness of the representation.

Given a filter at a given orientation and scale, a linear convolution model provides a translation-invariant representation. Such invariance can be extended to rotations and scalings by choosing to multiplex these sets of filters at different orientations and spatial scales. Ideally, the parameters of edges would vary in a continuous fashion to a full relative translation, rotation, and scale invariance. However this is difficult to achieve in practice and some compromise has to be found. Indeed, although orthogonal representations are popular in computer vision owing to their computational tractability, it is desirable in our context that we have a relatively high overcompleteness in the representation to achieve this invariance. For a given set of 256×256 images, we first chose to have 8 dyadic levels (i.e., by doubling the scale at each level) with 24 different orientations. Orientations are measured as nonoriented angles in radians, by convention in the range from $-\pi/2$ to $\pi/2$ (but not including $-\pi/2$) with respect to the x-axis. Finally, each image is transformed into a pyramid of coefficients. This pyramid consists of approximately $4/3 \times 256^2 \approx 8.7 \times 10^4$ pixels multiplexed on 8 scales and 24 orientations, that is, approximately 16.7×10^6 coefficients, an overcompleteness factor of about 256. This linear transformation is represented by a pyramid of coefficients, see Figure 14.2(b).

14.3.2 The SparseLets Framework

The resulting dictionary of edge filters is overcomplete. The linear representation would thus give a dense, relatively inefficient representation of the distribution of edges, see Figure 14.2(b). Therefore, starting from this linear representation, we searched instead for the most sparse representation. As we saw above in Section 14.2, minimizing the ℓ_0 pseudo-norm (the number of nonzero coefficients) leads to an expensive combinatorial search with regard to the dimension of the dictionary (it is NP-hard). As proposed by Perrinet [26], we may approximate a solution to this problem using a greedy approach. Such an approach is based on the physiology of V1. Indeed, it has been shown that inhibitory interneurons decorrelate excitatory cells to drive sparse code formation [55, 64]. We use this local architecture to iteratively modify the linear representation [42].

In general, a greedy approach is applied when the optimal combination is difficult to solve globally, but can be solved progressively, one element at a time. Applied to our problem, the greedy approach corresponds to first choosing the single filter Φ_i that best fits the image along with a suitable coefficient a_i, such that the single source $a_i\Phi_i$ is a good match to the image. Examining every filter Φ_j, we find the filter Φ_i with the maximal correlation coefficient ("matching" step), where

$$i = \text{argmax}_j \left(\left\langle \frac{\mathbf{I}}{\| \mathbf{I} \|}, \frac{\Phi_j}{\| \Phi_j \|} \right\rangle \right) \tag{14.7}$$

$\langle \cdot, \cdot \rangle$ represents the inner product and $\| \cdot \|$ represents the L_2 (Euclidean) norm. The index ("address") i gives the position (x and y), scale and orientation of the edge. We saw above that as filters at a given scale and orientation are generated by a translation, this operation can be efficiently computed using a convolution, but we keep this notation for its generality. The associated coefficient is the scalar projection

$$a_i = \left\langle \mathbf{I}, \frac{\Phi_i}{\| \Phi_i \|^2} \right\rangle \tag{14.8}$$

Second, knowing this choice, the image can be decomposed as

$$\mathbf{I} = a_i \Phi_i + R \tag{14.9}$$

where R is the residual image ("pursuit" step). We then repeat this two-step process on the residual (i.e., with $\mathbf{I} \leftarrow R$) until some stopping criterion is met. Note also that the norm of the filters has no influence in this algorithm on the matching step or on the reconstruction error. For simplicity and without loss of generality, we will hereafter set the norm of the filters to 1: $\forall j, \| \Phi_j \| = 1$ (i.e., the spectral energy sums to 1). Globally, this procedure gives us a sequential algorithm for reconstructing the signal using the list of sources (filters with coefficients), which greedily optimizes the ℓ_0 pseudo-norm (i.e., achieves a relatively sparse representation given the stopping criterion). The procedure is known as the *matching pursuit* (MP) algorithm [65], which has been shown to generate good approximations for natural images [26, 29].

We have included two minor improvements over this method: first, we took advantage of the response of the filters as complex numbers. As stated above, the modulus gives a response independent of the phase of the filter, and this value was used to estimate the best match of the residual image with the possible dictionary of filters (matching step). Then, the phase was extracted as the argument of the corresponding coefficient and used to feed back onto the image in the pursuit step. This modification allows for a phase-independent detection of edges, and therefore for a richer set of configurations, while preserving the precision of the representation.

Second, we used a "smooth" pursuit step. In the original form of the MP algorithm, the projection of the matching coefficient is fully removed from the image, which allows for the optimal decrease of the energy of the residual and allows for the quickest convergence of the algorithm with respect to the ℓ_0 pseudo-norm (i.e., it rapidly achieves a sparse reconstruction with low error). However, this efficiency comes at a cost, because the algorithm may result in nonoptimal representations because of choosing edges sequentially and not globally. This is often a problem when edges are aligned (e.g., on a smooth contour), as the different parts will be removed independently, potentially leading to a residual with gaps in the line. Our goal here is not necessarily to get the fastest decrease of energy, but rather to provide with the best representation of edges along contours. We therefore used a more conservative approach, removing only a fraction (denoted by α) of the energy at each pursuit step (for MP, $\alpha = 1$). Note that in that case, Eq. (14.6) has to be modified to account for the α parameter:

$$E_N = 1 - \alpha \cdot (2 - \alpha) \cdot \sum_{1 \leq k \leq N} \frac{a_k^2}{\| \mathbf{I} \|^2} \tag{14.10}$$

We found that $\alpha = 0.8$ was a good compromise between rapidity and smoothness. One consequence of using $\alpha < 1$ is that, when removing energy along contours, edges can overlap; even so, the correlation is invariably reduced. Higher and smaller values of α were also tested, and gave representation results similar to those presented here.

In summary, the whole coding algorithm is given by the following nested loops in pseudo-code:

1) draw a signal $\mathbf{I}$ from the database; its energy is $E = \| \mathbf{I} \|^2$,
2) initialize sparse vector $\mathbf{s}$ to zero and linear coefficients $\forall j, a_j = < \mathbf{I}, \Phi_j >$,
3) while the residual energy $E = \| \mathbf{I} \|^2$ is above a given threshold do:
   a. select the best match: $i = \text{ArgMax}_j |a_j|$, where $| \cdot |$ denotes the modulus,
   b. increment the sparse coefficient: $s_i = s_i + \alpha \cdot a_i$,
   c. update residual image: $\mathbf{I} \leftarrow \mathbf{I} - \alpha \cdot a_i \cdot \Phi_i$,
   d. update residual coefficients: $\forall j, a_j \leftarrow a_j - \alpha \cdot a_i < \Phi_i, \Phi_j >$,
4) the final set of nonzero values of the sparse representation vector $\mathbf{s}$ constitutes the list of edges representing the image as the list of couples $\pi_i = (i, s_i)$, where i represents an edge occurrence as represented by its position, orientation and scale and s_i the complex-valued sparse coefficient.

This class of algorithms gives a generic and efficient representation of edges, as illustrated by the example in Figure 14.3(a). We also verified that the dictionary used here is better adapted to the extraction of edges than Gabor's [42]. The performance of the algorithm can be measured quantitatively by reconstructing the image from the list of extracted edges. All simulations were performed using `Python` (version 2.7.8) with packages `NumPy` (version 1.8.1) and `SciPy`

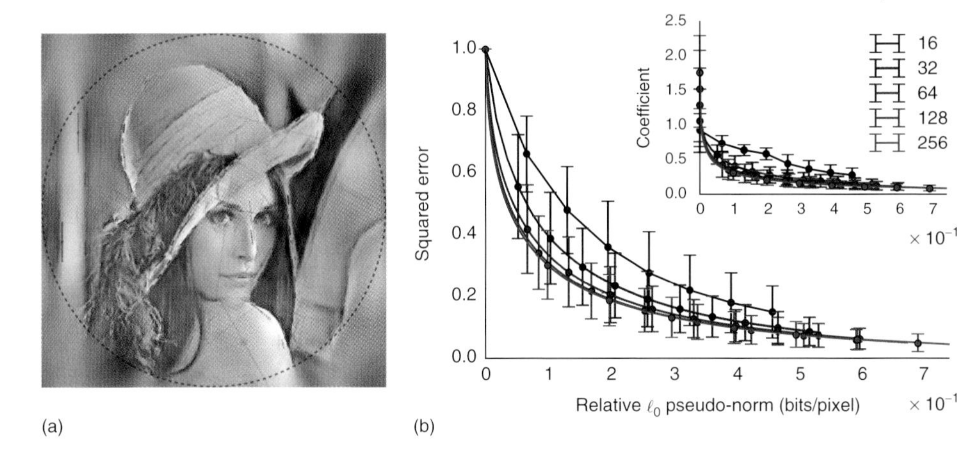

Figure 14.3 SparseLets framework. (a) An example reconstructed image with the list of extracted edges overlaid. As in Ref. [68], edges outside the circle are discarded to avoid artifacts. Parameters for each edge are the position, orientation, scale (length of bar), and scalar amplitude (transparency) with the phase (hue). We controlled the quality of the reconstruction from the edge information such that the residual energy is less than 3% over the whole set of images, a criterion met on average when identifying 2048 edges per image for images of size 256×256 (i.e., a relative sparseness of $\approx$ 0.01% of activated coefficients). (b) Efficiency for different image sizes as measured by the decrease of the residual's energy as a function of the coding cost (relative ℓ_0 pseudo-norm). (b, inset) This shows that as the size of images increases, sparseness increases, validating quantitatively our intuition on the sparse positioning of objects in natural images. Note, that the improvement is not significant for a size superior to 128. The SparseLets framework thus shows that sparse models can be extended to full-scale natural images, and that increasing the size improves sparse models by a degree of order (compare a size of 16 with that of 256).

(version 0.14.0) [66] on a cluster of Linux computing nodes. Visualization was performed using `Matplotlib` (version 1.3.1) [67].[4)]

14.3.3
Efficiency of the SparseLets Framework

Figure 14.3(a), shows the list of edges extracted on a sample image. It fits qualitatively well with a rough sketch of the image. To evaluate the algorithm quantitatively, we measured the ratio of extracted energy in the images as a function of the number of edges on a database of 600 natural images of size 256×256,[5)] see Figure 14.3(b). Measuring the ratio of extracted energy in the images, $N = 2048$ edges were enough to extract an average of approximately 97% of the energy of

4) These python scripts are available at *https://github.com/meduz/SparseEdges* and documented at *https://pythonhosted.org/SparseEdges*.
5) This database is publicly available at *http://cbcl.mit.edu/software-datasets/serre/SerreOlivaPoggio PNAS07*.

images in the database. To compare the strength of the sparseness in these full-scale images compared to the image patches discussed above (see Section 14.2), we measured the sparseness values obtained in images of different sizes. For these to be comparable, we measured the efficiency with respect to the relative ℓ_0 pseudo-norm in bits per unit of image surface (pixel): this is defined as the ratio of active coefficients times the numbers of bits required to code for each coefficient (i.e., $\log_2(M)$, where M is the total number of coefficients in the representation) over the size of the image. For different image and framework sizes, the lower this ratio, the higher the sparseness. As shown in Figure 14.3(b), we indeed see that sparseness increases relative to an increase in image size. This reflects the fact that sparseness is not only local (few edges coexist at one place) but is also spatial (edges are clustered, and most regions are empty). Such a behavior is also observed in V1 of monkeys as the size of the stimulation is increased from a stimulation over only the classical receptive field to 4 times around it [15].

Note that by definition, our representation of edges is invariant to translations, scalings, and rotations in the plane of the image. We also performed the same edge extraction where images from the database were perturbed by adding independent Gaussian noise to each pixel such that signal-to-noise ratio was halved. Qualitative results are degraded but qualitatively similar. In particular, edge extraction in the presence of noise may result in false positives. Quantitatively, one observes that the representation is slightly less sparse. This confirms our intuition that sparseness is causally linked to the efficient extraction of edges in the image.

To examine the robustness of the framework and of sparse models in general, we examined how results changed on changing the parameters for the algorithm. In particular, we investigated the effect of filter parameters B_f and B_θ. We also investigated how the overcompleteness factor could influence the result. We manipulated the number of discretization steps along the spatial frequency axis N_f (i.e., the number of layers in the pyramid) and orientation axis N_θ. The results are summarized in Figure 14.4 and show that an optimal efficiency is achieved for certain values of these parameters. These optimal values are in the order of that found for the range of selectivities observed in V1. Note that these values may change across categories. Further experiments should provide an adaptation mechanism to allow finding the best parameters in an unsupervised manner.

These particular results illustrate the potential of sparse models in computer vision. Indeed, one main advantage of these methods is to explicitly represent edges. A direct application of sparse models is the ability of the representation to reconstruct these images and therefore to use it for compression [26]. Other possible applications are image filtering or edge manipulation for texture synthesis or denoising [69]. Recent advances have shown that such representations could be used for the classification of natural images (see Chapter 13 or, for instance, Ref. [70]) or of medical images of emphysema [71]. Classification was also used in a sparse model for the quantification of artistic style through sparse coding analysis in the drawings of Pieter Bruegel the Elder [72]. These examples illustrate the different applications of sparse representations and in the following we illustrate some potential perspectives to further improve their representation efficiency.

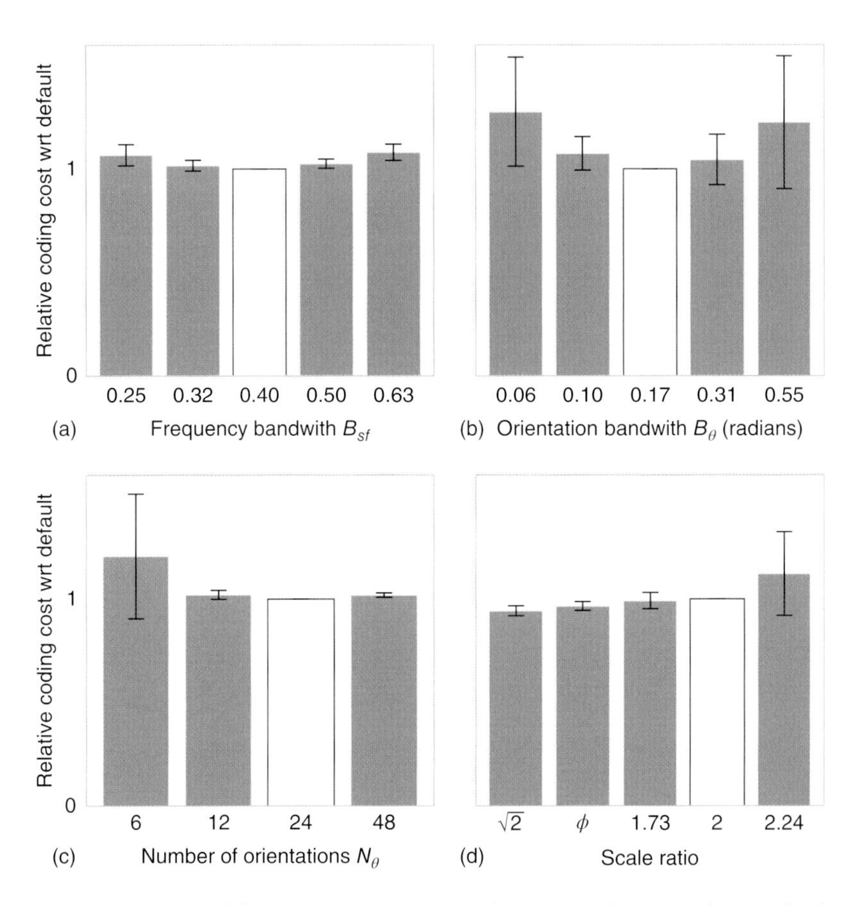

Figure 14.4 Effect of filters' parameters on the efficiency of the SparseLets framework. As we tested different parameters for the filters, we measured the gain in efficiency for the algorithm as the ratio of the code length to achieve 85% of energy extraction relative to that for the default parameters (white bar). The average is computed on the same database of natural images and error bars denote the standard deviation of gain over the database. First, we studied the effect of the bandwidth of filters respectively in the (a) spatial frequency and (b) orientation spaces. The minimum is reached for the default parameters: this shows that default parameters provide an optimal compromise between the precision of filters in the frequency and position domains for this database. We may also compare pyramids with different number of filters. Indeed from Eq. (14.4), efficiency (in bits) is equal to the number of selected filters times the coding cost for the address of each edge in the pyramid. We plot here the average gain in efficiency which shows an optimal compromise respectively (c) the number of orientations and (d) the number of spatial frequencies (scales). Note first that with more than 12 directions, the gain remains stable. Note also that a dyadic scale ratio (that is, of 2) is efficient but that other solutions—such as using the golden section ϕ—prove to be significantly more efficient, although the average gain is relatively small (inferior to 5%).

14.4
SparseEdges: Introducing Prior Information

14.4.1
Using the Prior in First-Order Statistics of Edges

In natural images, it has been observed that edges follow some statistical regularities that may be used by the visual system. We first focus on the most obvious regularity which consists in the anisotropic distribution of orientations in natural images (see Chapter 4 for another qualitative characterization of this anisotropy). Indeed, it has been observed that orientations corresponding to cardinals (i.e., to verticals and horizontals) are more likely than other orientations [73, 74]. This is due to the fact that our point of view is most likely pointing toward the horizon while we stand upright. In addition, gravity shaped our surrounding world around horizontals (mainly the ground) and verticals (such as trees or buildings). Psychophysically, this prior knowledge gives rise to the oblique effect [75]. This is even more striking in images of human scenes (such as a street, or inside a building) as humans mainly build their environment (houses, furnitures) around these cardinal axes. However, we assumed in the cost defined above (see Eq. (14.2)) that each coefficient is independently distributed.

It is believed that a homeostasis mechanism allows one to optimize this cost knowing this prior information [29, 76]. Basically, the solution is to put more filters where there are more orientations [73] such that coefficients are uniformly distributed. In fact, as neural activity in the assembly actually represents the sparse coefficients, we may understand the role of homeostasis as maximizing the average representation cost $\mathcal{C}(\mathbf{a}|\Phi)$. This is equivalent to saying that homeostasis should act such that at any time, and invariantly to the selectivity of features in the dictionary, the probability of selecting one feature is uniform across the dictionary. This optimal uniformity may be achieved in all generality by using an equalization of the histogram [7]. This method may be easily derived if we know the probability distribution function dP_i of variable a_i (see Figure 14.5(a)) by choosing a nonlinearity as the cumulative distribution function (see Figure 14.5(b)) transforming any observed variable $\overline{a}_i$ into

$$z_i(\overline{a}_i) = P_i(a_i \leq \overline{a}_i) = \int_{-\infty}^{\overline{a}_i} dP_i(a_i) \tag{14.11}$$

This is equivalent to the change of variables which transforms the sparse vector $\mathbf{a}$ to a variable with uniform probability distribution function in $[0, 1]^M$ (see Figure 14.5(c)). This equalization process has been observed in the neural activity of a variety of species and is, for instance, perfectly illustrated in the compound eye of the fly's neural response to different levels of contrast [76]. It may evolve dynamically to slowly adapt to varying changes, for instance to luminance or contrast values, such as when the light diminishes at twilight. Then, we use these point nonlinearities z_i to sample orientation space in an optimal fashion (see Figure 14.5(d)).

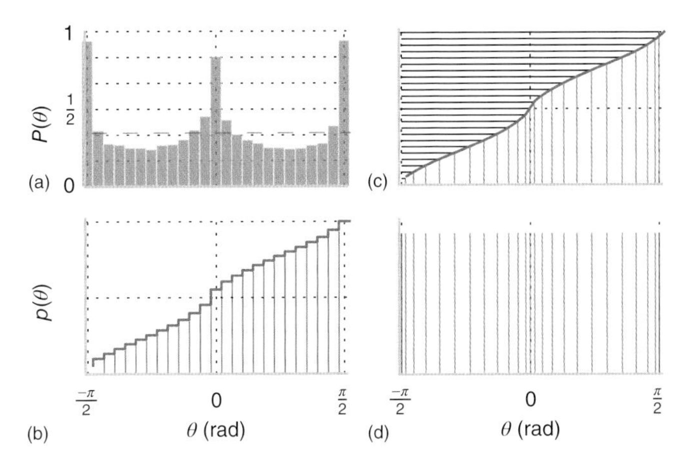

Figure 14.5 Histogram equalization. From the edges extracted in the images from the natural scenes database, we computed sequentially (clockwise, from the bottom left): (a) the histogram and (b) cumulative histogram of edge orientations. This shows that as was reported previously (see, for instance, Ref. [74]), cardinals axis are overrepresented. This represents a relative inefficiency as the representation in the SparseLets framework represents a priori orientations in a uniform manner. A neuromorphic solution is to use histogram equalization, as was first shown in the fly's compound eye by Laughlin [76]. (c) We draw a uniform set of scores on the y-axis of the cumulative function (black horizontal lines), for which we select the corresponding orientations (red vertical lines). Note that by convention these are wrapped up to fit in the $(-\pi/2, \pi/2]$ range. (d) This new set of orientations is defined such that they are a priori selected uniformly. Such transformation was found to well describe a range of psychological observations [73] and we will now apply it to our framework.

This simple nonparametric homeostatic method is applicable to the SparseLets algorithm by simply using the transformed sampling of the orientation space. It is important to note that the MP algorithm is nonlinear and the choice of one element at any step may influence the rest of the choices. In particular, while orientations around cardinals are more prevalent in natural images (see Figure 14.6(a)), the output histogram of detected edges is uniform (see Figure 14.6(b)). To quantify the gain in efficiency, we measured the residual energy in the SparseLets framework with or without including this prior knowledge. Results show that for a similar number of extracted edges, residual energy is not significantly changed (see Figure 14.6(c)). This is again due to the exponential convergence of the squared error [[51], p. 422] on the space spanned by the representation basis. As the tiling of the Fourier space by the set of filters is complete, one is assured of the convergence of the representation in both cases. However thanks to the use of first-order statistics, the orientation of edges are distributed such as to maximize the entropy, further improving the efficiency of the representation.

This novel improvement to the SparseLets algorithm illustrates the flexibility of the MP framework. This proves that by introducing the prior on first-order statistics, one improves the efficiency of the model for this class of natural images. Of course, this gain is only valid for natural images and would disappear

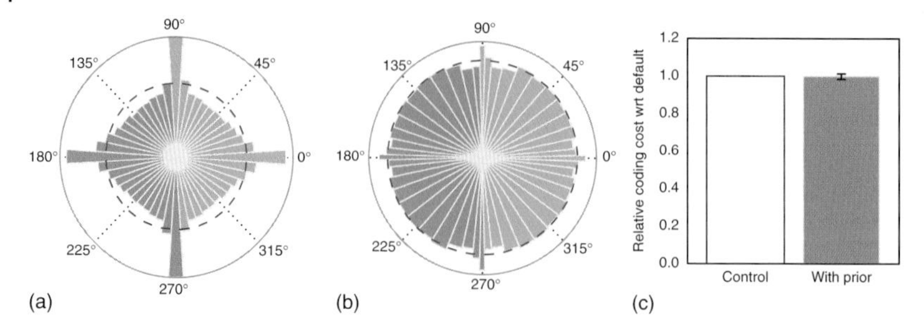

Figure 14.6 Including prior knowledge on the orientation of edges in natural images. (a) Statistics of orientations in the database of natural images as shown by a polar bar histogram (the surface of wedges is proportional to occurrences). As for Figure 14.5(a), this shows that orientations around cardinals are more likely than others (the dotted red line shows the average occurrence rate). (b) We show the histogram with the new set of orientations that has been used. Each bin is selected approximately uniformly. Note the variable width of bins. (c) We compare the efficiency of the modified algorithm where the sampling is optimized thanks to histogram equalization described in Figure 14.5 as the average residual energy with respect to the number of edges. This shows that introducing a prior information on the distribution of orientations in the algorithm may also introduce a slight but insignificant improvement in the sparseness.

for images where cardinals would not dominate. This is the case for images of close-ups (microscopy) or where gravity is not prevalent such as aerial views. Moreover, this is obviously just a first step as there is more information from natural images that could be taken into account.

14.4.2 Using the Prior Statistics of Edge Co-Occurrences

A natural extension of the previous result is to study the co-occurrences of edges in natural images. Indeed, images are not simply built from independent edges at arbitrary positions and orientations but tend to be organized along smooth contours that follow, for instance, the shape of objects. In particular, it has been shown that contours are more likely to be organized along co-circular shapes [77]. This reflects the fact that in nature, round objects are more likely to appear than random shapes. Such a statistical property of images seems to be used by the visual system as it is observed that edge information is integrated on a local "association field" favoring colinear or co-circular edges (see Chapter 13, Section 5, for more details and a mathematical description). In V1 for instance, neurons coding for neighboring positions are organized in a similar fashion. We have previously seen that statistically, neurons coding for collinear edges seem to be anatomically connected [12, 13] while rare events (such as perpendicular occurrences) are functionally inhibited [13].

Using the probabilistic formulation of the edge extraction process (see Section 14.2), one can also apply this prior probability to the choice of mechanism (matching) of the MP algorithm. Indeed at any step of the edge extraction

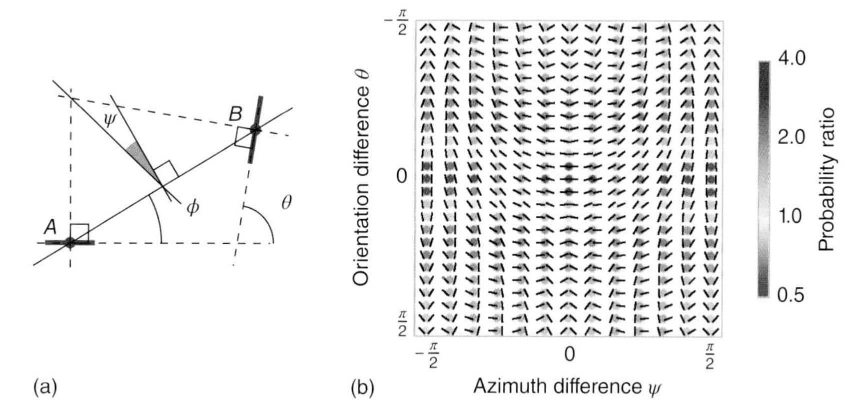

Figure 14.7 Statistics of edge co-occurrences. (a) The relationship between a pair of edges can be quantified in terms of the difference between their orientations θ, the ratio of scale σ relative to the reference edge, the distance $d = \| \vec{AB} \|$ between their centers, and the difference of azimuth (angular location) ϕ of the second edge relative to the reference edge. In addition, we define $\psi = \phi - \theta/2$ as it is symmetric with respect to the choice of the reference edge, in particular, $\psi = 0$ for co-circular edges. (b) The probability distribution function $p(\psi, \theta)$ represents the distribution of the different geometrical arrangements of edges' angles, which we call a "chevron map." We show here the histogram for natural images, illustrating the preference for colinear edge configurations. For each chevron configuration, deeper and deeper red circles indicate configurations that are more and more likely with respect to a uniform prior, with an average maximum of about 4 times more likely, and deeper and deeper blue circles indicate configurations less likely than a flat prior (with a minimum of about 0.7 times as likely). Conveniently, this "chevron map" shows in one graph that natural images have on average a preference for colinear and parallel angles (the horizontal middle axis), along with a slight preference for co-circular configurations (middle vertical axis).

process, one can include the knowledge gained by the extraction of previous edges, that is, the set $\mathcal{I} = \{\pi_i\}$ of extracted edges, to refine the log-likelihood of a new possible edge $\pi_* = (*, a_*)$ (where $*$ corresponds to the address of the chosen filter, and therefore to its position, orientation, and scale). Knowing the probability of co-occurrences $p(\pi_*|\pi_i)$ from the statistics observed in natural images (see Figure 14.7), we deduce that the cost is now at any coding step (where $\mathbf{I}$ is the residual image—see Eq. (14.9)):

$$\mathcal{C}(\pi_*|\mathbf{I}, \mathcal{I}) = \frac{1}{2\sigma_n^2} \| \mathbf{I} - a_* \Phi_* \|^2 - \eta \sum_{i \in \mathcal{I}} \| a_i \| \cdot \log\ p(\pi_*|\pi_i) \tag{14.12}$$

where η quantifies the strength of this prediction. Basically, this shows that, similarly to the association field proposed by Grossberg [78] which was subsequently observed in cortical neurons [79] and applied by Field *et al.* [80], we facilitate the activity of edges knowing the list of edges that were already extracted. This comes as a complementary local interaction to the inhibitory local interaction implemented in the pursuit step (see Eq. (14.9)) and provides a quantitative algorithm

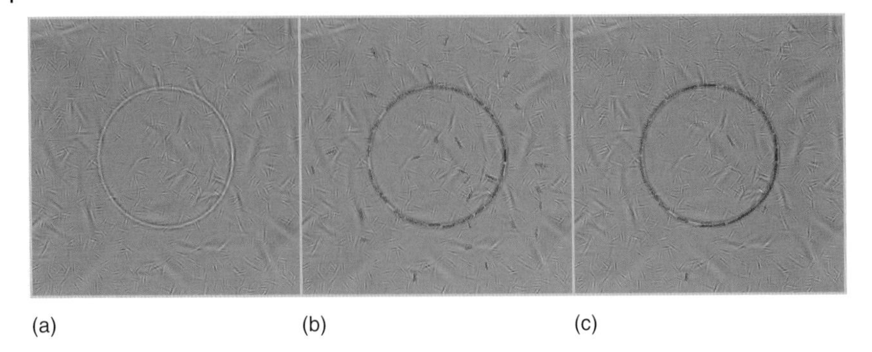

Figure 14.8 Application to rapid contour segmentation. We applied the original sparse edge framework to (a) the synthetic image of a circle embedded in noise. This noise is synthesized by edges at random positions, orientations and scales with a similar first-order statistics as natural scenes. (b) We overlay in red the set of edges which were detected by the SparseLets framework. (c) Then, we introduce second-order information in the evaluation of the probability in the sparse edges framework (with $\eta = 0.15$). This modifies the sequence of extracted edges as shown in blue. There is a clear match of the edge extraction with the circle, as would be predicted by psychophysical results of hand segmentation of contours. This shows that second-order information as introduced in this feed-forward chain may be sufficient to account for contour grouping and may not necessitate a recursive chain rule such as implemented in Ref. [68].

to the heuristics proposed in Ref. [42]. Note that although this model is purely sequential and feed-forward, this results possibly in a "chain rule" as when edges along a contour are extracted, this activity is facilitated along it as long as the image of this contour exists in the residual image. Such a "chain rule" is similar to what was used to model psychophysical performance [68] or to filter curves in images [81]. Our novel implementation provides a rapid and efficient solution that we illustrate here on a segmentation problem (see Figure 14.8).

Indeed, sparse models have been shown to foster numerous applications in computer vision. Among these are algorithms for segmentation in images [82] or for classification [83, 84]. We may use the previous application of our algorithm to evaluate the probability of edges belonging to the same contour. We show in Figure 14.8 the application of such a formula (in panel c vs classical sparse edge extraction in panel b) on a synthetic image of a circle embedded in noise (panel a). It shows that, while some edges from the background are extracted in the plain SparseLets framework (panel b), edges belonging to the same circular contour pop out from the computation similar to a chain rule (panel c). Note that contrary to classical hierarchical models, these results have been obtained with a simple layer of edge detection filters which communicate through local diffusion. An important novelty to note in this extension of the SparseLets framework is that there is no recursive propagation, as the greedy algorithm is applied in a sequential manner. These types of interaction have been found in area V1. Indeed, the processing may be modulated by simple contextual rules such as favoring colinear versus co-circular edges [85]. Such type of modulation opens up a wide range of potential applications to computer vision such as robust segmentation and algorithms for the autonomous classification of images [70]. More generally, it shows that simple feed-forward algorithms such as the one we propose may be sufficient to account for the sparse representation of images in lower-level visual areas.

14.5 Conclusion

In this chapter, we have shown sparse models at increasing structural complexities mimicking the key features of the primary visual cortex in primates. By focusing on a concrete example, the SparseLets framework, we have shown that sparse models provide an efficient framework for biologically inspired computer vision algorithms. In particular, by including contextual information, such as prior statistics on natural images, we could improve the efficiency of the sparseness of the representation.

Such an approach allows to implement a range of practical concepts (such as the good continuity of contours) in a principled way. Indeed, we based our reasoning on inferential processes such as they are reflected in the organization of neural structures. For instance, there is a link between co-circularity and the structure of orientation maps [86]. This should be included in further perspectives of these sparse models.

As we have seen, the (visual) brain is not a computer. Instead of using a sequential stream of semantic symbols, it uses statistical regularities to derive predictive rules. These computations are not written explicitly as it suffices that they emerge from the collective behavior in populations of neurons. As such, these rules are massively parallel, asynchronous, and error prone. Luckily, such neuromorphic computing architectures are beginning to emerge—yet, we lack a better understanding of how we may implement computer vision algorithms on such hardware.

In conclusion, this drives the need for a more biologically driven computer vision algorithm and a better understanding of V1. However, such knowledge is largely incomplete [87] and we need to develop a better understanding of results from electrophysiology. A promising approach in that sense is to include model-driven stimulation of physiological studies [88, 89] as they systematically test neural computations for a given visual task.

Acknowledgments

The author was supported by EC IP project FP7-269921, "BrainScaleS". Correspondence and requests for materials should be addressed to the author (email:Laurent.Perrinet@univ-amu.fr).

References

1. Friston, K., Adams, R.A., Perrinet, L., and Breakspear, M. (2012) Perceptions as hypotheses: saccades as experiments. *Front. Psychol.*, **3**, doi: 10.3389/fpsyg.2012.00151
2. Hubel, D. and Wiesel, T. (1968) Receptive fields and functional architecture of monkey's striate cortex. *J. Physiol.*, **195** (1), 215–244, *http://www.ncbi.nlm.nih.gov/pmc/articles/PMC1557912/*.
3. Barth, A.L. and Poulet, J.F.A. (2012) Experimental evidence for sparse firing in the neocortex. *Trends Neurosci.*, **35** (6), 345–355, doi: 10.1016/j.tins.2012.03.008
4. Willmore, B.D., Mazer, J.A., and Gallant, J.L. (2011) Sparse coding in striate and extrastriate visual cortex. *J. Neurophysiol.*, **105** (6), 2907–2919, doi: 10.1152/jn.00594.2010
5. Marr, D. (1983) *Vision: A Computational Investigation into the Human Representation and Processing of Visual Information*, Henry Holt and Company, Inc., *http://www.worldcat.org/isbn/0716715678*.
6. Attneave, F. (1954) Some informational aspects of visual perception. *Psychol. Rev.*, **61** (3), 183–193, *http://view.ncbi.nlm.nih.gov/pubmed/13167245*.
7. Atick, J.J. (1992) Could information theory provide an ecological theory of sensory processing? *Netw. Comput. Neural Syst.*, **3** (2), 213–252.
8. Wolfe, J., Houweling, A.R., and Brecht, M. (2010) Sparse and powerful cortical spikes. *Curr. Opin. Neurobiol.*, **20** (3), 306–312, doi: 10.1016/j.conb.2010.03.006
9. Field, D.J. (1994) What is the goal of sensory coding? *Neural Comput.*, **6** (4), 559–601, doi: 10.1162/neco.1994.6.4.559
10. Froudarakis, E., Berens, P., Ecker, A.S., Cotton, R.J., Sinz, F.H., Yatsenko, D., Saggau, P., Bethge, M., and Tolias, A.S. (2014) Population code in mouse V1 facilitates readout of natural scenes through increased sparseness. *Nat. Neurosci.*, **17** (6), 851–857, doi: 10.1038/nn.3707
11. Palm, G. (2013) Neural associative memories and sparse coding. *Neural Netw.*, **37**, 165–171, doi: 10.1016/j.neunet.2012.08.013
12. Bosking, W.H., Zhang, Y., Schofield, B., and Fitzpatrick, D. (1997) Orientation selectivity and the arrangement of horizontal connections in tree shrew striate cortex. *J. Neurosci.*, **17** (6), 2112–2127, *http://www.jneurosci.org/cgi/content/abstract/17/6/2112*.
13. Hunt, J.J., Bosking, W.H., and Goodhill, G.J. (2011) Statistical structure of lateral connections in the primary visual cortex. *Neural Syst. Circ.*, **1** (1), 3, doi: 10.1186/2042-1001-1-3
14. Field, D.J. (1987) Relations between the statistics of natural images and the response properties of cortical cells. *J. Opt. Soc. Am. A*, **4** (12), 2379–2394.
15. Vinje, W.E. and Gallant, J.L. (2000) Sparse coding and decorrelation in primary visual cortex during natural vision. *Science*, **287** (5456), 1273–1276, doi: 10.1126/science.287.5456.1273
16. DeWeese, M.R., Wehr, M., and Zador, A.M. (2003) Binary spiking in auditory cortex. *J. Neurosci.*, **23** (21), 7940–7949, *http://www.jneurosci.org/content/23/21/7940.abstract*.
17. Baudot, P., Levy, M., Marre, O., Monier, C., Pananceau, M., and Frégnac, Y. (2013) Animation of natural scene by virtual eye-movements evokes high precision and low noise in V1 neurons. *Front. Neural Circ.*, **7**, doi: 10.3389/fncir.2013.00206
18. Rao, R.P.N. and Ballard, D.H. (1999) Predictive coding in the visual cortex: a functional interpretation of some extra-classical receptive-field effects. *Nat. Neurosci.*, **2** (1), 79–87, doi: 10.1038/4580
19. Spratling, M.W. (2011) A single functional model accounts for the distinct properties of suppression in cortical area V1. *Vision Res.*, **51** (6), 563–576, doi: 10.1016/j.visres.2011.01.017
20. Perrinet, L.U., Adams, R.A., and Friston, K.J. (2014) Active inference, eye movements and oculomotor delays. *Biol. Cybern.*, pp. 1–25, doi: 10.1007/s00422-014-0620-8

21. Olshausen, B.A. and Field, D.J. (1996) Natural image statistics and efficient coding. *Netw. Comput. Neural Syst.*, **7** (6583), 333–339, doi: 10.1038/381607a0
22. Olshausen, B.A. and Field, D.J. (1997) Sparse coding with an overcomplete basis set: a strategy employed by V1? *Vision Res.*, **37** (23), 3311–3325, doi: 10.1016/S0042-6989(97)00169-7
23. Bell, A.J. and Sejnowski, T.J. (1997) The 'independent components' of natural scenes are edge filters. *Vision Res.*, **37** (23), 3327–3338.
24. Fyfe, C. and Baddeley, R. (1995) Finding compact and sparse-distributed representations of visual images. *Netw. Comput. Neural Syst.*, **6** (3), 333–344, doi: 10.1088/0954-898X/6/3/002
25. Zibulevsky, M. and Pearlmutter, B.A. (2001) Blind source separation by sparse decomposition. *Neural Comput.*, **13** (4), 863–882.
26. Perrinet, L., Samuelides, M., and Thorpe, S. (2004) Coding static natural images using spiking event times: do neurons cooperate? *IEEE Trans. Neural Netw.*, **15** (5), 1164–1175, doi: 10.1109/TNN.2004.833303, special issue on 'Temporal Coding for Neural Information Processing'
27. Rehn, M. and Sommer, F.T. (2007) A model that uses few active neurons to code visual input predicts the diverse shapes of cortical receptive fields. *J. Comput. Neurosci.*, **22** (2), 135–146.
28. Doi, E., Balcan, D.C., and Lewicki, M.S. (2007) Robust coding over noisy overcomplete channels. *IEEE Trans. Image Process.*, **16** (2), 442–452.
29. Perrinet, L.U. (2010) Role of homeostasis in learning sparse representations. *Neural Comput.*, **22** (7), 1812–1836, doi: 10.1162/neco.2010.05-08-795, *http://invibe.net/LaurentPerrinet/Publications/Perrinet10shl.*
30. Lewicki, M.S. and Sejnowski, T.J. (2000) Learning overcomplete representations. *Neural Comput.*, **12** (2), 337–365.
31. Smith, E.C. and Lewicki, M.S. (2006) Efficient auditory coding. *Nature*, **439** (7079), 978–982, doi: 10.1038/nature04485
32. Baddeley, R., Abbott, L.F., Booth, M.C.A., Sengpiel, F., Freeman, T., Wakeman, E.A., and Rolls, E.T. (1997) Responses of neurons in primary and inferior temporal visual cortices to natural scenes. *Proc. R. Soc. B: Biol. Sci.*, **264** (1389), 1775–1783, doi: 10.1098/rspb.1997.0246, *http://citeseer.nj.nec.com/19262.html*
33. Hebb, D.O. (1949) *The Organization of Behavior: A Neuropsychological Theory*, John Wiley & Sons, Inc., New York.
34. Zylberberg, J., Murphy, J.T., and DeWeese, M.R. (2011) A sparse coding model with synaptically local plasticity and spiking neurons can account for the diverse shapes of V1 simple cell receptive fields. *PLoS Comput. Biol.*, **7** (10), e1002 250, doi: 10.1371/journal.pcbi.1002250
35. Hunt, J.J., Dayan, P., and Goodhill, G.J. (2013) Sparse coding can predict primary visual cortex receptive field changes induced by abnormal visual input. *PLoS Comput. Biol.*, **9** (5), e1003 005, doi: 10.1371/journal.pcbi.1003005
36. Blättler, F. and Hahnloser, R.H.R. (2011) An efficient coding hypothesis links sparsity and selectivity of neural responses. *PLoS ONE*, **6** (10), e25 506, doi: 10.1371/journal.pone.0025506
37. Seriès, P., Latham, P.E., and Pouget, A. (2004) Tuning curve sharpening for orientation selectivity: coding efficiency and the impact of correlations. *Nat. Neurosci.*, **7** (10), 1129–1135, doi: 10.1038/nn1321
38. Ganguli, S. and Sompolinsky, H. (2012) Compressed sensing, sparsity, and dimensionality in neuronal information processing and data analysis. *Annu. Rev. Neurosci.*, **35** (1), 485–508, doi: 10.1146/annurev-neuro-062111-150410, *http://keck.ucsf.edu/surya/12.CompSense.pdf.*
39. Elad, M. (2010) *Sparse and Redundant Representations: From Theory to Applications in Signal and Image Processing*, 1st edn, Springer-Verlag, *http://www.amazon.com/exec/obidos/redirect?tag=citeulike07-20&path=ASIN/144197010X.*
40. Benoit, A., Caplier, A., Durette, B., and Herault, J. (2010) Using human visual system modeling for bio-inspired low

level image processing. *Comput. Vis. Image Underst.*, **114** (7), 758–773, doi: 10.1016/j.cviu.2010.01.011
41. Serre, T. and Poggio, T. (2010) A neuromorphic approach to computer vision. *Commun. ACM*, **53** (10), 54–61, doi: 10.1145/1831407.1831425
42. Fischer, S., Redondo, R., Perrinet, L.U., and Cristóbal, G. (2007) Sparse approximation of images inspired from the functional architecture of the primary visual areas. *EURASIP J. Adv. Signal Process.*, **2007** (1), 090 727–090 122, doi: 10.1155/2007/90727
43. Perrinet, L.U. (2008) Adaptive sparse spike coding: applications of neuroscience to the compression of natural images, in *Optical and Digital Image Processing Conference 7000 - Proceedings of SPIE, 7 - 11 April 2008*, vol. **7000** (ed. G.C. Peter Schelkens), SPIE.
44. Perrinet, L.U. (2007) Topics in dynamical neural networks: from large scale neural networks to motor control and vision, in *The European Physical Journal (Special Topics)*, vol. **142**, Springer-Verlag, Berlin and Heidelberg, pp. 163–225, doi: 10.1016/j.jphysparis.2007.10.011
45. Rissanen, J. (1978) Modeling by shortest data description. *Automatica*, **14**, 465–471.
46. Olshausen, B.A. (2002) Sparse codes and spikes, in *Probabilistic Models of the Brain: Perception and Neural Function* (eds R.P.N. Rao, B.A. Olshausen, and M.S. Lewicki), MIT Press, pp. 257–272.
47. Tikhonov, A.N. (1977) *Solutions of Ill-Posed Problems*, Winston & Sons, Washington, DC, *http://www.amazon.com/exec/obidos/redirect?tag=citeulike07-20&path=ASIN/0470991240.*
48. Bethge, M., Rotermund, D., and Pawelzik, K. (2003) Second order phase transition in neural rate coding: binary encoding is optimal for rapid signal transmission. *Phys. Rev. Lett.*, **90** (8), 088 104.
49. Nikitin, A.P., Stocks, N.G., Morse, R.P., and McDonnell, M.D. (2009) Neural population coding is optimized by discrete tuning curves. *Phys. Rev. Lett.*, **103** (13), 138 101.
50. Akaike, H. (1974) A new look at the statistical model identification. *IEEE Trans. Autom. Control*, **19**, 716–723.
51. Mallat, S. (1998) *A Wavelet Tour of Signal Processing*, 2nd edn, Academic Press.
52. Pece, A.E.C. (2002) The problem of sparse image coding. *J. Math. Imaging Vis.*, **17**, 89–108.
53. Lee, D.D. and Seung, S.H. (1999) Learning the parts of objects by non-negative matrix factorization. *Nature*, **401**, 788–791.
54. Ranzato, M.A., Poultney, C.S., Chopra, S., and LeCun, Y. (2007) Efficient learning of sparse overcomplete representations with an Energy-Based model. *Adv. Neural Inf. Process. Syst.*, **19**, 1137–1144.
55. Bolz, J. and Gilbert, C.D. (1989) The role of horizontal connections in generating long receptive fields in the cat visual cortex. *Eur. J. Neurosci.*, **1** (3), 263–268.
56. Oja, E. (1982) A simplified neuron model as a principal component analyzer. *J. Math. Biol.*, **15**, 267–273.
57. Liu, J. and Jia, Y. (2014) Hebbian-based mean shift for learning the diverse shapes of V1 simple cell receptive fields. *Chin. Sci. Bull.*, **59** (4), 452–458, doi: 10.1007/s11434-013-0041-4
58. Peharz, R. and Pernkopf, F. (2012) Sparse nonnegative matrix factorization with l0-constraints. *Neurocomputing*, **80**, 38–46, doi: 10.1016/j.neucom.2011.09.024
59. Charles, A.S., Garrigues, P., and Rozell, C.J. (2012) A common network architecture efficiently implements a variety of Sparsity-Based inference problems. *Neural Comput.*, **24** (12), 3317–3339, doi: 10.1162/neco_a_00372, *http://citeseerx.ist.psu.edu/viewdoc/download?doi=10.1.1.310.2639&rep=rep1&type=pdf.*
60. Aharon, M., Elad, M., and Bruckstein, A.M. (2006) On the uniqueness of overcomplete dictionaries, and a practical way to retrieve them. *Linear Algebra Appl.*, **416** (1), 48–67, doi: 10.1016/j.laa.2005.06.035
61. Daugman, J.G. (1980) Two-dimensional spectral analysis of cortical receptive field profiles. *Vision Res.*, **20**

(10), 847–856, doi: 10.1016/0042-6989(80)90065-6
62. Field, D.J. (1999) Wavelets, vision and the statistics of natural scenes. *Philos. Trans. R. Soc. London, Ser. A*, **357** (1760), 2527–2542, doi: 10.1098/rsta.1999.0446
63. Fischer, S., Sroubek, F., Perrinet, L.U., Redondo, R., and Cristóbal, G. (2007) Self-invertible 2D log-Gabor wavelets. *Int. J. Comput. Vision*, **75** (2), 231–246, doi: 10.1007/s11263-006-0026-8
64. King, P.D., Zylberberg, J., and DeWeese, M.R. (2013) Inhibitory interneurons decorrelate excitatory cells to drive sparse code formation in a spiking model of V1. *J. Neurosci.*, **33** (13), 5475–5485, doi: 10.1523/jneurosci.4188-12.2013
65. Mallat, S. and Zhang, Z. (1993) Matching Pursuit with time-frequency dictionaries. *IEEE Trans. Signal Process.*, **41** (12), 3397–3414.
66. Oliphant, T.E. (2007) Python for scientific computing. *Comput. Sci. Eng.*, **9** (3), 10–20, doi: 10.1109/MCSE.2007.58
67. Hunter, J.D. (2007) Matplotlib: a 2D graphics environment. *Comput. Sci. Eng.*, **9** (3), 90–95, doi: 10.1109/MCSE.2007.55
68. Geisler, W.S., Perry, J.S., Super, B.J., and Gallogly, D.P. (2001) Edge co-occurence in natural images predicts contour grouping performance. *Vision Res.*, **41** (6), 711–724, doi: 10.1016/s0042-6989(00)00277-7
69. Portilla, J. and Simoncelli, E.P. (2000) A parametric texture model based on joint statistics of complex wavelet coefficients. *Int. J. Comput. Vision*, **40** (1), 49–70, doi: 10.1023/a:1026553619983
70. Perrinet, L.U. and Bednar, J.A. (2014) Edge co-occurrences are sufficient to categorize natural versus animal images. *Scientific Reports* (2015), doi: 10.1038/srep11400.
71. Nava, R., Marcos, J.V., Escalante-Ramírez, B., Cristóbal, G., Perrinet, L.U., and Estépar, R.S.J. (2013) *Advances in Texture Analysis for Emphysema Classification*, Lecture Notes in Computer Science, vol. **8259**, Chapter 27, Springer-Verlag, Berlin and Heidelberg, pp. 214–221, doi: 10.1007/978-3-642-41827-3_27
72. Hughes, J.M., Graham, D.J., and Rockmore, D.N. (2010) Quantification of artistic style through sparse coding analysis in the drawings of Pieter Bruegel the Elder. *Proc. Natl. Acad. Sci.*, **107** (4), 1279–1283, doi: 10.1073/pnas.0910530107
73. Ganguli, D. and Simoncelli, E. (2010) Implicit encoding of prior probabilities in optimal neural populations, in *Advances in Neural Information Processing Systems*, vol. **23** (eds J. Lafferty, C.K.I. Williams, J. Shawe-Taylor, R.S. Zemel, and A. Culotta), pp. 658–666, *http://www.cns.nyu.edu/pub/lcv/ganguli10c-preprint.pdf.*
74. Girshick, A.R., Landy, M.S., and Simoncelli, E.P. (2011) Cardinal rules: visual orientation perception reflects knowledge of environmental statistics. *Nat. Neurosci.*, **14** (7), 926–932, doi: 10.1038/nn.2831
75. Keil, M.S. and Cristóbal, G. (2000) Separating the chaff from the wheat: possible origins of the oblique effect. *J. Opt. Soc. Am. A Opt. Image Sci. Vis.*, **17** (4), 697–710, doi: 10.1364/josaa.17.000697
76. Laughlin, S. (1981) A simple coding procedure enhances a neuron's information capacity. *Z. Naturforsch. Sect. C: Biosci.*, **36** (9-10), 910–912, *http://view.ncbi.nlm.nih.gov/pubmed/7303823.*
77. Sigman, M., Cecchi, G.A., Gilbert, C.D., and Magnasco, M.O. (2001) On a common circle: natural scenes and Gestalt rules. *Proc. Natl. Acad. Sci. U.S.A.*, **98** (4), 1935–1940, doi: 10.1073/pnas.031571498
78. Grossberg, S. (1984) *Outline of A Theory of Brightness, Color, and form Perception*, vol. **20**, Elsevier, pp. 59–86, doi: 10.1016/s0166-4115(08)62080-4
79. von der Heydt, R., Peterhans, E., and Baumgartner, G. (1984) Illusory contours and cortical neuron responses. *Science (New York)*, **224** (4654), 1260–1262, doi: 10.1126/science.6539501
80. Field, D.J., Hayes, A., and Hess, R.F. (1993) Contour integration by the human visual system: evidence for a local "association field". *Vision Res.*,

33 (2), 173–193, doi: 10.1016/0042-6989(93)90156-Q, *http://view.ncbi.nlm.nih.gov/pubmed/8447091.*

81. August, J. and Zucker, S. (2001) A Markov process using curvature for filtering curve images, in *Energy Minimization Methods in Computer Vision and Pattern Recognition*, Lecture Notes in Computer Science, vol. **2134** (eds M. Figueiredo, J. Zerubia, and A. Jain), Springer-Verlag, Berlin and Heidelberg, pp. 497–512, doi: 10.1007/3-540-44745-8_33.
82. Spratling, M.W. (2013) Image segmentation using a sparse coding model of cortical area V1. *IEEE Trans. Image Process.*, **22** (4), 1631–1643, doi: 10.1109/tip.2012.2235850
83. Spratling, M.W. (2013) Classification using sparse representations: a biologically plausible approach. *Biol. Cybern.*, **108** (1), 61–73, doi: 10.1007/s00422-013-0579-x
84. Dumoulin, S.O., Hess, R.F., May, K.A., Harvey, B.M., Rokers, B., and Barendregt, M. (2014) Contour extracting networks in early extrastriate cortex. *J. Vis.*, **14** (5), 18, doi: 10.1167/14.5.18
85. McManus, J.N.J., Li, W., and Gilbert, C.D. (2011) Adaptive shape processing in primary visual cortex. *Proc. Natl. Acad. Sci. U.S.A.*, **108** (24), 9739–9746, doi: 10.1073/pnas.1105855108
86. Hunt, J.J., Giacomantonio, C.E., Tang, H., Mortimer, D., Jaffer, S., Vorobyov, V., Ericksson, G., Sengpiel, F., and Goodhill, G.J. (2009) Natural scene statistics and the structure of orientation maps in the visual cortex. *Neuroimage*, **47** (1), 157–172, doi: 10.1016/j.neuroimage.2009.03.052
87. Olshausen, B.A. and Field, D.J. (2005) How close are we to understanding V1? *Neural Comput.*, **17** (8), 1665–1699, doi: 10.1162/0899766054026639, *http://portal.acm.org/citation.cfm?id=1118017.*
88. Sanz-Leon, P., Vanzetta, I., Masson, G.S., and Perrinet, L.U. (2012) Motion Clouds: model-based stimulus synthesis of natural-like random textures for the study of motion perception. *J. Neurophysiol.*, **107** (11), 3217–3226, doi: 10.1152/jn.00737.2011
89. Simoncini, C., Perrinet, L.U., Montagnini, A., Mamassian, P., and Masson, G.S. (2012) More is not always better: adaptive gain control explains dissociation between perception and action. *Nat. Neurosci.*, **15** (11), 1596–1603, doi: 10.1038/nn.3229

15
Biologically Inspired Keypoints

Alexandre Alahi, Georges Goetz, and Emmanuel D'Angelo

15.1
Introduction

Computer vision researchers have spent the past two decades handcrafting or learning the best image representations, that is, features, to extract semantically meaningful information from images and videos. The main motivation is to transform the visual signal into a more compact and robust representation in order to accurately recognize scenes, objects, retrieve similar contents, or even reconstruct the original 3D contents from 2D observations. A large number of applications can benefit from this transformation, ranging from human behavior understanding for security and safety purposes, panorama stitching, augmented reality, robotic navigation, to indexing and browsing the deluge of visual contents on the web.

Meanwhile, the neuroscience community is also studying how the human visual system encodes visual signals to allow the creation of advanced visual prosthesis for restoration of sight to the blind [1]. Although both communities address similar problems, they tackle their challenges independently, from different perspectives. Reducing the knowledge gap between these two worlds could contribute to both communities. Finding an image descriptor whose interpretation matches the accuracy (and/or robustness) of human performance has yet to be achieved. How similar are the image encoding strategies used in computer vision to our current understanding of the retinal encoding scheme? Can we improve their performance by designing descriptors that are biologically inspired?

Concurrently, progress in computational power and storage enables machine learning techniques to learn complex models and outperform handcrafted methods [2]. Features learned with convolutional neural networks (CNN) over a large dataset lead to impressive results for object and scene classification [3]. Can we design neuroscience experiments to validate the learned representations? Finally, theoretical models discussed in neuroscience can now easily be tested and evaluated on large-scale databases of billions of images.

Biologically Inspired Computer Vision: Fundamentals and Applications, First Edition.
Edited by Gabriel Cristóbal, Laurent Perrinet, and Matthias Keil.

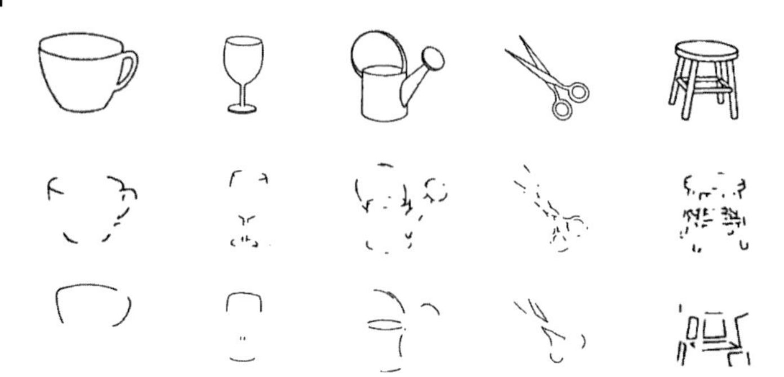

Figure 15.1 Illustration of the importance of corners to recognize objects as opposed to edges (illustration from Biederman [5]).

Neisser [4] modeled human vision as two stages: preattentive and attentive. The former extracts keypoints and the latter models the relationships among them. Fifty years later, this model is still inspiring the designs of computer vision algorithms to solve the visual search task. Accordingly, we humans do not scan a scene densely to extract meaningful information. Our eyes jump on a sparse number of salient visual parts to recognize a scene or an object. The corresponding eye movements are referred to as *saccades*. Where are these key salient locations and how can we robustly identify them?

Biederman [5] performed an interesting experiment showing the importance of corners as opposed to points on edges to recognize objects (see Figure 15.1). One can look at edges as roads on a map and corners as the route choices. With only access to information about the route choices, one can easily reconstruct the full map by linearly connecting the dots. Conversely, if we only share partial roads, it is not easy to reconstruct the full map. In other words, to recognize an object, it is actually intuitive to consider its highly informative/significant local parts, for example, corners. Such an assumption is in line with the claim that a keypoint should capture a distinctive and informative information of an object. Keypoints (Figure 15.2) are also seen as image locations that can be reliably tracked [6]. Image patches with large gradients are better candidates as keypoints than textureless ones.

Figure 15.2 Illustration of the keypoints extracted over an image using the implementation of the SIFT [7] algorithm available in Ref. [8]. Each circle represent a keypoint. The radius of the circle corresponds to the image patch used to describe the keypoint. The radius within each circle describes the orientation of the keypoint. More details are available in Section 15.4

In this chapter, we describe methods to extract and represent keypoints. We propose to highlight the design choices that are not contradictory to our current understanding of the human visual system (HVS). We are far from claiming any biological findings, but it is worth mentioning that state-of-the-art keypoint extraction and representation can be motivated from a biological standpoint. The chapter is structured as follows: first, we briefly define some terminologies for the sake of clarity. Then, we present a quick overview of our current understanding of the HVS. We highlight models and operations within the front end of the visual system that can inspire the computer vision community to design algorithms for the detection and encoding of keypoints. In Sections 15.4 and 15.5, we present details behind the state-of-the-art algorithms to extract and represent keypoints, although it is out of the scope of this chapter to go into intricate details. We nonetheless emphasize the design choices that *could* mimic some operations within the human visual system. Finally, we present a new evaluation framework to get additional insight on the performance of a descriptor. We show how to reconstruct a keypoint descriptor to qualitatively analyze its behavior. We conclude by showing how to design a better image classifier using the reconstructed descriptors.

15.2 Definitions

Several terminologies have been used in various communities to describe similar concepts. For the sake of clarity, we briefly define the following terms:

Definition 1 *A **keypoint** is a salient image point that visually stands out and is likely to remain stable under any possible image transformation such as illumination change, noise, or affine transformation to name a few.*

The image *patch* surrounding the keypoint is described by compiling a vector given various image operators. The patch size is given by a scale space analysis that is in practice 1 – 10% of the image size. In the literature, a keypoint can also be referred to as an *interest point*, *local feature*, *landmark*, or an *anchor point*. The terms keypoint *detector* or *extractor* have often been used interchangeably, although the latter is probably more correct.

The equivalent biological processes of extraction and encoding of keypoints take place in the front end of the HVS, possibly as early as in the retina.

Definition 2 *The **front end** of the visual system consists of its first few layers of neurons, from the retina photoreceptors to the primary visual cortex located in the occipital lobe of the brain.*

Definition 3 *The region over which a retinal ganglion cell (RGC) responds to light is called the* receptive field *of that cell (see Figure 15.5) [9].*

The terms *biologically inspired, biomimicking*, or *biologically plausible* have similar meanings and will sometimes be used interchangeably. In this chapter, we will be drawing analogies with the behavior of the front end of the visual system in general, and the retina in particular.

15.3
What Does the Frond-End of the Visual System Tell Us?

15.3.1
The Retina

The retina is a thin layer of tissue which lines the back of the eye (see Chapters 1 and 2). It is responsible for transducing visual information into neural signals that the brain can process. A rough analogy would be that the retina is the "CCD camera" of the visual system, even though it does much more than capture a per-pixel representation of the visual world [10]. In recent years, neuroscientists have taken significant steps toward understanding the operations and calculations which take place within the human retina, that is, how images *could be* transmitted to the brain. The mammalian retina itself has a layered structure, in which only three types of cells, the rod and cone photoreceptors, as well as the much rarer intrinsically photosensitive RGCs, are sensitive to light [11–13]. Intrinsically, photosensitive RGCs are not thought to be involved in the image-forming pathways of the visual system, and instead are thought to play a role in synchronizing the mammalian circadian rhythm with environmental time. We therefore ignore them when describing the image encoding properties of the retina.

15.3.2
From Photoreceptors to Pixels

The photoreceptors can loosely be thought of as the "pixels" of the visual system (see Figure 15.3). They are indirectly connected to the RGCs whose axons form the optic nerve via a finely tuned network of neurons, known as the *horizontal, bipolar, and amacrine* cells. The bipolar cells are responsible for vertical connections through the retinal network, with horizontal and amacrine cells providing lateral connections for comparison of neighboring signals. RGCs output action potentials (colloquially referred to as "spikes", which they "fire") to the brain. They consist of binary, on or off signals, which the RGCs can send at up to several hundred hertz. By the time visual signals leave the retina via the optic nerve, a number of features describing the image have been extracted, with remarkably complex operations sometimes taking place: edge detection across multiple spatial scales [14], but also motion detection [15], or even segregation of object and background motion [16]. At the output of the retina, at least 17 different types of ganglion cells have extracted different spatial and temporal features of the photoreceptor activity patterns [17] over regions of the retina that have little to no overlap for a given ganglion cell type, thereby efficiently delivering information to the subsequent stages

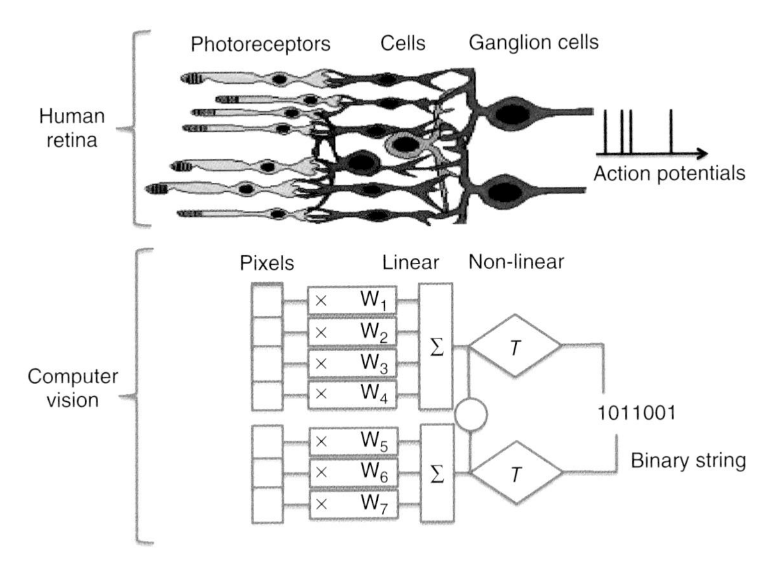

Figure 15.3 From the retina to computer vision: the biological pathways leading to action potentials can be emulated by a linear nonlinear cascade. [Upper part of the image is a courtesy of the book Avian Visual Cognition by R. Cook]

of the visual system [18]. It is worth noting that the retina does not respond to static images and instead transmits an "event-based" representation of the world to the visual cortex, where these events correspond in their simplest form to transitions from light to dark (OFF transitions), dark to light (ON transitions), or to the more complex signals mentioned in the above.

15.3.3 Visual Compression

The visual system in general, and the retina, in particular, is an exquisitely and precisely organized system, from spatial and temporal points of view. While there are more than 100 million photoreceptors in the human eye, including approximately 5 million cone photoreceptors, the optic nerve only contains about 1 million axons, indicating that a large part of compression of the visual information takes place before it even leaves the eye. In all but the most central regions of the retina—the fovea—several photoreceptors influence each ganglion cell.

15.3.4 Retinal Sampling Pattern

Anatomically, the retina is segmented into four distinct concentric regions, defined by their distance from the central region of the retina, known as the foveola (see Figure 15.4(a)). High-resolution vision is provided by the foveola. The fovea surrounds the foveola and contains the RGCs which process the visual

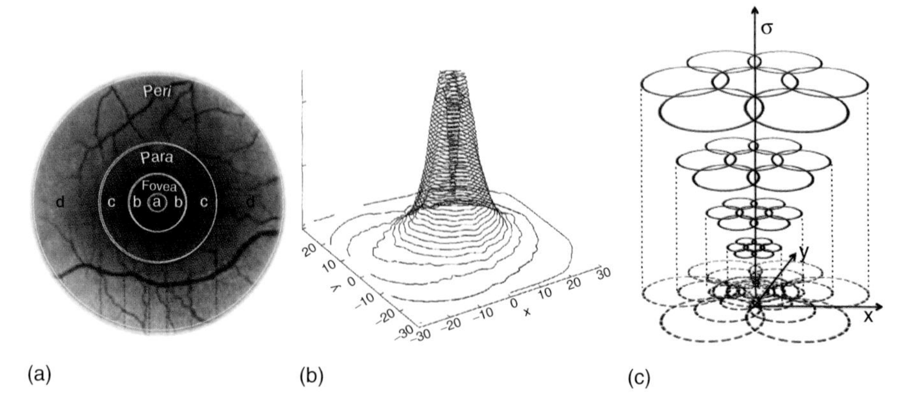

Figure 15.4 In the primate visual system, the fovea is the region of highest visual acuity. The RGCs connected to these photoreceptors are displaced in a belt which surrounds the foveola. (a) Illustration of the anatomical description of the retinal areas. The mean radial values of each of the retina areas are fovea: 1 mm; para: 3 mm; and peri: 5 mm. (b) The density of the RGC exponentially increases when moving toward the foveola [19]. The parafoveal area provides an area of intermediary visual acuity, while the low-resolution perifoveal area provides the visual system with a highly compressed, but also highly sensitive representation of the world. (c) In each region of the retina, a multiscale representation of the visual system further takes place, with different functional classes of RGCs extracting information at different spatial scales.

information captured by the cone photoreceptors located in the foveola. The foveola is indeed completely devoid of cells other than cone photoreceptors and Müller cells, which provide structural integrity to the tissue. An intermediate area, the parafoveal region, retains a relatively high RGC density and visual acuity. Both of these decrease to a minimum in the high-eccentricity perifoveal areas. In the periphery, rod photoreceptors form the majority of light-sensitive cells. Combined with a high convergence of the visual pathways, which results in high amplification of the visual signals, this makes the peripheral retina highly sensitive and efficient in low-illumination conditions, even if acuity leaves to be desired in it.

In the fovea, there is a one-to-one connection between individual photoreceptors and RGCs [20]. As the distance to the foveola increases, RGCs connect to more and more photoreceptors and consequently, image compression increases and visual acuity diminishes (Figure 15.4(b)). In the most peripheral regions of the retina, the midget ganglion cells, which are the smallest RGCs and thought to be responsible for the low-acuity visual pathways, can be connected to upward of 15 individual cone photoreceptors [9].

15.3.5
Scale-Space Representation

The fine-to-coarse evolution of acuity across the visual field suggests that the retina captures a multiscale representation of the world. The local spatial

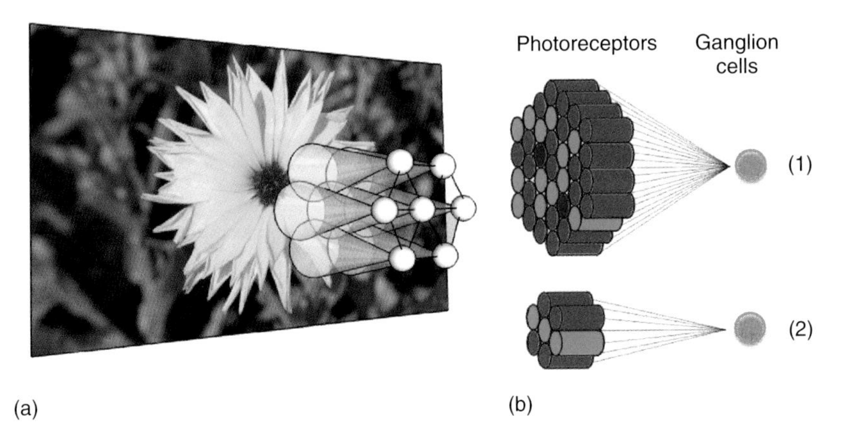

Figure 15.5 (a) Schematic representation of seven ganglion cells and their receptive fields, sampling from neighboring regions of an image (image courtesy of [21]). (b) In the perifoveal area, RGC-receptive fields are large and consist of many photoreceptors (1). As eccentricity decreases, the number of photoreceptors influencing RGCs decreases (2).

organization of the retina further reinforces this idea. At any point in the retina, each cone photoreceptor connects to multiple RGCs via a network of bipolar and amacrine cells. Different functional types of RGCs connect to different numbers of photoreceptors, resulting in a local multiscale representation of the visual signals [9], with so-called midget cells extracting information at finer visual scales, while so-called parasols encode a coarser description of the images (see Figures 15.4(c) and 15.5).

15.3.6
Difference of Gaussians as a Model for RGC-Receptive Fields

Receptive fields of RGCs consist of a central disk, the "center," and a surrounding ring, the "surround." The center and the surround are antagonistic, in that they respond to light increments of opposite polarity. An RGC that responds to light transitions from dark to light over its center will respond to transitions from light to dark over its surround. Such a cell is called an *ON-center OFF-surround* RGC, often conveniently shortened to "OFF RGC". Conversely, OFF-center ON-surround, or ON RGCs, also exist. The center–surround shape of an RGC-receptive field is classically modeled as a difference of Gaussians (DoG) [22, 23] (see Section 15.4 for more details on the DoG). While this model fails to capture the fine structure of receptive fields [9], it appears to accurately describe the relative weighting of center and surround cells at coarser scales.

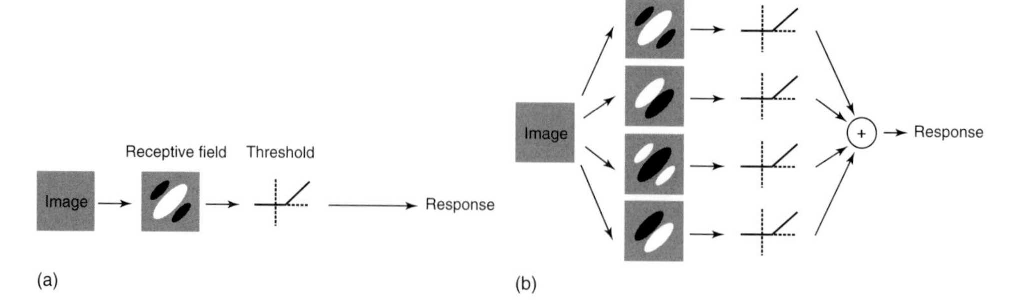

Figure 15.6 Illustration of the simple and complex cells proposed by Movhson *et al.* [25]. Image courtesy of [26]. (a) Simple cells are modeled using a linear filter followed by a rectifying nonlinearity. (b) For complex cells, models start with linear filtering by a number of receptive fields followed by rectification and summation.

15.3.7 A Linear Nonlinear Model

While differences of Gaussians can describe retinal ganglion-cell-receptive fields relatively well on average, the retina does not behave like a simple linear system in general. A classical computational description of the retina [24] consists of a linear – nonlinear filter cascade known as the *LN model* (see Figure 15.6(a)), sometimes extended to take into account couplings between adjacent neurons and the non-Poisson stochastic firing patterns of the RGCs. The extended version of the LN model is known as the generalized linear model (GLM). In the LN model, a spatial filter mask extracts a region of interest over the image, and the result is transformed into a stochastic neural firing rate by the nonlinearity. This filter is typically chosen to be an exponential. In computer vision, it is possible to adopt a similar approach for extracting and representing features, as shown on Figure 15.3.

Today, researchers are probing the retina at the single-photoreceptor level [9], measuring cells' responses to various stimuli, and trying to develop a precise understanding of how the retina extracts visual information across functional cell types.

15.3.8 Gabor-Like Filters

Further in the visual system, Hubel and Wiesel, Nobel Prize winners, discovered through seminal studies [27] that in the visual cortex of the cat, there exist cells that respond to very specific sets of stimuli. They separated neurons in the visual cortex into two different functional groups. For *simple cells,* responses can be predicted using a single linear filter for their receptive field, followed by a unique nonlinearity. For *complex cells* the response cannot be predicted using a single linear – nonlinear cascade (see Figure 15.6(b)).

Simple cells have the following properties:

- The spatial regions of the visual field over which inputs to the cell are excitatory and inhibitory are distinct.
- Spatially distinct regions of the receptive field add up linearly.
- Excitatory and inhibitarory regions balance each other out.

As a consequence, it is possible to predict the response of these cells remarkably well, using a simple linear – nonlinear model.

In 1987, Jones and Palmer performed extensive characterization of the functional receptive fields of simple cells, and found remarkable correspondence between measured receptive fields and two dimensional Gabor filters [28], thereby suggesting that the visual system adopts a sparse and redundant representation of visual signals (see Chapter 14). Building on these findings, Olshausen and Field demonstrated in 1996 that a learning algorithm that attempts to find a sparse linear code for natural scenes will lead to Gabor-like filters, similar to those found in simple cells of the visual cortex [29] (see Chapter 14). Gabor filters have since been widely adopted by the computer vision community.

15.4 Bioplausible Keypoint Extraction

The computer vision community has developed many algorithms over the past decades to identify keypoints within an image. In 1983, Burt and Adelson [30] built a Laplacian pyramid for coarse-to-fine extraction of keypoints. Since then, new methods based on multiscale Laplacian ofGaussians (LoG) [31], DoG [32], or Gabor filters [33] have been proposed. DoG is an approximation of LoG which was proposed to speed up long computation times associated with LoGs. Grossberg *et al.* [34] extended the DoG to directional differences-of-offset-Gaussians (DOOG) to improve the extraction step.

A popular method for keypoint extraction consists of computing the first-order statistics from zero-crossing of DoGs at different scales [35]. The scale-invariant feature transform (SIFT) [7] uses such an approach and is still one of the most used detectors after 15 years of research in the field.

15.4.1 Scale-Invariant Feature Transform

The SIFT consists of the following steps:

1) Compute the DoG of the image (see Figure 15.7).
2) Keypoints are the local maxima/minima of the DoG in scale space.
3) Use a Taylor expansion to interpolate an accurate position of the keypoint at a subpixel accuracy. It is similar to fitting a 3D curve to get location in terms of (x, y, σ) where x, y is the pixel location and σ the Gaussian variance (referred to as *scale*).)

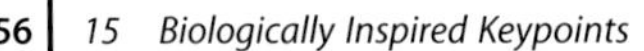

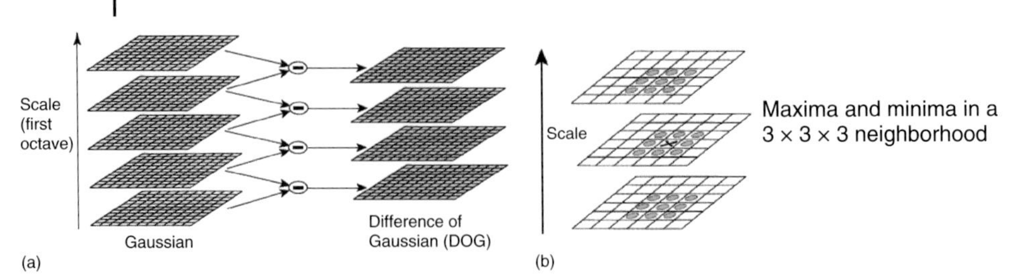

Figure 15.7 Illustration of the scale-space computation with DoG used in SIFT (a). We also illustrate the $3 \times 3 \times 3$ neighborhood used to compute the local maxima/minima (b). Image courtesy of [7].

4) Discard keypoints along edges by discarding points with high ratio between the principal curvatures. The principal curvatures can be computed from the Hessian matrix.

Such a method leads to keypoints that are scale invariant. In general, basic keypoint extraction methods require the computation of image derivatives in both the x and y direction. The image is convolved with derivatives of Gaussians. Then, the Hessian matrix is formed, representing the second-order derivative of the image. It describes the local curvature of the image. Finally, the maxima of the determinant of the Hessian matrix are interpolated. A very popular fast implementation of such an algorithm was proposed by Bay *et al.*, and was called *speeded-up robust features* (SURF) [36].

15.4.2 Speeded-Up Robust Features

The algorithm behind SURF is the following:

1) Compute the Gaussian second-order derivative of the image (instead of the DoG in SIFT).
2) A non-maximum suppression algorithm is applied in $3 \times 3 \times 3$ neighborhoods (position and scale).
3) The maxima of the determinant of the Hessian matrix are interpolated in space and scale.

Both SIFT and SURF find extrema at subsampled pixels that can adjust accuracy at larger scales. Agrawal *et al.* [37] use simplified center–surround filters (squares or octagons instead of circles) to approximate the Laplacian in an algorithm referred to as the *center surround extrema* (CenSurE) (see Figure 15.8).

15.4.3 Center Surround Extrema

The CenSurE algorithm is as follows:

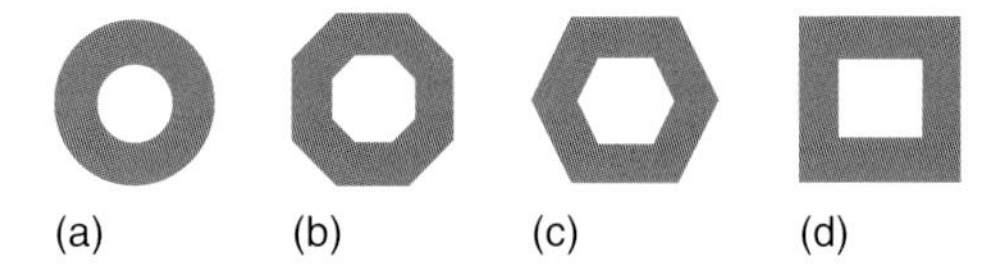

Figure 15.8 Illustration of various type of center–surround filters used in the CenSure method [37]. (a) Circular filter. (b) Octagonal filter. (c) Hexagonal filter. (d) Box filter.

1) Octagonal center–surround filters of seven different sizes are applied to each pixel.
2) A non-maximal suppression algorithm is performed in a $3 \times 3 \times 3$ neighborhood.
3) Points that lie along the lines are rejected if the ratio of the principal curvature is too big. The filtering is done using the Harris method [38].

Although CenSurE was designed for speed, simpler and faster methods exist that also use circular patterns to evaluate the likelihood of having a keypoint. Rosten and Drummond [39] proposed a fast computational technique called *features from accelerated segment test* (*FAST*).

15.4.4 Features from Accelerated Segment Test

The FAST uses binary comparisons with each pixel in a circular ring around the candidate pixel. It uses a threshold to determine whether a pixel is less than the center pixel. The second derivative is not computed as it is in many state-of-the-art methods. As a result, it can be sensitive to noise and scale.

In brief, state-of-the-art keypoint extractor algorithms rely on methods for which parallels can be drawn with the way the HVS operates. These methods mainly study the gradient of the image with LoG, its approximated version with DoG, or center–surround filters. Gabor-like keypoint detectors were proposed in 1992 to compute local energy of the signal [40, 41]. In this framework, a keypoint is the maximum of the first- and second-order derivatives with respect to position. A multiscale strategy is suggested in Refs [42, 43]. Note that Gabor functions are more often used to represent the keypoints as opposed to extracting them.

15.5 Biologically Inspired Keypoint Representation

15.5.1 Motivations

In computer vision, the keypoint location itself is not informative enough to solve retrieval or recognition tasks. A vector is computed for each keypoint to provide information about the image patch surrounding the keypoint. The patch can be described densely with Gabor-like filters or sparsely with Gaussian kernels.

15.5.2 Dense Gabor-Like Descriptors

It is quite common to use the response to specific filters to describe a patch. The same filters can be applied for many different orientations. In the 1990s, Freeman and Adelson [44] used steerable filters to evaluate the response of an arbitrary orientation by interpolation from a small set of filters. As mentioned in Section 15.3, the cells' responses in the HVS can be seen as kernels. While there is no general consensus in the research community on the precise nature of the kernels, there is a convergence toward kernels that are roughly shaped like Gabor functions, derivatives, or differences of round or elongated Gaussian functions.

Over the past decades, the most common descriptors have been those modeling the gradient responses over a set of orientations. Indeed, the most popular keypoint descriptor has been the SIFT representation [7] which uses a collection of histogram of oriented gradients (HOG). Such a feature counts occurrences of gradient orientations in an image patch. In practice, the orientations are quantized into eight bins. Each bin can be seen as the response to a filter, that is, a simple cell. Figure 15.9 illustrates how a HOG can be seen as the responses of a set of simple cells.

15.5.2.1 Scale-Invariant Feature Transform Descriptor

The SIFT description vector is obtained as follows:

1) Compute the gradient magnitude and direction for every pixel of the patch surrounding the keypoint. Note that a Gaussian weighting function is applied to emphasize the importance of pixels near the keypoint location.
2) The image patch is segmented into $4 \times 4 = 16$ cells. The eight-bin HOG feature is computed for each cell.
3) The vector is sorted given the highest orientation and normalized to be less sensitive to illumination changes.

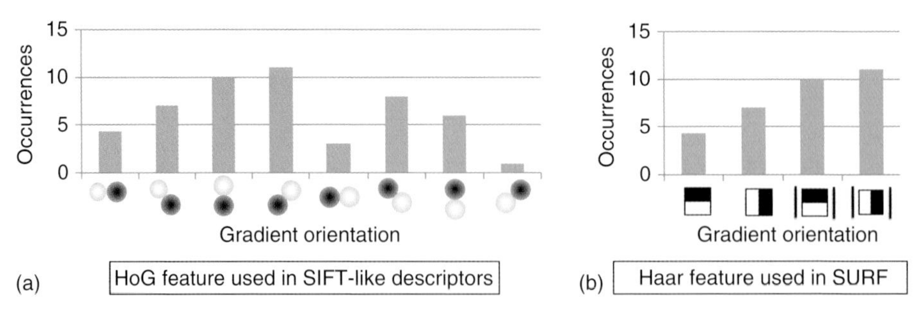

Figure 15.9 Illustration of the filter responses used to describe SIFT [7] (a) and (b) SURF [36] keypoints.

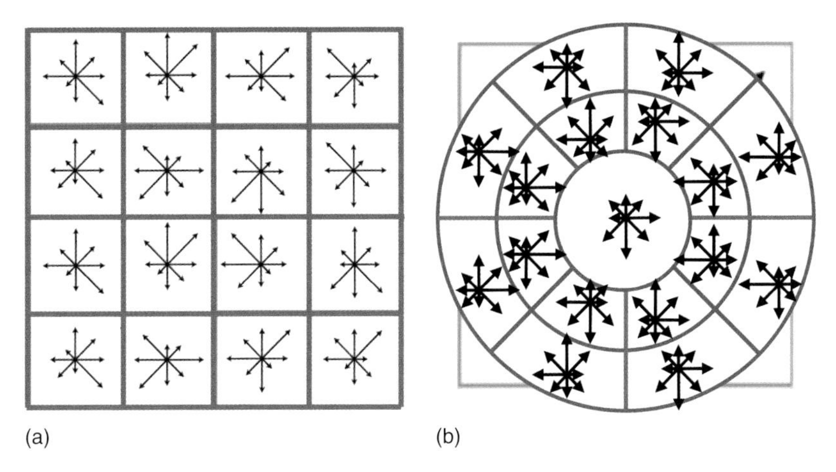

Figure 15.10 Comparison between the (a) SIFT [7] and (b) GLOH [46] strategies to segment an image patch into cells of HOG.

Different variants of the SIFT descriptor have been proposed in the literature such as PCA-SIFT [45], or GLOH [46]. The former reduces the descriptor size from 128 to 36 dimensions using principal component analysis (PCA). The latter uses a log-polar binning structure instead of the four quadrants used by SIFT (see Figure 15.10). GLOH uses spatial bins of radius 6, 11, and 15, with eight angular bins, leading to a 272-dimensional histogram. The descriptor size is further reduced from 272 to 128 using PCA.

Although it is well accepted in the computer vision community that HOG-like descriptors are powerful features, their performance varies with respect to the number of quantization levels, sampling and overlapping strategies (e.g., polar [46] vs rectangular [7]), and whether the gradient is signed or unsigned. Bay *et al.* [36] propose to approximate the HOG representation with Haar-like filters to describe SURF keypoints.

15.5.2.2 **Speeded-Up Robust Features Descriptor**

The SURF description vector is obtained as follows:

1) A Gaussian-weighted Haar wavelet response in the x and y directions is applied every 30°to identify the orientation of the keypoint.
2) The region surrounding the keypoint is broken down into a 4×4 grid. In each cell, the response in the x and y directions to Haar-wavelet features are added together. The sum of the absolute values is computed to form a $4 \times 4 \times 4 = 64$ element vector.
3) This vector is then normalized to be invariant to changes in contrast.

Figure 15.9 illustrates the SIFT and SURF description vectors.

15.5.3 Sparse Gaussian Kernels

In the signal processing community, researchers have recently started trying to reconstruct signals given a sparsity constraint with respect to a suitable basis. Interestingly, many signals and phenomena can be represented sparsely, with only a few nonzero (or active) elements in the signal. Images are in fact sparse in the gradient domain. They are considered as piecewise smoothed signals. As a result, we do not need to compute the responses of filters over all locations.

Calonder *et al.* [47] showed that computing the Gaussian difference between a few random points within the image patch of a keypoint is good enough to describe it for keypoint matching purposes. In addition, the Gaussians do not need to be concentric. They further use the sign of the difference to represent the keypoint, leading to a very compact binary descriptor. This method belongs to the family of descriptors coined *local binary descriptors* (LBDs).

15.5.3.1 Local Binary Descriptors

Local Binary Descriptors perform the following steps N times:

1) Compute the Gaussian average at two locations x_i and x_j within the image patch with variance σ_i and σ_j, respectively.
2) Compute the difference between these two measurements
3) Binarize the resulting difference by retaining the sign only.

There are several considerations in the design of LBD. How should we select the Gaussians (locations and variances)? How many times should we perform the steps (i.e., what is a good size for N)? What is the optimal sampling of the image patch to encode as much information as possible with the lowest number of comparison (i.e., bits)?

Interestingly, LBD can be compared to the linear nonlinear filter cascades present in the HVS (see Figure 15.3). Alahi *et al.* [48] propose an LBD inspired by the retinal sampling grid. There are many possible sampling grids to compare pairs of Gaussian responses: for instance, BRIEF [47] and ORB [49] use random Gaussian locations with the same fixed variance. BRISK [50] uses a circular pattern with various Gaussian sizes. Alahi *et al.* proposed the fast retina keypoint (FREAK) [48] which uses a retinal sampling grid, as illustrated in Figure 15.11. They heuristically show that the matching performance increases when a retinal sampling grid is used as compared to previous strategies (see Figure 15.12).

15.5.3.2 Fast Retina Keypoint Descriptor

The FREAK uses a retinal pattern to promote a coarse-to-fine representation of a keypoint. Figure 15.11 illustrates the topology of the Gaussians kernels. Each circle represents the 1-standard deviation contour of a Gaussian applied to the corresponding sampling points. The pattern is made of one central point and seven concentric circles of six points each, with circles being shifted by 30°with respect to the contiguous ones. The FREAK pattern can be seen as a multiscale description

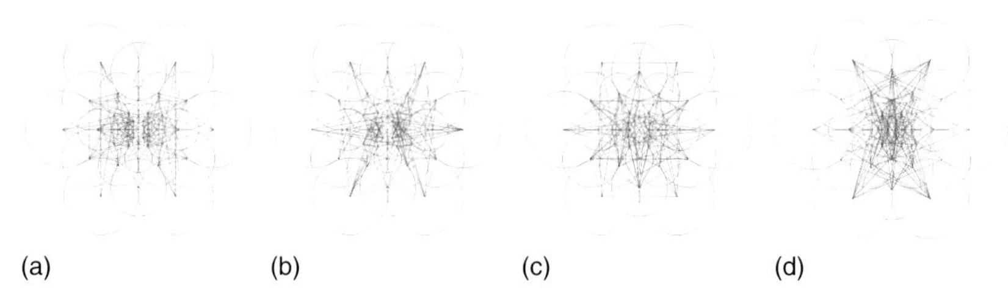

Figure 15.11 Illustration of the retinal sampling grid used in FREAK [48]. Each line represents the pairs of Gaussians selected to measure the sign of their difference. The first cluster mainly involves perifoveal receptive fields and the last ones foveal ones.

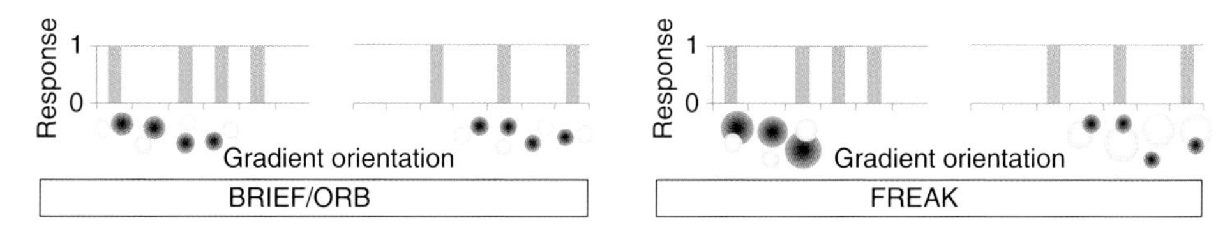

Figure 15.12 Illustration of the filter responses used to describe the binaries' descriptors (BRIEF [47], ORB [49], and FREAK [48]).

of the keypoint image patch. The coarse-to-fine representation allows the keypoint to be matched in a cascade manner, that is, segment after segment, to speed up the retrieval step. Given the retinal pattern, thousands of pairs of Gaussian responses can be measured. Alahi *et al.* [48] use a learning step similar to ORB [49] to select a restricted number of pairs. The selection algorithm is as follows:

1) Form a matrix D representing several hundred thousands of keypoints. Each row corresponds to a keypoint represented with its large descriptor made of all possible pairs in the retina sampling pattern illustrated in Figure 15.11. There are more than 900 possible pairs with 4 dozens of Gaussians.
2) Compute the mean of each column. In order to have a discriminant feature, high variance is desired. A mean of 0.5 leads to the highest variance of a binary distribution.
3) Order the columns with respect to the highest variance.
4) Keep the best column (mean of 0.5) and iteratively add the remaining columns having low correlation with the selected columns.

Figure 15.11 illustrates the pairs selected by grouping them into four clusters (128 pairs per group)[1]. Note that using more than 512 pairs did not improve the performance of the descriptor. The first cluster involves more of the peripheral Gaussians and the last one, the central ones. The remaining clusters highlight activities in the remaining area of the retina (i.e., the parafoveal are). The obtained clusters seem to match the behavior of the human eye. We first use the perifoveal

1) Groups of 128 pairs are considered here because of hardware requirements. The Hamming distance used in the matching process is computed by segments of 128 bits thanks to the SSE instruction set.

receptive fields to estimate the location of an object of interest. Then, validation is performed with the more densely distributed receptive fields in the foveal area. Although the used feature selection algorithm is greedy, it seems to match our understanding of the model of the human retina. Furthermore, it is possible to take advantage of the coarse-to-fine structure of the FREAK descriptor to match keypoints.

15.5.3.3 Fast Retina Keypoint Saccadic Matching

To match FREAK keypoints, Alahi *et al.* [48] have proposed to mimic saccadic search by parsing the descriptor in several steps. They start by matching with the first 128 bits of the FREAK descriptor representing coarse information. If the distance is smaller than a threshold, they further continue the comparison with the rest of the binary string to analyze finer information. As a result, a cascade of comparisons is performed decreasing the time required for the matching step with respect to other LBD. More than 90% of the candidates are discarded with the first 128 bits of the FREAK descriptor using a threshold large enough to speed up matching but not to alter the final result. Figure 15.13 illustrates the saccadic match. For visualization purposes, an object of interest is represented with a single FREAK descriptor of the size of its bounding circle (Figure 15.13(a)). Then, a new image is searched for the same object. All candidate image regions (a regular grid of keypoints at three different scales) are also described with a single descriptor of the size of the candidate region. The first cascade (Figure 15.13(b) and (d)) (first 16 bytes) discards many candidates and selects very few of them to compare with the remaining bytes. In Figure 15.13(e), the last cascade (Figure 15.13c) has correctly

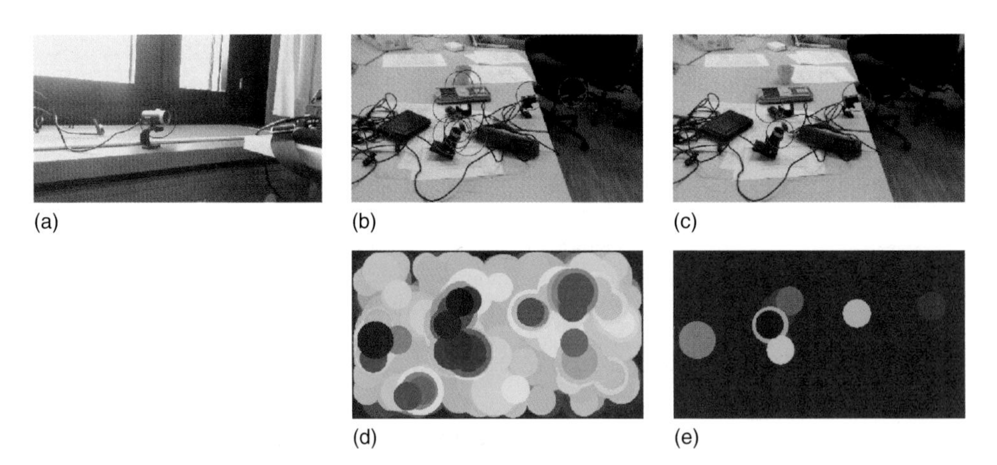

Figure 15.13 Illustration of the FREAK cascade matching. (a) An object of interest. (b) Ten best matches after the first cascade. (c) Three best matches after the last cascade. (d) Heatmap after the first cascade. (e) Heatmap after the last cascade. In (b) and (d), the distance of all regions to the object of interest is illustrated in color jet. In (c), we can see that the object of interest is located in the few estimated regions. These regions can simulate the saccadic search of the human eyes.

selected the locations of our object of interest despite the changes of illuminations and viewpoints.

15.5.4
Fast Retina Keypoint versus Other Local Binary Descriptors

All LBDs use Gaussian responses in order to be less sensitive to noise. BRIEF [47] and ORB [49] use the same kernel for all points in the patch. To match the retina model, FREAK uses different kernels for every sample point, similar to BRISK [50]. The difference from BRISK is the exponential change in size and the overlapping kernels. It has been observed that overlapping between points of different concentric circles improves performance. One possible reason is that with the presented overlap, more information is captured. An intuitive explanation could be formulated as follows.

Let us consider the intensities measured at the receptive fields A, B, and C where $I_A > I_B$, $I_B > I_C$, and $I_A > I_C$.

If the fields do not overlap, then the last test $I_A > I_C$ is does not add any discriminant information. However, if the fields overlap partially, new information can be encoded. In general, adding redundancy allows us to use fewer receptive fields. This has been a strategy employed in compressed sensing (CS) or dictionary learning for many years. According to Olshausen and Field [51], such redundancies also exist in the receptive fields of the visual cortex.

15.6
Qualitative Analysis: Visualizing Keypoint Information

15.6.1
Motivations

In the previous sections, we presented several methods to extract and represent keypoints. It is beyond the scope of this chapter to present all possible techniques. Tuytelaars and Mikolajczyk [46] extensively describe the existing keypoint extractor techniques and their desired properties for computer vision applications such as repeatability, locality, quantity, accuracy, efficiency, invariance, and robustness. For instance, the repeatability score is computed as follows:

$$repeatability = \frac{\#correspondences}{\min(n_a, n_b)}, \tag{15.1}$$

where # refers to the number of correspondences; n_a and n_b are the number of detected keypoints from an image a and its transformed image b, respectively. It measures whether a keypoint remains stable over any possible image transformation.

The performance of keypoint descriptors is solely compared on the basis of their performance in full grown, end-to-end machine vision tasks such as image registration or object recognition, as in Ref. [49]. While this approach is essential for developing working computer vision applications, it provides little hindsight on the type and quantity of local image content that is embedded in the keypoint under test.

We dedicate the last section of this chapter to the presentation of two lines of work that bring a different viewpoint on local image descriptors. In the first part, we see a method to invert binarized descriptors, thus allowing a qualitative analysis of the differences between BRIEF and FREAK. The second part presents an experiment that combines descriptor inversion, crowd intelligence, and some psychovisual representations to build better image classifiers.

Eventually, these research projects will yield a better methodology for developing image part descriptors that are not only accurate but also mimic the generalization properties experienced every day in the human mind.

15.6.2
Binary Feature Reconstruction: From Bits to Image Patches

In the preceding sections of the chapter, we learned that BRIEF and FREAK are LBDs: they produce a *bitstream* that should unambiguously *describe* local image content in the *vicinity* of a given keypoint. They differ only in the patch sampling strategy. While BRIEF uses fixed width Gaussians randomly spread over the patch, FREAK, on the other hand, relies on the retinal pattern and its Gaussian measurements are concentric and wider away from the patch center.

15.6.2.1 Feature Inversion as an Inverse Problem

While a detailed description of the solver used to invert binary descriptors is beyond the scope of the current chapter, we describe in this paragraph the mathematical modeling process that led to its development so that the reader can apply it to their own reconstruction problems.

Let us consider that we are given an image patch of size $\sqrt{N} \times \sqrt{N}$. We want to describe it using an LBD (be it BRIEF or FREAK) of length M bits. Empirically, we obtain a component of the feature descriptor by

1) picking a location within the patch;
2) overlapping this area with a Gaussian (green area in Figure 15.14);
3) multiplying point-to-point, the value of the Gaussian with the pixel values in the patch;
4) taking the mean of these premultiplied values;
5) repeating this process with another location and another Gaussian (red area in Figure 15.14);
6) computing the difference between the green mean and the red mean;
7) assigning $(+1)$ for the current feature vector component if the difference is positive and (-1), otherwise.

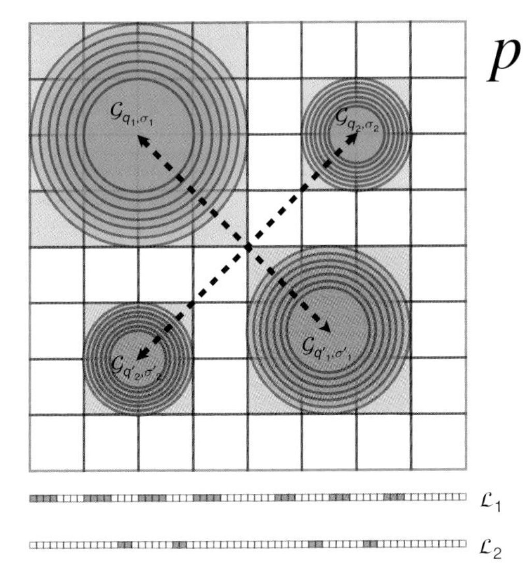

Figure 15.14 Example of a local descriptor for an 8×8 pixels patch and the corresponding sensing matrix. Only two measurements of the descriptor are depicted; each one is produced by subtracting the Gaussian mean in the lower (red) area from the corresponding upper (green) one. All the integrals are normalized by their area to have values in [0, 1]. Below this are the corresponding vectors.

If we reshape the image patch into a column of size N, steps 1–4 are mathematically represented by the dot product $\langle \mathcal{G}_{q,\sigma}, \boldsymbol{p} \rangle$ between a normalized Gaussian $\mathcal{G}_{q,\sigma} \in \mathbb{R}^N$ of width σ centered in $\boldsymbol{q} \in \mathbb{R}^2$ and the patch vector $\boldsymbol{p} \in \mathbb{R}^N$. Computing steps 1–5 for different pairs of points inside a patch and taking the difference depicted in step 6 is fully represented by the linear operator $\mathcal{L}$ defined by

$$\begin{aligned}\mathcal{L} : \mathbb{R}^N &\to \mathbb{R}^M \\ \boldsymbol{p} &\mapsto (\langle \mathcal{G}_{q_i,\sigma_i}, \boldsymbol{p} \rangle - \langle \mathcal{G}_{q'_i,\sigma'_i}, \boldsymbol{p} \rangle)_{1 \le i \le M},\end{aligned} \tag{15.2}$$

Since $\mathcal{L}$ is linear, it can be represented as a matrix $\boldsymbol{\mathcal{L}} \in \mathbb{R}^{M \times N}$ (Figure 15.14 bottom) applied to the input vectorized patch $\boldsymbol{p}$ and obtained by stacking M rows $\boldsymbol{\mathcal{L}}_i$:

$$\boldsymbol{\mathcal{L}}_i = \mathcal{G}_{q_i,\sigma_i} - \mathcal{G}_{q'_i,\sigma'_i}, \; 1 \le i \le M. \tag{15.3}$$

The final binary descriptor is obtained by the composition of this sensing matrix with a componentwise quantization operator $\mathbb{B}$ defined by $\mathbb{B}(\mathbf{x})_i = \text{sign} x_i$, so that, given a patch $\boldsymbol{p}$, the corresponding LBD reads as

$$\overline{\boldsymbol{p}} := \mathscr{B}(\boldsymbol{\mathcal{L}} \boldsymbol{p}) \quad \in \{-1, +1\}^M.$$

Even without the binarization step, the reconstruction problem for *any* LBD operator is *ill-posed* because there are typically many fewer measurements M than there are pixels in the patch. In Reference [52] $32 \times 32 = 1024$ pixel image patches are described using $M = 256$ measurements only. Classically, this problem can be made tractable by using a *regularized inverse problem* approach: adding regularization constraints allows the incorporation of a priori knowledge about the desired solution that will compensate for the lost information. The reconstructed patches will be obtained by solving a problem made of two constraints:

- a data term that will tie the reconstruction to the input binary descriptor;
- a regularization term that will enforce some prior knowledge about the desirable properties of the reconstruction.

In order to avoid any bias introduced by a specific training set, a natural candidate for the regularization constraint is the sparsity of the reconstructed patches in some wavelet frame $\boldsymbol{W}$. This requires that the patches have only a few nonzero coefficients after analysis in this wavelet frame. A simple way to enforce this constraint is to add a penalty term that grows with the number of nonzero coefficients. Mathematically, counting coefficients can be performed by measuring the *cardinality* (written $\sharp\{.\}$) of the set they belong to. For the case at hand, let us decompose these different steps:

1) A candidate patch $\mathbf{x}$ in a wavelet frame $\boldsymbol{W}$ that is obtained by computing the matrix-vector product $\boldsymbol{y} = \boldsymbol{W}\boldsymbol{x}$ is analyzed.
2) The set of interest is made of the components y_i of the vector $\mathbf{y}$ that are nonzero.
3) We count how many of these are there: $\sharp\{|y_i| \neq 0\}$.

Variational frameworks are designed to minimize the *norm* of error vectors instead of the cardinality of a set. This is why these frameworks use instead the ℓ_0-norm of a vector, which is defined in terms of the number of its nonzero coefficients:

$$\| \mathbf{y} \|_0 = \sharp\{y_i \,:\, |y_i| \neq 0\} = \| \boldsymbol{W}\boldsymbol{x} \|_0. \tag{15.4}$$

This assumption may seem abstract, but it amounts to say that all the spatial frequencies are not present in an image and that most of the information is carried by only a subset of them. We experience this every day, as it is one of the principles behind compression schemes such as JPEG.

As a last step, defining a normalization constraint is required. From Eq. (15.3), it is clear that an LBD is a *differential* operator by nature. Thus, the original values of the pixels are definitely lost and we can reconstruct their variations only. This ambiguity can be solved by arbitrarily fixing the dynamic range of the pixels to the interval $[0, 1]$ and constraining the reconstructed patches to belong to the set $\mathcal{S}$ of the patches of mean 0.5.

Finally, the reconstruction problem can be written as

$$\hat{\boldsymbol{p}} = \underset{\mathbf{x}\in\mathbb{R}^N}{\arg\min}\; \underbrace{\mathcal{J}(\mathbf{x})}_{\text{Data term}} \quad \text{subject to } \underbrace{\| \boldsymbol{W}\boldsymbol{x} \|_0 \leq K \;\text{ and }\; \boldsymbol{x} \in \mathcal{S}}_{\text{Regularization term}}. \tag{15.5}$$

The data term $\mathcal{J}$ first applies the linear operator $\boldsymbol{\mathcal{L}}$ to the current reconstruction estimate $\boldsymbol{x}$, binarizes the result and counts how many bits are different between

$\mathbb{B}(\mathcal{L}\boldsymbol{x})$ and the input descriptor $\boldsymbol{p}$. Eq. (15.4) is a special instance of a Lasso-type program [53] that can be solved with an iterative procedure (see Refs. [54, 55], and Chapter 14, Section 1).

15.6.2.2 Interest of the Retinal Descriptor

This has taken us a long way from keypoints and biology. We come back to our matter by picking a solver for Eq. (15.4) (see Ref. [54]). Does the resolution of the problem 15.5 exhibit meaningful differences when applied to regular keypoints or to bioinspired ones? Let us proceed to some experiments using BRIEF (for the regular one) and FREAK (for the bioinspired one).

We start by decomposing a well known image into nonoverlapping patches. Computing BRIEF and FREAK descriptors then reconstructing each patch yields the results shown in Figure 15.15. These results are strikingly different. BRIEF allows to immediately recognize the original image of Lena. The edges, however, are larger and more blurred: each reconstructed patch behaves like a large, blurred sketch of the original image content. Using FREAK instead, the reconstructed image looks like a weird patchwork. Carefully observing what happens in the center, however, shows that the original image directions are accurately reconstructed.

When considering the retinal pattern from Figure 15.11, this effect is quite intuitive. In this sampling, the outer cells have a coarser resolution than the innermost ones. Thus, the image geometry encoded in a FREAK descriptor has a very coarse scale near the orders of the patch and becomes more and more sensitive when moving to the center. Owing to the random nature of the measurement positions

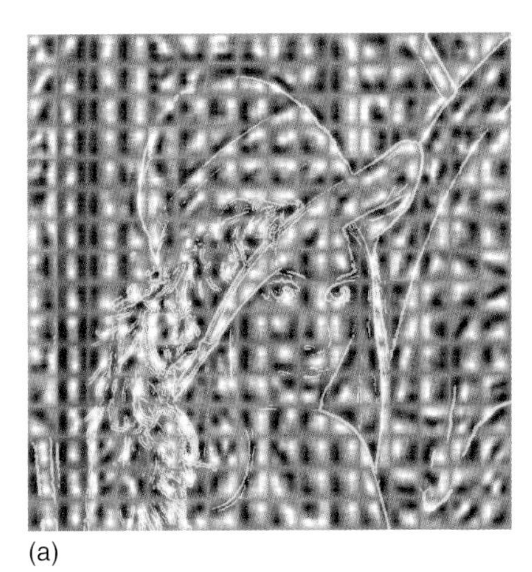

(a)

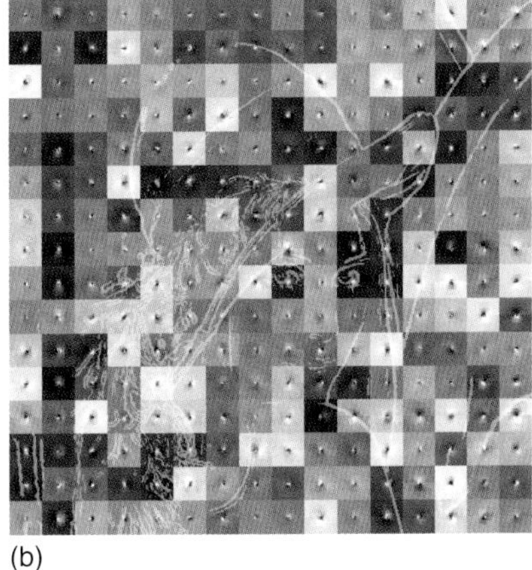

(b)

Figure 15.15 Reconstruction of Lena from binary LBDs. There is no overlap between the patches used in the experiment, thus the blockwise aspect. Yellow lines are used to depict some of the edges of the original image. In both cases, the orientation selected for the output patch is consistent with the original gradient direction. (a) BRIEF allows to recover large blurred edges, while (b) FREAK concentrates most of the information in the center of the patch.

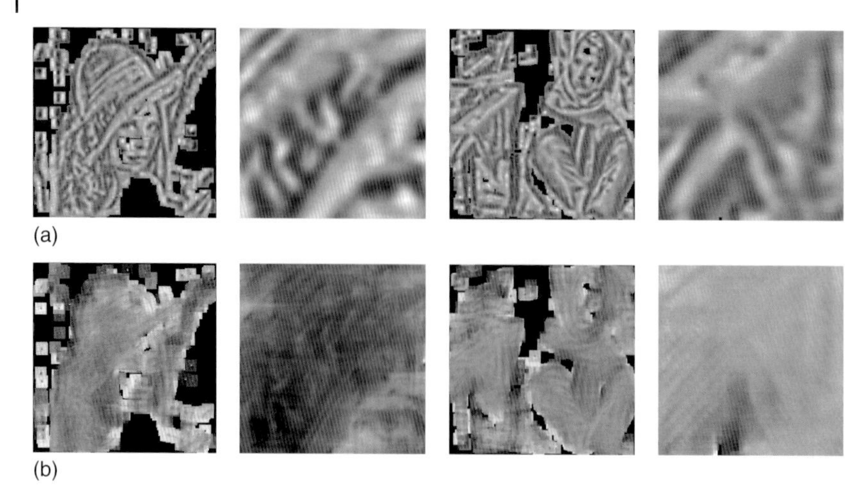

Figure 15.16 Reconstruction of LBDs centered on FAST keypoints only. Columns: Lena image, its zoom version; Barbara image, and its zoom version. (a) using BRIEF. (b) using FREAK. Since the detected points are usually very clustered, there is a dense overlap between patches, yielding a reconstruction that is visually better than the one in Figure 15.15.

inside a patch, BRIEF does consider all the content equally and is not biased toward its center.

Because FREAK focalizes its attention to an area near the center of the patch, it encodes finer details in the descriptor than BRIEF. This is confirmed by the experiment shown in Figure 15.16. In this second experiment, the patches to reconstruct are first selected by the FAST keypoint detector. Since FAST points tend to aggregate near corners and edges, there is a large overlap between the patches that discards the block artifacts from Figure 15.15. The capability of FREAK to encode fine edge information is well illustrated by this example. The fine pattern of the tissue can be correctly perceived while it is blurred out by BRIEF. Even the more variable orientations of the hat feathers of Lena can be reconstructed.

15.6.3
From Feature Visualization to Crowd-Sourced Object Recognition

A key difference between the human mind and machine vision classifiers is the ability of humans to easily generalize their knowledge to never-seen instances of objects. A kind of brute force approach to tackle this problem is, of course, to make training sets bigger, but this approach will be inherently limited by practical cost issues and by more subtle effects of low occurrence rates for some objects of interest. In Reference [56], Vondrick *et al.* propose an original way around this problem that associates the results of a feature inversion algorithm and human psychophysics tools to train and build better classifiers that are capable of recognizing objects *without ever seeing them* during the training phase.

How did they manage to do so? Their work is based on the following observation: if we generate hundreds or thousands of noise images of a given size, then some of them will contain accidental shapes that *look like* actual objects or geometric figures that can be detected by human observers. This principle is used, for example, in studies of the HVS [57].

Of course, generating random images and waiting for humans to see anything inside would also be time- and money-consuming, and this would still require features to be computed and selected from these random images. Furthermore, these random noise images might not be very efficient, as they would probably lack geometric features such as edges when generated as pixelwise random noise. However, given a feature inversion algorithm, it is possible to reconstruct image parts from random feature vectors and submit these images to human viewers instead.

Thus, Vondrick *et al.* selected the popular HOG [58] and CNN [59] object descriptors that are used in most state-of-the-art object visual classifiers. They then took the following steps:

1) *Random realizations* of HOG and CNN [59] feature descriptors were generated.
2) A *feature inversion* algorithm was fed with these random vectors. The feature inversion method that they used is described in Ref. [60].
3) These reconstruction results were submitted to *human viewers* on a large-scale sample using Amazon Mechanical Turk.[2] These workers were tasked with *identifying the category* (if any) of the object inside each picture.

Finally, they trained a visual object classifier that approximates the average human worker decision between object categories and pure noise. This is an *imaginary* classifier, as it has never seen a real object in its training phase but only hallucinated objects instead.

The reference images used by the imaginary classifiers for different object classes can be seen in Figure 15.17. It is interesting to note that the usual geometric shape of the object is retrieved in each case: cars and televisions have corners and are roughly rectangular, while bottles, persons, or fire hydrants are vertically elongated. Since CNN also encode the original color of the images, it is remarkable that the fire hydrant model selected from the human workers turned out to be red, as one might expect.

Table 15.1 shows the average precision of these object classifiers on the PASCAL VOC 2011 challenge compared to the output of a purely random classifier. In most cases, the imaginary classifier significantly outperforms the random classifier, showing that relevant information was transferred from the human viewers to the automatic system. Vondrick *et al.* also managed to train classifiers as a compromise between an actual training set and hallucinated images. When the size of the training set grows, the benefits of the imaginary classifiers vanish. This is,

2) Amazon Mechanical Turk is a crowd-sourcing Internet marketplace that enables researchers to coordinate the use of human intelligence to perform large-scale tasks at a reduced labor cost.

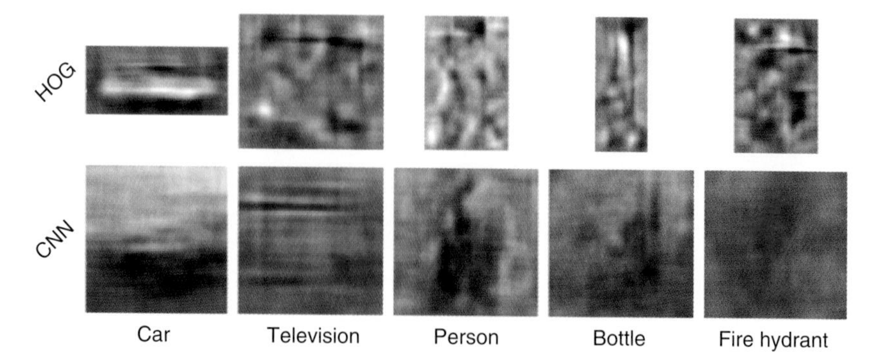

Figure 15.17 Decision boundaries acquired from Mechanical Turk workers after presenting them inverted random image descriptors. Illustration reprinted from Vondrick *et al.* [56].

Table 15.1 Average precision of imaginary classifiers (trained without any real object images) on the PASCAL VOC 2011 dataset. Boldface numbers emphasize cases where the imaginary classifier clearly outperforms a random classifier. Reprinted from Vondrick *et al.* [56].

	Car	Person	Fire hydrant	Bottle	TV
HOG	**22.9**	**45.5**	0.8	**15.9**	**27.0**
CNN	**27.5**	**65.6**	**5.9**	6.0	**23.8**
Chance	7.3	32.3	0.3	4.5	2.6

however, a promising first step to building better object recognition systems from a smaller number of examples and future research may succeed in transferring the generalization skills of the human mind to automatic systems.

As a closing remark, recall that CNN were designed to mimic the human visual cortex [59]. Using the training method proposed by Vondrick *et al.*, we can then build an image recognition system whose behavior reproduces that of the human brain and which exploits information from a human crowd although it has never seen a real image.

15.7 Conclusions

Keypoint extraction and representation techniques are usually the first step toward more challenging computer vision tasks such as object retrieval or 3D reconstruction. Such a low-level feature extraction step becomes more practical if its computational cost is low. Such a condition has led to the recent designs of fast algorithms for computing binary descriptors that are consistent with our current understanding of the front end of the visual system. However, we would like to point out to the reader that many other informative statistical approaches exist to describe an image patch, fusing several low-level features such as the covariance descriptor [61] or image moments [62], to name a few. These methods are typically computationally more costly than the binary descriptors. In the last section of this chapter, we saw that such binary features can be inverted to yield plausible visual reconstructions. This process can be used not only to better understand differences between descriptors but also to open up novel and promising approaches to developing complete machine vision systems.

We would like to conclude this chapter by sharing some links on available codes to run the presented methods in this chapter. First, codes are available in matlab [63] or C/C++/python [8] to extract and represent keypoints with the presented algorithms. Second, the source code to generate the experiments of the feature inversion algorithms are available online as well [64] and [65]. The detailed mechanics of both techniques are presented in Refs [54, 66] for LBD and Ref. [60] for HOG/CNN. The LBD reconstruction framework makes heavy use of results and tools from an emerging field called *compressive sensing* (*CS*), and in particular 1-bit CS. The Rice University maintains a comprehensive list of CS resources at the address [67], including 1-bit CS references. Finally, deriving actual computer vision algorithms from psychovisual studies is an arduous task although the whole field of research is impregnated with David Marr's primal sketch. Readers interested in an alternative way of linking psychological theories of perception and mathematical algorithms can find some alternative inspiration in A. Desolneux's work regrouped in Ref. [68]. This work tries to estimate salient structures from noise, but also relies on the Gestalt theory (as formalized by Wertheimer and Kanizsa) to model the human perception. This allows the authors to derive robust, parameter-less algorithms to detect low-level features in images.

References

1. Jepson, L.H., Hottowy, P., Mathieson, K., Gunning, D.E., Dabrowski, W., Litke, A.M., and Chichilnisky, E.J. (2013) Focal electrical stimulation of major ganglion cell types in the primate retina for the design of visual prostheses. *J. Neurosci.*, **33** (17), 194–205.
2. Krizhevsky, A., Sutskever, I., and Hinton, G.E. (2012) Imagenet classification with deep convolutional neural networks. Advances in Neural Information Processing Systems, pp. 1097–1105.
3. Russakovsky, O., Deng, J., Su, H., Krause, J., Satheesh, S., Ma, S., Huang, Z., Karpathy, A., Khosla, A., Bernstein, M., Berg, A.C., and Fei-Fei, L. (2015) ImageNet Large Scale Visual Recognition Challenge. *International Journal of Computer Vision (IJCV)*, 1–42, doi: 10.1007/s11263-015-0816-y.
4. Neisser, U. (1964) *Visual Search*, Scientific American.
5. Biederman, I. (1987) Recognition-by-components: a theory of human image understanding. *Psychol. Rev.*, **94** (2), 115.
6. Shi, J. and Tomasi, C. (1994) Good features to track. 1994 IEEE Computer Society Conference on Computer Vision and Pattern Recognition, 1994. Proceedings CVPR'94, IEEE, pp. 593–600.
7. Lowe, D.G. (1999) Object recognition from local scale-invariant features. The Proceedings of the 7th IEEE International Conference on Computer vision, 1999, vol. 2, IEEE, pp. 1150–1157.

8. Opencv's Features2d Framework, *http://docs.opencv.org/* (accessed 30 April 2015).
9. Field, G.D., Gauthier, J.L., Sher, A., Greschner, M., Machado, T.A., Jepson, L.H., Shlens, J., Gunning, D.E., Mathieson, K., Dabrowski, W. et al. (2010) Functional connectivity in the retina at the resolution of photoreceptors. *Nature*, **467**, 673–677.
10. Gollisch, T. and Meister, M. (2010) Eye smarter than scientists believed: neural computations in circuits of the retina. *Neuron*, **65** (2), 150–164.
11. Granit, R. (1947) *Sensory Mechanisms of the Retina*, Oxford University Press, London.
12. Stryer, L. (1988) Molecular basis of visual excitation, *Cold Spring Harbor Symposia on Quantitative Biology*, vol. **53**, Cold Spring Harbor Laboratory Press, pp. 283–294.
13. Berson, D.M., Dunn, F.A., and Takao, M. (2002) Phototransduction by retinal ganglion cells that set the circadian clock. *Science*, **295** (5557), 1070–1073.
14. Barlow, H.B., Fitzhugh, R., and Kuffler, S.W. (1957) Change of organization in the receptive fields of the cat's retina during dark adaptation. *J. Physiol.*, **137**, 338–354.
15. Borst, A. and Euler, T. (2011) Seeing things in motion: models, circuits, and mechanisms. *Neuron*, **71**, 974–994.
16. Ölveczky, B., Baccus, S., and Meister, M. (2003) Segregation of object and background motion in the retina. *Nature*, **423**, 401–408.
17. Field, G.D. and Chichilnisky, E.J. (2007) Information processing in the primate retina: circuit and coding. *Annu. Rev. Neurosci.*, **30**, 1–30.
18. Doi, E., Gauthier, J.L., Field, G.D., Shlens, J., Sher, A., Greschner, M., Machado, T.A., Jepson, L.H., Mathieson, K., Gunning, D.E., Litke, A.M., Paninski, L., Chichilnisky, E.J., and Simoncelli, E.P. (2012) Efficient coding of spatial information in the primate retina. *J. Neurosci.*, **32** (46), 16256–16264.
19. Garway-Heath, D.F., Caprioli, J., Fitzke, F.W., and Hitchings, R.A. (2000) Scaling the hill of vision: the physiological relationship between light sensitivity and ganglion cell numbers. *Invest. Ophthalmol. Visual Sci.*, **41** (7), 1774–1782.
20. Kolb, H. and Marshak, D. (2003) The midget pathways of the primate retina. *Doc. Ophthalmol.*, **106** (1), 67–81.
21. Ebner, M. and Hameroff, S. (2011) Lateral information processing by spiking neurons: a theoretical model of the neural correlate of consciousness. *Comput. Intell. Neurosci.*, **2011**, 247879.
22. Dacey, D., Packer, O.S., Diller, L., Brainard, D., Peterson, B., and Lee, B. (2000) Center surround receptive field structure of cone bipolar cells in primate retina. *Vision Res.*, **40**, 1801–1811.
23. Enroth-Cugell, C. and Robson, J.G. (1966) The contrast sensitivity of retinal ganglion cells of the cat. *J. Physiol.*, **187**, 517–552.
24. Pillow, J.W., Shlens, J., Paninski, L., Sher, A., Litke, A.M., Chichilnisky, E.J., and Simoncelli, E.P. (2008) Spatio-temporal correlations and visual signalling in a complete neuronal population. *Nature*, **454**, 995–999.
25. Movshon, J.A., Thompson, I., and Tolhurst, D. (1978) Spatial and temporal contrast sensitivity of neurones in areas 17 and 18 of the cat's visual cortex. *J. Physiol.*, **283** (1), 101–120.
26. Carandini, M. (2006) What simple and complex cells compute. *J. Physiol.*, **577** (2), 463–466.
27. Hubel, D.H. and Wiesel, T.N. (1962) Receptive fields, binocular interaction and functional architecture in the cat's visual cortex. *J. Physiol.*, **160** (1), 106.
28. Jones, J.P. and Palmer, L.A. (1987) An evaluation of the two-dimensional Gabor filter model of simple receptive fields in the cat striate cortex. *J. Neurophysiol.*, **58** (6), 1233–1258.
29. Olshausen, B.A. and Field, D.J. (1996) Emergence of simple-cell receptive field properties by learning a sparse code for natural images. *Nature*, **381**, 607–609.
30. Burt, P.J. and Adelson, E.H. (1983) The Laplacian pyramid as a compact image code. *IEEE Trans. Commun.*, **31** (4), 532–540.
31. Sandon, P.A. (1990) Simulating visual attention. *J. Cognit. Neurosci.*, **2** (3), 213–231.

32. Gaussier, P. and Cocquerez, J.P. (1992) Neural networks for complex scene recognition: simulation of a visual system with several cortical areas. International Joint Conference on Neural Networks, 1992. IJCNN, vol. 3, IEEE, pp. 233–259.
33. Azzopardi, G. and Petkov, N. (2013) Trainable COSFIRE filters for keypoint detection and pattern recognition. *IEEE Trans. Pattern Anal. Mach. Intell.*, **35** (2), 490–503.
34. Grossberg, S., Mingolla, E., and Todorovic, D. (1989) A neural network architecture for preattentive vision. *IEEE Trans. Biomed. Eng.*, **36** (1), 65–84.
35. Brecher, V.H., Bonner, R., and Read, C. (1991) Model of human preattentive visual detection of edge orientation anomalies. Orlando'91, Orlando, FL, International Society for Optics and Photonics, pp. 39–51.
36. Bay, H., Tuytelaars, T., and Van Gool, L. (2006) SURF: speeded up robust features, *Computer Vision–ECCV 2006*, Springer-Verlag, pp. 404–417.
37. Agrawal, M., Konolige, K., and Blas, M.R. (2008) CenSurE: center surround extremas for realtime feature detection and matching, *Computer Vision–ECCV 2008*, Springer-Verlag, pp. 102–115.
38. Harris, C. and Stephens, M. (1988) A combined corner and edge detector. Alvey Vision Conference, vol. 15, Manchester, UK, p. 50.
39. Rosten, E. and Drummond, T. (2006) Machine learning for high-speed corner detection, *Computer Vision–ECCV 2006*, Springer-Verlag, pp. 430–443.
40. Heitger, F., Rosenthaler, L., Von Der Heydt, R., Peterhans, E., and Kübler, O. (1992) Simulation of neural contour mechanisms: from simple to end-stopped cells. *Vision Res.*, **32** (5), 963–981.
41. Rosenthaler, L., Heitger, F., Kübler, O., and von der Heydt, R. (1992) Detection of general edges and keypoints, *Computer Vision ECCV'92*, Springer-Verlag, pp. 78–86.
42. Robbins, B. and Owens, R. (1997) 2D feature detection via local energy. *Image Vision Comput.*, **15** (5), 353–368.
43. Manjunath, B., Shekhar, C., and Chellappa, R. (1996) A new approach to image feature detection with applications. *Pattern Recognit.*, **29** (4), 627–640.
44. Freeman, W.T. and Adelson, E.H. (1991) The design and use of steerable filters. *IEEE Trans. Pattern Anal. Mach. Intell.*, **13** (9), 891–906.
45. Ke, Y. and Sukthankar, R. (2004) PCA-SIFT: a more distinctive representation for local image descriptors. Proceedings of the 2004 IEEE Computer Society Conference on Computer Vision and Pattern Recognition, 2004. CVPR 2004, vol. 2, IEEE, pp. II–506.
46. Tuytelaars, T. and Mikolajczyk, K. (2008) Local invariant feature detectors: a survey. *Found. Trends® in Comput. Graph. and Vision*, **3** (3), 177–280.
47. Calonder, M., Lepetit, V., Strecha, C., and Fua, P. (2010) BRIEF: binary robust independent elementary features, *Computer Vision–ECCV 2010*, Springer-Verlag, pp. 778–792.
48. Alahi, A., Ortiz, R., and Vandergheynst, P. (2012) FREAK: fast retina keypoint, 2012 IEEE Conference on Computer Vision and Pattern Recognition (CVPR), IEEE, pp. 510–517.
49. Rublee, E., Rabaud, V., Konolige, K., and Bradski, G. (2011) ORB: an efficient alternative to SIFT or SURF. 2011 IEEE International Conference on Computer Vision (ICCV), IEEE, pp. 2564–2571.
50. Leutenegger, S., Chli, M., and Siegwart, R.Y. (2011) BRISK: binary robust invariant scalable keypoints. 2011 IEEE International Conference on Computer Vision (ICCV), IEEE, pp. 2548–2555.
51. Olshausen, B.A. and Field, D.J. (2004) What is the other 85% of V1 doing?. *Prob. Syst. Neurosci.*, **4** (5), 182–211.
52. Leutenegger, S., Chli, M., and Siegwart, Y. (2011) BRISK: binary robust invariant scalable keypoints. 2011 IEEE International Conference on Computer Vision (ICCV), pp. 2548–2555.
53. Tibshirani, R. (1994) Regression shrinkage and selection via the lasso. *J. R. Stat. Soc. Ser. B*, **58**, 267–288.
54. d'Angelo, E., Jacques, L., Alahi, A., and Vandergheynst, P. (2014) From bits to images: inversion of local binary descriptors. *IEEE Trans. Pattern Anal.*

Mach. Intell., **36** (5), 874–887, doi: 10.1109/TPAMI.2013.228.
55. Jacques, L., Laska, J.N., Boufounos, P.T., and Baraniuk, R.G. (2011) Robust 1-bit compressive sensing via binary stable embeddings of sparse vectors. *IEEE Trans. Inf. Theory*, **59** (4), 2082–2102.
56. Vondrick, C., Pirsiavash, H., Oliva, A., and Torralba, A. (2014) Acquiring Visual Classifiers from Human Imagination, *http://web.mit.edu/vondrick/imagination/* (accessed 30 April 2015).
57. Beard, B.L. and Ahumada, A.J. Jr. (1998) Technique to extract relevant image features for visual tasks. SPIE Proceedings, vol. 3299, pp. 79–85, doi: 10.1117/12.320099.
58. Dalal, N. and Triggs, B. (2005) Histograms of oriented gradients for human detection. IEEE Computer Society Conference on Computer Vision and Pattern Recognition, 2005. CVPR 2005, vol. 1, pp. 886–893, doi: 10.1109/CVPR.2005.177.
59. LeCun, Y., Bottou, L., Bengio, Y., and Haffner, P. (1998) Gradient-based learning applied to document recognition. *Proc. IEEE*, **86** (11), 2278–2324, doi: 10.1109/5.726791.
60. Vondrick, C., Khosla, A., Malisiewicz, T., and Torralba, A. (2013) HOGgles: visualizing object detection features. 2013 IEEE International Conference on Computer Vision (ICCV), pp. 1–8, doi: 10.1109/ICCV.2013.8.
61. Tuzel, O., Porikli, F., and Meer, P. (2006) Region covariance: a fast descriptor for detection and classification. *Computer Vision–ECCV 2006*, Springer-Verlag, pp. 589–600.
62. Hwang, S.K., Billinghurst, M., and Kim, W.Y. (2008) Local descriptor by Zernike moments for real-time keypoint matching. Congress on Image and Signal Processing, 2008. CISP'08, vol. 2, IEEE, pp. 781–785.
63. MathWorks Matlab r2014b Computer Vision Toolbox, *http://www.mathworks.com/products/computer-vision/* (accessed 30 April 2015).
64. GitHub LBD Reconstruction Code, *https://github.com/sansuiso/LBDReconstruction* (accessed 30 April 2015).
65. Source Code to Invert Hog, *https://github.com/CSAILVision/ihog* (accessed 30 April 2015).
66. d'Angelo, E., Alahi, A., and Vandergheynst, P. (2012) Beyond bits: reconstructing images from local binary descriptors. 2012 21st International Conference on Pattern Recognition (ICPR), pp. 935–938.
67. Rice Compressive Sensing Resource, *http://dsp.rice.edu/cs.*
68. Desolneux, A., Moisan, L., and Morel, J.M. (2008) *From Gestalt Theory to Image Analysis*, Springer-Verlag.

Part IV
Applications

16
Nightvision Based on a Biological Model

Magnus Oskarsson, Henrik Malm, and Eric Warrant

16.1
Introduction

The quest to understand the biological principles that exquisitely adapt organisms to their environments and that allow them to solve complex problems, has led to unexpected insights into how similar problems can be solved technologically. These biological principles often turn out to be refreshingly simple, even "ingenious," seen from a human perspective, making them readily transferable to man-made devices. This mimicry of biology – known more broadly as "biomimetics" – has already led to a great number of technological innovations, from Velcro (inspired by the adhesive properties of burdock seed capsules) to convection-based air conditioning systems in skyscrapers (inspired by the nest ventilation system of the African fungus-growing termite). In this chapter, we describe a recent biomimetic advance inspired by the visual systems of nocturnal insects.

Constituting the great majority of all known species of animal life on Earth, the insects have conquered almost every imaginable habitat (excluding oceans). And despite having eyes, brains, and nervous systems only a fraction the size of our own, insects rely on vision to find food, to navigate and home, to locate mates, escape predators and migrate to new habitats. Even though insects do not see as sharply as we do, many experience many more colors, can see polarized light, and can clearly distinguish extremely rapid movements. Remarkably, this is even true of nocturnal insects, whose visual world at night is around 100 million times dimmer than that experienced by their day-active relatives. Research over the past 15–20 years – particularly on fast-flying and highly aerodynamic moths and bees and on ball-rolling dung beetles – indicates that nocturnal insects that rely on vision for the tasks of daily life invariably see extremely well (reviewed by Warrant and Dacke in Ref. [1]). We now know, for instance, that nocturnal insects are able to distinguish colors [2, 3], detect faint movements [4], and avoid obstacles during flight by analyzing optic flow patterns [5], to learn visual landmarks [6–8], to orient to the faint polarization pattern produced by the moon [9], and to navigate using the faint stripe of light provided by the Milky Way [10].

Biologically Inspired Computer Vision: Fundamentals and Applications, First Edition.
Edited by Gabriel Cristóbal, Laurent Perrinet, and Matthias Keil.

To see so well at night, these insects – whose small eyes struggle to capture sufficient light to see – must overcome a number of fundamental physical limitations in order to extract reliable information from what are inherently dim and very noisy visual images. Recent research has revealed that nocturnal insects employ several neural strategies for this purpose. It is these strategies that have provided the inspiration for a new night-vision algorithm that dramatically improves the quality of video recordings made in very dim light.

Most people have experienced problems with taking pictures and recording video under poor lighting conditions. Low contrast and grainy images are often the result. Most of today's digital cameras are based on the same principles that animal eyes are built on. The light that hits the camera is focused by an optical system, typically involving one or many lenses. The light is focused on a sensor that transforms the photon energy into an electrical signal. These sensors are typically CCD or CMOS arrays, arranged in a grid of pixels. The electric current is then transformed into a digital signal that can be stored in memory.

Since the underlying principles of both animal and camera vision are similar, it is natural to try to mimic the neural processes of nocturnal animals in order to construct efficient computer vision algorithms. In the following sections, we describe both the underlying biological principles and our computer vision approach in detail. Parts of the work in this chapter have previously been presented in [11–14].

16.1.1 Related Work

Enhancing low-light video in most cases involves some form of amplification of the signal in order to increase the contrast. This is – as described in the following sections – also true for our approach. This amplification will introduce more apparent noise in the signal, and the most difficult part of the video enhancement process is the reduction of noise in the signal.

There exists a multitude of noise reduction techniques that apply weighted averaging for noise reduction purposes. Many authors have additionally realized the benefit of trying to reduce the noise by filtering the sequences along motion trajectories in spatiotemporal space and in this way, ideally, avoid motion blur and unnecessary amounts of spatial averaging. These noise reduction techniques are usually referred to as *motion-compensated (spatio-)temporal filtering*. In Ref. [15], means along the motion trajectories are calculated while in Refs [16] and [17], weighted averages, dependent on the intensity structure and noise in a small neighborhood, are applied. In Ref. [18], so-called linear minimum mean square error filtering is used along the trajectories. Most of the motion-compensated methods use some kind of block-matching technique for the motion estimation that usually, for efficiency, employs rather few images, commonly two or three. However, in Refs [16, 18] and [19], variants using optical flow estimators are presented. A drawback of the motion-compensating methods is that the filtering relies on a good motion estimation to give a good output, without excessive blurring. The motion estimation is especially complicated for sequences severely degraded by noise. Different approaches have been applied to deal with this problem, often by simply reducing the influence of the temporal filter in difficult areas. This, however, often leads to disturbing noise at, for example, object contours.

Some of the most successful approaches to image denoising over the recent years belong to the group of block-matching algorithms. These include the nonlocal means algorithms [20] and the popular BM3D approach [21, 22]. BM3D is the algorithm that we have tried which has the most comparable result to our own.

Another class of noise reducing video processing methods uses a cascade of directional filters and in this way analyzes the spatiotemporal intensity structure in the neighborhood of each point. The filtering and smoothing is done primarily in the direction that corresponds to the filter that gives the highest output, cf. Refs [23, 24] and [25]. These methods work well for directions that coincide with the fixed filter directions but have a pronounced degradation in the output in directions in-between the filter directions. For a review of a large number of spatiotemporal noise reduction methods, see Ref. [26].

An interesting family of smoothing techniques for noise reduction are the ones that solve an edge-preserving anisotropic diffusion equation on the images. This approach was pioneered by Perona and Malik [27] and has had many successors, including the work by Weickert [28]. These techniques have also been extended to 3D and spatiotemporal noise reduction in video, cf. Refs [29] and [30]. In Refs [29] and [31], the so-called *structure tensor* or *second-moment matrix* is applied in a similar manner to our approach in order to analyze the local spatiotemporal intensity structure and steer the smoothing accordingly. The drawbacks of techniques based on diffusion equations include the fact that the solution has to be found using an often time-consuming iterative procedure. Moreover, it is very difficult to find a suitable stopping time for the diffusion, at least in a general and automatic manner. These drawbacks make these approaches, in many cases, unsuitable for video processing.

A better approach is to apply single-step structure-sensitive adaptive smoothing kernels. The bilateral filters introduced by Tomasi and Manduchi [32] for 2D images falls within this class. Here, edges are maintained by calculating a weighted average at every point using a Gaussian kernel, where the coefficients in the kernel are attenuated on the basis of how different the intensities are in the corresponding pixels compared to the center pixel. This makes local smoothing very dependent on the correctness in intensity in the center pixel, which cannot be assumed in images heavily disturbed by noise.

An approach that is closely connected to both bilateral filtering and to anisotropic diffusion techniques based on the structure tensor is the structure-adaptive anisotropic filtering presented by Yang *et al.* [33]. We base our spatiotemporal smoothing on the ideas in this paper, but extend it to our special setting for video and very low light levels. For a study of the connection between anisotropic diffusion, adaptive smoothing, and bilateral filtering, see Ref. [34].

Another group of algorithms connected to our work are in the field of high dynamic range (HDR) imaging, cf. Ref. [35, 36] and the references therein. The aim of HDR imaging is to alleviate the restriction caused by the low dynamic range of ordinary CCD cameras, that is, the restriction to the ordinary 256 intensity levels for each color channel. Most HDR imaging algorithms are based on using multiple exposures of a scene with different settings for each exposure and then using different approaches for storing and displaying this extended image data. However, the HDR techniques are not especially aimed at the kind of low-light-level data that we are targeting in this work, where the utilized dynamic range in the input data is in the order of 5–20 intensity levels and the SNR is extremely low.

There are surprisingly few published studies that particularly target noise reduction in low-light-level video. Two examples are the method by Bennett and McMillan [37] and the technique presented by Lee *et al.* [38]. In Ref. [38], very simple operations are combined in a system presumably developed for easy hardware implementation in, for example, mobile phone cameras and other compact digital video cameras; this method cannot handle the high levels of noise that we target in this work. The approach taken by Bennett and McMillan [37] for low-dynamic-range image sequences is more closely connected to our technique. Their virtual exposure framework includes the bilateral ASTA-filter (adaptive spatiotemporal accumulation) and a tone mapping technique. The ASTA-filter, which changes to relatively more spatial filtering in favor of temporal filtering when motion is detected, is in this way related to the biological model described in the following sections. However, as bilateral filters are applied, the filtering is edge sensitive and the temporal bilateral filter is additionally used for local motion detection. The filters apply novel dissimilarity measures to deal with the noise sensitivity of the original bilateral filter formulation. The filter size applied at each point is decided by an initial calculation of a suitable amount of amplification using tone mapping of an isotropically smoothed version of the image. A drawback of the ASTA-filter is that it requires global motion detection as a preprocessing step to be able to deal with moving camera sequences. The sequence is then warped to simulate a static camera sequence.

16.2 Why Is Vision Difficult in Dim Light?

From the outset, it is important to point out that the nocturnal visual world is essentially identical to the diurnal visual world. The contrasts and colors of objects are almost identical. The only real distinguishing difference is the mean level of light intensity, which can be up to 11 orders of magnitude dimmer at night [39, 40], depending on the presence or absence of moonlight and clouds, and whether the habitat is open or closed (i.e., beneath the canopy of a forest). It is this difference that severely limits the ability of a visual system to distinguish the colors and contrasts of the nocturnal world. Indeed, many animals, especially diurnal animals, distinguish very little at all. In the end, the greatest challenge for an eye that views a dimly illuminated object is to absorb sufficient photons of light to reliably discriminate it from other objects [6, 40–42]. The main role of all photoreceptors is to transduce incoming light into an electrical signal whose amplitude is proportional to the light's intensity. The mechanism of transduction involves a complex chain of biochemical events within the photoreceptor that uses the energy of light to open ion channels in the photoreceptor membrane, thereby allowing the passage of charged ions and the generation of an electrical response. Photoreceptors are even able to respond to single photons of light with small but distinct electrical responses known as *bumps* (as they are called in the invertebrate literature: Figure 16.1a,b). At higher intensities, the bump responses fuse to create a graded response whose duration and amplitude is proportional to the duration and amplitude of the light stimulus. At very low light levels, a light stimulus of constant intensity will be coded as a train of bumps generated in the retina at a particular rate, and at somewhat higher light levels, the constant intensity will be coded by a graded potential of particular amplitude. At the level of the photoreceptors, the reliability of vision is determined by the repeatability of this response: for repeated presentations of the same stimulus, the reliability of vision is maximal if the rate of bump generation, or the amplitude of the graded response, remains exactly the same for each presentation. In practice, this is never the case, especially in dim light.

Why is this so? The basic answer is that the visual response (and as a consequence, its repeatability) is degraded by visual "noise." Part of this noise arises from the stochastic nature of photon arrival and absorption (governed by Poisson statistics): each sample of N absorbed photons (or signal) has a certain degree of uncertainty (or noise) associated with ($\pm\sqrt{N}$ photons). The relative magnitude of this uncertainty (i.e., $N/\sqrt{N} = \sqrt{N}$) is greater at lower rates of photon absorption, and these quantum fluctuations set an upper limit to the visual signal-to-noise ratio [45–47]. As light levels fall, fewer photons are absorbed, the noise relative to the signal is greater, and less of it can be seen. This photon "shot noise" limits detection reliability and is equally problematic for artificial imaging systems, such as cameras, as it is for eyes.

There are also two other sources of noise that further degrade visual discrimination by photoreceptors in dim light. The first of these, referred to as "transducer noise," arises because photoreceptors are incapable of producing an identical electrical response, of fixed amplitude, latency, and duration, to each (identical) photon of absorbed light. This source of noise, originating in the biochemical processes leading to signal amplification, degrades the reliability of vision, particularly at slightly brighter light levels when photon shot noise no longer dominates [43, 48, 49].

The second source of noise, referred to as "dark noise," arises because of the occasional thermal activation of the biochemical pathways responsible for transduction, which even occur in perfect darkness [50]. These activations produce "dark events," electrical responses that are indistinguishable from those produced by real photons, and these are more frequent at higher retinal temperatures. At very low light levels this dark noise can significantly contaminate visual signals. In insects and crustaceans, dark events seem to be rare, only around 10 every hour at 25°C, at least in those species where it has been investigated [43, 51, 52]. But

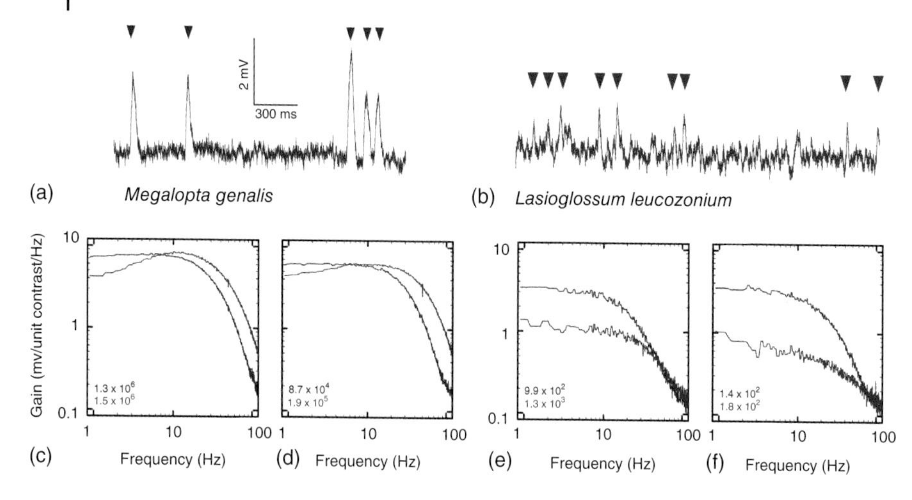

Figure 16.1 Adaptations for nocturnal vision in the photoreceptors of the nocturnal sweat bee *Megalopta genalis*, as compared to photoreceptors in the closely related diurnal sweat bee *Lasioglossum leucozonium*. (a) and (b) Responses to single photons (or "photon bumps": arrowheads) recorded from photoreceptors in *Megalopta* (a) and *Lasioglossum* (b) Note that the bump amplitude is larger, and the bump time course much slower, in *Megalopta* than in *Lasioglossum*. (c) – (f) Average contrast gain as a function of temporal frequency in *Megalopta* (blue curves, $n = 8$ cells) and *Lasioglossum* (red curves, $n = 8$ cells) at different adapting intensities, indicated as "effective photons" per second in each panel for each species (for each species, each stimulus intensity was calibrated in terms of "effective photons," i.e., the number of photon bumps per second the light source elicited, thereby eliminating the effects of differences in the light gathering capacity of the optics between the two species, which is about 27×: Ref. [43]). In light-adapted conditions (c) and (d), both species reach the same maximum contrast gain per unit bandwidth, although *Lasioglossum* has broader bandwidth and a higher corner frequency (the frequency at which the gain has fallen off to 50% of its maximum). In dark-adapted conditions (e) and (f), *Megalopta* has a much higher contrast gain per unit bandwidth. All panels adapted with kind permission from Ref. [44].

in nocturnal toad rods, the rate is much higher – 360 per hour at 20°C [53] – and this sets the ultimate limit to visual sensitivity [54, 55].

16.3 Why Is Digital Imaging Difficult in Dim Light?

As discussed in the introduction, animal eyes and digital cameras share much of their underlying functions, from the optics to the sensor. A digital camera's CCD or CMOS sensor is the corresponding element to the retina and photoreceptors in biology, and the basic purpose is the same – to convert light photons into an electrical signal that can be measured and interpreted. Many of the noise sources described in the previous section are therefore the same or analogous to the ones found in digital cameras. We now briefly discuss the specific characteristics of different types of noise that are present in digital images and relate them to their biological counterparts. For an excellent review of noise in digital cameras based on CCD sensors, see Ref. [56].

The incoming light to the digital camera, of course, suffers from the same physical limitations as the incoming light to an eye. In addition, as described in the previous section, photon shot noise is particularly problematic at low light levels. The incoming light to a sensor array is converted to an electrical signal by the semiconductor. This signal is amplified on the chip to produce a measurable voltage. The "transducer noise" found in photoreceptors has its technical counterparts in "fixed pattern noise," "blooming," and "read noise." Fixed pattern noise is, as its name indicates, a variation in response of different spatial pixels to uniform illumination (due to variations in quantum efficiency and charge collection volume). Blooming is the process when the illumination of nearby pixels influences a pixel's own response. This effect can in many cases be ignored. The read noise is generated by the amplification at each pixel. It can often be modeled as zero, that is, noise that is independent of the number of collected electrons. This noise dominates shot noise at low signal levels.

In the previous section, the "dark noise" in photoreceptors was described. This thermal effect is also present in digital sensors and is called *dark current noise.* Thermal energy generates free electrons in silicon and this will generate signals that are indistinguishable from the true signal. The effect of dark current noise is more severe at higher temperatures.

Finally, in order to produce a digital image, the electrical signal is quantized into a digital signal with a fixed number of bits. Depending on the chosen bit depth, this process can influence the output more or less, but it is often reasonable to model it as an addition of a zero mean, uniformly distributed, noise term.

From this discussion, it is clear that animals have to address many of the same noise issues as we have in digital cameras. This is especially true at low light levels.

16.4 Solving the Problem of Imaging in Dim Light

As we have seen above, the paucity of photons in dim light causes fundamental limitations to imaging reliability in both eyes and cameras. We next describe how these limitations can be overcome in visual systems, by describing the neural adaptations that have evolved for improving visual reliability in dim light. The adaptations that have evolved over millions of years of evolution are surprisingly elegant and simple and essentially involve two main processing steps: (i) an enhancement of the neural image created in the retina and (ii) a subsequent optimal filtering of this image in space and time at a higher level in the visual system. These two processing steps are the ones we have now employed in the development of a new night-vision algorithm that dramatically improves the quality of video recordings made in very dim light.

We now give an overview of the basic steps in our algorithm. The details of the different steps and their biological motivations are given in the subsequent

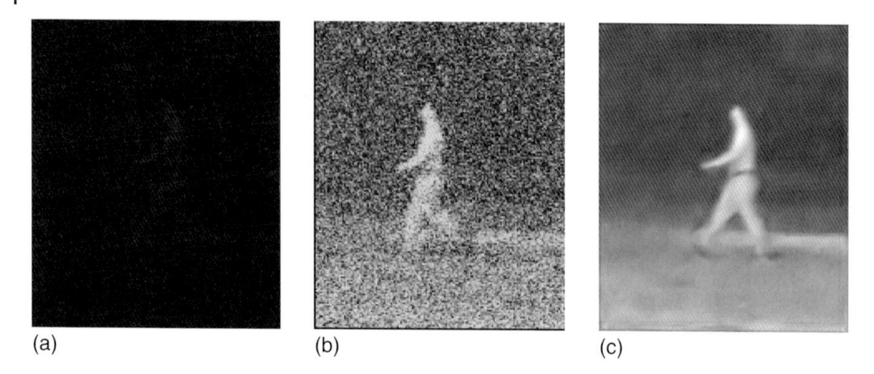

Figure 16.2 (a) One frame from a very dark input sequence. We have little contrast in the image and the details are hard to discern. (b) One frame of the amplified sequence. Here we have structures clearly visible, but the noise in the image has also been amplified. (c) One frame from the final output of Algorithm 16.1. The structures are preserved and the noise has been highly suppressed.

sections. We start with a low-light video sequence $I_{\text{in}}(x)$, where $x \in D \subset \mathbb{R}^3$ is a spatiotemporal point or voxel. One frame of an example low-light sequence is given in Figure 16.2 (a). The first step in our algorithm is the amplification of the signal. This is done by applying a one-dimensional nonlinear amplification function $T(s)$ on the input gray values (or color values) so that $I_T(x) = T(I_{\text{in}}(x))$. The function $T(s)$ could be fixed or adapted to the input values. Some examples are discussed in Section 16.4.1. Since we not only amplify the signal but also the noise, we obtain in general a very noisy output video sequence. One frame of an amplified sequence is given in Figure 16.2 (b). Now comes the difficult part, namely, to reduce the noise but keep the signal intact. In order to do this we must in some way distinguish the noise from the signal.

We try to mimic the ideas from the summation principles found in nocturnal insects, where the signal is retrieved through summing in space and time. The details are described in Section 16.4.2.

Almost all noise reduction algorithms are based on averaging the signal. But we do not want to average pixels or voxels that have different signal value, that is, we do not want to average over edges in the signal structure. Since we are dealing with a video sequence, edges appear for two different reasons, (i) due to differences in physical appearance of objects, such as boundaries of objects or texture and color changes and (ii) due to motion of either the camera or objects.

We consider the image sequence as a spatiotemporal volume, as depicted to the left in Figure 16.3. Our goal is to construct smoothing summation kernels that adapt to the local spatiotemporal intensity structure. These kernels should be large in directions where there are no edges and small in directions where we have strong edges. To the right in Figure 16.3 the kernels for two different points are shown. The man walking gives rise to an intensity streak in the image stack. The estimated kernels adapt to this local spatiotemporal structure so that we do not get smoothing over edges. Notice that the kernel at the point with a

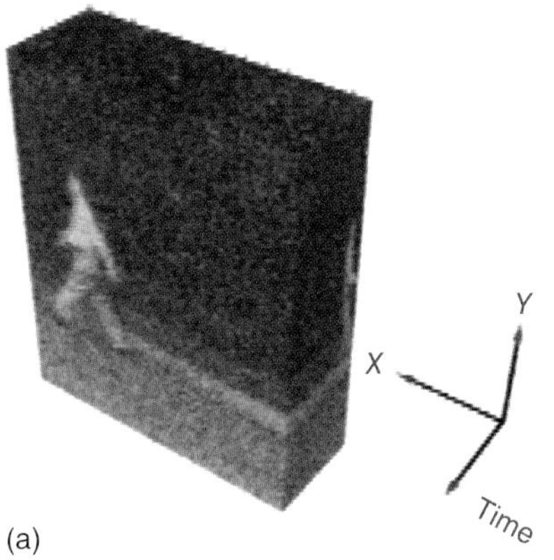

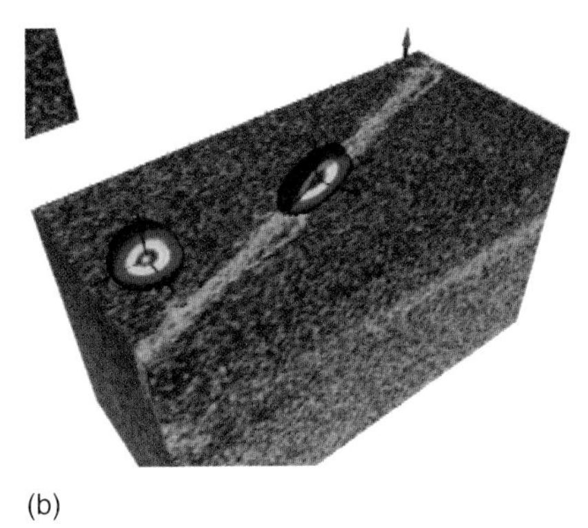

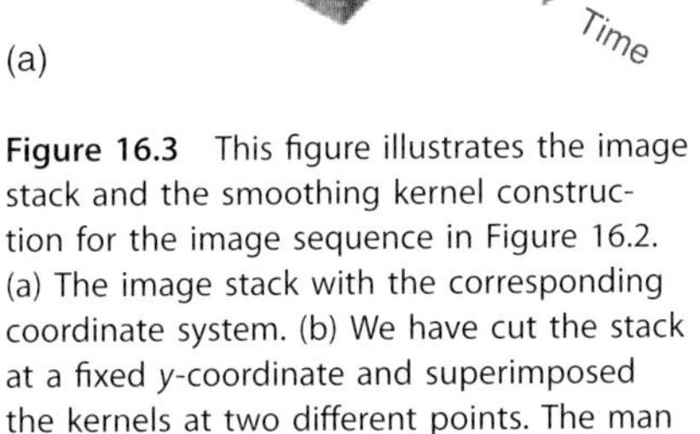

Figure 16.3 This figure illustrates the image stack and the smoothing kernel construction for the image sequence in Figure 16.2. (a) The image stack with the corresponding coordinate system. (b) We have cut the stack at a fixed y-coordinate and superimposed the kernels at two different points. The man walking gives rise to an intensity streak in the image stack. The estimated kernels adapt to this local spatiotemporal structure so that we do not get smoothing over edges. Notice that the kernel at the point with the more structurally isotropic neighborhood is quite isotropic, whereas the kernel at the point with the man walking is elongated in the direction of the intensity streak.

more structurally isotropic neighborhood is quite isotropic, whereas the kernel at the point with the man walking is elongated in the direction of the intensity streak.

We use the so-called *structure tensor* to estimate the local spatiotemporal structure at each point and use this to construct the smoothing kernels. This will also include calculating the eigenvalues and eigenvectors that we use to orient and stretch our kernels. The details are given in Section 16.4.2.2. When we have constructed the kernels we can estimate the new gray values by integration (or summation as we are dealing with discrete structures). The summation is done over a local space–time neighborhood for each point x_0, typically $13 \times 13 \times 13$ voxels. The different steps are summarized in Algorithm 16.1.

16.4.1 Enhancing the Image

In this section, we describe the first step in our method. We start with a biological motivation and then discuss the intensity transformation step in detail.

16.4.1.1 Visual Image Enhancement in the Retina

As we have mentioned previously, most animal photoreceptors are capable of detecting single photons of light, responding to them with small but discrete voltage responses that in insects are known as "bumps" (Figure 16.1(a) and (b)). Research has revealed that these bumps are much larger in nocturnal and

Algorithm 1 Low-Light Video Enhancement

1: Given a low-intensity input video sequence:

$$I_{\text{in}}(\mathbf{x}), \mathbf{x} \in D \subset \mathbb{R}^3.$$

2: Apply an amplifying intensity transformation $T : \mathbb{R} \to \mathbb{R}$,

$$I_T(\mathbf{x}) = T(I_{\text{in}}(\mathbf{x})).$$

3: **for** each $\mathbf{x}_0 \in D$ and a neighborhood $\Omega(\mathbf{x}_0)$

4: Calculate the structure tensor $\mathbf{J}(\mathbf{x}_0)$.

5: Calculate the eigenvalues and eigenvectors of $\mathbf{J}(\mathbf{x}_0)$.

6: Construct the summation kernel $k : \mathbb{R}^2 \to \mathbb{R}$,

$$k(\mathbf{x}_0, \mathbf{x}), \mathbf{x} \in \Omega(\mathbf{x}_0).$$

7: Integrate the output intensity:

$$I_{\text{out}}(\mathbf{x}_0) = \iiint\limits_{\Omega(\mathbf{x}_0)} k(\mathbf{x}_0, \mathbf{x}) I_T(\mathbf{x}) \mathrm{d}\mathbf{x}.$$

8: **end for**

crepuscular insects than in diurnal insects [44, 57], revealing an intrinsic benefit for vision in dim light. From electrophysiological recordings of photoreceptors in two closely related nocturnal and diurnal halictid bee species – the nocturnal *Megalopta genalis* (Figure 16.1(a)) and the diurnal *Lasioglossum leucozonium* (Figure 16.1(b)) – bump amplitude in the nocturnal species was found to be much greater than in the diurnal species [44]. These larger bumps have also been reported from other nocturnal and crepuscular arthropods (e.g., crane flies, cockroaches, and spiders [57–60]); and in locusts – which are considered to be diurnal insects – bump size can even vary at different times of the day, becoming significantly larger at night [61]. Larger bumps indicate that the photoreceptor's gain of transduction is greater, and thus in *Megalopta* the gain of transduction is greater than in *Lasioglossum*. This higher transduction gain manifests itself as a higher "contrast gain," which means that the photoreceptor has a greater voltage response per unit change in light intensity (or contrast). At lower light levels, the contrast gain is up to five times higher in *Megalopta* than in *Lasioglossum*, which results in greater signal amplification and the potential for improved visual reliability in dim light (Figure 16.1(c)–(f)). One problem with having a higher gain is that it not only elevates the visual signal it also elevates several sources of visual noise by the same amount. Thus, for this strategy to pay off, subsequent stages of processing are needed to reduce the noise. As we will see in the following, this processing involves a summation of photoreceptor responses in space and time.

In addition to a higher contrast gain, the photoreceptors of nocturnal insects tend also to be significantly slower than those of their day-active relatives. Despite compromising temporal resolution, slower vision in dim light (analogous to having a longer exposure time on a camera) is beneficial because it increases the visual signal-to-noise ratio and improves contrast discrimination at lower temporal frequencies by suppressing photon noise at frequencies that are too high to be reliably resolved [64, 65]. Temporal resolution can be measured in several ways, for instance, by measuring the ability of a photoreceptor to follow a light source whose intensity modulates sinusoidally over time: a photoreceptor that can follow a light source modulating at a high frequency is considered to be fast. Thus, in the frequency domain, slower vision is equivalent to saying that the temporal bandwidth is narrow, or more precisely, that the temporal corner frequency is low – this is often defined as the frequency where the response has fallen to 50% of its maximum value (i.e., by 3 dB), and lower values indicate slower vision. In *Megalopta* it is around 7 Hz in dark-adapted conditions. In diurnal *Lasioglossum* the dark-adapted corner frequency is nearly three times the value found in *Megalopta*, around 20 Hz, a value that is nonetheless considerably less than that typical of the diurnal, highly maneuverable and rapidly flying higher flies (50–107 Hz; [57]). The difference in temporal properties between the two bee species are most likely due to different photoreceptor sizes, and different numbers and types of ion channels in the photoreceptor membrane [57, 66–69].

16.4.1.2 Digital Image Enhancement

The procedure of intensity transformation is also commonly referred to as *tone mapping*. Tone mapping could actually be performed either before or after the noise reduction, with a similar output, as long as the parameters for the smoothing kernels are chosen to fit the current noise level.

In the virtual exposures method of Bennett and McMillan [37], a tone mapping procedure is applied where the actual mapping is a logarithmic function similar to the one proposed by Drago *et al.* [70]. The tone mapping procedure also contains additional smoothing using spatial and temporal bilateral filters and an attenuation of details, found by the subtraction of a filtered image from a nonfiltered image. We instead choose to do all smoothing in the separate noise reduction stage and here concentrate on the tone mapping.

The tone-mapping procedure of Bennett and McMillan involves several parameters, both for the bilateral smoothing filters and for changing the acuteness of two different mapping functions – one for the large-scale data and one for the attenuation of the details. These parameters have to be set manually and will not adapt if the lighting conditions change in the image sequence. Since we aim for an automatic procedure we instead opt for a modified version of the well-known procedure of histogram equalization, cf. Ref. [71]. Histogram equalization is parameter-free and increases the contrast in an image by finding a tone mapping that evens out the intensity histogram of the input image as much as possible. In short, histogram equalization works in the following way. We assume that we have an input image I_{in} and an output image I_{out}, both defined with continuous gray values $0 \cdots 1$, and where the gray value distribution of I_{in} is $g(s)$. The output image is obtained by transforming the input image with a gray level transformation $T(s)$ so that the output distribution is flat, that is, with the distribution $h(s) = 1$. If we further assume that $T(s)$ is an increasing function, with $T(0) = 0$ and $T(1) = 1$ the following equation will hold:

$$\int_0^s g(t)dt = \int_0^{T(s)} h(t)dt = \int_0^{T(s)} 1dt = T(s) \tag{16.1}$$

So by applying the transformation

$$T(s) = \int_0^s g(t)dt \tag{16.2}$$

we get an output image with a completely flat distribution. This derivation works exactly for continuous images, but as real images are discrete, the integration is replaced by a summation of the discrete histogram of the input image and the resulting image only has an approximately flat distribution. Furthermore, for many images, histogram equalization gives a too extreme mapping, which, for example, saturates the brightest intensities so that structure information here is lost. It also heavily changes the local intensity distribution in ways that change the appearance of the images. We therefore apply contrast-limited histogram equalization as presented by Pizer *et al.* in Ref. [72], but without the tiling that applies different mappings to different areas (tiles) in the image. In the contrast-limited histogram equalization, a parameter, the clip-limit β, sets a limit on the derivative of the slope of the mapping function. If the mapping function, found by histogram equalization exceeds this limit, the increase in the critical areas is spread equally over the mapping function. An example can be seen in Figure 16.2 (b). As illustrated in the image, this process amplifies the signal, but just as in the biological system from the previous section, it is clear that we need to address the amplified noise in the signal.

16.4.2
Filtering the Image

16.4.2.1 Spatial and Temporal Summation in Higher Visual Processing

As mentioned in Section 16.2, even though slow photoreceptors of high contrast gain are potentially beneficial for improving visual reliability in dim light, these benefits may not be realized because of the contamination of visual signals by various sources of visual noise. The solution to overcome the problems of noise contamination is to neurally sum visual signals in space and time [41, 73–77].

We have already discussed summation of photons in time above: when light gets dim, the slower photoreceptors of nocturnal animals can improve visual reliability by integrating signals over longer periods of time [41, 64, 65]. Even slower vision could be obtained by neurally integrating (summing) signals at a higher level in the visual system. However, temporal summation has a cost: the perception of fast-moving objects is seriously degraded. This is potentially disastrous for a fast-flying nocturnal animal (such as a nocturnal bee or moth) that needs to negotiate obstacles. Not surprisingly, temporal summation is more likely to be employed by slow-moving animals. Summation of visual signals in space can also improve image quality. Instead of each visual channel collecting photons in isolation (as in bright light), the transition to dim light could activate specialized, laterally spreading neurons which couple the channels together into groups. Each of the summed groups – themselves now defining the channels – could collect considerably more photons over a much wider visual angle, albeit with a simultaneous and unavoidable loss of spatial resolution. Despite being much brighter, the image would become necessarily coarser. In *Megalopta*, such laterally spreading neurons have been found in the first optic ganglion (lamina ganglionaris), the first visual processing station in the brain. The bee's four classes of lamina monopolar cells (or L-fibers, L1–L4) – which are responsible for the analysis of photoreceptor signals arriving from the retina – are housed within each neural "cartridge" of the lamina, a narrow cylinder of lamina tissue that resides below each ommatidium. Compared to the L-fibers of the diurnal honeybee *Apis mellifera*, those of *Megalopta* have lateral processes that extensively spread into neighboring cartridges (Figure 16.4(a)): cells L2, L3, and L4 spread to 12, 11, and 17 lamina cartridges respectively, while the homologous cells in *Apis* spread respectively to only 2, 0, and 4 cartridges [62, 78]. Similar laterally spreading neurons have also been found in other nocturnal insects, including cockroaches [79], fireflies [80], and hawkmoths [81].

Spatial and temporal summation strategies have the potential to greatly improve the visual signal-to-noise ratio in dim light (and thereby the reliability of vision) for a narrower range of spatial and temporal frequencies [77, 82, 83]. However, despite the consequent loss in spatial and temporal resolution, summation would tend to average out the noise (which is uncorrelated between channels) or significantly reduce its amplitude. Thus, summation would maximize nocturnal visual reliability for the slower and coarser features of the world. Those features that are faster and finer – and inherently noisy – would be filtered out. However, it is, of course, far better to see a slower and coarser world, than nothing at all.

These conclusions can be visualized theoretically [77]. Both *Megalopta* (Figure 16.4(b)) and *Apis* (Figure 16.4(c)) are able to resolve spatial details in a scene at much lower intensities with summation than without it [82]. These theoretical results assume that both bees experience an angular velocity during flight of 240 deg/s, a value that has been measured from high-speed films of *Megalopta* flying at night. At the lower light levels where *Megalopta* is active, the optimum visual performance shown in Figure 16.4(b) is achieved with an integration time of about 30 ms and summation from about 12 ommatidia (or cartridges). This integration time is close to the photoreceptor's dark-adapted value [6], and the extent of predicted spatial summation is very similar to the number of cartridges to which the L2 and L3 cells actually branch [62], thus strengthening the hypothesis that the lamina monopolar cells are involved in spatial summation. Interestingly, even in the honeybee *Apis*, summation can improve vision in dim light (Figure 16.4(c)). The Africanized subspecies, *Apis mellifera scutellata*, and the closely related south-east Asian giant honeybee

Apis dorsata, both forage on large pale flowers during dusk and dawn, and even throughout the night, if a moon half-full or larger is present in the sky. This ability can be explained only if bees optimally sum photons over space and time [84], and this is also revealed in Figure 16.4(c) (for an angular velocity of 240 deg/s). At the lower light levels where *Apis* is active, the optimum visual performance shown in Figure 16.4(c) is achieved with an integration time of about 18 ms and summation from about three or four cartridges. As in *Megalopta*, this integration time is close to the photoreceptor's dark-adapted value, and the extent of predicted spatial summation is again very similar to the number of cartridges to which the L2 and L3 cells actually branch.

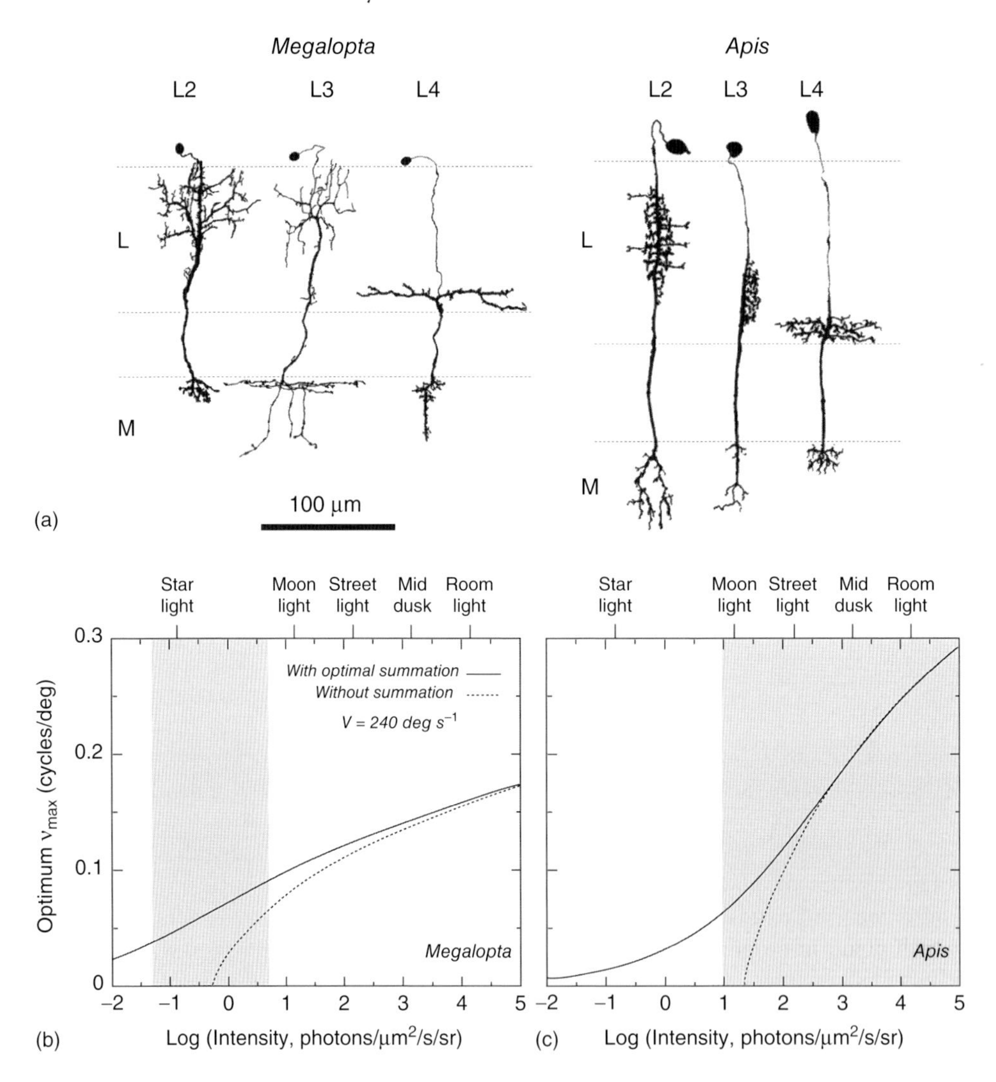

Figure 16.4 Spatial summation in nocturnal bees. (a) Comparison of the first-order interneurons – L-fiber types L2, L3, and L4 – of the *Megalopta genalis* female (i) and the worker honeybee *Apis mellifera* (ii). Compared to the worker honeybee, the horizontal branches of L-fibers in the nocturnal halictid bee connect to a much larger number of lamina cartridges, suggesting a possible role in spatial summation. L = lamina, M = medulla. Reconstructions from Golgi-stained frontal sections. Adapted from Refs [62, 63]. (b) and (c) Spatial and temporal summation modeled at different light intensities in *Megalopta genalis* and (b) *Apis mellifera* (c) for an image velocity of 240 deg/s (measured from *Megalopta genalis* during a nocturnal foraging flight [6]). Light intensities are given for 540 nm, the peak in the bee's spectral sensitivity. Equivalent natural intensities are also shown. The finest spatial detail visible to flying bees (as measured by the maximum detectable spatial frequency, ν_{max}) is plotted as a function of light intensity. When bees sum photons optimally in space and time (solid lines) vision is extended to much lower light intensities (nonzero ν_{max}) compared to when summation is absent (dashed lines). Note that nocturnal bees can see in dimmer light than honeybees. Gray areas denote the light intensity window within which each species is normally active (although honeybees are also active at intensities higher than those presented on the graph). Figure reproduced from Ref. [40] with permission from Elsevier.

16.4.2.2 Structure Tensor Filtering of Digital Images

We now present our summation method inspired by the biological principles from the previous section. It is in parts based on the structure-adaptive anisotropic image filtering by Yang *et al.* [33] but in order to improve its applicability to video data and make it suitable for our low-light-level vision objective, it involves a number of modifications and extensions.

A new image $I_{out}(\mathbf{x}_0)$ is obtained, by applying at each spatiotemporal point $\mathbf{x}_0 = (x_0, y_0, t_0)$, a kernel $k(\mathbf{x}_0, \mathbf{x})$ to the original image $I_{in}(\mathbf{x}_0)$ such that

$$I_{out}(\mathbf{x}_0) = \frac{1}{\mu(\mathbf{x}_0)} \iiint_{\Omega(\mathbf{x}_0)} k(\mathbf{x}_0, \mathbf{x}) I_{in}(\mathbf{x}) d\mathbf{x} \quad (16.3)$$

where

$$\mu(\mathbf{x}_0) = \iiint_{\Omega(\mathbf{x}_0)} k(\mathbf{x}_0, \mathbf{x}) d\mathbf{x} \quad (16.4)$$

is a normalizing factor. The normalization makes the sum of the kernel elements equal to 1 in all cases, so that the mean image intensity does not change. The area Ω over which the integration, or in the discrete case, summation, is made is chosen as a finite neighborhood centered around x_0, typically 13^3 voxels.

Since we want to adapt the filtering to the spatiotemporal intensity structure at each point, in order to reduce blurring over spatial and temporal edges, we calculate a kernel $k(x_0, x)$ individually for each point x_0. The kernels should be wide in directions of homogeneous intensity and narrow in directions with important structural edges. To find these directions, the intensity structure is analyzed by the so-called *structure tensor* or *second-moment matrix*. The structure tensor has been applied in image analysis in numerous papers, for example, Refs [85–87].

The tensor $\mathbf{J}_\rho(\mathbf{x}_0)$ is defined in the following way:

$$\mathbf{J}_\rho(\nabla I(\mathbf{x}_0)) = G_\rho \star (\nabla I(\mathbf{x}_0)\nabla I(\mathbf{x}_0)^T) \tag{16.5}$$

where

$$\nabla I(\mathbf{x}_0) = \left[\frac{\partial I}{\partial x_0}\frac{\partial I}{\partial y_0}\frac{\partial I}{\partial t_0}\right]^T \tag{16.6}$$

is the spatiotemporal intensity gradient of I at the point x_0. G_ρ is the Gaussian kernel function

$$G_\rho(\mathbf{x}) = \frac{1}{\mu}e^{-\frac{1}{2}(\frac{x^2+y^2+t^2}{\rho^2})} \tag{16.7}$$

where μ is the normalizing factor. The notation $\star$ means elementwise convolution of the matrix $\nabla I(\mathbf{x}_0)\nabla I(\mathbf{x}_0)^T$ in a neighborhood centered at $\mathbf{x}_0$. It is this convolution that gives us the smoothing in the direction of gradients, which is the key to the noise insensitivity of this method. The numerical estimates of the time derivatives in equation 16.6 are in our implementations taken symmetrically around the given frame for the batch version of our method. We have also implemented an online version where we take the time derivatives only backward from the current frame.

Eigenvalue analysis of $\mathbf{J}_\rho$ will now give us the structural information that we seek. The eigenvector $\mathbf{v}_1$, corresponding to the smallest eigenvalue λ_1, will be approximately parallel to the direction of minimum intensity variation while the other two eigenvectors will be orthogonal to this direction. The magnitude of each eigenvalue will be a measure of the amount of intensity variation in the direction of the corresponding eigenvector. For a deeper discussion on eigenvalue analysis of the structure tensor, see Ref. [88].

The basic form of the kernels $k(\mathbf{x}_0, x)$ that are constructed at each point $\mathbf{x}_0$ is that of a Gaussian function,

$$k(\mathbf{x}_0, \mathbf{x}) = e^{-(1/2)(\mathbf{x}-\mathbf{x}_0)^T R\Sigma^2 R^T(\mathbf{x}-\mathbf{x}_0)} \tag{16.8}$$

including a rotation matrix R and a scaling matrix Σ. The rotation matrix is constructed from the eigenvectors $\mathbf{v}_i$ of $\mathbf{J}_\rho$,

$$R = \begin{bmatrix}\mathbf{v}_1 & \mathbf{v}_2 & \mathbf{v}_3\end{bmatrix} \tag{16.9}$$

while the scaling matrix has the following form:

$$\Sigma = \begin{bmatrix}\frac{1}{\sigma(\lambda_1)} & 0 & 0\\ 0 & \frac{1}{\sigma(\lambda_2)} & 0\\ 0 & 0 & \frac{1}{\sigma(\lambda_3)}\end{bmatrix} \tag{16.10}$$

The function $\sigma(\lambda_i)$ is a decreasing function that sets the width of the kernel along each eigenvalue direction. The theory in Ref. [33] is mainly developed for 2D images and measures of corner strength and of anisotropism, both involving ratios of the maximum and minimum eigenvalues, are there calculated at every point x_0. An extension of this to the 3D case is then discussed. However, we have not found these two measures to be adequate for the 3D case because they focus too much on singular corner points in the video input and to a large extent disregard the linear and planar structures that we want to preserve in the spatiotemporal space. For example, a dependence of the kernel width in the temporal direction on the eigenvalues corresponding to the spatial directions does not seem appropriate in a static background area. We instead simply let an exponential function depend directly on the eigenvalue λ_i in the current eigenvector direction $\mathbf{v}_i$ in the following way:

$$\sigma(\lambda_i, \mathbf{x}_0) = \begin{cases} \Delta\sigma e^{-\frac{\lambda_i}{d}+\lambda_{\min}} + \sigma_{\min}, & \lambda_i > \lambda_{\min} \\ \sigma_{\max}, & \lambda_i \leq \lambda_{\min} \end{cases} \tag{16.11}$$

where $\Delta\sigma = (\sigma_{\max} - \sigma_{\min})$, so that σ attains its maximum $\sigma_{\max}$ below $\lambda = \lambda_{\min}$ and asymptotically approaches its minimum $\sigma_{\min}$ when $\lambda \rightarrow \infty$. The parameter d scales the width function along the λ-axis and has to be set in relation to the current noise level. Since the part of the noise that stems from the quantum nature of light, that is, the photon shot noise, depends on the brightness level, it is signal-dependent and the parameter d should ideally be set locally. However, we have noticed that when the type of camera and the type of tone mapping is fixed, a fixed value of d usually works for a large part of the dynamic range. When changing the camera and tone mapping approach, a new value of d has to be found for optimal performance.

When the widths $\sigma(\lambda_i, \mathbf{x}_0)$ have been calculated and the kernel subsequently constructed according to 16.8, Eq. 16.3 is used to calculate the output intensity I_{out} of the smoothing stage at the current pixel $\mathbf{x}_0$.

In biological vision, edges are often perceived stronger than they actually are. This effect, called *lateral inhibition*, is due to the ability of an excited neuron to reduce the response of its neighboring neurons. We have mimicked this behavior in an edge-sharpening version of our algorithm. In order to do this, we simply replace the Gaussian kernel with the following base kernel:

$$k_{hb}(\mathbf{x}) = \frac{1}{\mu} e^{-(1/2)((\mathbf{x}^T\mathbf{x})/(\sigma^2))}(1 - \alpha\,\mathbf{x}^T\,\mathbf{x}) \tag{16.12}$$

which is stretched and rotated in the same way as before. In the top row of Figure 16.5, the difference between the two kernels is shown. The bottom row shows the output of our algorithm using the two different kernels.

16.5 Implementation and Evaluation of the Night-Vision Algorithm

We have implemented Algorithm 16.1 based on the methods described in Sections 16.4.1.2 and 16.4.2.2. We show results from running our method in

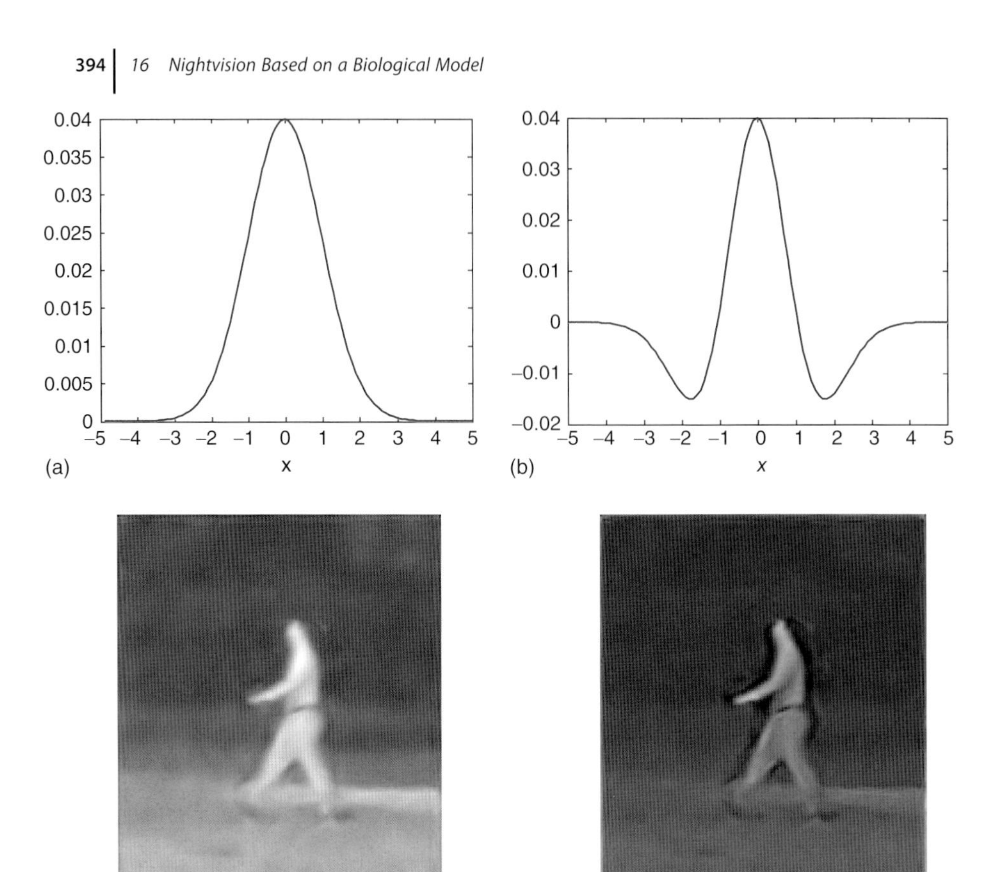

Figure 16.5 This shows the impact of using the lateral inhibition technique described in Section 16.4.2.2. The top row shows the basic kernels in one dimension used in the summation. (a) A standard Gaussian. (b) A sharpening version. The bottom row shows frames from the output of our algorithm. (c) The result of running the algorithm with a Gaussian kernel. (d) The result using the sharpening kernel.

Section 16.5.4, but we start by discussing a number of technical and implementation details. These relate to computational aspects, noise estimation, and automatic parameter selection, and how we handle color data.

16.5.1 Adaptation of Parameters to Noise Levels

The amount of noise in an image sequence changes depending on the brightness level and the signal-to-noise ratio (SNR) decreases the darker the area gets. Since we want the algorithm to adapt to changing light levels, both spatially within the same image and temporally in the image sequence, the width function needs to depend on the noise level and adapt to the SNR. This adaptation is governed by the measure d. In this section, we briefly discuss how the noise in the eigenvalues can be modeled and this in turn gives us a way of choosing the parameters in the algorithm in an appropriate way. This is similar to the type of strategy advocated theoretically (and supported experimentally from recordings in vertebrate visual systems) in Ref. [89].

If we use the very simple model of the image sequence as a signal with added independent and identically distributed Gaussian noise of zero mean, we can say the following about the noise in the estimated eigenvalues. First of all, the sequence is filtered with Gaussian filters as well as differentiation filters to estimate the gradients in the sequence at every point. The noise in the estimated gradients will hence also be Gaussian but with a different variance. The structure tensor is calculated from the estimated gradients. As the structure tensor is made up of the outer product of the gradients, when we filter the tensor elementwise, the tensor elements will be a sum of squares of Gaussian-distributed variables.

We model the noise in the eigenvalues as gamma distributed. This is not the exact distribution as the calculation of the eigenvalues and the structure tensor includes a number of nonlinear steps, but we have seen in our experiments that we capture enough of the statistics using this approximation. In Figure 16.6, histograms of the eigenvalues of the structure tensors for an image are shown in blue. We can fit a gamma probability density function (PDF) to this histogram in a least squares manner. The resulting gamma distribution is shown in red. In our experience, the eigenvalues generally follow a gamma distribution well.

From the gamma distribution means, $\overline{\lambda}_i$, and variances, $s^2_{\lambda_i}$, can be estimated. These parameters can be used to set the parameters, that is, d, of the algorithm.

16.5.2 Parallelization and Computational Aspects

The visual system of most animals is based on a highly parallel architecture. On looking at the overview of Algorithm 16.1, two things are clear. The algorithm is linear in the number of voxels in the input image sequence and it is highly parallelizable. The output of each voxel depends only on a neighborhood of that given voxel. The most expensive parts are the calculation of the structure tensor, including the gradient calculation, the elementwise smoothing, and the actual filtering, or summation. This is a task for which the graphics processing units of modern graphics cards are very well suited.

We have implemented the whole adaptive enhancement methodology as a combined CPU/GPU algorithm. All image pre- and postprocessing is performed on the CPU. This includes image input/output and the amplification step. If we use histogram equalization as a contrast amplification, this requires summation over all pixels. This computation is not easily adapted to a GPU, as the summation would have to be done in multiple passes. However, as these steps constitute a small part of the execution time, a simpler CPU implementation is adequate here. In summary, by exploiting the massively parallel architecture of modern GPUs, we obtain interactive frame rates on a single nVidia GeForce 8800-series graphics card.

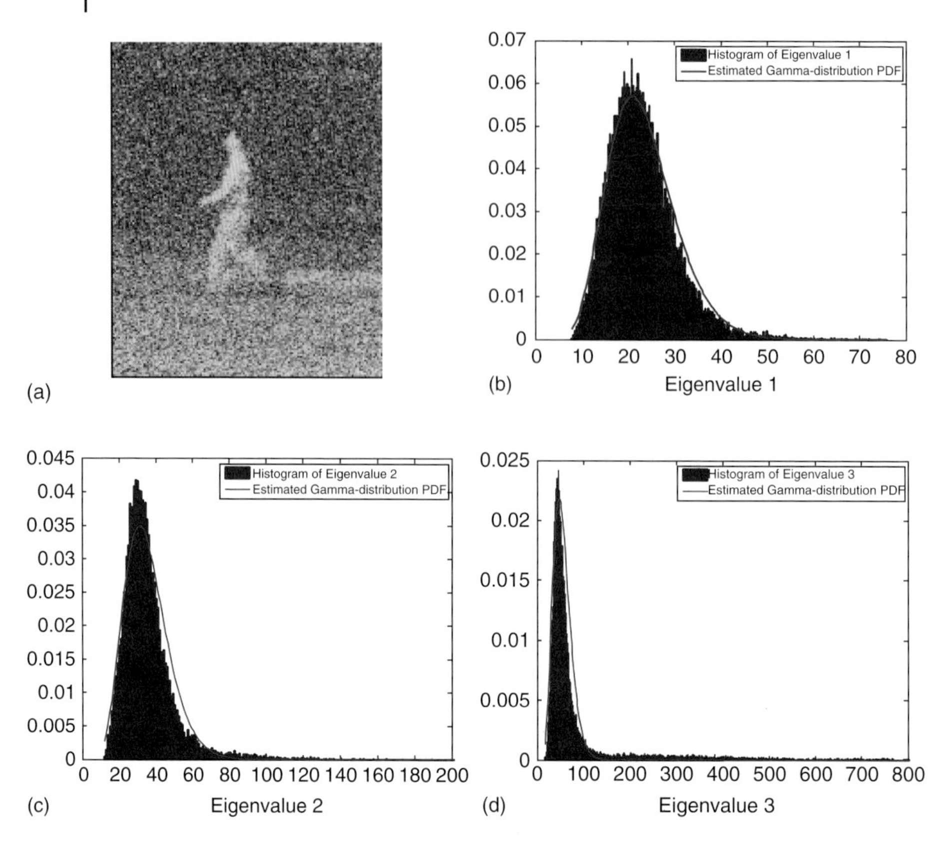

Figure 16.6 (a) One frame of an amplified noisy image sequence. (b)–(d) Histograms of eigenvalues 1–3 in blue. Also shown in red are fitted gamma distributions to the data.

16.5.3 Considerations for Color

The discussion so far has dealt with intensity images. We now discuss some special aspects of the algorithm when it comes to processing color images.

In applying the algorithm to RGB color image data, one could envision a procedure where the color data in the images are first transformed to another color space including an intensity channel, for example, the HSV color space, cf. Ref. [71]. The algorithm could then be applied unaltered to the intensity channel, while smoothing of the other two channels could either be performed with the same kernel as in the intensity channel or by isotropic smoothing. The HSV image would then be transformed back to the RGB color space.

However, in very dark color video sequences there is often a significant difference in the noise levels in the different input channels: for example, the blue channel often has a relatively higher noise level. It is therefore essential that it is possible to adapt the algorithm to this difference. To this end, we chose to calculate the structure tensor $\mathbf{J}_\rho$, and its eigenvectors and eigenvalues, in the intensity channel, which we simply define as the mean of the three color channels. The widths of the kernels are then adjusted separately for each color channel by using a different value of the scaling parameter d for each channel. This gives a clear improvement of the output with colors that are closer to the true chromaticity values and with less false color fluctuations than in the above-mentioned HSV approach.

When acquiring raw image data from a CCD or CMOS sensor, the pixels are usually arranged according to the so-called Bayer pattern. It has been shown, cf. Ref. [90], that it is efficient and suitable to perform the interpolation from the Bayer pattern to three separate channels, so-called *demosaicing*, simultaneously to the denoising of the image data. We apply this approach here for each channel, by setting to zero the coefficients in the summation kernel $k(\mathbf{x}_0, \mathbf{x})$ corresponding to pixels where the intensity data is not available and then normalizing the kernel. A smoothed output is then calculated for both the noisy input pixels and the pixels where data are missing.

16.5.4 Experimental Results

In this section, we describe some experimental results. We have tested our algorithm on a number of different types of sequences involving static and moving scenes and with large or little motion of the camera. We have also used a number of different cameras, ranging from cheap consumer cameras to high-end machine vision cameras. We have compared our method to a number of different methods described in the related work section, but most algorithms are not targeted at low-light vision and cannot handle the resulting levels of noise. The algorithm that gives comparable results to ours is BM3D for video, Ref. [21]. We have used the implementation from Ref. [91].

In Figure 16.7, the results of running our algorithm are shown for a number of different input sequences. Also shown is the result of running BM3D. In Figure 16.8, a close-up is shown for two of the frames from Figure 16.7. One can see that BM3D gives very sharp results with very little noise left. In this sense the quality is slightly better than our approach. However, owing to the high noise levels, two types of artifacts can be noticed in the BM3D images: the block structure is visible and we get hallucination effects due to the codebook nature of BM3D. This results in small objects appearing or disappearing.

In order to test the noise reduction part of our method, we conducted a semisynthetic experiment in the following way. We recorded a video sequence under good lighting conditions. This resulted in a sequence with very little noise. We then synthetically constructed a low-light video sequence by dividing the 8-bit gray values by a constant K (e.g., 50), added Gaussian noise with standard deviation σ (e.g., 1), and finally truncated the gray values to integer values. We then amplified the sequence by multiplying by K. This gave us a highly noisy sequence, with known ground truth. One frame of the dark sequence can be seen in second row

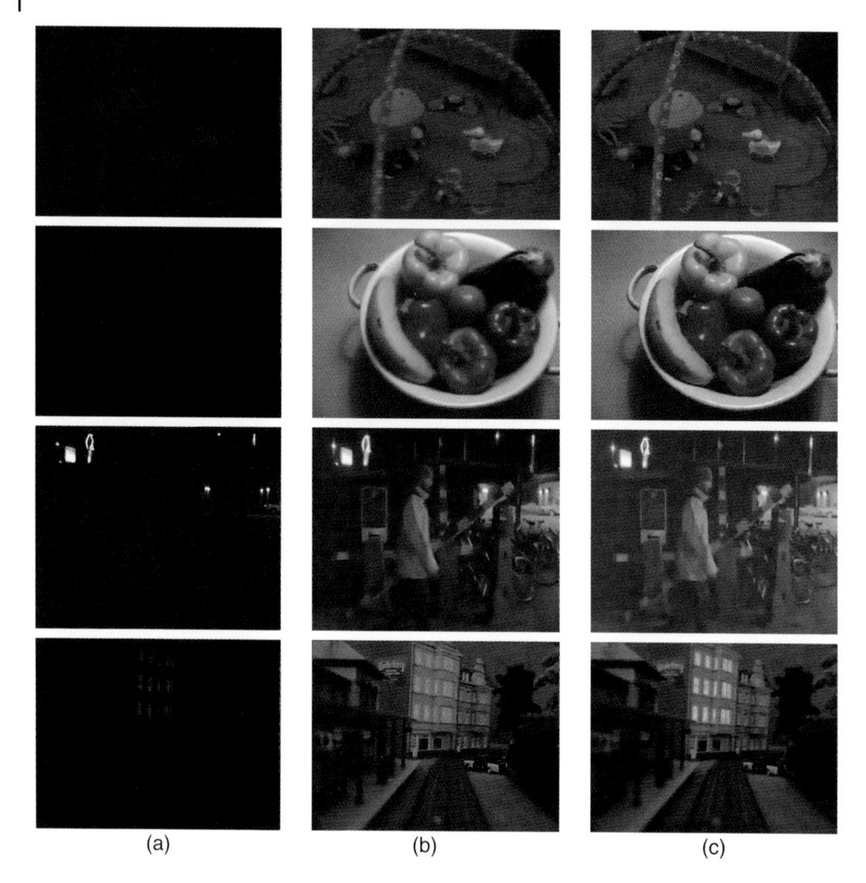

Figure 16.7 Illustration of results from four different input sequences. (a) One frame of the dark input sequence. (b) One frame of our approach and (c) one frame of the result of running BM3D. The two top rows were taken by a standard compact digital camera, a Canon IXUS 800 IS. The third row was taken by a handheld consumer camcorder, a Sony DCR-HC23. The bottom row was taken by an AVT Marlin machine vision camera. The two top and the bottom sequences have large camera motions and the two bottom sequences have additional motion in the scene.

Figure 16.7. We ran our noise reduction step and compared it to BM3D. The resulting peak signal-to-noise ratio (PSNR) and the structure similarity (SSIM) index [92] can be seen in Table 16.1 together with the values for the initial noisy amplified signal. The results for our method and BM3D are very similar. The difference is that our method gives less sharp results, while BM3D gives more blocking artifacts. In terms of computational complexity, the two methods are comparable. As described in Section 16.5.2, our smoothing method is linear in the number of processed voxels, but the factor is highly dependent on the size of the summation neighborhood. This is also true for BM3D which is highly dependent on the size of the block matching neighborhood.

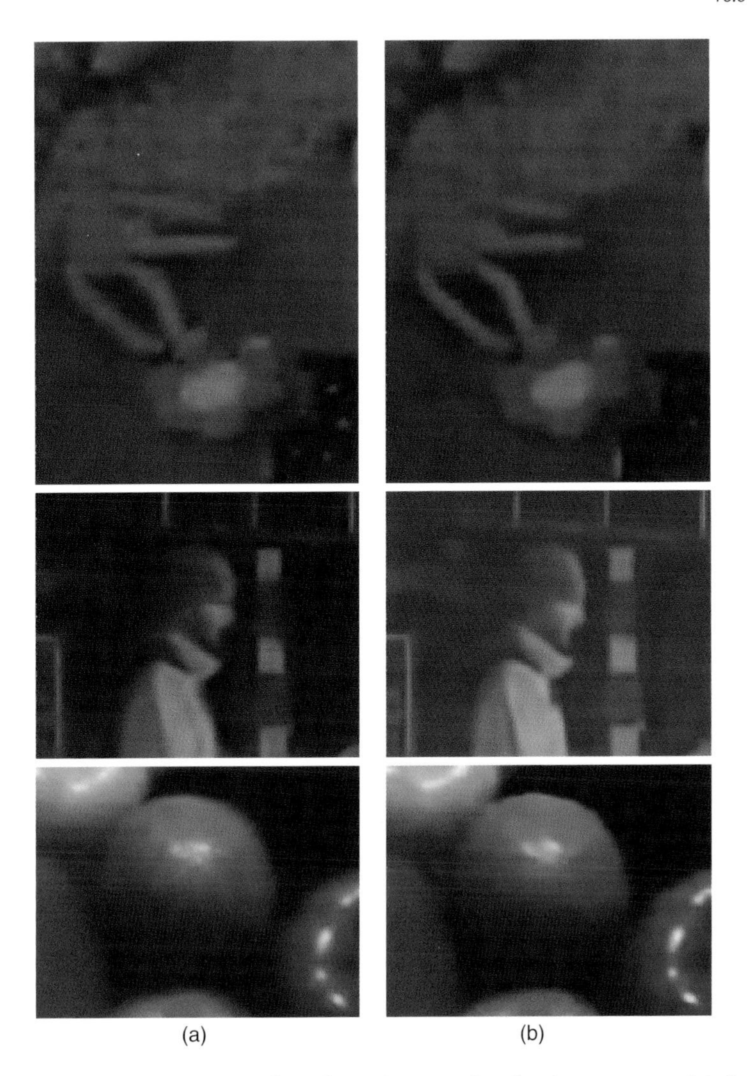

Figure 16.8 A close-up of one frame from the output sequences. (a) The result using our approach and (b) the result using BM3D. One can see that BM3D gives very sharp edges but in some cases, it hallucinates and removes parts due to the codebook nature of BM3D. One can also see some blocking artifacts due to the block structure of BM3D.

16.6 Conclusions

The colors and contrasts of the nocturnal world are just as rich as those found in the diurnal world, and many animals – both vertebrate and invertebrate – have evolved visual systems to exploit this abundant information. Via a combination of highly sensitive optical eye designs and unique alterations in the morphology,

Table 16.1 Peak signal to noise-ratio (PSNR) and structural similarity (SSIM) index for the semisynthetic test sequence.

	PSNR	SSIM
Noisy input	14.7	0.072
Proposed method	27.5	0.777
BM3D	28.6	0.788

circuitry, and physiology of the retina and higher visual centers, nocturnal animals are capable of advanced and reliable vision at night. In the case of the nocturnal bee *Megalopta genalis*, greatly enlarged corneal facets and rhabdoms and slow photoreceptors with high contrast gain ensure that visual signal strength is maximal as it leaves the eye and travels to the lamina. Even though it remains to be shown conclusively, anatomical and theoretical evidence suggests that once the visual signals from large groups of ommatidia reach the lamina, they are spatially (and possibly temporally) summed by the second-order monopolar cells, resulting in an enhanced signal and reduced noise. The greatly improved SNR that this strategy could afford, while confined to a narrower range of spatial and temporal frequencies, would ensure that nocturnal visual reliability is maximized for the slower and coarser features of the world. Those features that are faster and finer – and inherently noisy – would be filtered out. But slower and coarser features, in contrast, would be seen more clearly. Using these biological principles as inspiration, we have developed a night-vision algorithm that processes and enhances video sequences captured in very dim light. This algorithm, similarly to a nocturnal visual system, relies on initial amplification of image signals and a local noise-reducing spatiotemporal summation that is weighted by the extent of image motion occurring in the same image locality. The algorithm dramatically increases the visibility of image features, including the preservation of edges, forms, and colors. This could be used in a number of applications, that is, improving videos taken at night or in dim rooms. It could also be used for surveillance purposes. One of our goal application areas has been in improving vehicular video data, which in turn could be used for example, for pedestrian detection, obstacle avoidance, or traffic sign detection.

Our method provides not only a significant result in the field of dim-light image processing but it also strengthens the evidence that summation strategies are essential for reliable vision in nocturnal animals.

Acknowledgment

The authors are indebted to Toyota Motor Engineering & Development Europe, the US Air Force Office of Scientific Research (AFOSR), and the Swedish Research Council for their financial support.

References

1. Warrant, E.J. and Dacke, M. (2011) Vision and visual navigation in nocturnal insects. *Annu. Rev. Entomol.*, **56**, 239–254.
2. Kelber, A., Balkenius, A., and Warrant, E.J. (2002) Scotopic colour vision in nocturnal hawkmoths. *Nature*, **419**, 922–925.
3. Somanathan, H., Borges, R.M., Warrant, E.J., and Kelber, A. (2008) Nocturnal bees learn landmark colours in starlight. *Curr. Biol.*, **18** (21), R996–R997.
4. Theobald, J., Warrant, E.J., and O'Carroll, D.C. (2010) Wide-field motion tuning in nocturnal hawkmoths. *Proc. R. Soc. London, Ser. B*, **277**, 853–860.
5. Baird, M., Kreiss, E., Wcislo, W.T., Warrant, E.J., and Dacke, M. (2011) Nocturnal insects use optic flow for flight control. *Biol. Lett.*, 7, 499–501.
6. Warrant, E.J. (2004) Vision in the dimmest habitats on earth. *J. Comp. Physiol. A*, **190**, 765–789.
7. Somanathan, H., Borges, R.M., Warrant, E.J., and Kelber, A. (2008) Visual ecology of Indian carpenter bees I: light intensities and flight activity. *J. Comp. Physiol. A*, **194**, 97–107.
8. Reid, S.F., Narendra, A., Hemmi, J.M., and Zeil, J. (2011) Polarised skylight and the landmark panorama provide night-active bull ants with compass information during route following. *J. Exp. Biol.*, **214**, 363–370.
9. Dacke, M., Nilsson, D.E., Scholtz, C.H., Byrne, M., and Warrant, E.J. (2003) Insect orientation to polarized moonlight. *Nature*, **424**, 33.
10. Dacke, M., Baird, E., Byrne, M., Scholtz, C.H., and Warrant, E.J. (2013) Dung beetles use the Milky Way for orientation. *Curr. Biol.*, **23**, 298–300.
11. Malm, H. and Warrant, E. (2006) Motion dependent spatiotemporal smoothing for noise reduction in very dim light image sequences. *Proceedings International Conference on Pattern Recognition*, Hong Kong, pp. 954–959.
12. Malm, H., Oskarsson, M., Warrant, E., Clarberg, P., Hasselgren, J., and Lejdfors, C. (2007) Adaptive enhancement and noise reduction in very low light-level video. IEEE 11th International Conference on Computer Vision, 2007. ICCV 2007, IEEE, pp. 1–8.
13. Malm, H., Oskarsson, M., and Warrant, E. (2012) Biologically inspired enhancement of dim light video, *Frontiers in Sensing*, Springer-Verlag, pp. 71–85.
14. Warrant, E., Oskarsson, M., and Malm, H. (2014) The remarkable visual abilities of nocturnal insects: neural principles and bioinspired night-vision algorithms. *Proc. IEEE*, **102** (10), 1411–1426.
15. Kalivas, D. and Sawchuk, A. (1990) Motion compensated enhancement of noisy image sequences. *Proceedings IEEE International Conference Acoustics Speech, and Signal Processing*, Alburquerque, NM, pp. 2121–2124.
16. Özkan, M., Sezan, M., and Tekalp, A. (1993) Adaptive motion-compensated filtering of noisy image sequences. *IEEE Tran. Circuits Syst. Video Technol.*, **3** (4), 277–290.
17. Miyata, K. and Taguchi, A. (2002) Spatio-temporal separable data-dependent weighted average filtering for restoration of the image sequences. *Proceedings IEEE International Conference on Acoustics, Speech and Signal Processing*, vol. 4, Alburquerque, NM, pp. 3696–3699.
18. Sezan, M., Özkan, M., and Fogel, S. (1991) Temporally adaptive filtering of noisy image sequences. *Proceedings IEEE International Conference on Acoustics, Speech and Signal Processing*, vol. 4, Toronto, Canada, pp. 2429–2432.
19. Dugad, R. and Ahuja, N. (1999) Video denoising by combining kalman and wiener estimates. *Proceedings International Conference on Image Processing*, vol. 4, Kobe, Japan, p. 152–156.
20. Buades, A., Coll, B., and Morel, J.M. (2005) A non-local algorithm for image denoising. IEEE Computer Society Conference on Computer Vision and Pattern Recognition, 2005. CVPR 2005, vol. 2, IEEE, pp. 60–65.
21. Dabov, K., Foi, A., and Egiazarian, K. (2007) Video denoising by sparse 3d transform-domain collaborative filtering.

Proceedings of the 15th European Signal Processing Conference, vol. 1, p. 7.

22. Dabov, K., Foi, A., Katkovnik, V., and Egiazarian, K. (2007) Image denoising by sparse 3-d transform-domain collaborative filtering. *IEEE Trans. Image Process.*, **16** (8), 2080–2095.
23. Martinez, D. and Lim, J. (1985) Implicit motion compensated noise reduction of motion video signals. *Proceedings IEEE International Conference on Acoustics, Speech and Signal Processing*, Tampa, FL, pp. 375–378.
24. Arce, G. (1991) Multistage order statistic filters for image sequence processing. *IEEE Trans. Signal Proc.*, **39** (5), 1146–1163.
25. Ko, S.J., and Forest, T. (1993) Image sequence enhancement based on adaptive symmetric order statistics. *IEEE Trans. Circuits Syst. II: Analog Digital Signal Proc.*, **40** (8), 504–509.
26. Brailean, J., Kleihorst, R., Efstratiadis, S., Katsaggelos, A., and Lagendijk, R. (1995) Noise reduction filters for dynamic image sequences: a review. *Proc. IEEE*, **83** (9), 1272–1289.
27. Perona, P. and Malik, J. (1990) Scale-space and edge detection using anisotropic diffusion. *IEEE Trans. Pattern Anal. Mach. Intell.*, **12** (7), 629–639.
28. Weickert, J. (1998) *Anisotropic Diffusion in Image Processing*, Teubner-Verlag, Stuttgart.
29. Uttenweiler, D., Weber, C., Jähne, B., Fink, R., and Scharr, H. (2003) Spatiotemporal anisotropic diffusion filtering to improve signal-to-noise ratios and object restoration in fluorescence microscopic image sequences. *J. Biomed. Opt.*, **8** (1), 40–47.
30. Lee, S. and Kang, M. (1998) Spatio-temporal video filtering algorithm based on 3-d anisotropic diffusion equation. *Proceedings International Conference on Image Processing*, vol. 2, pp. 447–450.
31. Weickert, J. (1999) Coherence-enhancing diffusion filtering. *Int. J. Comput. Vision*, **31** (2-3), 111–127.
32. Tomasi, C. and Manduchi, R. (1998) Bilateral filtering for gray and color images. *Proceedings of the 6th International Conference Computer Vision*, pp. 839–846.
33. Yang, G., Burger, P., Firmin, D., and Underwood, S. (1996) Structure adaptive anisotropic image filtering. *Image Vision Comput.*, **14**, 135–145.
34. Barash, D., and Comaniciu, D. (2004) A common framework for nonlinear diffusion, adaptive smoothing, bilateral filtering and mean shift. *Image Vision Comput.*, **22**, 73–81.
35. Kang, S., Uytendaele, M., Winder, S., and Szeliski, R. (2003) High dynamic range video. SIGGRAPH '03: ACM SIGGRAPH 2003 Papers, pp. 319–325.
36. Reinhard, E., Ward, G., Pattanaik, S., and Debevec, P. (2005) *High Dynamic Range Imaging: Acquisition, Display, and Image-Based Lighting*, Morgan Kaufmann Publishers.
37. Bennett, E. and McMillan, L. (2005) Video enhancement using per-pixel virtual exposures. *Proceedings SIGGRAPH*, Los Angeles, CA, pp. 845–852.
38. Lee, S.W., Maik, V., Jang, J., Shin, J., and Paik, J. (2005) Noise-adaptive spatio-temporal filter for real-time noise removal in low light level images. *IEEE Tran. Consum. Electron.*, **51** (2), 648–653.
39. Martin, G.R. (1990) *Birds By Night*, T. and A.D. Poyser, London.
40. Warrant, E.J. (2008) Nocturnal vision, in *The Senses: A Comprehensive Reference (Vol. 2: Vision II)* (eds T. Albright and R. Masland), Academic Press, pp. 53–86.
41. Laughlin, S.B. (1990) Invertebrate vision at low luminances, in *Night Vision* (eds R.F. Hess, L.T. Sharpe, and K. Nordby), Cambridge University Press, pp. 223–250.
42. Warrant, E.J. (2006) *The Sensitivity of Invertebrate Eyes to Light* (E.J. Warrant and D.-E. Nilsson Invertebrate Vision), Cambridge University Press, Cambridge, pp. 167–210.
43. Lillywhite, P.G. and Laughlin, S.B. (1979) Transducer noise in a photoreceptor. *Nature*, **277**, 569–572.
44. Frederiksen, R., Wcislo, W.T., and Warrant, E.J. (2008) Visual reliability and information rate in the retina of a nocturnal bee. *Curr. Biol.*, **18**, 349–353.

45. Rose, A. (1942) The relative sensitivities of television pickup tubes, photographic film and the human eye. *Proc. Inst. Radio Eng. N. Y.*, **30**, 293–300.
46. de Vries, H.L. (1943) The quantum character of light and its bearing upon threshold of vision, the differential sensitivity and visual acuity of the eye. *Physica*, **10**, 553–564.
47. Land, M.F. (1981) Optics and vision in invertebrates, in *Handbook of Sensory Physiology*, V. VII/6B (ed. H. Autrum), Springer-Verlag, Berlin, Heidelberg and New York, pp. 471–592.
48. Lillywhite, P.G. (1977) Single photon signals and transduction in an insect eye. *J. Comp. Physiol.*, **122**, 189–200.
49. Laughlin, S.B. and Lillywhite, P.G. (1982) Intrinsic noise in locust photoreceptors. *J. Physiol.*, **332**, 25–45.
50. Barlow, H.B. (1956) Retinal noise and absolute threshold. *J. Opt. Soc. Amer.*, **46**, 634–639.
51. Dubs, A., Laughlin, S.B., and Srinivasan, M.V. (1981) Single photon signals in fly photoreceptors and first order interneurons at behavioural threshold. *J. Physiol.*, **317**, 317–334.
52. Doujak, F.E. (1985) Can a shore crab see a star? *J. Exp. Biol.*, **166**, 385–393.
53. Baylor, D.A., Matthews, G., and Yau, K.W. (1980) Two components of electrical dark noise in toad retinal rod outer segments. *J. Physiol.*, **309**, 591–621.
54. Aho, A.C., Donner, K., Hydén, C., Larsen, L.O., and Reuter, T. (1988) Low retinal noise in animals with low body temperature allows high visual sensitivity. *Nature*, **334**, 348–350.
55. Aho, A.C., Donner, K., Helenius, S., Larsen, L.O., and Reuter, T. (1993) Visual performance of the toad (*Bufo bufo*) at low light levels, retinal ganglion cell responses and prey-catching accuracy. *J. Comp. Physiol. A*, **172**, 671–682.
56. Healey, G.E. and Kondepudy, R. (1994) Radiometric CCD camera calibration and noise estimation. *IEEE Trans. Pattern Anal. Mach. Intell.*, **16** (3), 267–276.
57. Laughlin, S.B. and Weckström, M. (1993) Fast and slow photoreceptors: a comparative study of the functional diversity of coding and conductances in the Diptera. *J. Comp. Physiol. A*, **172**, 593–609.
58. Laughlin, S.B., Blest, A.D., and Stowe, S. (1980) The sensitivity of receptors in the posterior median eye of the nocturnal spider *Dinopis*. *J. Comp. Physiol.*, **141**, 53–65.
59. Heimonen, K., Salmela, I., Kontiokari, P., and Weckström, M. (2006) Large functional variability in cockroach photoreceptors: optimization to low light levels. *J. Neurosci.*, **26**, 13454–13462.
60. Pirhofer-Walzl, K., Warrant, E.J., and Barth, F.G. (2007) Adaptations for vision in dim light: impulse responses and bumps in nocturnal spider photoreceptor cells (*Cupiennius salei* Keys). *J. Comp. Physiol. A*, **193**, 1081–1087.
61. Horridge, G., Duniec, J., and Marčelja, L. (1981) A 24-hour cycle in single locust and mantis photoreceptors. *J. Exp. Biol.*, **91** (1), 307–322.
62. Greiner, B., Ribi, W.A., Wcislo, W.T., and Warrant, E.J. (2004) Neuronal organisation in the first optic ganglion of the nocturnal bee *Megalopta genalis*. *Cell Tissue Res.*, **318**, 429–437.
63. Ribi, W. (1975) The first optic ganglion of the bee. *Cell Tissue Res.*, **165** (1), 103–111.
64. Hateren, J.Hv. (1992) Real and optimal neural images in early vision. *Nature*, **360**, 68–70.
65. Hateren, J.Hv. (1993) Spatiotemporal contrast sensitivity of early vision. *Vision Res.*, **33**, 257–267.
66. Laughlin, S.B. (1996) Matched filtering by a photoreceptor membrane. *Vision Res.*, **36**, 1529–1541.
67. Weckström, M. and Laughlin, S.B. (1995) Visual ecology and voltage-gated ion channels in insect photoreceptors. *Trends Neurosci.*, **18** (1), 17–21.
68. Niven, J.E., Vähäsöyrinki, M., Kauranen, M., Hardie, R.C., Juusola, M., and Weckström, M. (2003) The contribution of shaker K+ channels to the information capacity of *Drosophila* photoreceptors. *Nature*, **421** (6923), 630–634.

69. Niven, J.E., Anderson, J.C., and Laughlin, S.B. (2007) Fly photoreceptors demonstrate energy-information trade-offs in neural coding. *PLoS Biol.*, **5** (4), e116.
70. Drago, F., Myszkowski, K., Annen, T., and Chiba, N. (2003) Adaptive logarithmic mapping for displaying high contrast scenes. *Proceedings EUROGRAPHICS*, vol. 22, pp. 419–426.
71. Gonzalez, R. and Woods, R. (1992) *Digital Image Processing*, Addison-Wesley.
72. Pizer, S., Amburn, E., Austin, J., Cromartie, R., Geselowitz, A., Geer, T., teer Haar Romeny, B., Zimmerman, J., and Zuiderveld, K. (1987) Adaptive histogram equalization and its variations. *Comput. Vision Graph. Image Process.*, **39**, 355–368.
73. Snyder, A.W. (1977) Acuity of compound eyes, physical limitations and design. *J. Comp. Physiol.*, **116**, 161–182.
74. Snyder, A.W., Laughlin, S.B., and Stavenga, D.G. (1977) Information capacity of eyes. *Vision Res.*, **17**, 1163–1175.
75. Snyder, A.W., Stavenga, D.G., and Laughlin, S.B. (1977) Spatial information capacity of compound eyes. *J. Comp. Physiol.*, **116**, 183–207.
76. Laughlin, S.B. (1981) Neural principles in the peripheral visual systems of invertebrates, in *Handbook of Sensory Physiology*, V. VII/6B (ed. H. Autrum), Springer-Verlag, Berlin, Heidelberg and New York, pp. 133–280.
77. Warrant, E.J. (1999) Seeing better at night: life style, eye design and the optimum strategy of spatial and temporal summation. *Vision Res.*, **39**, 1611–1630.
78. Greiner, B., Ribi, W.A., and Warrant, E.J. (2005) A neural network to improve dim-light vision? Dendritic fields of first-order interneurons in the nocturnal bee *Megalopta genalis. Cell Tissue Res.*, **323**, 313–320.
79. Ribi, W.A. (1977) Fine structure of the first optic ganglion (lamina) of the cockroach, Periplaneta Americana. *Tissue Cell*, **9**, 57–72.
80. Ohly, K.P. (1975) The neurons of the first synaptic regions of the optic neuropil of the firefly *Phausius splendidula L.* (Coleoptera). *Cell Tissue Res.*, **158**, 89–109.
81. Strausfeld, N.J. and Blest, A.D. (1970) Golgi studies on insects I. The optic lobes of *Lepidoptera. Philos. Trans. R. Soc. London, Ser. B*, **258**, 81–134.
82. Theobald, J.C., Greiner, B., Wcislo, W.T., and Warrant, E.J. (2006) Visual summation in night-flying sweat bees: a theoretical study. *Vision Res.*, **46**, 2298–2309.
83. Klaus, A. and Warrant, E.J. (2009) Optimum spatiotemporal receptive fields for vision in dim light. *J. Vision*, **9** (4), 1–16.
84. Warrant, E.J., Porombka, T., and Kirchner, W.H. (1996) Neural image enhancement allows honeybees to see at night. *Proc. R. Soc. London, Ser. B*, **263**, 1521–1526.
85. Jähne, B. (1993) *Spatio-Temporal Image Processing*, Springer-Verlag.
86. Bigun, J. and Granlund, G.H. (1987) Optimal orientation detection of linear symmetry. *Proceedings of the IEEE 1st International Conference on Computer Vision*, London, Great Britain, pp. 433–438.
87. Knutsson, H. (1989) Representing local structure using tensors. The 6th Scandinavian Conference on Image Analysis, Oulu, Finland, pp. 244–251.
88. Haussecker, H. and Spies, H. (1999) *Handbook of Computer Vision and Applications*, vol. 2, Chap. Motion, Academic Press, pp. 125–151.
89. Atick, J.J. (1992) Could information theory provide an ecological theory of sensory processing? *Netw. Comput. Neural Syst.*, **3** (2), 213–251.
90. Hirakawa, K. and Parks, T. (2006) Joint demosaicing and denoising. *IEEE Trans. Image Process.*, **15** (8), 2146–2157.
91. *http://www.cs.tut.fi/foi/GCF-BM3D/.*
92. Wang, Z., Bovik, A.C., Sheikh, H.R., and Simoncelli, E.P. (2004) Image quality assessment: from error visibility to structural similarity. *IEEE Trans. Image Process.*, **13** (4), 600–612.

17
Bioinspired Motion Detection Based on an FPGA Platform

Tim Köhler

17.1
Introduction

Visual detection of motion is one of the computer vision methods that has been in high demand. Several applications of motion detection can be found in mobile robotics, for example, obstacle avoidance or flight stabilization of unmanned air vehicles. In the case of small and light robots, particularly, computer vision processing has been very limited because of power consumption and weight. As flying insects are able to solve the same tasks with very limited processing and power resources, biologically inspired methods promise to provide very efficient solutions. However, even if a biological solution for a computer vision task is found, in an application, the manner in which this solution is implemented is also of critical importance.

In this chapter, the combination of (a) a bioinspired motion detection method and (b) a very versatile design flow is presented. The implementation described here uses specific devices called *field-programmable gate arrays* (*FPGA*s). Such a hardware design process allows an implementation which is very well adapted to an algorithm and, thus, can lead to a very efficient implementation.

This chapter's scope covers topics discussed at greater length in other chapters. For more details on compound- and biosensors, see Chapters 6 and 12 which focuses on the topic of motion detection.

Section 17.2 provides a short motivation for hardware implementation of biologically inspired computer vision in motion detection. Section 17.3 gives an introduction to insect motion detection. The results of behavioral biology and neural biology studies are summarized therein. Section 17.4 presents an overview of findings on robotic implementations of biological insect motion detection. This includes solutions using FPGA and, therefore, a short introduction to FPGAs as well. The main part of this chapter is Section 17.5. in which an FPGA-based module implementing bioinspired motion detection is described. Section 17.6 shows experimental results of the motion detection module. Finally, Section 17.7 discusses the implementation and the experimental results and Section 17.8 concludes the chapter.

Biologically Inspired Computer Vision: Fundamentals and Applications, First Edition.
Edited by Gabriel Cristóbal, Laurent Perrinet, and Matthias Keil.

17.2
A Motion Detection Module for Robotics and Biology

Visual detection of motion is realized in biology in a very efficient and reliable manner. Insects such as flies are capable of detecting an approaching object very quickly. On the other hand, the incorporated resources for this motion detection are quite limited. For example, the housefly *Musca domestica* has compound eyes with only about 3000 single facets (called *ommatidia*). Furthermore, the neural processing for the detection of motion is also limited (see Section 17.3). Taking all these factors into account, the fly has the ability for quick visual detection of motion with little processing, its low weight, small size, and a very small energy consumption.

This efficient biological motion detection[1] gives a very interesting blueprint for technical motion detection applications. So-called *micro air vehicles* (*MAV*), for example, face problems comparable to those of flying insects (flight stabilization, obstacle detection). They cannot use powerful computers to solve these tasks because of size and weight limitations. Having an efficient implementation derived from biology could solve such problems. Therefore, implementing the biological findings in engineering in an efficient manner is needed. There are many examples of such implementations that have been published (see Refs [1, 2], and [3]).

A second motivation can be found in biorobotics in general: the flow of scientific findings is not always in the direction from biology to robotics. Instead, the opposite direction can also be of importance . In biology, explanatory models can be set up on the basis of behavioral experiments. Insect motion detection, as an example, was studied in the 1950s by Hassenstein and Reichardt [4]. Such behavioral studies can lead to models that explain the observed behaviors. But to be certain that a stated model is a correct explanation, the opposite direction need to be tested; that is, using a synthetic implementation of the proposed model to perform the tasks studied in the behavioral experiments. Such a synthetic implementation could be realized and tested in computer simulations. However, to arrive at acceptable results, the test conditions often need to be as realistic as possible (real environment, comparable optical parameters such as spatial and temporal resolution). This is why a robotic implementation of a model is necessary. Of course, a verification via such a synthetic model alone is still not be proof regarding the correctness of the proposed model. But if the synthetic implementation fails to show the behavior noticed in the behavioral experiments – that is, even under optimal conditions, it is very likely that the proposed model would fail in the real biological counterpart, too, and therefore it is not valid. The option of carrying out neural measurements in living insects is chosen in neural biology. However, even with this method, finding all three parts, namely, preprocessing, motion detection,

1) The motion detection described here is done visually, that is, an optical flow is actually detected. Although this optical flow is sometimes (especially in experimental setups) generated just visually without any motion, the method and its application is often referred to as *motion detection*.

and postprocessing, may not be feasible. Thus, there is sufficient motivation for a computer vision implementation both in robotics and in biology.

17.3
Insect Motion Detection Models

The insect visual system can be separated into two parts: first, there is the so-called compound eye with optical and receptive neural components and, second, there is the further neural processing. A sketch of a compound eye can be seen in Figure 17.1(a) and a single photoreception unit (called *ommatidium*) in Figure 17.1(b). This structure is an example; different variations of this can also be found. One variation is optical superposition (no complete separation of the ommatidia by pigment cells); another is the number of photoreceptor cells per ommatidium. Some more details can be found in Ref. [5].

A diagram of both parts (compound eye and further neural processing) is depicted in Figure 17.2. Most of the motion detection components of insect vision are located or supposed to be located in the further neural processing, that is,. in the *lamina*, the *lobula plate*, and the *medulla* [7, 8].

Insect motion detection has been studied from the 1950s. One of the first findings was the identification of a separation in local detection components, on the one hand, and an integration of multiple local detection signals, on the other. Through integration, a detection signal for a larger region covering the whole field of view is generated. The single local detection components are referred to as *elementary motion detectors* (*EMD*) in the following.

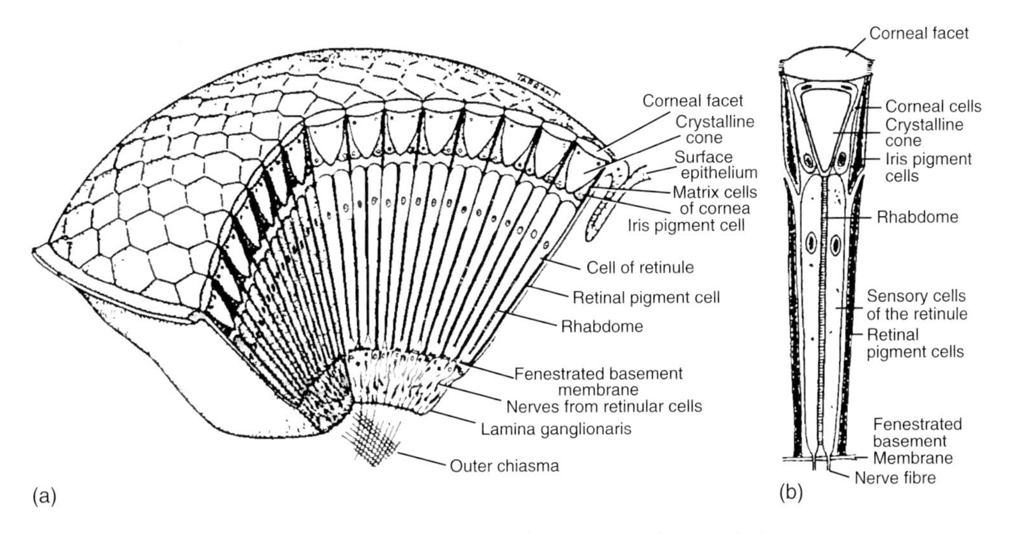

Figure 17.1 (a) The section of a compound eye. Exemplarily, the ommatidia are optically separated here. This is not the case for all complex eyes. (b) The section of an ommatidium (again with an optical separation from the neighbor ommatidia). (Adapted from Ref. [[6], pages 155 and 167]).

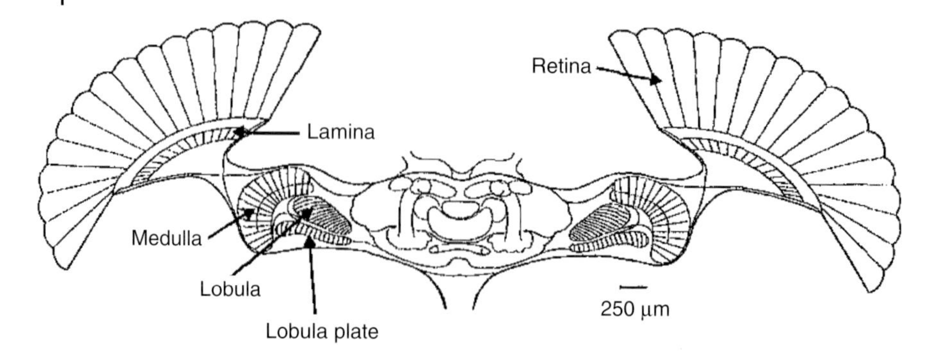

Figure 17.2 Schematic horizontal cross section through the head of a fly including the major parts of its head ganglion (adapted from Ref. [7], Figure 1, with kind permission from Springer Science and Business Media).

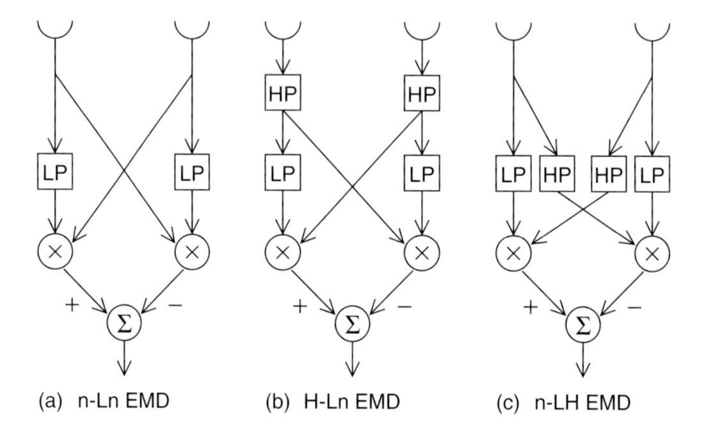

Figure 17.3 (a), (b), and (c) Three versions of EMDs. The "∪" on top denotes a photoreceptor. "LP" and "HP" stand for lowpass filter and highpass filter, respectively. The correlation component "×" can be realized by a multiplication and "Σ" by a summation (see text). The labels below the figures are used to refer to these models. The labels conform to a naming scheme in which the first letter represents the input stage (in this case, with a highpass input filter "H" or with no preprocessing "n"). The following two letters stand for the components between the preprocessing and the correlation inputs: The first letter identifies the component before the first correlation input (vertical arms in the figures, in these three models, all are lowpass filters "L"). The second letter represents the component in front of the second correlation inputs (the crossed arms in the figures, highpass filters "H" or no filter operations "n"). Such symmetrical models are sometimes referred to as "full-detector," while one half of these detectors (i.e., just one correlation element and no summation) are sometimes called "half-detector" (figures taken from Refs. [10, 11]).

Figure 17.3 illustrates three different versions of EMD . A feature common to all three is that the signals of only two photoreceptors (or two groups of photoreceptors) are used to detect motion components along one (called *preferred*) motion vector from one of the receptors (or the receptor groups) to the other [9]. For motion (components) in the direction from the first to the second receptor (in the figures, from the left to the right receptor) the detector generates a positive output value. For motion in the opposite direction, the output is negative.

The simplest EMD version can be seen in Figure 17.3(a). Here, the signals of the two photoreceptors are delayed by a lowpass filter component. The delayed signal of one receptor (i.e., the output of one of the lowpass filters) is correlated with the undelayed signal of the other receptor (see the review of Egelhaaf and Borst [12]). As correlation function, a simple multiplication leads to results that are comparable to biological data. Finally, the difference between the two correlation results needs to be computed.

The resulting behavior is as follows (see Figure 17.4): if an object moves from a position in front of the left photoreceptor to a position in front of the right receptor then the corresponding reflected brightness values will be perceived first by the left receptor and after a certain delay (depending on the object's motion speed), by the right receptor. If the motion speed matches the delay in the left receptor's lowpass filter, then the corresponding brightness values are correlated at one of the multiplication components. This correspondence leads to high correlation results, while the second multiplication component is supplied with uncorrelated

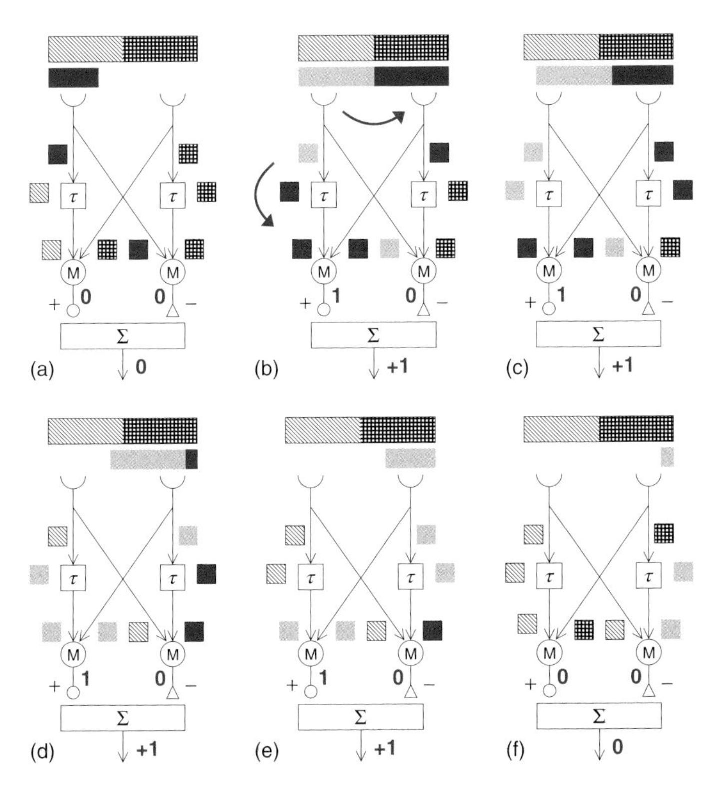

Figure 17.4 Functionality of the EMD. Exemplarily, a red and blue object moves from the left-hand side to the right-hand side. Depicted are six states (a)–(f). The "τ" denotes a delay element (see text) and "M" a correlation.

brightness values (leading to small correlation values). As a result, the small correlation value is subtracted from the large correlation value and the detector's output value is large.

In case of motion in the opposite direction, a large value is subtracted from a small value leading to a large negative output. Finally, in case of no motion (in the direction between the two receptors), both correlation results are equal and the detector's output will be zero. The two computation parts up to one correlation (left-hand side, right-hand side) are sometimes called "half-detectors" and could be used alone if only one motion direction is to be detected. If the delay induced by the motion velocity and the lowpass filters' delays do not match, then the correlation result will be less than optimal and the absolute value of the detector's output will be less than the maximum output. At very slow or very fast motion speed, the detector's output might not be distinguishable from zero.

As Harris *et al.* [13] have shown, the simple EMD's response differs from empirical data. This is especially the case for specific pattern types and motion conditions (see Ref. [14]). A proposed improvement is the introduction of a first-order highpass filter to suppress constant illumination differences as depicted in Figure 17.3(b) (see Ref. [15]).

A further variation was proposed by Kirschfeld [16] (see also [15]) and can be seen in Figure 17.3(c). Two of the advantages of this version are, first, again a better correspondence to biological data under specific conditions and, second, a stronger response compared to a H-Ln EMD (see caption of Figure 17.3) with the same time constants.

There are further variants of EMD types, some of which have been published quite recently. For example, Ref. [17] presents two versions with a reduced sensitivity to contrast changes. This property is especially interesting for motion detection in mobile robots.

A feature common for all these variations is that a single EMD uses just two photoreceptors (or two local photoreceptor groups). However, to generate a reliable detector response the output values of multiple EMDs are combined. A simple realization is a summation of the single EMD responses – which is sufficient in terms of an acceptable accordance with biological data and sufficient for many robotic applications. Two effects of such an aggregation are an increased reliability and a better ratio of motion-induced detector response to nonsystematic responses (signal-to-noise ratio, SNR). However, probably the main advantage of a combination of several (locally detecting) EMDs is the detectability of (a) regional/global motion and (b) complex motion types. In Figure 17.5 three examples are depicted.

As an alternative to a simple summation of the single EMD responses [18] suggest a nonlinear passive membrane model. Extensions or modifications such as this—also of the EMDs' filters or correlation—have been proposed several times after the original publications by Hassenstein and Reichardt. Often they lead to detector responses which fit better to measurements of the biological counterpart. Another example of this is a weighting of the single EMD responses (leading to so-called "matched filters") which was proposed by Franz and Krapp [19].

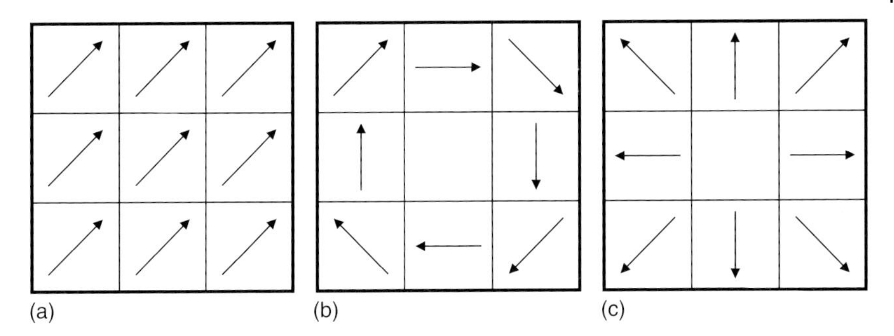

Figure 17.5 By the aggregation of several EMD responses to EMD arrays (covering up to the whole field of view), specific types of motion can be detected. For the sake of simplicity, all examples show a quadratic field of view of up to nine EMDs. The arrows depict single EMDs with the respective preferred direction. (a) The EMD array responds strongest to its preferred motion direction from the bottom left to the top right. (b) The array has a clockwise rotation as preferred optical flow . (c) The EMD array responds strongest to approaching objects, leading to "looming" optical flow (figures taken from Refs [10, 11]).

Expected steady-state responses to a constantly moving sine pattern for the H-Ln- and the n-LH-EMD type (see caption of Figure 17.3) were derived by Borst *et al.* [15]. The estimations are based on the ideal behavior of first-order, lowpass and first-order, highpass filters. The expected steady-state response of a H-Ln-EMD to a constantly moving sine pattern is derived by Borst *et al.* [15] as shown in Eq. (17.1). Given is the average value of multiple detectors, each with a different position, that is, a different phase φ. The stimulus is generated by a sine pattern with period length λ moved with constant velocity v. The resulting stimuli at the receptors have a frequency of $\omega = 2\pi \cdot v/\lambda$. The time constants τ_l and τ_h are those of the lowpass filters and the highpass filters, respectively. Note that I, the constant DC component of the stimulus (depending on the constant average of the illumination and the constant average of the pattern brightness), is not included in the given equation. Thus, the detector's output signal does not depend on these constant properties. Of course, the amplitude of the input signal (ΔI) has a major influence on the detector's response.

$$\langle R_i \rangle_\varphi = \Delta I^2 \frac{\tau_h \omega \cdot \tau_l \tau_h \omega^2}{(1 + \tau_l^2 \omega^2) \cdot (1 + \tau_h^2 \omega^2)} \cdot \sin\left(\frac{2\pi \Delta\varphi}{\lambda}\right) \tag{17.1}$$

In Eq. (17.2), the expected steady-state response of the n-LH type detector array when excited with the same constantly moving sine pattern can be seen. Both theoretical steady-state responses are compared with measurements in the experiments Section 17.6 below.

$$\langle R_i \rangle_\varphi = \Delta I^2 \frac{\tau_h \omega \cdot (1 + \tau_l \tau_h \omega^2)}{(1 + \tau_l^2 \omega^2) \cdot (1 + \tau_h^2 \omega^2)} \cdot \sin\left(\frac{2\pi \Delta\varphi}{\lambda}\right) \tag{17.2}$$

These two equations are needed in the experimental Section 17.6. Further explanations and theoretical ideal responses are given by Borst *et al.* [15].

17.4
Overview of Robotic Implementations of Bioinspired Motion Detection

For the implementation of bioinspired algorithms in general, the methods and types of hardware used could be like those for any other application. However, as often one motivation to use a biologically inspired algorithm is power/weight efficiency, the hardware basis for such an implementation needs to be selected accordingly.

There are generally two options for algorithmic implementations:

1) Implementation in software
2) Implementation in hardware.

In case of the EMD-based motion detection, there are some publications presenting implementations of option (1). Zanker and Zeil [20] propose a software implementation which runs on a standard personal computer. With respect to the power and/or weight limitations of several EMD-applications (especially micro air vehicles, but also typical mobile robots), the need for a PC is often a knockout criterion. An alternative software implementation was presented by Ruffier *et al.* [21]. Regarding power/weight limitations, this implementation of EMDs is less demanding because it is based on a microcontroller.

Algorithmic implementations in hardware can be separated into two groups:

1) Implementation on dedicated analog, digital, or mixed-signal hardware
2) Implementation on reconfigurable hardware.

Designing hardware dedicated for a specific algorithm allows a very good (usually seen as optimal) realization with respect to power-efficiency, weight-efficiency, or a certain combination of both. This is especially the case while designing one's own semiconductor chip (integrated circuit (IC)) specifically for this algorithm. Such ICs are called "application-specific integrated circuits (ASIC)" and have the drawback of a relatively complex and time-consuming design and manufacturing process. An EMD-ASIC was presented by Harrison [22–24]. Using standard ICs instead of own ASICs (but in a dedicated circuit) is generally easier to realize. Such an implementation can still be quite power-efficient and was proposed by Franceschini *et al.* [25].

Using reconfigurable hardware can be seen as an intermediate solution between software- and (dedicated) hardware-implementations. Here, the hardware can be adapted to an application (in certain limits). This adaptation (or "configuration") can be optimized for a specific algorithm and for a specific design goal (e.g., power-efficiency). Such a solution can be, for example, more power-efficient than a microcontroller-based (software) implementation but will usually be not as power-efficient as an optimized ASIC. The actual advantage over a microcontroller-based and disadvantage compared to an ASIC-based implementation depends on the specific application and algorithm and can even vanish.

17.4.1 Field-Programmable Gate Arrays (FPGA)

There are different kinds of reconfigurable ICs available today. Some reconfigurable analog ICs are called “field-programmable analog arrays (FPAAs)”. FPAA-based implementations of EMDs have been presented in Refs [21] and [10]. In these implementations, analog implies that the signal processing is carried out using continuous values (signals as continuous voltage or continuous current levels) and (with some limitations) continuous in time. By these means, theoretically, any small value changes at any point in time for any small duration could lead to a response of the system. The advantage of such an analog processing is that discretization effects that may occur when generating digital (e.g., binary-represented) values are avoided. More specifically, an analog processing is potentially closer to the continuous biological processing. However, today the analog-to-digital conversion can be chosen in a way that avoids discretization effects in practice. Much more common and used in a vast number of applications are the digital reconfigurable ICs called “field-programmable gate array (FPGA)”. Today, there are many more manufacturers of FPGAs and many more different FPGAs available on the market than are analog-reconfigurable IC types and manufacturers.

FPGAs consist of general building blocks which can be connected quite arbitrarily. Each building block usually consists of a programmable logic function and one or a few general memory elements. The programmable logic function is implemented by one or a few lookup tables (LUT) which map a limited number of digital input signals to a digital output. With these two components (logic function and general memory), in principle, any digital logic circuit can be implemented. Usually, memories are used to store machine states or variables and logic functions are used for any combinatorial circuits, including arithmetic functions. If logic capacity or memory size of one building block is not sufficient for a needed functionality, then several blocks are combined. Depending on the manufacturer of the FPGA, these general building blocks (or groups of these building blocks) are called, for example, “logic cell (LC),” “logic element (LE),” “adaptive logic modules (ALM),” “slice,” or “configurable logic block (CLB).”

Very different FPGA-based implementations of EMDs have been proposed so far. Aubepart *et al.* [1] shows a very fast solution (frame rates of 2.5 kHz up to 5 kHz) but with only 12 photodiodes as receptors. Zhang *et al.* [26] presents a high-resolution array working on image streams with 256 × 256 pixels at 350 fps. However, this setup uses an FPGA that is connected (via a PCI bus) with a personal computer (AMD Opteron). Thus, it is not designed for stand-alone use and has a high power consumption. In contrast, Ref. [3] propose a solution designed for micro air vehicles. The image data is also processed with 350 fps. The resolution is a bit worse with 240×240 pixels but the whole system has a weight of only 193 g and a power consumption of only 3.67 W. In fact, there are many systems called “micro air vehicle” with a total weight of just 10 g or less which clearly could not carry this module. However, in the context of, for example, robust, fast, long-distance, and/or long endurance unmanned air vehicles (capable of operating radiuses of several kilometers or top speeds of more than 100 km/h) the term "micro air vehicle" is used for larger systems, too.

17.5 An FPGA-Based Implementation

An FPGA-based implementation of EMD motion detection is discussed in more detail in this section. This FPGA implementation was designed to be used as a stand-alone module on different mobile robots. In the following subsection, the hardware basis is described. Following this, in Section 17.5.2, the configuration of the FPGA is presented. This configuration realizes the EMD model properties, while the hardware basis is independent of this specific algorithm. Parts of this work have been published in Refs [10, 11].

17.5.1 FPGA-Camera-Module

As this implementation was designed to be used alternately on different midsize mobile robots (weight around ≈8 kg, size down to about half of a standard PC case), some requirements were given: (i) the solution needed to be usable stand-alone (and not, for example only together with a PC), (ii) the weight and size needed to be limited for the intended robot platforms, and (iii) the power consumption needed to be as small as possible to drain the batteries as little as possible. Furthermore, later changes of the EMD implementation (e.g., an implementation of other variants of EMD or EMD integration) should be possible. Hence, as hardware basis, a stand-alone module with an FPGA and an onboard camera was chosen.

The module consists of an off-the-shelf FPGA board and a dedicated extension board, both stacked together. In Figure 17.6(a) the FPGA board can be seen. Its dimensions are 98 mm × 57 mm. The integrated FPGA is a "Stratix-II-S60" of the manufacturer Altera. With 24,176 adaptive logic modules (ALMs) it was a midsize FPGA at rollout. Besides the FPGA, some peripheral components and the connectors, also 32 MB DDR-II SDRAM are mounted on the board (see Figure 17.6(b)). However, this memory is not needed for the EMD implementation.

The second board of the FPGA-camera module is attached via two of the FPGA board connectors. It holds an application-specific circuit to add the missing components needed for the motion detection application. These components are (1) a CMOS camera, (2) a *digital-to-analog converter (DAC)*, and (3) user interface and serial interface components. The board is depicted in Figure 17.7.

The CMOS camera is a Kodak KAC-9630. This is a high-speed camera with 127×100 pixels and a maximum frame rate of 580 fps. The selection of a low-resolution but high-speed camera type is for the purpose of mimicking (up to some degree) the properties of insect vision. The number of ommatidia in insects

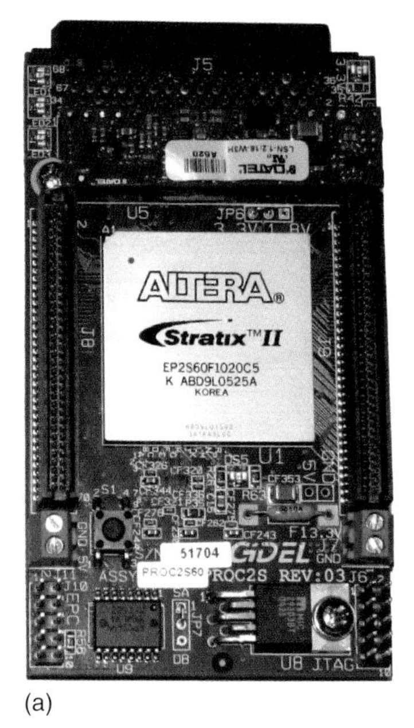

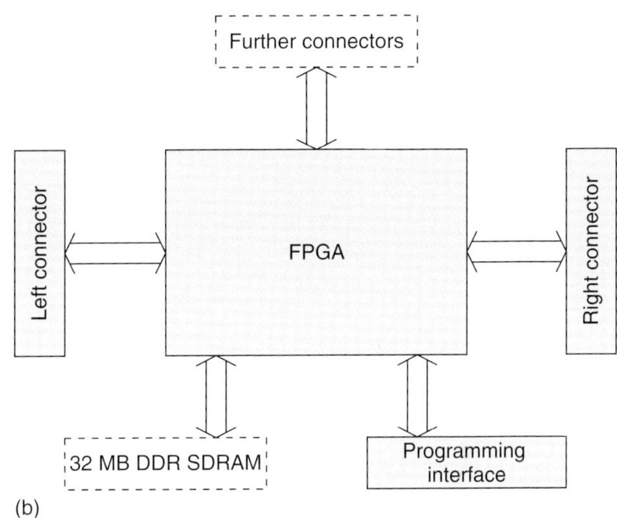

(a) (b)

Figure 17.6 (a) A photo of the FPGA board. On the back side (not visible in the photo) further connectors are mounted [10]. (b) A sketch of the main components and connections of the FPGA board. The dashed parts are not used in this application. The programming interface allows a reconfiguration (programming) of the FPGA by an external PC. However, this is used for development only. To use the module (especially stand-alone), at power-up, a pre-preprogrammed configuration is loaded into the FPGA.

Figure 17.7 Photo of the extension board (front view). On the left-hand side the dot-matrix display and the DIP switch serving as a simple user interface can be seen. The camera module is mounted in the center. On the right-hand side the serial interface components are located. The DAC is mounted below the camera module [10].

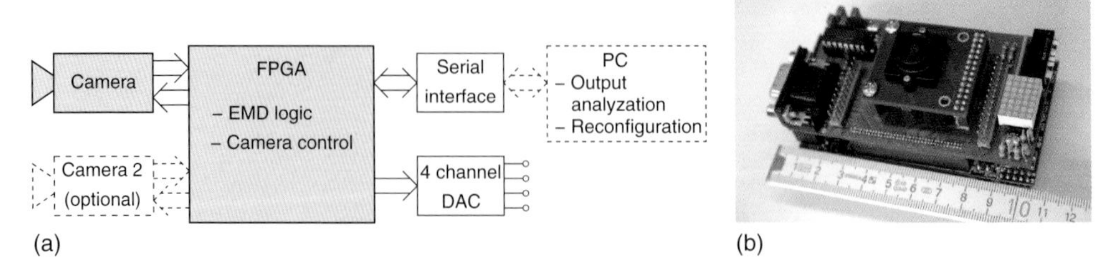

Figure 17.8 (a) A schematic of the EMD module (both boards). The dashed components are not part of the module and need to be attached via the extension board connectors. The main components are shaded. (b) A photo of the whole EMD module (camera side, ruler units are cm) [11].

range from some hundreds (e.g., common fruit fly *Drosophila melanogaster*: 700) up to some thousands. The spatial resolution (inter-ommatidia angle) can vary from tens of degrees down to, for example, an ommatidia angle of 0.24° in the acute zone in the dragonfly *Anax junius* [5]. Together with the properties of the (exchangeable) lens (horizontal angle of view of 40°), both the number of camera pixels and the inter-pixel angle of 0.315° are appropriate for most insect species.

With the four-channel digital-to-analog converter, an analog output of four independent motion detector arrays can be generated. Details are given in the following section. A simple user interface allows changing of the main settings and display of currently selected modes and important parameters such as brightness setting. Further settings, parameter output, and detector output are accessible via a serial interface.

The components of the whole setup are sketched in Figure 17.8(a). A second camera could be attached at an extra connector. The whole module can be seen in Figure 17.8(b).

17.5.2 A Configurable Array of EMDs

The motion detection computation is carried out by the FPGA. A schematic of the configured FPGA-design is shown in Figure 17.9. The two components "camera controller" and "I^{2}C driver" (in the figure on left-hand side) allow a manual or automatic setting of camera parameters, that is, the integration time and the analog camera gain.

The components realizing the actual EMD-array computation are depicted in the center of the figure. This computation is basically sequential, with the EMDs being updated one after the other. However, the computations are carried out in a pipeline (from top to bottom in the figure) parallelizing the lowpass filter, highpass-filter, correlation, and integration in dedicated parts of the FPGA. The pixel data stream from the camera is sampled in a framebuffer. Subsequently, the filter updates are computed. The new filter output values (stored in "HP memory"

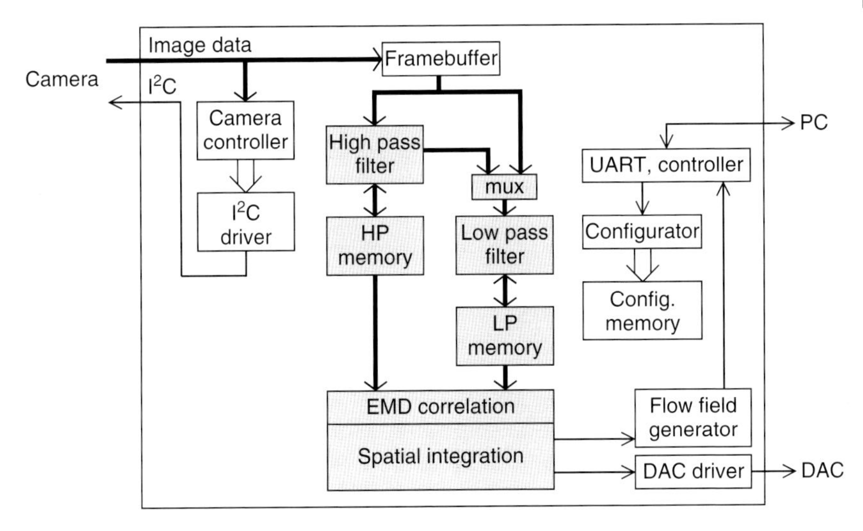

Figure 17.9 The core functionality of the EMD camera module is implemented inside the FPGA. The main FPGA-internal function blocks are shown in this schematic. Thick arrows indicate high bandwidth (image) data. Shaded components realize the EMD computation itself adapted from Ref. [11]. ©IOP Publishing. Reproduced by permission of IOP Publishing. All rights reserved.

and "LP memory") of the two "half-detectors" are correlated and their difference is calculated and summed up for the whole EMD array.

The integration result is (on the right in Figure 17.9) fed to (1) an analog output channel and (2) the serial interface components. Depending on the selected mode, current detector responses are sent out via the serial interface. Alternatively, camera images or other processing results can be transmitted, for example, to an attached PC.

Independent detector responses are accessible via the four analog output channels. Each detector can be supplied with EMD responses of the whole field of view or just a part of it. The camera images are separated into 5 × 5 sub-images (see Figure 17.10(a)) for this purpose. For each subimage, up to 24 × 20 EMDs are computed. The preferred detection direction for each of the 5 × 5 subarrays can be configured in one of eight 45°-steps or "off." Thereby, each of the four detectors can be configured as, for example, a horizontal/vertical/diagonal linear motion detector or as a detector for complex motion types as sketched in Figure 17.5 (e.g., rolling or looming optical flow).

As described in Section 17.3 the optical parameters of the compound eyes of different species vary a lot. Different inter-ommatidia angles are especially relevant for motion detection as they determine the (minimum) inter-receptor angle of EMDs in the different species. To allow tests with EMDs with different inter-receptor angles, a parameter (called *inter-receptor spacing* a_{xx}) is adjustable in this EMD-camera module. The effect of the a_{xx} parameter is depicted in Figure 17.10(b).

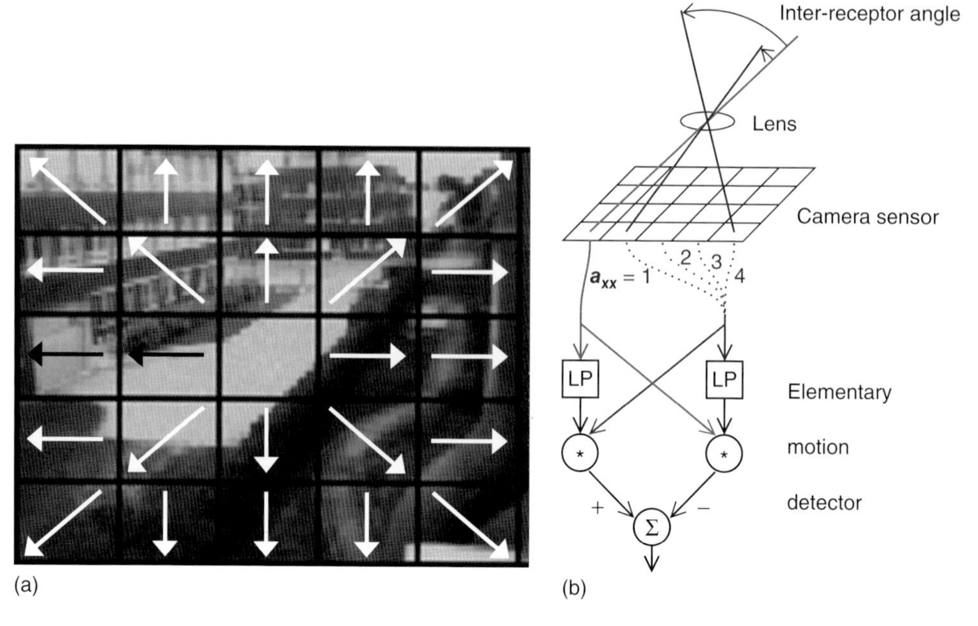

Figure 17.10 (a) A photo taken by the EMD camera module and supplemented by the visualization of an EMD array configuration example (looming detector). (b) The effect of the inter-receptor spacing. The parameter a_{xx} is adjustable in the range from 0 (EMD inoperable) to 10. The corresponding inter-receptor angle depends on the installed lens [11]. © IOP Publishing. Reproduced by permission of IOP Publishing. All rights reserved.

Further configurable options are the lowpass filter and highpass filter time constants and the EMD type (n-Ln-, H-Ln-, n-LH-EMD, see caption of Figure 17.3). The camera parameters (integration time, gain) and the EMD update rate can be changed, too. Finally, the output via the serial interface can be selected, for example, EMD array output, image data, flowfield data. The latter produces a two-dimensional optical flow vector per 5×5 subimage.

The measured power consumption of the whole module is in a range of 1 – 2 W. The higher consumption was generated with a special FPGA configuration having several features, which especially used the DDR memory. A standard design without external memory but with most of the features and running at an update rate of 100 Hz (camera frame rate 100 fps, which is also usable in darker indoor environments) led to the measurement of around 1 W.

This is a relatively low consumption caused by using an FPGA-based design with a rather slow-clocked pipeline design. The camera data is read at a rate of 10 MHz what determines the (slow) speed of the whole pipeline. Because of the EMD-serial computation, only a small part of the FPGA is used (13% of the FPGA logic resources).

17.6 Experimental Results

In this section, experimental results of the presented FPGA-based implementation are shown. The purpose is to show the general behavior of the module. More elaborate and extensive test results can be found in Ref. [10]. There, for example, variations of all parameters are also discussed.

Two test setups were chosen. In the first one, the response of a simple, single-direction, sensitive EMD array to a linear optical flow in the "preferred" detection direction is studied. In this setup, a printed sine pattern was placed in front of the EMD camera module. The module was mounted on a gantry crane with four degrees of freedom: three translatory axes (x, y, z) and one rotatory axis (around the z axis). In the first setup the crane moved only along one translatory axis. The motion axis was in parallel to the sine gradient of the test pattern (and in parallel to the "preferred" detection of the EMD array).

With constant motion speed, the response of the EMD array is also constant. This can be seen in Figure 17.11(a) where the responses for different motion velocities (x axis of the plot) are given. The variation of the motion detector response (minimum and maximum values are given for each velocity) is caused by the practical setup (especially motion speed deviation and illumination differences

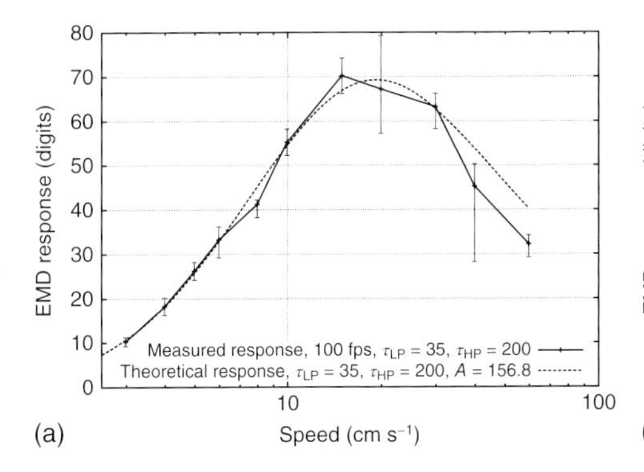

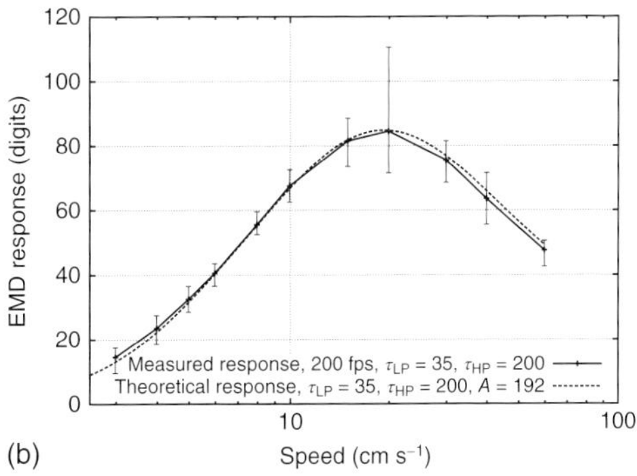

Figure 17.11 Results of the first test setup. In both plots, EMD array responses of several runs are shown. In each run, the camera module was moved linearly with constant speed in front of the test pattern (see text). The motion velocity was changed from run to run. The average, minimum, and maximum EMD array responses are given in the plots for the chosen velocity. The average EMD response is plotted as a solid curve over the motion speed (x-axis in log scale). The EMD camera module was configured as a Harris–O'Carroll detector with a lowpass time constant τ_l of 35 ms and a highpass time constant τ_h of 200 ms. In addition, the theoretically expected responses are plotted as dashed curves [as derived by Borst *et al.* [15]]. A free amplitude parameter A is manually adapted to the measurement data. Therefore, just the curve's shape, peak position, and the relative values – not the absolute ones – can be evaluated. (a) In the plot, the frame rate has been set to 100 fps. (b) A frame rate of 200 fps has been chosen.

over the whole pattern). The theoretical EMD responses (with adapted amplitude) are also given in the figure.

In the runs used for the data presented in Figure 17.11(a) the framerate of the camera was set to 100 fps. In the second plot, Figure 17.11(b), responses are shown of runs with an increased framerate of 200 fps. As can be seen, the strong diversion between theoretical and measured responses for the higher velocities in Figure 17.11(a) are not measurable with 200 fps framerate. This indicates an effect of aliasing.

In the second test setup, the EMD camera module was moved along all four axes of the gantry crane (separately, one axis after the other, see Figure 17.12(a)). As test pattern, a checkerboard was used this time. The input image of the test pattern can be seen in Figure 17.12(b). Another difference between the first and the second test setup is the configuration of the EMD array under test. While in the first test only one simple linear detector field was configured and tested, in the second setup, four different EMD array configurations were tested in parallel. The four configurations were (1) a linear vertical detector field, (2) a horizontal detector, (3) a looming detector, and (4) a rotation detector field (see Figure 17.13(a)). The motions carried out by the gantry crane can be recognized by the axes' velocities plotted in Figure 17.12(a).

The responses of the four EMD arrays are shown in Figure 17.13(b). Basically, the response of the four detector fields is the strongest to the motion that

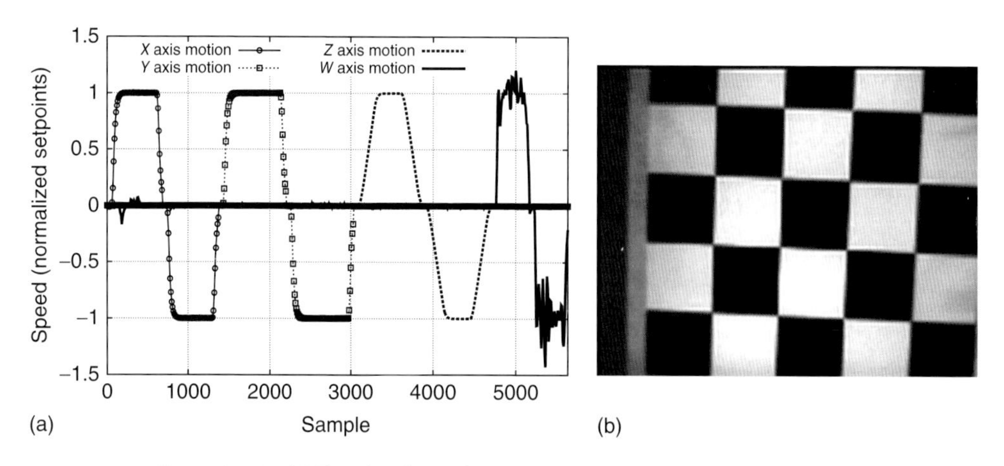

Figure 17.12 (a) The plot shows the motion speed of the four gantry axes in the second test. The data was generated during the same run as used for the data plotted in Figure 17.13(b). The *x*-, *y*-, and *z*-axes of the gantry are parallel to the camera image's vertical, horizontal, and optical axes, respectively. The w-axis of the gantry is parallel to the camera's optical axis. (b) This shows an image of the printed checkerboard test pattern taken by the camera of the EMD module. The integration time and analog gain of the camera were chosen to get a high contrast (both figures taken from Refs [10, 11]). See text and Figure 17.13 for further details.

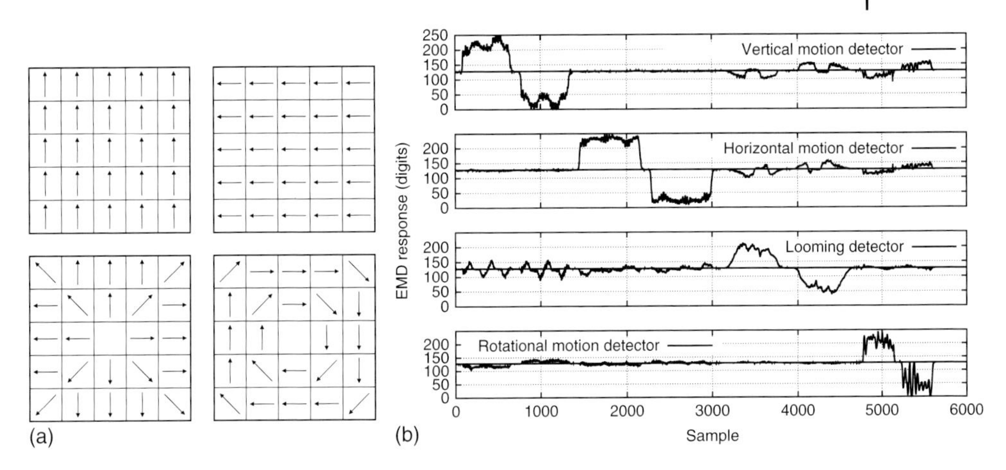

Figure 17.13 (a) The configured preferred directions of the four detectors used in the second test setup are shown (vertical, horizontal, looming, and rotational). (b) The EMD output of the four simultaneously computing detectors during the second test can be seen (both figures taken from Refs [10, 11]). Different movement patterns relative to a checkerboard (Figure 17.12(b)) were carried out sequentially (Figure 17.12(a)). First, there was a vertical translation, followed by a horizontal translation. The third pattern was generated by moving toward and away from the checkerboard (looming) and the last movement pattern corresponded to a rotation around the optical axis. The four detectors selectively respond to their preferred movement pattern and respond only weakly or not at all to their non-preferred patterns. See text and Figure 17.12 for further details.

generates their preferred optical flow. During the motion, the response can vary with the pattern (see the vertical detector response which depends on the number of horizontal edges visible in the camera image). Furthermore, the response can vary depending on the motion speed (compare Figure 17.13(b) and Figure 17.12(a) for the rotation part).

17.7 Discussion

The FPGA-based EMD camera module presented was designed as a versatile motion detection module for robotic applications, as well as for biological experiments. Hence, the number of EMDs needs to be high enough, the EMD parameters and the array parameters need to be adaptable, and the module needs to be usable stand-alone. Especially compared to previous publications (see Section 17.4.1) these conditions are met very well.

While the number of EMDs per array in the solution described above (and proposed in Ref. [11]) is less than in the setup presented in Ref. [3], the power consumption is also lower. Furthermore, the focus of the module described in this chapter differs from the design in Ref. [3]. As described above, versatility was one main "must" here: being able to switch online between different types of EMD, different time constants, and different preferred flow fields for the four EMD arrays. As shown in Section 17.6, the general detection behavior is as expected for the EMDs and choosing different configurations is also possible.

The comparison of the two designs proposed in Refs [3] and [11] shows very well the advantages and the possibilities of FPGA-based implementations. Depending on the application, a design can be optimized for speed and computational power, for example, with several parallel processing units running at high frequencies. Alternatively, an FPGA design can be optimized for a low power consumption (given a minimum computational power needed for the application). For example, the design shown in Figure 17.9 uses a pipeline structure with parallel computation of the filters. However, the EMDs are computed serially one after the other. Thanks to the optimized computation within the pipeline (as it is possible within an FPGA), the overall computation speed is still enough for computing the whole array serially before the next frame data needs to be computed. The gain from this design is that only 13% of the available FPGA logic elements and only 39% of FPGA-internal memory is used. Therefore, the module consumes only 1–2 W of power.

17.8 Conclusion

In this chapter, mobile robotics was introduced as an application field of bioinspired computer vision methods. In many robotic applications, the resources available for the computer vision task are limited. Most often, the limitations are given in power consumption, space, and/or weight. As could be shown in this chapter, choosing a biologically inspired method can be a solution to meeting such requirements. Some of the tasks to be accomplished by a mobile robot have also to be solved in biology (e.g., by an insect). Further, quite often the solution found in biology is also very limited in resource consumption.

Also discussed in this chapter were different ways of implementing biologically inspired methods. As chosen for the implementation presented here, FPGAs combine several advantages. Using FPGAs for the implementation of a biologically inspired computer vision method can lead to a close-to-optimum solution with a good trade-off between effort and costs on the one hand and achieved performance, low power consumption, size, and weight, on the other.

Some more details of the implementation and tests presented here can be found in the publications by Koehler et al. [11] and [10]. Köhler [10] also describes an FPAA-based implementation of EMD and gives a comparison of different implementation methods.

Acknowledgments

The author would like to thank Ralf Möller and the Computer Engineering Group at Bielefeld University, Germany for their support while carrying out most of the design and experiment work presented in this chapter. A special thank goes to Frank Röchter for the work carried out as part of his Diploma thesis and to Jens P. Lindemann for his support in biologically related questions.

Furthermore, the author would like to thank Frank Kirchner and the Robotics Innovation Center (RIC) of the German Research Center for Artificial Intelligence (DFKI) for their support while writing this chapter.

In Section 17.5, an implementation developed in multiple Diploma and Bachelor theses of different authors and further work by the author (all at the Computer Engineering Group at Bielefeld University, Germany) is presented. Further tests were carried out at the Robotics Innovation Center (RIC) at the German Research Center of Artificial Intelligence (DFKI). Parts of this work have been published in Refs [10] and [11]. Thanks to IOP Publishing and to the Andere Verlag for giving their agreement.

References

1. Aubepart, F., Farji, M.E., and Franceschini, N. (2004) FPGA implementation of elementary motion detectors for the visual guidance of micro-air-vehicles. 2004 IEEE International Symposium on Industrial Electronics, vol. 1, pp. 71–76.
2. Aubépart, F. and Franceschini, N. (2007) Bio-inspired optic flow sensors based on FPGA: application to micro-air-vehicles. *Microprocess. Microsyst.*, **31**, 408–419, doi: *http://dx.doi.org/10.1016/j.micpro.2007.02.004*, special Issue on Sensor Systems.
3. Plett, J. *et al.* (2012) Bio-inspired visual ego-rotation sensor for MAVs. *Biol. Cybern.*, **106**, 51–63, doi: 10.1007/s00422-012-0478-6.
4. Hassenstein, B. and Reichardt, W. (1956) Systemtheoretische Analyse der Zeit-, Reihenfolgen-, und Vorzeichenauswertung bei der Bewegungsperzeption des Rüsselkäfers *Chlorophanus*. *Z. Naturforsch.*, **11** (9-10), 513–524.
5. Land, M.F. (1997) Visual acuity in insects. *Annu. Rev. Entomol.*, **42**, 147–177.
6. Duke-Elder, S.S. (1958) *System of Ophthalmology*, The C.V. Mosby Company, St. Louis, MO.
7. Borst, A. and Haag, J. (1996) The intrinsic electrophysiological characteristics of fly lobula plate tangential cells: I. Passive membrane properties. *J. Comput. Neurosci.*, **3** (4), 313–336.
8. Egelhaaf, M. (2006) The neural computation of visual motion information, in *Invertebrate Vision*, Chapter 10 (eds E. Warrant and D.-E. Nilsson), Cambridge University Press, pp. 399–461.
9. Reichardt, W. (1961) Autocorrelation, a principle for the evaluation of sensory information by the central nervous system, in *Sensory Communication* (ed. W.A. Rosenblith), MIT Press, New York, pp. 303–317.
10. Köhler, T. (2009) *Analog and Digital Hardware Implementations of Biologically Inspired Algorithms in Mobile Robotics*, Der Andere Verlag.
11. Köhler, T. *et al.* (2009) Bio-inspired motion detection in an FPGA-based smart camera module. *Bioinspiration Biomimetics*, **4** (1), 015008.

12. Egelhaaf, M. and Borst, A. (1989) Transient and steady-state response properties of movement detectors. *J. Opt. Soc. Am. A*, **6** (1), 116–126.
13. Harris, R.A., O'Carroll, D.C., and Laughlin, S.B. (1999) Adaptation and the temporal delay filter of fly motion detectors. *Vision Res.*, **39** (16), 2603–2613.
14. Harris, R.A. and O'Carroll, D.C. (2002) Afterimages in fly motion vision. *Vision Res.*, **42** (14), 1701–1714.
15. Borst, A., Reisenman, C., and Haag, J. (2003) Adaptation of response transients in fly motion vision. II: model studies. *Vision Res.*, **43** (11), 1309–1322.
16. Kirschfeld, K. (1972) The visual system of Musca: studies on optics, structure and function, in *Information Processing in the Visual Systems of Arthropods* (ed. R. Wehner), Springer-Verlag, Berlin, Heidelberg, New York, pp. 61–74.
17. Babies, B. *et al.* (2011) Contrast-independent biologically inspired motion detection. *Sensors*, **11**, 3303–3326.
18. Lindemann, J.P. *et al.* (2005) On the computations analyzing natural optic flow: quantitative model analysis of the blowfly motion vision pathway. *J. Neurosci.*, **25**, 6435–6448, doi: 10.1523/JNEUROSCI.1132-05.2005.
19. Franz, M.O. and Krapp, H.G. (2000) Wide-field, motion sensitive neurons and matched filters for optic flow fields. *Biol. Cybern.*, **83**, 185–197.
20. Zanker, J.M. and Zeil, J. (2002) An analysis of the motion signal distributions emerging from locomotion through a natural environment, in *Biologically Motivated Computer Vision: Second International Workshop, BMCV 2002, Tübingen, Germany, November 22-24, 2002. Proceedings*, vol. 2525/2002 (eds H. Bulthoff, S.-W. Lee, T. Poggio, and C. Wallraven), Springer-Verlag GmbH, pp. 146–156.
21. Ruffier, F. *et al.* (2003) Bio-inspired optical flow circuits for the visual guidance of micro air vehicles. Proceedings of the 2003 International Symposium on Circuits and Systems, 2003. ISCAS '03, vol. 3 (III), pp. 846–849.
22. Harrison, R.R. (2000) An analog VLSI motion sensor based on the fly visual system. PhD thesis, California Institute of Technology, Pasadena, CA, *http://www.ece.utah.edu/~harrison/thesis.pdf* (accessed 14 May 2015).
23. Harrison, R.R. (2004) A low-power analog VLSI visual collision detector, in *Advances in Neural Information Processing Systems*, vol. 16 (eds S. Thrun, L. Saul, and B. Scholkopf), MIT Press, Cambridge, MA.
24. Harrison, R.R. (2005) A biologically inspired analog IC for visual collision detection. *IEEE Trans. Circ. Syst.*, **52** (11), 2308–2318.
25. Franceschini, N., Pichon, J., and Blanes, C. (1992) From insect vision to robot vision. *Philos. Trans. R. Soc. London, Ser. B*, **337** (1281), 283–294.
26. Zhang, T. *et al.* (2008) An FPGA implementation of insect-inspired motion detector for high-speed vision systems. IEEE International Conference on Robotics and Automation (ICRA), Pasadena, CA, pp. 335–340.

18
Visual Navigation in a Cluttered World

N. Andrew Browning and Florian Raudies

18.1
Introduction

In this chapter, we explore the biological basis for visual navigation in cluttered environments and how this can inform the development of computer vision techniques and their applications. In uncluttered environments, navigation can be preplanned when the structure of the environment is known, or simultaneous localization and mapping (SLAM) can be used when the structure is unknown. In cluttered environments where the identity and location of objects can change, and therefore they cannot be determined in advance or expected to remain the same over time, navigation strategies must be reactive. Reactive navigation consists of a body (person, animal, vehicle, etc.) attempting to traverse through a space of unknown static and moving objects (cars, people, chairs, etc.) while avoiding collision with those objects. Visual navigation in cluttered environments is thereby primarily a problem of reactive navigation.

The first step to reactive navigation is the estimation of self-motion, answering the question: where am I going? Visual estimation of self-motion can be used for collision detection and time-to-contact (TTC) estimation. The second step is object detection (but not necessarily identification), determining the location of objects that pose an obstacle to the desired direction of travel. Visual information from a monocular view alone is typically insufficient to determine the distance to objects during navigation; however, TTC can be easily extracted from either optic flow or from the frame-to-frame expansion rate of the object. TTC can be used to provide constraints for the estimation of relative distance and for reaction time. A steering decision (or trajectory generator) must bring this information together to select a route through the cluttered environment toward a goal location. We provide mathematical problem formulations throughout, but do not provide derivations. Instead, we encourage the reader to go to the original articles as listed in the references.

Throughout the chapter, we use a systems model (ViSTARS) to frame discussion of how – at a system level – the brain combines information at different stages of processing into a coherent behavioral response for reactive navigation. The

Biologically Inspired Computer Vision: Fundamentals and Applications, First Edition.
Edited by Gabriel Cristóbal, Laurent Perrinet, and Matthias Keil.

motivation for this is twofold: (i) to provide a concrete description of the computational processing from input to output and to demonstrate the concepts in question, as opposed to providing an abstract mathematical formulation or a biological description of a cell response in a limited environment and (ii) to demonstrate how the concepts discussed in this chapter can be applied to robotic control. ViSTARS has been shown to fit a wide range of primate neurophysiological and human behavioral data and has been used as the basis for robotic visual estimation and control for reactive obstacle avoidance and related functions.

18.2 Cues from Optic Flow: Visually Guided Navigation

The concept of *Optic Flow* was formulated by Gibson during World War II and published in 1950 to describe how visual information *streams* over the eye as an animal moves through its environment. Brain representations of optic flow have been discovered in a variety of species. For example, in primates, optic flow is represented as a vector field at least as early as primary visual cortex (V1) and in rabbits is computed directly in the retinal ganglion cells.

Primate visual processing is often characterized as having two distinct pathways (two-stream hypothesis), the ventral pathway and the dorsal pathway. The ventral, or parvocellular, pathway projects from retina through lateral geniculate nucleus (LGN) to V1 and to inferior temporal lobe (IT). This pathway is primarily concerned with the identity of objects (the what stream). The dorsal, or magnocellular, pathway projects from retina through LGN to V1, medial temporal lobe, and medial superior temporal area (MST) and is primarily concerned with the location and motion of objects (the where stream). Closer inspection of the cells in magnocellular pathway indicates that this pathway may be primarily concerned with (or to be more accurate, modeled with concepts from) the estimation and interpretation of optic flow. Table 18.1 illustrates possible roles for the primate magnocellular pathway in visual navigation.

Optic flow is, by Gibson's original definition, continuous in time; however, we generally do not consider optic flow fields during saccadic eye movements (or rapid camera panning) to be informative since produced flows are dominated by rotational flow induced by eye movements. Moreover, optic flow is more commonly abstracted as a vector field that is discretized in time and space. As such, optic flow encodes the movement of pixels or features between captured moments (or video frames). While this abstraction has been vital to the analysis of optic flow, and is used in this chapter, the reader should remember that biological visual systems are additionally capable of integrating over longer periods of time and representing motion along curved trajectories, for example, through a polar representation. Moreover, biological visual systems are asynchronous. They do not have a fixed or regular integration time step. However, until asynchronous sensors are widely available, it is unlikely that any near-term application of optic flow will utilize this aspect of the biological visual system.

Table 18.1 Primate magnocellular brain areas and their possible role in visual navigation.

Magnocellular pathway brain area	Possible role
Retina-LGN (parasol ganglion cells)	Contrast normalization; intensity change detection
Primary visual cortex (V1)	Speed and direction of motion (vector field, optic flow)
Area MT	Motion differentiation (object boundaries) and motion pooling (aperture resolved, vector field, optic flow)
Area MST	Object trajectory, heading, and time to contact (TTC)

Optic flow is defined as the change of patterns of light on the retina caused by the relative motion between the observer and the world; the structure and motion of the world is carried in the pattern of light [1]. Optic flow is usually estimated through detection and tracking of changes in brightness in an image stream. This is a nontrivial task for which various biological and computational solutions exist [2, 3]. Computation of optic flow would be a chapter in itself, and for the current discussion of visual navigation, we assume that a vector field representation of optic flow is available and follows the following definition for translational self-motion:

$$\begin{pmatrix} \dot{x}_T \\ \dot{y}_T \end{pmatrix} = \frac{1}{Z}\begin{pmatrix} -f & 0 & x \\ 0 & -f & y \end{pmatrix}\begin{pmatrix} v_x \\ v_y \\ v_z \end{pmatrix} = \frac{v_z}{Z}\begin{pmatrix} x - x_{FOE} \\ y - y_{FOE} \end{pmatrix} \quad \text{with} \quad \begin{pmatrix} x_{FOE} \\ y_{FOE} \end{pmatrix} = \frac{f}{v_z}\begin{pmatrix} v_x \\ v_y \end{pmatrix}. \tag{18.1}$$

where f is the focal length of the camera, x and y are the image coordinates of a point in space (X,Y,Z) projected on to the image plane, x_{FOE} and y_{FOE} are the image coordinates of the focus of expansion (FOE) of the vector field (which corresponds to heading during forward translation), v_x is velocity (in space, not in the image) along the x axis, v_y is the velocity along the y axis, v_z is the velocity along the z-axis, Z is the position of the object measured along the z-axis (depth), and $\dot{x}_T$ and $\dot{y}_T$ are the image velocity in x and y coordinates, respectively.

Optic flow provides a rich set of cues for visually guided navigation, including:

1) Self-motion of the observer
2) TTC with surfaces/objects in the environment
3) Object segmentation and trajectories
4) Object deformations and morphing

Animals move to actively obtain visual information, and this information is utilized through movements of head, eyes, and body, which in turn generates new visual information. Many robots and unmanned vehicles are built with the same capabilities, primarily to optimize the use of sensor and computational resources through active vision [4]. In this chapter, we simplify the discussion of optic flow analysis by assuming that the eye is fixed in body-centric coordinates in order that we can focus on reactive visual navigation rather than eye or head movements. We justify this simplification through the fact that the effect of rotations can be estimated, and removed, either with extraretinal signals or from optic flow itself provided there is sufficient depth structure.

If we assume that there is no rotation of the eye/head/sensor, then the relative motion between the eye and the world is caused only by body motion. This body motion can be modeled through a 3D translational and 3D rotational velocity vector. Optic flow is estimated from the projection of those translational and rotational vectors on to the retina/sensor. Analysis of optic flow allows for the estimation of the 3D translational direction and 3D rotation velocity from the visual input stream. Note that 3D translational speed cannot be recovered due to the inherent size/distance ambiguity of a 2D projection of a 3D world, the object could be small and close by or large and far away. However, TTC can be recovered from the image stream and provides a behaviorally relevant measure of the relative distance to objects in the scene. Note that some authors (e.g., [5] Eq. (2.13)) derive temporal measures of distance (range/range rate) in their analysis but refer to them as relative depth rather than TTC. This omission is perhaps because they did not perform the final computation to put the measure into standard units of time or perhaps to avoid potential confusion with the definition of TTC (see Section 4).

Accurate TTC requires an estimate of the current direction of travel, which for the purposes of this chapter we define as heading. Heading can be estimated from optic flow or from an inertial sensor, although it should be noted that in practice mapping inertial heading estimates into the image frame often adds significant synchronization and calibration issues. TTC from optic flow provides a dense map of the relative distances to surfaces and objects in the scene. By dense, we mean that each sample location in the image is assigned a TTC value. For most applications, sample location is synonymous with pixel location. This dense map can serve as a rough sketch of objects, the volume of a scene, and provides information about empty, drivable space – this is also the basis of *structure from motion* algorithms, which are not discussed here. In applications where the speed of the sensor is known (e.g., a robot which measures its own speed), TTC multiplied by own speed provides an estimate of the true distance to static objects. Depending on what assumptions can be made with respect to object speed or size, it is also possible to estimate the distance to *moving* objects in some circumstances.

Optic flow can be used to segment both static and moving objects. In the moving object case, the directions and speeds of the vector field in the region of the object are inconsistent with the surrounding vector field generated by self-motion. Static object boundaries typically coincide with depth discontinuities – an object is in front of (or behind) another object. Due to the projective geometry of camera or models for image formation of the eye close versus far objects will have different image velocities. For instance, if there are two objects, one at 5 m distance and one at 10 m distance, and the camera moves to the left, the optic flow vectors from the closer object will have twice the speed compared to the speed of vectors from the far object. These speed discontinuities (that do not affect vector direction/angle) in optic flow indicate a depth discontinuity in the real world. For instance, a desk in front of the wall or a mug on the table introduces such discontinuities when moving relative to them. This phenomenon is commonly called motion parallax – closer objects move faster on your retina than those further away. A speed discontinuity in the optic flow can be detected and interpreted as object boundary. Motion parallax can be used to gain a depth ordering between objects which appear overlapping or connected within the visual field. Thus, optic flow not only facilitates the detection of edges of objects but also allows for the creation of a depth ordering among connected objects. Information about objects and their depth ordering helps navigation based on readily available and always up-to-date information about the visual world. Depth ordering is not the same as estimating relative depth from TTC but is rather a quasi-independent depth measure, which provides a more robust representation of object ordering in the environment requiring only relatively accurate speed estimation and, unlike TTC, does not require knowledge of current heading.

For ease of discussion, in this chapter, we focus only on rigid (nondeforming) independently moving objects on straight-line trajectories. These simplifications are valid for the vast majority of visual navigation tasks involving vehicles and world structure and can be mitigated for tasks involving biological objects. These visual variables, self-motion, TTC, and object boundaries can be integrated to generate motor plans and actions for steering toward a goal, avoiding obstacles, or pursuing a moving goal (or target). For reactive navigation, biological systems tightly couple perception and action through reflexive visual responses and *visual navigation in a cluttered world.*

18.3
Estimation of Self-Motion: Knowing Where You Are Going

Knowing your current direction of travel is essential for navigation. During reactive navigation, objects are only obstacles if they impede your desired trajectory; otherwise, they are benign and sometimes helpful environmental structure.[1] Whether the current objective of the navigation is defined visually (e.g., the objective is to dock with, or intercept, a particular object) or nonvisually (e.g., to go north) current direction of travel must be known in order to measure deviation from that objective. We define the current direction of travel as *self-motion, egomotion,* or *heading.* For our purposes here (and as is common in the biological literature), we use these terms interchangeably while understanding that, depending on the vehicle in question, heading and direction of travel are not always the same. Moreover, we limit the definition to straight-line trajectories. A detailed discussion of the estimation of curvilinear trajectories, which we define as *path,* is beyond the scope of this chapter and is only briefly touched upon.

Heading in a rigid, static world is represented in optic flow through the FOE [1]. The FOE is the point in the image space from which all vectors originate.

1) In more complex environments where objects can pose a threat based on their identity (e.g., a sniper scope), then additional mechanisms are required to detect, identify, and characterize the threat.

Figure 18.1 (a) Example optic flow field overlaid video from car driving forward past traffic on a public road, black circle indicates the focus of expansion and current heading. Image brightness has been increased and optic flow vectors have been sub-sampled and scaled for illustration. (b) Close up centered around the focus of expansion. The time to contact of the car, which we are heading towards, is encoded by the ratio of the optic flow vector magnitude to the distance of the origin of that vector from the FOE.

During forward translation, with a forward-facing camera, the FOE will usually be within the borders of the image frame. However, there are cases where the FOE may be located outside the borders of the image frame. In addition to the FOE, there is a corresponding focus of contraction (FOC) that occurs at the location at which all vectors converge, which for forward translation is usually behind you. Figure 18.1 shows a typical optic flow field for forward translation with a forward-facing camera. Since the FOE coincides with heading, we define the core problem of heading estimation as the localization of the FOE. Mathematically, knowledge of the FOE is also required to convert from an observed optic flow field to a TTC representation.

Optic flow fields corresponding to observer translations have been used as visual stimuli in many neurophysiological studies to determine if optic flow and the FOE could be used by primates for navigation. In primates, neurons in dorsal medial superior temporal area (MSTd) were found to be selective for radial expansion patterns – or in other words, they respond preferentially to translations with a particular heading [6, 7]. In this section, we begin with mathematical definitions before describing methods for estimating heading.

We assume that optic flow has been extracted from an image sequence. Formally, optic flow is defined as the vector field $(\dot{x}(x, y), \dot{y}(x, y)) \in \Re^2$ with x and y denoting Cartesian coordinates of the image position. Here, we focus on a model which describes this optic flow. First, we assume that the eye can be modeled by a pinhole camera with the focal length f that projects 3D points $P = (X, Y, Z)^t$ onto the 2D image plane $\vec{x} = (x, y)^t = f/Z \cdot (X, Y)^t$. When modeling the human vision system, this is a simplistic assumption, resulting in some error, but works well for digital cameras, which are used in most real-world applications. Second, we assume that the observer is moving through a rigid environment, which is characterized by the distances *Z(x,y)* as seen through the pinhole camera at image location x and y. Third, we assume the observer's movement is described by the instantaneous 3D linear velocity $\vec{v} = (v_x, v_y, v_z)^t$ and 3D rotational velocity $\vec{\omega} = (\omega_x, \omega_y, \omega_z)^t$. Formally, the model for optic flow is the visual image motion equation [5]

$$\begin{pmatrix} \dot{x} \\ \dot{y} \end{pmatrix} = \underbrace{\frac{1}{Z}\overbrace{\begin{pmatrix} -f & 0 & x \\ 0 & -f & y \end{pmatrix}}^{=:A}\begin{bmatrix} v_x \\ v_y \\ v_z \end{bmatrix}}_{\text{translational flow}} + \underbrace{\frac{1}{f}\overbrace{\begin{pmatrix} x \cdot y & -(f^2 + x^2) & f \cdot y \\ (f^2 + y^2) & -x \cdot y & -f \cdot x \end{pmatrix}}^{=:B}\begin{bmatrix} \omega_x \\ \omega_y \\ \omega_z \end{bmatrix}}_{\text{rotational flow}}. \tag{18.2}$$

Each image location (x, y) is assigned a flow vector $(\dot{x}, \dot{y})$, which is a velocity vector in the 2D image plane. These velocities express the movement of light intensity patterns within the image. When the observer moves in a static, rigid environment, Eq. (18.2) describes all these velocities. For a digital camera, we often associate the image locations (x, y) with pixel locations and the flow vectors $(\dot{x}, \dot{y})$ with "pixel velocities." Analysis of Eq. (18.2) allows us to understand the impact of specific translations and rotations on optic flow fields, as shown in Figure 18.2.

Equation (18.2) allows us to understand how we can estimate the FOE (heading) from observed optic flow. If we simplify the problem by assuming that there are no rotations between measurements, and that we want an estimate of the FOE in image coordinates, then we can define a least-squares solution to find the intersection point of all of the motion vectors on the image plane. This problem can be posed using the standard matrix formulation of the form $Ax = y$ and solved in the least-squares sense using the pseudoinverse, $x_{LS} = (A^T A)^{-1} A^T y$. If we define $m = \frac{\dot{y}}{\dot{x}}$, $n = \frac{\dot{x}}{\dot{y}}$, $b = mx - y$, and $c = ny - x$, then we can construct the solution as follows:

$$\begin{bmatrix} (m_1 - m_2) & (n_1 - n_2) \\ (m_1 - m_3) & (n_1 - n_3) \\ \vdots & \vdots \\ (m_{k-1} - m_k) & (n_{k-1} - n_k) \end{bmatrix} \begin{bmatrix} x_{FOE} \\ y_{FOE} \end{bmatrix} = \begin{bmatrix} (b_1 - b_2) + (c_1 - c_2) \\ (b_1 - b_3) + (c_1 - c_3) \\ \cdots \\ (b_{k-1} - b_k) + (c_{k-1} - c_k) \end{bmatrix} \tag{18.3}$$

where $\dot{x}$ and $\dot{y}$ are the observed velocities at the location defined by x, and y, k is the total number of optic flow vectors. The visual angle of the FOE on the image plane corresponds to the current bearing angle, and this is often sufficient as a heading estimate for many applications, including TTC estimation which requires only the pixel coordinates of the FOE. This mathematical formulation continues to work when the FOE is outside the image plane, in practice, however, error will increase as the FOE gets further from the center of the image plane.

In a more general sense, we can formulate an optimization problem to estimate the translational and rotational motion and environmental structure that caused the observed optic flow pattern. Equation (18.4) poses one such nonlinear

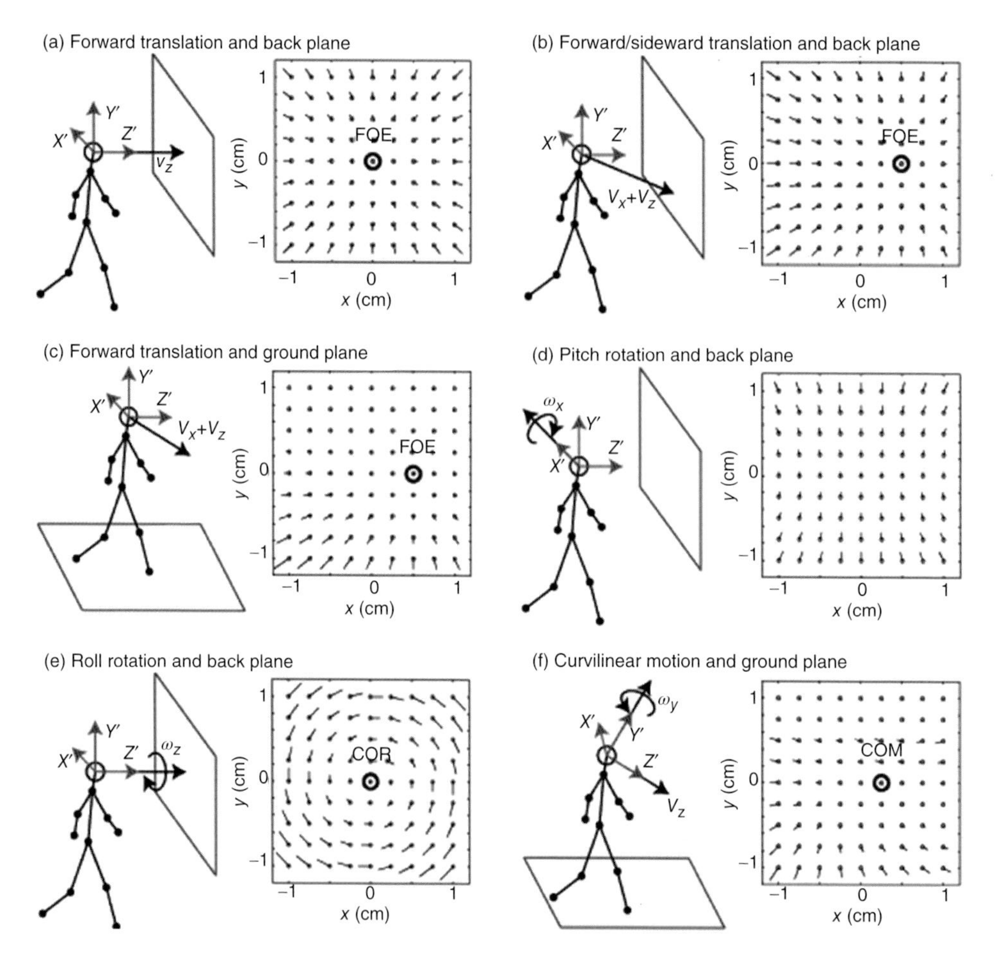

Figure 18.2 Characteristic optic flow patterns generated for observer motions toward a back plane or when walking along a ground plane. a) Flow for straight ahead translation toward a back plane at 10 m distance. b) Flow for the translation components vx = 0.45m/sec, vy = 0m/sec, and vz = 0.89m/sec toward a back plane at 10 m distance. c) The same translation as in b) but now above a ground plane. The angle between the line of sight and ground plane is 30 degrees and the distance of the ground plane measured along this line of sight is -10 meters. In these three cases a)-c) the focus of expansion (FOE) indicates the direction of travel in image coordinates. d) Flow for the pitch rotation of ωx = 3°/sec. e) Flow for the roll rotation of ωz = 9°/sec. In this case, the center of rotation (COR) appears at the image coordinates (0, 0). f) Flow for curvilinear motion with the translation components vx = 0.71m/sec, vy = 0m/sec, and vz = 0.71m/sec, and the rotational components ωx = 0°/sec, ωy = 3°/sec, and ωy = 0°/sec above the ground plane like in c). The center of motion (COM) marks the position of zero length vectors. All flow patterns are generated using the model Eq. (18.2).

optimization problem that can be solved through gradient descent [8]:

$$(\hat{\vec{v}}, \hat{\vec{\omega}}, \hat{Z}) = \arg\ \min_{\vec{v},\vec{\omega},Z} \int_{\Omega_{\vec{x}}} \left\| \begin{pmatrix} \dot{x} \\ \dot{y} \end{pmatrix} - \frac{1}{Z} A\vec{v} - B\vec{\omega} \right\|_2^2 d\vec{x}. \tag{18.4}$$

In Eq. (18.4), we use the estimated optic flow $(\dot{x}, \dot{y})$ and subtract the model flow defined by Eq. (18.2). This results in a residual vector, which we take the Euclidean norm $\|\cdot\|^2$of. Assuming that the self-motion parameters are the same within the entire image plane we integrate over the area $\Omega_{\vec{x}}$. This means, we have to estimate the model parameters $\vec{v}$, $\vec{\omega}$, and Z, whereas parameter Z can be different for each image location. We minimize the residual value, the difference between estimate and model, if we look for the minimum in the space of these parameters.

More details on the estimation of self-motion and a comparison between different methods are given in [9].

As noted previously, bioinspired methods are supported by neurophysiological data from primate brain area MST [6, 7]. Rather than taking a gradient descent approach or analytical treatment, the solution is found by sampling all reasonable motions $\vec{v}$ and $\vec{\omega}$ together with all reasonable depths Z and to evaluate a bioinspired likelihood function for these parameters [10, 11]. Each of these constructed sample optic flow patterns defines a *template* which represents travel in a particular direction. Self-motion is estimated by finding the template that best matches the observed optic flow field. The match score can be defined as a simple inner product between the template and the observed optic flow, normalized by the overall energy (sum of vector magnitudes) of the observed optic flow, or by using a standard correlation measure. Other hybrid approaches use optimization and sampling strategies [12–14]. Many biological template match methods reduce the large sample space by sampling (generating templates for) only a subset of possible self-motions. For example, the model in [10] samples only fixating self-motion. The ViSTARS model [15–18] computes heading in model MSTd using an inner product between the optic flow estimate in MT with a set of templates represented in the weight matrices connecting MT with MSTd. Each template corresponds to motion in a particular direction, with approximately 1° spacing horizontally and approximately 5° spacing vertically. In later work, ViSTARS was updated to define the templates to simultaneously calculate TTC and heading through the same computation. This has the benefit of eliminating all peripheral scaling bias[2)] and thereby equally weighting all components of the optic flow field in the heading estimate. ViSTARS can tolerate small amounts of uncorrected rotation in the optic flow field but requires additional information about eye/camera rotations to estimate heading while the camera is panning. The template matching in ViSTARS is highly robust to noise in the optic flow field through the spatiotemporal smoothing that occurs at each processing stage of the model. This system-level smoothing, in space and time, emerges naturally from model neurons parameterized based on the documented properties of V1, MT, and MST cells. ViSTARS

2) Peripheral scaling bias is the phenomenon where during forward translation motion vector magnitude further from the FOE is larger than motion vector magnitude closer to the FOE.

produces heading estimates that are consistent with human and primate heading estimation data, including in the presence of independently moving objects (which violate the mathematical assumptions of a static world), and are consistent with the temporal evolution of signals in MT and MSTd. The ViSTARS templates have recently been extended to incorporate spiral templates which are a closer match to primate neural data and enable the explanation of human curvilinear path estimation data [18].

Royden [19] presented a model of heading estimation based on a differential motion representation in primate MT. In a rigid world, differencing neighboring vectors in space results in a new vector with an arbitrary magnitude but that still converges on the FOE [20]. The chief benefit of using differential motion vectors is the simultaneous removal of rotation from the optic flow field. From the differential motion field, either a template match or a least-squares solution can be used to find the FOE. Heading estimates from Royden's modeling work yield similar heading estimation performance to humans in similar environments including in the presence of independently moving objects.

In summary, heading is directly observable from an optic flow field through estimation of the location of the FOE. There are a number of mathematical abstractions used throughout the biological and engineering literature; however, they can generally be placed in one of two groups: template matching and linear/nonlinear optimization problems (for a summary review, see [9]).

18.4
Object Detection: Understanding What Is in Your Way

Reactive navigation is by definition a behavioral response to the existence of unexpected objects along the desired trajectory. Therefore, object detection is a major component of reactive navigation. Note though that object *identification and classification* are not necessary. The fact that there is an object on the current trajectory is sufficient information to initiate an avoidance maneuver. In some applications, certain objects may pose a more serious obstacle than others, which requires some element of object identification or classification, but that is beyond the scope of this chapter. When a camera moves through the environment, optic flow can be extracted from the image plane. In a rigid, static world, the direction of all of those vectors has a common origin at the FOE, but the magnitude is a function of relative distance.[3] The boundaries of obstacles can be found by looking for depth discontinuities in the optic flow field – regions where neighboring pixels have significantly different vector magnitudes. Note that the word *object* is defined here in terms of how it impacts navigation, it may consist of multiple individual objects, or it may be a part of a larger object. Moving obstacles will also exhibit motion discontinuities, and the direction of the motion vectors corresponding to the moving object will point to a unique

3) More accurately, it is a function of the reciprocal of TTC.

Table 18.2 Analytical models for five segmentation cues for objects cues using optic flow. The optic flow of the object is denoted by $(\dot{x}_O, \dot{y}_O)$, and the optic flow of the background is denoted by $(\dot{x}_B, \dot{y}_B)$. The sign of these defined expressions has the interpretation as indicated in the left column.

Cue	Model term
Accretion ($\Delta > 0$) and deletion ($\Delta < 0$)	$\Delta = \Delta_B - \Delta_O$ with $\Delta_{B/I} = \begin{pmatrix} \dot{x}_{B/O} \\ \dot{y}_{B/O} \end{pmatrix}^t \begin{pmatrix} \cos(\beta) \\ \sin(\beta) \end{pmatrix}$ (18.5)
Expansion ($\delta > 0$) and contraction ($\delta < 0$)	$\delta = \delta_O - \delta_B$ with $\delta_{O/I} = \partial_x \dot{x}_{B/O} + \partial_y \dot{y}_{B/O}$ (18.6)
Acceleration ($\alpha > 0$) and deceleration ($\alpha < 0$)	$\alpha = \alpha_O - \alpha_\beta$ with $\alpha_{B/I} = \begin{pmatrix} \ddot{x}_{O/I} \\ \ddot{y}_{O/I} \end{pmatrix}^t \begin{pmatrix} \cos(\beta) \\ \sin(\beta) \end{pmatrix}$ (18.7)
Local, spatial curvature: concave ($\vartheta > 0$) and convex ($n < 0$)	$\vartheta = \vartheta_O - \vartheta_B$ with $\vartheta_{B/O} = \dfrac{\begin{vmatrix} \dot{x}_{B/O} & \dot{x}_{B/O} \cdot \left(\partial_x \dot{x}_{B/O}\right) + \dot{y}_{B/O} \cdot (\partial_y \dot{x}_{B/O}) \\ \dot{y}_{B/O} & \dot{x}_{B/O} \cdot (\partial_x \dot{y}_{B/O}) + \dot{y}_{B/O} \cdot (\partial_y \dot{y}_{B/O}) \end{vmatrix}}{\sqrt{\dot{x}_{B/O}^2 + \dot{y}_{B/O}^2}^{\,3}}$ (18.8)
Local, temporal curvature: concave ($\gamma > 0$) and convex ($\gamma < 0$)	$\gamma = \gamma_O - \gamma_B$ with $\gamma_{B/O} = \dfrac{\dot{x}_{B/O} \cdot \ddot{y}_{B/O} - \dot{y}_{B/O} \cdot \ddot{x}_{B/O}}{\sqrt{\dot{x}_{B/O}^2 + \dot{y}_{B/O}^2}^{\,3}}$ (18.9)

object FOE specific to that object. This object FOE can be used to determine the relative trajectory of the object with respect to the observer's trajectory. Finally, the interior (space between edges) of objects will, by definition, be a region of the image that produces no strong discontinuities. In other words, the interior of an object will have uniform or slowly varying vector magnitudes.

In this section, we define the local segmentation cues: accretion/deletion, expansion/contraction, acceleration/deceleration, spatial curvature, and temporal curvature (Table 18.2, Figure 18.3).

Accretion and deletion are expressed with respect to the angle β, which is the angle of the normal to the object's edge in the 2D image plane. Accretion ($\Delta > 0$) denotes the uncovering of background by the object, and deletion happens when the object covers the background ($\Delta < 0$). The unit of accretion and deletion is m/s.

We define expansion and contraction using spatial derivatives of optic flow. For forward self-motion, an expansion ($\delta > 0$) indicates that the object is in front of the background. For a backward self-motion, the object appears behind the background, for example, when looking through an open door in the back wall. Assuming forward self-motion, a contraction ($\delta < 0$) indicates that the object is behind the background. The unit of expansion and contraction is 1/s or Hz.

Acceleration and deceleration use the temporal dependency of the optic flow computing second-order temporal derivatives. This cue indicates acceleration for

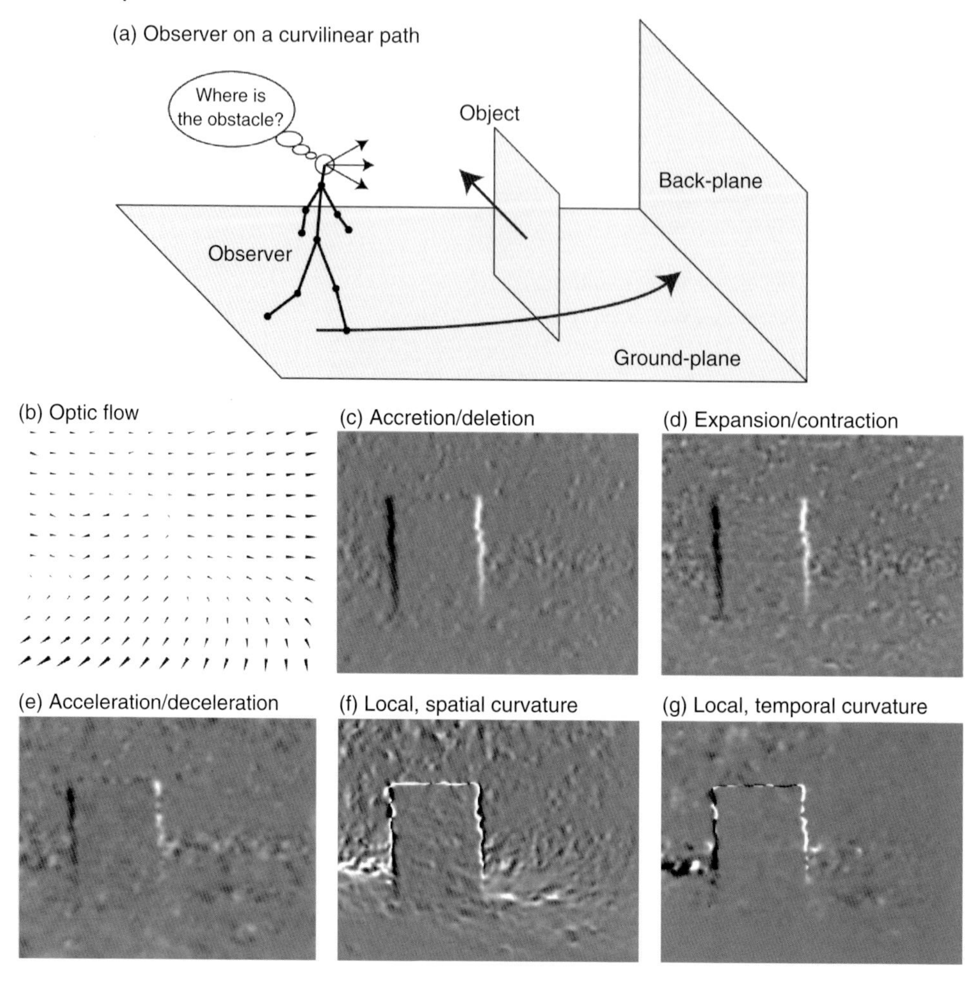

Figure 18.3 Segmentation cues for observer on a curvilinear path going around an independently moving object. a) Outline of motions. b) Estimated flow (Zach et al., [21]). c) Accretion (>0) and deletion (<0) cue. d) Expansion (>0) and contraction (<0) cue. e) Acceleration (>0) and deceleration (<0) cue. f) Local, spatial curvature cue. g) Local, temporal curvature cue. For curvature the sign indicates a concave (>0) or a convex (<0) curvature.

$\alpha > 0$ and it indicates deceleration for $\alpha < 0$. These accelerations and decelerations have the same interpretation as for objects in classical physics; however, here, the objects are patches of light in the image plane. The unit of this cue is m/s^2.

A combination of spatial derivatives of optic flow and the optic flow itself allows for the construction of a local curve, which approximates the characteristics of the optic flow, locally. This curvature is zero in cases where the spatial derivatives of the optic flow evaluate all to zero. For instance, translational self-motion toward a parallel backplane provides no curvature. However, if rotational self-motion is present, curvature is present. In general, the corresponding curve has a concave shape for $\vartheta > 0$ and a convex shape for $\vartheta < 0$. The unit of this spatial curvature is 1/m.

Instead of defining curvature in the image plane, an alternative is to define curvature in the temporal domain. Looking at one fixed location in the image plane, we define a curve through the two components of optic flow which change over time. This curve has a concave shape for $\gamma > 0$ and a convex shape for $\gamma < 0$. The unit for this temporal curvature is 1/m.

These five cues describe the segmentation between object and background [22]. They can also be derived from observed optic flow $(\hat{x}, \hat{y})$ through calculation of the spatial and temporal derivatives and the expressions from Eqs. (18.5) to (18.9) substituting the terms $(\dot{x}_{B/O}, \dot{y}_{B/O})$ (Figure 18.3). In practice, the use of derivatives on measured optic flow can result in noisy information.

Neurophysiological evidence suggests that subsets of cells in primate brain area MT respond to motion boundaries [23] through cells that respond most strongly to differences in estimated motion between two lobes of the receptive field. Differential motion cells are analogous to simple cells in V1 but look for differences in motion rather than differences in intensity. These differential motion cells project to ventral medial superior temporal (MSTv). The ViSTARS model segments objects using differential motion filters located in model MT, and the differential motion filters are grouped into object representations in model MSTv. These representations were demonstrated sufficient to match key human reactive navigation data.

Recent extensions to the ViSTARS model include the introduction of space-variant normalization across the differential motion filters. This provides more robust and reliable cell responses across the visual field and is more consistent with the space-variant representations in human and primate visual systems, for example, the log-polar representation in V1. This is implemented in such a way that each vector is normalized by its distance from the FOE; in other words, it is converted into expansion rate (mathematically defined as the reciprocal of TTC – see TTC section in the following text). Specifically, the optic flow vector field is converted to polar coordinates (r, θ), where r is the magnitude of the vector and theta is the angle from the origin (at the FOE). In each pixel location, the differences between a center (5×5 pixels) and surround filter (49×49 pixels) are calculated:

$$\theta_d = abs(\theta_c - \theta_s)/2\pi \tag{18.10}$$

$$r_d = [r_c - r_s]^+ \tag{18.11}$$

where subscripts d, c, and s denote the difference, center, and surround, respectively, and $[]^+$ denotes half-wave rectification (max(x,0)). The magnitude difference is then arbitrarily scaled by the distance from the FOE, D:

$$R_d = 1000\, r_d/D \tag{18.12}$$

The step defined by Eq. (18.12) incorporates an arbitrary scaling parameter (1000) applied to the expansion rate at that location in space; the scaling parameter should be chosen taking into account the frame rate of the system and the expected range of observed expansion rates. Finally, both the angle and magnitude are thresholded, and the conjunction is defined as the output of the differential motion filter O:

$$O = \begin{cases} 1 \text{ if } \theta_d > 0.25 \text{ and } R_d > 0.5m_R \\ 0 \text{ otherwise} \end{cases} \tag{18.13}$$

where m_R is the mean of all of the R_d values within the current image and the parameters 0.25 and 0.5 are chosen based on the desired sensitivity of the filter. Equations 18.10–18.13 are simplifications of the ViSTARS dynamical models of differential cells in MT (the full biological motivation and mathematical definition is described in [16]) modified through space-variant scaling to eliminate peripheral bias.

Peripheral bias is the phenomenon that motion vector magnitude further from the FOE is larger, for objects the same distance away, than for motion vectors closer to the FOE. Space-variant normalization (Eq. (18.12)) reduces the dynamic range required to analyze motion across the optic flow field by eliminating peripheral bias. The net effect of this space-variant normalization is that small, slow-moving objects can be accurately segmented from a moving wide field of view imager irrespective of where on the image plane they are projected (Figure 18.4).

In summary, knowing the object identity is not necessary for reactive navigation. Objects can be detected in optic flow through analysis of optic flow and specifically through differentiation of the motion field which results in motion boundaries corresponding to depth discontinuities in space.

Figure 18.4 ViSTARS can segment small, slow-moving objects from a moving wide field of view imager through space-variant differential motion processing. (a) The camera is moving forward at around 4 m/s (10 mph), two people are walking to the right (red box). (b)Space-variant differential motion accurately segments the people.

18.5
Estimation of TTC: Time Constraints from the Expansion Rate

As noted in previous sections, optic flow does not carry sufficient information to determine the distance to objects or surfaces in the observed scene. This is due to the size/distance ambiguity – a small object at a short distance is projected on to the retina/eye/sensor exactly the same as a large object at a far distance.

By contrast, the *expansion rate* of an object – when defined as a ratio = $\dot{w}/w$ with w being the size of the object in the image plane (width, height, area) – is the reciprocal of the ratio of distance to closing speed, which is the TTC (τ, TTC): mathematically, $\tau \equiv Z/\|v_z\| \equiv w/\dot{w}$. Note that although τ is often defined in the biological literature using angular size, this mathematical equivalence is only approximately true when using angular sizes. Moreover, the term "expansion rate" is often used to refer to the change in object size ($\dot{w}$) rather than as a ratio (as defined previously); this is not mathematically speaking a *rate* unless it is defined as a ratio against some other quantity (although one could argue that it is implicitly the ratio per unit of time between image frames). The definition of expansion rate as the ratio of "change in size" to "size" is critical to understanding and exploiting the relationship between TTC and object size growth on the image plane. It allows us to mathematically prove that expansion rate, E = 1/TTC which in turn allows us to infer, for example, that a 20% expansion rate per second is exactly equal to a 5 s TTC. These relationships are easily derived using perspective projection [17] and provide a robust representation for visual navigation.

The definition of TTC is the time until an object passes the infinite plane parallel to the image plane passing through the camera focal point. TTC does not indicate anything about the likelihood of a collision with an object. When combined with information on the trajectory of an object, for example, using bearing information to determine that an object is on a collision trajectory, TTC is the time-to-collision. Note the important difference between contact and collision in this context. When an object is not on a collision trajectory, TTC is the time to pass by the object. This mathematically abstract definition of TTC can be a source of confusion, which may be why some authors have used the term relative depth instead – particularly when talking about static environments.

Optic flow defines motion of a point on the image plane from one time to another relative to a FOE/contraction, and so TTC can be computed from optic flow as follows (Figure 18.5):

$$\tau = Z/\|v_z\| = (x - x_{FOE})/\dot{x} = (y - y_{FOE})/\dot{y} \tag{18.14}$$

Self-motion creates an FOE, which allows for a simple transformation from optic flow to a dense TTC (relative depth) map of environmental structure. For parallel trajectories, the FOE is at infinity, and in this case, TTC is also infinity. Before discussing the application of TTC for visual navigation, it is important to understand the assumptions that go into the derivation of TTC from visual information:

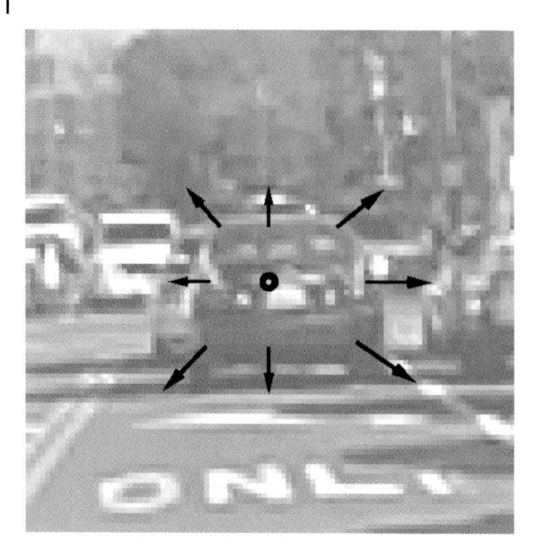

Figure 18.5 Close-up centered around the focus of expansion from Figure 18.1. The time to contact of the car, which we are heading toward, is encoded by the ratio of the optic flow vector magnitude to the distance of the origin of that vector from the FOE.

1) The world is rigid (nondeforming).
2) The observer (camera) is traveling on a straight-line trajectory (between measurements).
3) The observer (camera) is traveling at a fixed speed (between measurements).
4) If the world contains one or more independently moving objects:
   a. The motion vectors from those objects are analyzed with respect to their own object's FOE.
   b. The object is moving on a straight-line trajectory relative to the observer.
   c. The object is traveling at a fixed speed relative to the observer.

Assumption 1 is violated when the field of view contains objects that are constantly changing their shape, such as jelly fish, or that are deforming due to external forces, such as trees blowing in the wind. For realistic viewing conditions, these motions rarely occur. Even if they appear, the motions introduced by such deformations are small compared to the motion due to translation of the sensor. As a result, the error introduced by these deformations is small. Moreover, if TTC is estimated in a filter, these deformation motions will tend to smooth out, and error will be extremely small compared to measurement error in general.

Assumption 2 is violated whenever the observer changes direction. The TTC coordinate frame is inherently body based, when the body turns, so does the coordinate frame – and therefore the plane to which TTC relates. This would be true of a distance metric also; however, a distance-based coordinate frame can be rotated relatively easily, and a direction-based coordinate frame, such as TTC, is more complex. As with assumption 1, if TTC is being estimated in a filter and the changes in trajectory are smooth, then the TTC estimates will automatically update as trajectory changes. This means that it is essential that TTC be conceptualized as a dynamic visual variable that changes in response to observer behaviors.

Assumption 3 is violated when the observer changes speed. This is obvious that if you start moving faster toward something, then the TTC changes; therefore, any process that estimates TTC must assume that there was a constant speed between the image measurements that were used to estimate optic flow. As with assumptions 1 and 2, provided TTC is estimated in a filter and speed changes are smooth, then TTC estimates will be automatically updated as the speed changes.

Assumption 4 relates to independently moving objects which create their own FOE. Just as the observer's FOE provides information about heading and TTC, so does the object's FOE:

IF the object's FOE lies within the boundary of the object
THEN the object is likely to be on a direct collision course
IF the object's FOE lies outside the boundary of the object
THEN the object is not likely to be on a collision course

However, reliably measuring the FOE of an object is nontrivial, particularly when the camera is on a moving, vibrating vehicle, and the object appears small in the image. As a result, it is often more reliable and robust to assume the FOE is at the center of the object to estimate TTC rather than estimate the FOE. The error introduced by this FOE assumption can be computed and it can be minimized by using a dense TTC map across the whole object and averaging to provide object TTC. The resulting TTC error is high for objects not on a collision course but error becomes close to zero for objects on a direct collision course. This assumption thereby provides accurate behaviorally relevant information for navigation but would not be suitable for the application of object tracking. Note that assumptions relating to fixed-line trajectories and fixed speeds of moving objects are the same as for the moving object case and the same explanations and mitigations apply. Rather than using the object's FOE to determine collision likelihood, it is often more reliable to use the 2D object trajectory on the image plane and an alternate (nonoptic flow based) analysis, for example, constant bearing strategy or an image movement-based tracking model.

TTC (and expansion/looming) responsive cells have been observed in many animals from locusts to frogs and to pigeons, with pigeons being the most studied. In primates, the receptive fields of MST cells are particularly suited for the estimation of TTC [17], although there is evidence for TTC response in higher areas as well.

The ViSTARS model incorporates TTC estimation directly into model MSTd cells, simultaneously solving for FOE and TTC. This is performed through a redefinition of the heading templates to normalize each motion vector by its distance from the corresponding template FOE. This allows the template to simultaneously locate the FOE and estimate the average TTC to objects within the span of the template. The templates are mathematically defined to equally weigh each optic flow vector, irrespective of its location on the image plane, as a function of expansion rate. This means that heading estimation equally weights the center region and the periphery, and that the template match score is proportional to the mean TTC of the object(s) within the receptive field of the cell [17].

The ViSTARS TTC template definition provides all of the information necessary to reproject the optic flow into a dense TTC map, which can then be fed directly into a steering field. Simply multiply each component of the winning template (the template with the maximal response when matched against optic flow), with the observed optic flow, and the result is a spatial representation of the world where the magnitude of the response at each location is proportional to the expansion rate at that location. In other words, the winning template provides a scale factor for each pixel location that will convert the optic flow representation into an expansion rate (1/TTC) representation. When combined with object boundaries from object segmentation (see previous section), this provides a complete representation of space to facilitate reactive obstacle avoidance.

In summary, TTC can be extracted directly from optic flow, assuming the location of the FOE is known. TTC, and its reciprocal expansion rate, provides relevant behavioral information as to the imminence of collision with any object and are a robust way to measure relative distances from optic flow. Moreover, representation of the world as a function of TTC (e.g., as expansion rate) directly facilitates reactive navigation behaviors.

18.6 Steering Control: The Importance of Representation

The biological basis of steering control in higher mammals has not yet been discovered. This chapter is primarily concerned about biological inspiration for applied visual reactive control. In this section, we discuss how the visual representations observed in primate brain (and discussed previously) can be efficiently combined for reactive navigation in a biologically plausible computational framework.

Steering is the manifestation of reactive navigation through behavioral modification. When an obstacle is detected, the observer must change trajectory to avoid it. When the current trajectory is not on an intercept path with the goal object/location/direction, the observer must change trajectory. Steering can therefore be thought of as a decision-making process or a cognitive process. The steering process must analyze the visual variables, heading, visual angle and TTC of the obstacle, and visual angle and TTC of the goal, to determine an appropriate trajectory avoiding collision with obstacles while navigating toward the goal.

The chosen representation for these visual variables constrains the space of possible solutions. It seems likely that the brain represents these variables in a spatial map [24]. In a spatial map, each unique position in space has a corresponding unique position in the map. In ViSTARS and related models, TTC is represented through the magnitude of activation at any given point in the 2D map. *Three* 2D maps are required for reactive navigation. The first represents heading, the second represents one or more obstacles, and the third represents the goal;

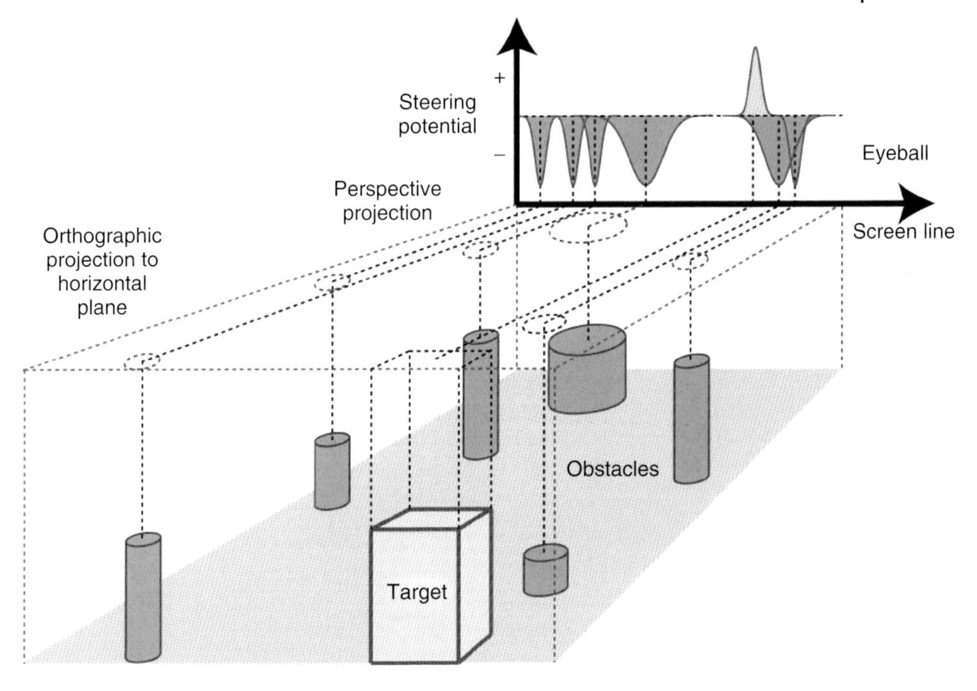

Figure 18.6 Illustration of how representations of the goal (target) and obstacles are combined into the steering field in ViSTARS. The obstacle representation is inverted to create minima in regions where navigation will result in a collision. This is added to the target representation to create a maximum at the point in space that will safely avoid the obstacles while still heading toward the goal.

more representations could be incorporated to fuse additional information such as attention and auditory information. The precursor model to ViSTARS [24] showed how three representations such as these could be minimally combined to produce a *steering field* [25], whereby the magnitude of activation at any given location defines the navigability of that location. The maxima in the representation defines the desired trajectory.

Figure 18.6 illustrates how the obstacle and goal representations are generated from projection of objects into the steering field in ViSTARS through a 1D example. A key requirement here is that the activation around an object's boundary should reduce in amplitude as a function of distance of this object boundary to the camera location in a logarithmic manner. This distance dependency is a key feature of ViSTARS object representations.

The obstacle representation is inverted to create minima in regions where navigation will result in a collision. The obstacle representation is summed with the goal representation to create the steering field (not shown) with maxima at locations that will safely avoid the obstacles. The global maximum in the steering field defines the trajectory that will head toward the goal while avoiding the obstacles, directly defining where the observer should steer toward.

When applied to the constraints of a robot, the control mechanism will be specific to the vehicle and application, but in most cases, the controller will make trajectory modifications to align the current heading with the closest maxima in the steering field. To provide temporal stability (damping) in the steering field, the current heading estimate can be added into the steering field (as it was in the ViSTARS publications), but in most cases, damping can be handled within the vehicle controller itself.

The ViSTARS steering field is similar to a potential field control law in that it creates attractors and repellers, which are used to generate trajectories. However, unlike a potential field controller, the steering field is based on a first person perspective view of the environment extracted in real time during navigation and does not require a priori information as to the location of objects within the environment. As such, it does not rely on any memorized locations and is entirely based on the current observation of obstacles and goal from an egocentric viewpoint. ViSTARS, and its precursor STARS, demonstrated how trajectories generated with this steering field were qualitatively and quantitatively similar to those produced by humans in similar environments.

It is worth noting that [26] utilizes a very similar steering field in a spike-latency model for sonar-based navigation in an algorithm called "Openspace," and so the representation of goals and obstacles as distributions across navigable space may also provide a multimodal reactive navigation capability.

18.7 Conclusions

This chapter provided an analysis of the relevant visual representations and variables required to support reactive visual navigation: self-motion, object boundaries, TTC/relative depth, and steering decisions. Likely, mappings of these representations to the magnocellular pathway of the primate visual system are shown in Table 18.1. Other animals, including insects, have different but often comparable circuits involved in the computation of these visual variables (see Chapter 17 about how elementary motion detectors are used by insects). A recent review [27] provides details of these representations in the visual system of honeybees. Biological systems tightly integrate visual perception with the motor system to allow for rapid reactive navigation in a changing world. Understanding the detailed representations within the primate brain may allow us to build robotic systems with the same capabilities. Basic biological capabilities like reactive navigation are fundamental for the integration of robotics into our daily lives and in the automation of highly dangerous tasks such as firefighting, search and rescue, and so on.

More information about the mathematical models (including code) described in this chapter, along with related work, can be found on *http://opticflow.bu.edu*, *http://vislab.bu.edu*, and *http://vislab.bu.edu/projects/vistars/*.

Acknowledgments

NAB was supported in part by the Office of Naval Research (ONR) N00014-11-1-0535 and Scientific Systems Company Inc. FR was supported in part by CELEST, a National Science Foundation Science (NSF) of Learning Center (NSF SMA-0835976), NSF BCS-1147440, the ONR N00014-11-1-0535, and ONR MURI N00014-10-1-0936.

References

1. Gibson, J.J. (1950) *The Perception of the Visual World*, Houghton Mifflin, Boston.
2. Raudies, F. (2013) Optic flow. *Scholarpedia*, **8** (7), 30724.
3. Scharstein, D. *http://vision.middlebury.edu/flow/eval/* (accessed 13 April 2015)
4. Aloimonos, J., Weiss, I., and Bandyopadhyay, A. (1988) Active vision. *Int. J. Comput. Vision*, **2**, 333–356.
5. Longuet-Higgins, H.C. and Prazdny, K. (1980) The interpretation of a moving retinal image. *Proc. R. Soc. Lond. B Biol. Sci.*, **208**, 385–397.
6. Tanaka, K., Hikosaka, K., Saito, H., Yukie, M., Fukada, Y., and Iwai, E. (1986) Analysis of local and wide-field movements in the superior temporal visual areas of the macaque monkeys. *J. Neurosci.*, **6** (1), 134–144.
7. Duffy, C.D. and Wurtz, R.H. (1995) Response of monkey MST neurons to optic flow stimuli with shifted centers of motion. *J. Neurosci.*, **15** (7), 5192–5208.
8. Soatto, S. (2000) Optimal structure from motion: local ambiguities and global estimates. *Int. J. Comput. Vision*, **39** (3), 195–228.
9. Raudies, F. and Neumann, H. (2012) A review and evaluation of methods estimating ego-motion. *Comput. Vision Image Understanding*, **116** (5), 606–633.
10. Perrone, J.A. and Stone, L.S. (1994) A model of self-motion estimation within primate extrastriate visual cortex. *Vision Res.*, **34** (21), 2917–2938.
11. Perrone, J.A. (1992) Model for the computation of self-motion in biological systems. *J. Opt. Soc. Am. A*, **9** (2), 177–192.
12. Jepson, A. and Heeger, D. (1992) Linear Subspace Methods for Recovering Translational Direction. Technical Report RBCV-TR-92-40, University of Toronto, Department of Computer Science, Toronto.
13. Lappe, M. and Rauschecker, J. (1993) A neural network for the processing of optic flow from ego-motion in man and higher mammals. *Neural Comput.*, **5**, 374–391.
14. Lappe, M. (1998) A model of the combination of optic flow and extraretinal eye movement signals in primate extrastriate visual cortex – neural model of self-motion from optic flow and extraretinal cues. *Neural Networks*, **11**, 397–414.
15. Browning, N.A., Mingolla, E., and Grossberg, S. (2009) A neural model of how the brain computes heading from optic flow in realistic scenes. *Cogn. Psychol.*, **59** (4), 320–356.
16. Browning, N.A., Grossberg, S., and Mingolla, E. (2009) Cortical dynamics of navigation and steering in natural scenes: motion-based object segmentation, heading, and obstacle avoidance. *Neural Networks*, **22** (10), 1383–1398.
17. Browning, N.A. (2012) A neural circuit for robust time-to-contact estimation based on primate MST. *Neural Comput.*, **24** (11), 2946–63.
18. Layton, O.W. and Browning, N.A. (2014) A unified model of heading and path perception in primate MSTd. *PLoS Comput. Biol.*, **10** (2), e1003476. doi: 10.1371/journal.pcbi.1003476
19. Royden, C.S. (2002) Computing heading in the presence of moving objects: a model that uses motion-opponent operators. *Vision Res.*, **42** (28), 3043–58.
20. Rieger, J. and Lawton, D.T. (1985) Processing differential image motion. *J. Opt. Soc. Am.*, **2** (2), 354–359.

21. Zach, C., Pock, T., and Bischof, H. (2007) A duality based approach for realtime TV-L optical flow. Hamprecht, F.A., Schnörr, C., and Jähne, B. (Eds.): *DAGM 2007*, Lecture Notes in Computer Science, **4713**, Springer, Heidelberg, pp. 214–223
22. Raudies, F. and Neumann, H. (2013) Modeling heading and path perception from optic flow in the case of independently moving objects. *Front. Behav. Neurosci.*, **7**, 23. doi: 10.3389/fnbeh.2013.00023
23. Born, R.T. and Bradley, D.C. (2005) Structure and function of visual area MT. *Annu. Rev. Neurosci.*, **28**, 157–189.
24. Elder, D.M., Grossberg, S., and Mingolla, E. (2009) A neural model of visually guided steering, obstacle avoidance, and route selection. *J. Exp. Psychol.: Hum. Percept. Perform.*, **35** (5), 1501.
25. Schöner, G., Dose, M., and Engels, C. (1995) Dynamics of behavior: theory and applications for autonomous robot architectures. *Rob. Autom Syst.*, **16**, 213–245.
26. Horiuchi, T.K. (2009) A spike-latency model for sonar-based navigation in obstacle fields. *IEEE Trans. Circuits Syst. I*, **56** (11), 2393–2401.
27. Srinivasan, M.V. (2011) Honeybees as a model for the study of visually guided flight, navigation and biologically inspired robotics. *Physiol. Rev.*, **91**, 413–460.

Index

Biologically Inspired Computer Vision: Fundamentals and Applications, First Edition.
Edited by Gabriel Cristóbal, Laurent Perrinet, and Matthias Keil.

t